Molecular Biology Biochemistry and Biophysics

27

Effects of Ionizing Radiation on DNA

Physical, Chemical and Biological Aspects

Edited by
J. Hüttermann W. Köhnlein R. Téoule
Coordinating Editor: A. J. Bertinchamps

With Contributions by
G. Ahnström J. Cadet T. Coquerelle W. A. Cramp A. Ehrenberg
J. P. Elliott A. Gräslund S. Gregoli U. Hagen J. N. Herak
G. Hotz J. Hüttermann W. Köhnlein A. R. Lehmann A. Ore
A. Rupprecht G. Scholes D. Schulte-Frohlinde C. von Sonntag
R. Téoule H. Traut A. Van de Vorst K. F. Weibezahn E. Westhof
C. M. Wheeler S. J. Wyard

With 74 Figures

Springer-Verlag Berlin Heidelberg New York 1978

This is publication no. 1202 of the Biology, Radiation Protection and Medical Research programme of the Commission of the European Communities

ISBN-13: 978-3-642-81198-2 e-ISBN-13: 978-3-642-81196-8
DOI: 10.1007/978-3-642-81196-8

Library of Congress Cataloging in Publication Data. Main entry under title: Effects of ionizing radiation on DNA. (Molecular biology, biochemistry, and biophysics; 27) "Publication No. 1202 of the Biology-medical research programme of the Commission of the European Communities." Bibliography: p. Includes index. 1. Radiogenetics. 2. Deoxyribonucleic acid. I. Bertinchamps, A.J., 1924— II. Ahnström, G. III. Commission of the European Communities. IV. Series.
QH465.R3E33 574.8'732 77-25857

Softcover reprint of the hardcover 1st edition 1978

2131/3130-543210

Foreword

This publication presents the culmination of collaborative effort between specialists in the wide range of disciplines concerned with the effects of ionizing radiations on nucleic acids.

The authors belong to a group formed under the aegis of the Commission of the European Communities some eight years ago with the object of facilitating the exchange of ideas and information between European scientists working in this field.

The Commission's aim was not to replace traditional information channels, but to strengthen the various links between the scientists in the group, as if they were working together in a single team.

In addition to the annual meetings of the group, contact was maintained by individual visits, exchanges of young scientists, experiments using mutual equipment, and the regular publication of a Newsletter, mainly to announce the availability of preprints.

Bruxelles, December 1977

A. J. Bertinchamps
Commission of the
European Communities.

Preface

Mankind today is faced with a choice of cardinal importance, namely, deciding on the source and quantity of energy to be made available for tomorrow's world. If, as is probable, the energy is to be mainly of nuclear origin, we must expect a marked expansion of the nuclear industry, which could well become the world's leading industry in the not-too-distant future. This would carry with it the problems caused by the presence of enormous quantities of radioactivity, despite the stringent precautions taken to date and those to be taken in future. Even without undertaking special studies, it is obvious that some degree of interaction between the world of the atom and the world of living matter will be inevitable.

The quantitative assessment of this interaction and tentative predictions about it are currently being discussed. No one disputes its reality. Experience shows the many facets of the problem, whose most formidable aspects - if we consider only the genetic and carcinogenetic effects - all have one common denominator, namely DNA.

It is obviously the effects of radiation on man which hold our attention most. Although these are being studied elsewhere and much pertinent information has been obtained via epidemiology and animal or cell radiobiology, the answers to the principal question, especially the dose-effect relationship at low doses seem, if only for statistical reasons, beyond the scope of these various disciplines. I think all radiobiologists agree, in consequence, that these problems can be solved only by possessing detailed knowledge of the action of radiation at molecular level. This action is not instantaneous, but a sequence of complex and interdependent events. Radiation energy decays according to a continuous process:

- the first stage comprises the events occurring between the absorption of energy and the formation of free radicals;
- the second stage is the chemical transformation of these radicals into stable radiation products;
- the third includes the effects caused by the incorporation of these abnormal products in biologically active systems.

Each stage evolves in time and space at its own particular level: the first, at atomic level, lasts somewhere in the region of 10^{-16} s; the third, at the level of the entire living organism, may last several years.

The free radicals, which possess paramagnetism, appear at an early stage in the chain of events. Because of their paramagnetic properties, electron spin resonance can be used to detect them, measure them, and possibly determine their structure.

After a succession of fast reactions involving the formation of secondary radicals, these unstable entities are deactivated and give rise to stable chemical products which, having lost their paramagnetism, can no longer be studied by electron spin resonance.

At this point, chemical methods take over for measuring and identifying the properties of the stable products. As we shall see, a large number of products resulting from the radiolysis of nucleic acid constituents have been studied by chromatography. In the third stage, the purely biological processes of the radiation effects are studied by a number of techniques geared to the phenomenon itself and to the material studied.

Each of the techniques used at the different stages of radiation damage involve certain restrictions as to the nature of the samples which may be studied. ESR as normally employed calls for dehydrated samples. Radiochemical techniques are usually applied to aqueous solutions. This raises the major question of the relevance of these models to living organisms.

It would be highly desirable to identify the interdependence and the interaction of all these products, especially those responsible for biological alterations and their radical precursors. This has been the aim of the authors but it has been only partially achieved. The interactions are too numerous and the mechanisms too complex for the reader to receive an exact and faithful picture and for the gap separating the experimental model from the living organism itself to be ignored.

The following pages must therefore be regarded only as an attempt, which perhaps deserves publicizing, despite its inadequacies, and may stimulate a similar effort at integration on a broader scale.

Although they have been unable to arrive at the comprehensive view they had hoped for, the authors nonetheless feel they have followed the right road towards solving a crucial problem.

The contents of the book are divided into three separate parts, corresponding to the distribution of radiation effects in time.

The authors wish to express their sincere thanks to A. Van de Vorst, on whose initiative the book was written, and to H. Schibilla, who, throughout the writing of it, unflaggingly attended to all the arrangements without which the work could not have been completed.

Bruxelles, December 1977

A. J. Bertinchamps
Commission of the
European Communities.

Contents

Section I: Physical Aspects

Section III: Biological Aspects

Editors

Albert J. Bertinchamps	Prof. Dr. med., coordinating editor, Commission des Communautés Européenes, Biologie, Radioprotection et Recherche médicale, 200, rue de la Loi, B-1049 Bruxelles
Jürgen Hüttermann	Prof. Dr. rer. nat., editor on Physical Aspects, Institut für Biophysik und physikalische Biochemie, Universität Regensburg, Universitätsstr. 31, D-8400 Regensburg 2
Wolfgang Köhnlein	Prof. Dr. rer. nat., editor on Biological Aspects, Institut für Strahlenbiologie, Universität Münster, Hittorfstr. 17, D-4400 Münster
Robert Téoule	Dr. s.c. phys., editor on Chemical Aspects, Département de Recherche Fondamentale, Laboratoire de Radiobiologie, Commissariat à l'Energie Atomique, Centre d'Etudes Nucléaires de Grenoble, 85X, F-38041 Grenoble Cédex

Contributors

Gunnar Ahnström	Prof. Dr. phil., Department of Radiation Chemistry, Wallenberg Laboratory, University of Stockholm, Lilla Frescati, S-104 05 Stockholm
Jean Cadet	Dr. s.c. phys. Département de Recherche Fondamentale, Laboratoire de Radiobiologie, CEA/CEN-G, 85 X, F-38041 Grenoble Cédex
Thérèse Coquerelle	Dr. med. Kernforschungszentrum Karlsruhe Institut für Strahlenbiologie, Postfach 3640, D-7500 Karlsruhe 1
William A. Cramp	Ph. D., Medical Research Council, Cyclotron Unit, Hammersmith Hospital, Ducane Road, GB-London W12 OHS
Anders Ehrenberg	Ph. D., Prof. of Biophysics, Department of Biophysics, University of Stockholm, Arrhenius Laboratory, S-104 05 Stockholm
Joan Pieter Elliott	B.A., Ph. D., Department of Physics, Guy's Hospital, Medical School, GB-London Bridge SE1 9RT
Astrid Gräslund	Ph. D., Department of Biophysics, University of Stockholm, Arrhenius Laboratory S-104 05 Stockholm
Silvano Gregoli	Dr. s.c. phys., Commission des Communautés Européenes, Biologie, Radioprotection et Recherche médicale, Laboratoire de Radiobiologie moléculaire, U.L.B., B-1640 Rhode-St-Genèse
Ulrich Hagen	Prof. Dr. med. Dr. rer. nat., Kernforschungszentrum Karlsruhe, Institut für Strahlenbiologie, Postfach 3640, D-7500 Karlsruhe 1
Janko N. Herak	Prof. Dr. rer. nat., Faculty of Pharmacy and Biochemistry, University of Zagreb, YU-41000 Zagreb

Gerhart Hotz	Prof. Dr. med., Kernforschungszentrum Karlsruhe, Institut für Strahlenbiologie, Postfach 3640, D-7500 Karlsruhe 1
Jürgen Hüttermann	Prof. Dr. rer. nat., Institut für Biophysik und physikalische Biochemie, Universität Regensburg, Universitätsstr. 31, Postfach 397, D-8400 Regensburg
Wolfgang Köhnlein	Prof. Dr. rer. nat., Institut für Strahlenbiologie, Universität Münster, Hittorfstr. 17, D-4400 Münster
Alan R. Lehmann	Ph. D., Medical Research Council, University of Sussex, Falmer, GB-Brighton BN1 9QG
Aadne Ore	Ph. D., Professor of Physics, Fisisk Institutt, Universitetet i Oslo, Box 1048 Blindern, N-Oslo 3
Allan Rupprecht	Dr. technol., Institute of inorganic and physical Chemistry, University of Stockholm, Arrhenius Laboratory, S-104 05 Stockholm
George Scholes	Ph. D., D. Sc., Reader in Radiation Chemistry, Radiation and Biophysical Chemistry Laboratory, University of Newcastle Upon Tyne, GB-Newcastle Upon Tyne NE1 7RU
Dietrich Schulte-Frohlinde	Prof. Dr. rer. nat., Institut für Strahlenchemie im Max-Planck-Institut für Kohlenforschung, Stiftstr. 34-36, D-4330 Mülheim a.d. Ruhr
Clemens von Sonntag	Priv.-Doz. Dr. rer. nat., Institut für Strahlenchemie im Max-Planck-Institut für Kohlenforschung, Stiftstr. 34-36, D-4330 Mülheim a.d. Ruhr
Robert Téoule	Dr. s.c. phys., Département de Recherche Fondamentale, Laboratoire de Radiobiologie, CEA/CEN-G, 85 X, F-38041 Grenoble Cédex
Horst Traut	Prof. Dr. rer. nat., Institut für Strahlenbiologie der Universität Münster, Hittorfstr. 17, D-4400 Münster
Albert van de Vorst	Prof. associé, Université de Liège, Département de Physique atomique et moléculaire, B-4000 Sart-Tilman par Liège 1

Karl Friedrich Weibezahn	Dr. rer. nat., Kernforschungszentrum Karlsruhe, Institut für Strahlenbiologie, Postfach 3640, D-7500 Karlsruhe 1
Eric Westhof	Dr. sc. phys., Institut für Biophysik und physikalische Biochemie, Universität Regensburg, Universitätsstr. 31, Postfach 397, D-8400 Regensburg 2
Christopher M. Wheeler	Ph. D., Radiation and Biophysical Chemistry Laboratory, University of Newcastle Upon Tyne, GB-Newcastle Upon Tyne NE1 7RU
Sidney J. Wyard	D. sc., Prof. of Physics applied to Medicine, Department of Physics, Guy's Hospital Medical School, GB-London Bridge SE1 9RT

Section I
Physical Aspects

Chapter 1

Structure and Electronic Properties of DNA

1.1 Nomenclature and Primary Structure of DNA

An understanding of the behavior of nucleic acids under the action of ionizing radiations necessitates a knowledge of those basic physico-chemical properties which govern the structure, the chemical reactivity, and the interactions of these polymers. The

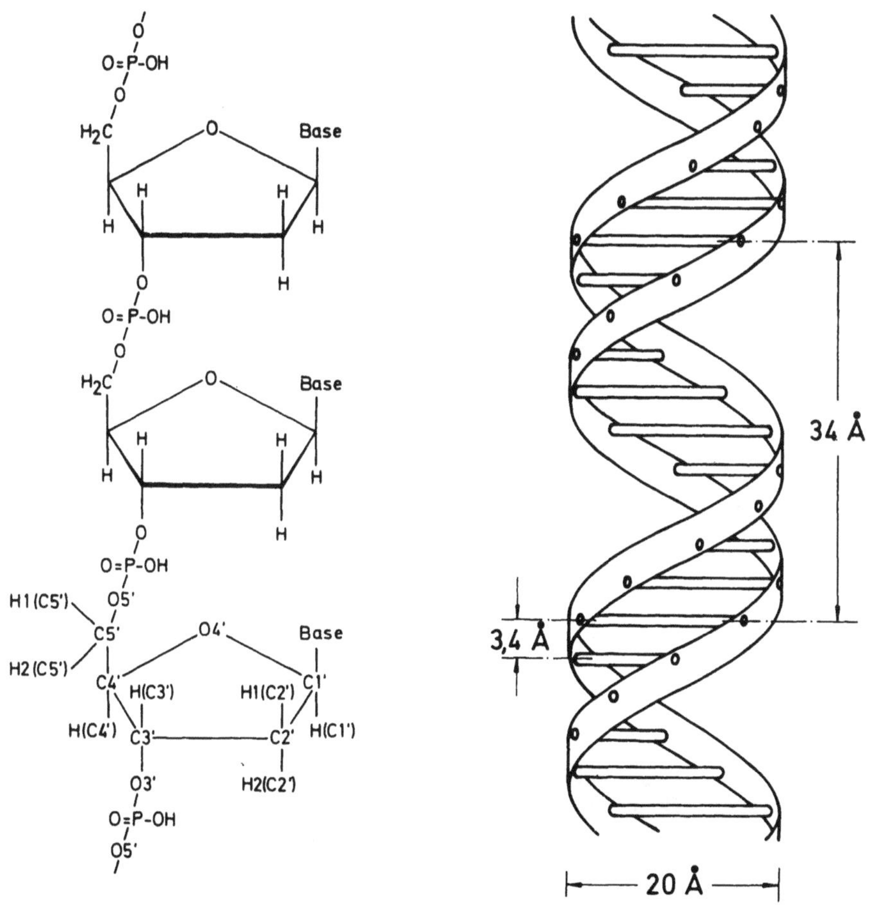

Fig. 1.1. The primary structure of DNA with a schematic representation of B-DNA. The ribbons represent the two phosphate-sugar chains and the horizontal rods the hydrogen-bonded base pairs (1 Å = 10^{-1} nm)

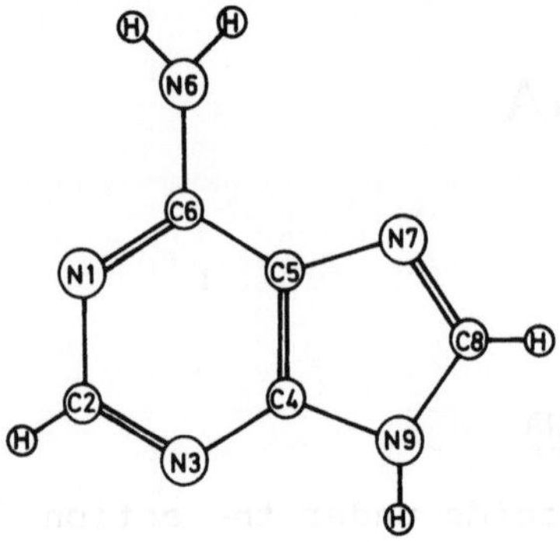

I

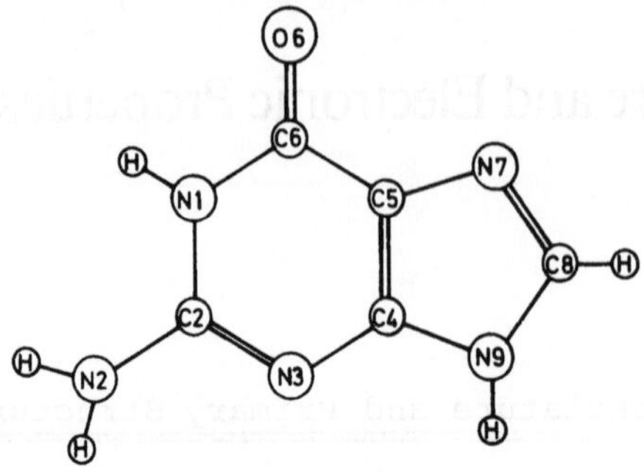

II

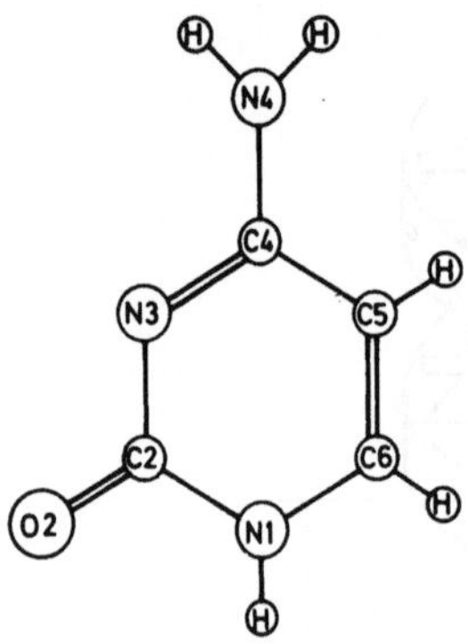

III

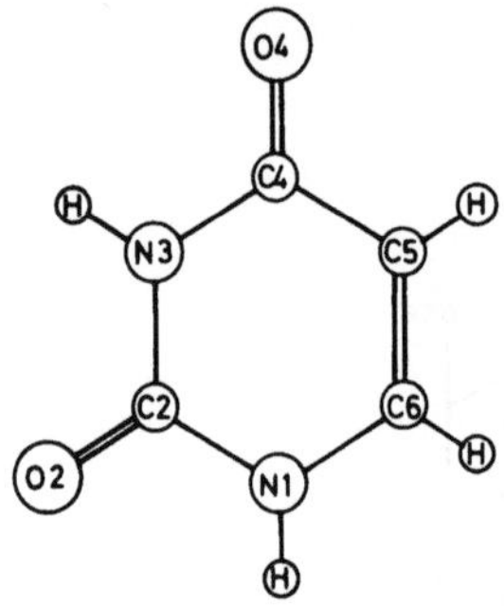

IV

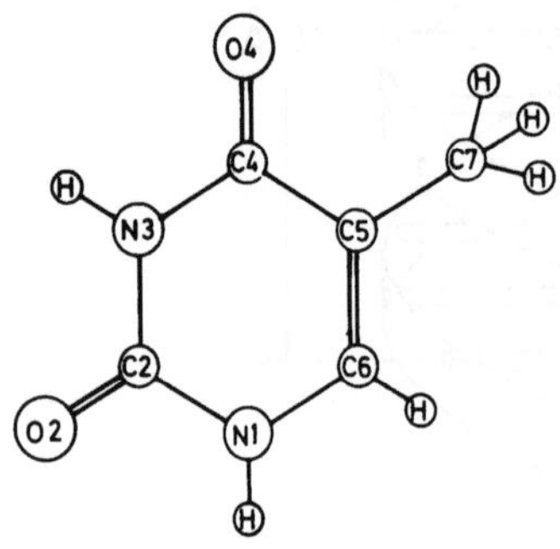

V

Fig. 1.2. Ring positions of the common purine and pyrimidine bases: *I* adenine, *II* guanine, *III* cytosine, *IV* uracil, and *V* thymine

chemical and physico-chemical aspects of nucleic acid structure have been discussed elsewhere in great detail (CHARGAFF and DAVIDSON, 1955; JORDAN, 1960; MICHELSON, 1963; BLOOMFIELD et al., 1974) and only a short summary is given, in which the number of references has been kept to a minimum.

The nucleic acids are linear, unbranched polymeric molecules built up by extended repetition of nucleotides (Fig. 1.1). Each of these is composed of a nitrogenous purine or pyrimidine base, a 5-carbon sugar which is 2'-deoxy-D-ribose for deoxyribonucleic acid (DNA) and D-ribose for ribonucleic acid (RNA), and a phosphate group. The base is joined to the C1' position of the sugar by a β-glycosidic linkage, while the phosphate is esterified at the C3' and C5' positions.

Four heterocyclic bases occur predominantly in the DNA. These are the purines adenine and guanine and the pyrimidines cytosine and 5-methyluracil or thymine in DNA. Thymine is replaced by uracil in RNA. The chemical structure and the numerical designation of the atoms of the bases are shown in Figure 1.2. The standard nomenclature generally used to identify the bases and the corresponding nucleosides and nucleotides is summarized in Table 1.1. In addition to the main components, other bases are found occasionally, in particular in RNA. These include a number of methyl derivatives of adenine and guanine like 6-methylaminopurine, 6,6-dimethylaminopurine, 2-methyladenine, 1-methylguanine, 2-methylamino-6-hydroxypurine, 2,2-dimethylamino-6-hydroxypurine. In the case of pyrimidines, two derivatives of cytosine are found in DNA: 5-methylcytosine in the DNA of higher animals as well as plants, and 5-hydroxymethylcytosine in T-even bacteriophages.

1.2 Dimensions and Conformations

In this chapter, the structural information obtained from X-ray diffraction studies on single crystals is summarized and, when possible, compared with that obtained in solutions. The reviews of VOET and RICH (1970), ARNOTT (1970), SUNDARALINGAM (1973), and TS'O (1974) should be consulted for further details. Single crystals of bases, nucleosides, and nucleotides are generally grown from aqueous solutions, solutions of hydrochloric or hydrobromic acid, or from aqueous mixtures containing ethanol or dimethylsulphoxide. The crystals may contain solvent molecules in the unit cell. Their influence on the bond distances and bond angles is very limited, whereas their presence may exert a strong influence on the hydrogen bonding schemes. The molecules of the unit cell usually form pairs in which the bases are bonded between themselves by at least two hydrogen bonds. In mixed crystals, the pairing involves different bases. These pairs will be discussed in I.1.4.1. A common feature of all crystal structures of nucleic acid bases is that a hydrogen atom bonded to a nitrogen or an oxygen almost always participates in hydrogen bonding (DONOHUE, 1952). Another feature of pyrimidine and purine crystal structures is the way the bases stack with partial overlap yielding specific stacking patterns (see I.1.4.2). In a crystal, the molecules must arrange themselves in such a way as to fill space optimally. This close-packing requirement together with the tendencies to form all possible hydrogen bonds and to stack the bases forces the molecules to form alternating hydrophobic and hydro-

Table 1.1. Nomenclature of the common bases, nucleosides and nucleotides[a]

Base	Abbre-viation	Nucleoside				Nucleotide			
		Ribo	Abbre-viation	Deoxyribo	Abbre-viation	Ribo	Abbre-viation	Deoxyribo	Abbre-viation
Adenine	Ade	Adenosine	Ado	Deoxyadeno-sine	dAdo	Adenosine-5'-monophosphate or 5'-adenylic acid	5'-AMP	Deoxadenosine-5'-monophosphate or 5'-deoxyadenylic acid	5'-dAMP
Guanine	Gua	Guanosine	Guo	Deoxyguano-sine	dGuo	Guanosine-5'-monophosphate or 5'-guanylic acid	5'-GMP	Deoxyguanosine-5'-monophosphate or 5'-deoxygua-nylic acid	5'-dGMP
Cytosine	Cyt	Cytidine	Cyd	Deoxycyti-dine	dCyd	Cytidine-5'-monophosphate or 5'-cytidylic acid	5'-CMP	Deoxycytidine-5'-monophosphate or 5'-deoxycyti-dylic acid	5'-dCMP
Uracil	Ura	Uridine	Urd	Deoxyuri-dine	dUrd	Uridine-5'-monophosphate or 5'-uridylic acid	5'-UMP	Deoxyuridine-5'-monophosphate or 5'-deoxyuri-dylic acid	5'-dUMP
Thymine	Thy	Ribosyl-	Thd	Thymidine	dThd	-	-	Thymidine-5'-monophosphate or 5'-thymidylic acid	5'-dTMP

[a] Recommendations from the IUPAC-IUB Commission on Biochemical Nomenclature (CBN) Abbreviations and Symbols for Nucleic Acids, Polynucleotides and their Constituents. J. Biol. Chem. 245, 5171-5176 (1970).

philic zones of bases and ribose moieties in most nucleoside crystal structures. The hydrophobic region is held together by a complex network of hydrogen bonds between the polar groups of the riboses and, when present, the molecules of crystallization. In the hydrophobic and nonpolar region, the Van der Waals forces predominate and promote the self-pairing of the bases (see I.1. 4.1). In I.1.4, it will be shown that similar interactions are responsible for the double-helical structure of DNA.

1.2.1 The Bases

The shortest bonds are those with a high double bond character and are therefore of primary importance for the sites of attack in the molecules (see I.1.3.1). In pyrimidines, the intercarbon bond with the shortest length is always the C5 to C6 bond (1.34 to 1.35 Å), whereas the shortest C-N bond is the N3 to C4 bond of cytosine (1.34 to 1.35 Å). In purines, the C2 to N3 bond of adenine is the shortest C-N bond. In adenine and guanine, the C8 to N7 bond is shorter than the C8 to N9 bond. The aromatic conjugated rings of the purines and pyrimidines are expected to be planar since small deviations from planarity rapidly decrease the resonance stabilization energy of such systems. The deviations from planarity of the atoms in the bases are usually between 0.01 Å and 0,1 Å, the exocyclic atoms being most deviated. These distortions of the molecules have been attributed to their response to the packing forces. The purines and pyrimidines illustrated in Figure 1.2 are presented in their stable tautomeric forms, i.e., the keto (lactam) form for the hydroxylated bases and the amino form for the aminated ones. The nitrogen to which the D-ribose or the D-deoxyribose is attached in the nucleosides and nucleotides (N1 in pyrimidines and N9 in purines) always has a hydrogen attached in the unsubstituted bases.

The sites of protonation of the bases usually involve a ring nitrogen atom and not the exocyclic groups. All charged cytosine molecules are protonated at N3 and the guanine ones at N7 of the imidazole ring. The adenine base contains two sites, the first one is the pyrimidine nitrogen N1 and the second one the imidazole nitrogen N7. The pyrimidine nitrogen N3 is protonated only in N7 substituted adenines (KISTENMACHER and SHIGEMATSU, 1975). Only one protonated uracil derivative has been analysed, 1-methyluracil hydrobromide, and it is protonated on the oxygen bonded to C4 (VOET and RICH, 1970). The most pronounced effect of protonation is adjustment of the ring bond angles. Predominantly, an enlargement of the internal angle at the site of protonation together with a decrease of the adjacent angles takes place (TRUS and MARSH, 1972). Such a trend can also be observed between a nitrogen with an extra-annular attachment, like N1 in guanine, and a nitrogen without an extra-annular attachment, like N3 in cytosine. The bond lengths are not substantially altered following protonation of the ring.

Spectroscopic studies in solution confirm the results of X-ray diffraction concerning the most stable tautomeric forms and the sites of protonation of the bases. In aqueous solutions, the ring nitrogens N1 of adenine, N3 of cytosine, and N7 of guanine add a proton in the pH range 2-4.5. In the pH region 9-10, the ring nitrogens N1 of guanine, N1 of uracil, and N1 of thymine lose a proton. In acidic solutions, adenine, cytosine, and guanine are therefore positively charged with uracil and thymine neutral. On the other hand, in alkaline solutions, adenine and cytosine are neutral with uracil, thymine, and guanine negatively charged. It is to be noted that the pK values of the bases are influenced by the ionic strength. Hence, the pK values of adenine and cytosine are increased by salts, while those of uracil, thymine and guanine are decreased by salts.

1.2.2 The Nucleosides and Nucleotides

The five-membered ring of the pentose unit is not planar but one or two atoms are displaced from the mean plane formed by the others. Usually, the plane is defined by atoms C1', O4', and C4' with atoms C2' and C3' displaced from it, one ∿ 0.5 Å and the other ∿ 0.1 Å. When the atom most out-of-plane is on the same side as C5' the conformation is called ENDO, otherwise it is EXO. Common nucleosides and nucleotides show either C2'-ENDO or C3'-ENDO conformation in the solid state (Fig. 1.3). The effect of the puckering on the bond distances is negligible; it is mainly

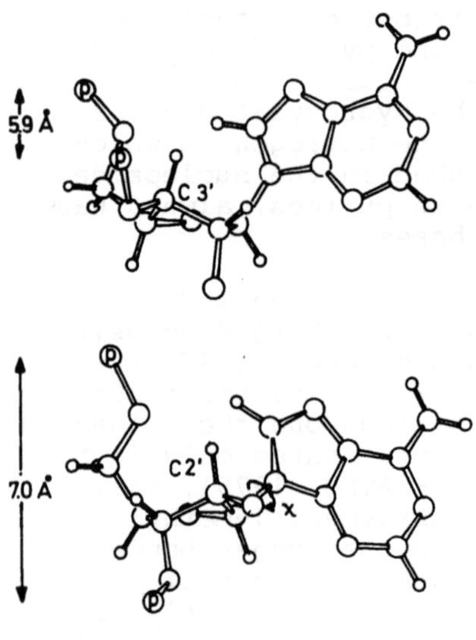

Fig. 1.3: The two preferred conformations for the nucleotides. The base is in the anti range (the glycosyl torsion angle is indicated by χ). In the upper confromation the sugar is in the C3'-ENDO pucker and in the lower conformation it is C2'-ENDO. The C4'-C5' bond is in the preferred gauche$^+$ conformation. (Redrawn after SUNDARALINGAM, 1975)

observed in the bond angles around C2' and C3'. NMR studies have shown that the furanose ring of the common pyrimidine and purine nucleosides (-tides) exists in solution in dynamic equilibrium between the C2'-ENDO and C3'-ENDO puckered forms (HRUSKA, 1973). A further major mode of internal motion of nucleosides and nucleotides is the rotation around the C4'-C5' bond. In the solid state and in solution the preferred conformation is gauche$^+$ where the O5' oxygen atom lies over the ribose ring. However, the other two staggered conformations are also found for the nucleosides.

Another parameter necessary to determine the conformational state of a nucleoside (-tide) is the orientation of the base relative to the pentose, since rotation about the C1'-N9 (or N1) bond is allowed. However, this rotation is not free but is restricted by steric interactions to two regions: the anti and the syn (DONOHUE and TRUEBLOOD, 1960). In the anti range the glycosyl torsion angle is centered around 30° and in the syn range around 240° (see Fig. 1.3). The glycosyl torsion angle is correlated with the conformation of the sugar ring. Usually, with the base in the anti range and the gauche$^+$ rotamer, the C3'-ENDO sugars are found with lower values of the glycosyl torsion angle than the C2'-ENDO ones. The nucleosides and nucleotides show a preference for the anti conformation. However, in the purine nucleosides, the syn conformation is often observed. SUNDARALINGAM (1973) has shown that the nucleotides are more "rigid" than the nucleosides, since they are only found in the anti conformation with an overwhelming preference for the gauche$^+$ conformation. The flexibility in the nucleotides seems to be restricted to the puckering of the pentose, which can either be C2'-ENDO or C3'-ENDO. These two preferred conformations of the nucleotide unit are represented in Figure 1.3. The sugar pucker makes the C2'-ENDO nucleotide longer by about 1 Å than the C3'-ENDO nucleotide (SUNDARALINGAM, 1975). In polynucleotides, the main source of flexibility for the folding of the backbone is provided by the internucleotide phosphodiester.

1.2.3 DNA

According to the model initially proposed by WATSON and CRICK (1953), a DNA molecule consists of two anti-parallel polynucleotide chains coiled as a double helix about a common axis (Fig. 1.1). The purine and pyrimidine bases are stacked in a parallel array in the core of the helix and are nearly perpendicular to the helix axis. Each adenine is hydrogen-bonded to a thymine in the other strand and each guanine to a cytosine. With this model of the secondary structure of DNA, most of the experimental results known at that time could be accomodated and especially the observation made by CHARGAFF (1951) on the presence of equimolar amounts of guanine and cytosine as well as adenine and thymine.

More refined X-ray diffraction studies of oriented fibres have shown, however, that DNA fibres can exist in three double-stranded helical structures: the A-DNA, the B-DNA, and the C-DNA. The B-DNA is generally considered the native form, while the C-DNA is closely related to the B-DNA. At low humidities and low ionic strength, the A-form is obtained, which at higher humidities transforms into B-DNA if excess salt is present. Under intermediate conditions of relative humidity and salt content, one sometimes observes the C-DNA conformation. ARNOTT (1976) concludes that the B-DNA conformation is favoured in polar conditions while the other conformations are preferred in less polar environments. He also suggests that the nucleotide sequence affects the structural behavior. Thus, the secondary structure of DNA may change under less polar conditions from the B- to the A- and then C-form only if there exists a sufficient number of purine-purine stacks involving the guanine base. The reluctance of A-T rich DNA to show the B → A transition is very probably due to the presence of nucleotide doublet sequence favoring the B-conformation and not to the composition of sequence. In the A-DNA, the deoxyribose ring exhibits the C3'-ENDO puckering and the base is in the anti range. On the other hand, B-DNA and C-DNA are the only polynucleotides with C2'-ENDO sugars (or more precisely C3'-EXO sugars). As in nucleosides and nucleotides, the B- and C-DNAs have a larger glycosyl torsion angle than the A-DNA since they have C2'-ENDO sugars. It is interesting to add that all RNA structures are very similar. The sugar puckering is C3'-ENDO and the base is in the anti range. Consequently, the ability of DNA to exist in two forms is brought about by the deoxydation at C2' (ARNOTT, 1970).

In all secondary structures of DNA, the two helices are right-handed and the chains are held together by H-bonds between complementary bases according to the Watson-Crick schemes. However, the pitch of the helix and the number of bases per turn vary. Also the plane defined by the base pairs is perpendicular to the helix axis only in B-DNA. In A-DNA, the normal of the base pairs is tilted by 20° to the helix axis.

DNA does not occur only as long double-stranded linear rods of restricted flexibility. In some organisms such as virus and bacteriophage single-stranded DNA is found. Tertiary structures such as supercoiled and open-cyclic molecules are obtained when double-stranded DNA is isolated in an undegraded form. In the supercoiled structure, the double-stranded DNA chain is covalently closed and twisted into a superhelix. An open-cyclic structure is obtained, when one strand is broken so that it can rotate round the other nonbroken cyclic strand. This form also occurs with both ends covalently bonded in the native form. Denaturation of double-stranded molecules by heat or alkaline titration splits the two strands into two single-stranded chains which are much more flexible and therefore more closely resemble random coils of higher density. The single chains may remain chained together forming catenanes when supercoiled DNA is denatured. Such helix → coil transitions are characterized by a transition temperature (or melting temperature), a decrease in the viscosity of the solution, and an increase in the UV absorption of the solution (see I.1.3.2).

The Watson-Crick double helix contains one narrow and one broad groove between the externally located sugar-phosphate chains. At physiological pH, the phosphates are negatively charged and, for this reason, DNA is usually found associated with various metal cations occurring naturally, with polycationic amines (spermine) in some viruses and bacteria, or with positively charged proteins like histone or protamine in the chromatin of eucaryotic cells. The molecular weights of native DNA molecules range from 10^6 to more than 10^9 daltons with contour lengths between 1 to 10^3 μm and with 5.10^3 to 5.10^6 nucleotide pairs.

1.3 Electronic Properties

1.3.1 Molecular Orbital Calculations

Twenty years ago, PULLMAN et al. (1956) applied, for the first time, the molecular orbital method of quantum chemistry to the study of the structure of purine and pyrimidine bases. The aim of such research is the correlation between certain electronic and energetic molecular indices and the corresponding physico-chemical properties.

In this section, attention is given to those electronic properties of the bases which are important for the interpretation of their relative behavior under the action of ionizing radiation (PULLMAN and PULLMAN, 1963). Among these are the electron donor and acceptor properties, which measure, respectively, the ionization potential and the electron affinity. In the Hückel approximation of the molecular orbital method, the appropriate indices are the energy of the highest filled molecular orbital for the electron donor capacity and the energy of the lowest empty molecular orbital for the electron acceptor ability. The calculations yield these energies in the form $E_i = \alpha + k_i\beta$, where α is the coulomb and β the resonance integral. The values of k_i are generally in the range 0 to 1.5 for the highest filled molecular orbital and 0 to -1.5 for the lowest empty molecular orbital. The closer to zero the values of the coefficients k_i for both orbitals become, the greater are the electron donor or the electron acceptor capacities of the molecule. The results for the five naturally occurring bases are summed up in Table 1.2. These reveal that the purines are better electron donors than the pyrimidines and that guanine should be the best donor among the five bases. Furthermore, cytosine and adenine are better electron acceptors than thymine and guanine, cytosine being the best acceptor of all bases.

Table 1.2. Energy coefficients (in β units) of molecular orbitals (PULLMAN and PULLMAN, 1963)

Base	Highest filled molecular orbital	Lowest empty molecular orbital
Adenine	0.49	-.87
Guanine	0.31	-1.05
Cytosine	0.60	-0.80
Uracil	0.60	-0.96
Thymine	0.51	-0.96

The irradiation of water leads to the formation of very reactive species, such as hydrogen atoms, $OH^\cdot$ radicals and solvated electrons. The reactivity of the bases toward these agents is of primary importance. Three indices are considered to be relevant: the electronic net charge, the mobile bond order, and the index of free valence. The electronic net charge measures the excess or the lack of charge on an atom brought about by the delocalization of the electrons in the molecule. The mobile bond order is a measure of the double bond character of a particular bond in a molecule. And the index of free valence indicates the binding power not used by an atom in the π-bonds of the molecule (see II.1). The reactivity towards free radicals, for example hydrogen atoms, is governed by the mobile bond order and the index of free valence. The reaction should occur at the bonds with highest mobile bond order and highest free valences on the atoms forming the bond. In the pyrimidines, this bond is the C5-C6 double bond and, in the purines, it is the N7-C8 double bond. Among the pyrimidines, uracil and thymine are found to be more reactive toward free radicals than cytosine. In uracil and cytosine, the carbon C5 is expected to be the site of attack and, in the purines, it should be carbon C8. Electrophilic attack, on the other hand, occurs at positions characterized by appropriate values of the electronic net charge and the index of free valence. If both parameters are greater for one atom than for the other atoms in a molecule, the former will be the site of electrophilic attack. The oxygen of cytosine is interesting because among all oxygens of the carbonyl groups of the bases, it has the highest negative charge and the highest free valence. Therefore, it is the most reactive exocyclic oxygen of the bases toward proton fixation.

These calculations have been performed on neutral molecules. The electronic indices are different, however, for the ionized molecules. For example, in the anions of uracil and cytosine, the centre of reactivity for proton addition is the carbon C6. Therefore, a hydrogen atom will attack carbon C5 in uracil and cytosine, whereas a proton is expected to add to carbon C6 in the electron adducts of these molecules (WESTHOF et al., 1975). In the case of thymine the site of hydrogenation of the neutral molecule is identical with that of protonation of the electron adduct (see II.2).

The position of the base pairs in the double helix is such that these reactive groups are turned outwards. The N7-C8 double bond of purines and the groups at C5 in pyrimidines stick out in the broad groove, while the carbonyl oxygens O2 of pyrimidines are exposed in the narrow groove.

Another important parameter concerning sensitivity toward radiation damage is the resonance energy. This index is a measure of the gain in energy of a molecule brought about by delocalization of the π-electrons. Indeed, it has been suggested that the higher stability of the aromatic molecules towards radiation damage is related to their resonance stabilization (DUCHESNE, 1957). There is a good correlation between the radioresistance of a molecule and its resonance energy per π-electron (VAN DE VORST, 1970). Thus the purines are found to have greater values for the resonance energy per π-electron than the pyrimidines, which explains the greater sensitivity to radiation damage of the latter (see II.1 and II.2).

1.3.2 Absorption Spectra

By virtue of their conjugated double-bond systems, the purine and pyrimidine bases absorb light in the region of 200-300 nm. This is illustrated in Figure 1.4. CLARK and TINOCO (1965) classify the UV absorption bands of the nucleic acid bases into three groups. The first band includes the longest wavelength of absorption (260-280 nm). Guanine and cytosine absorb toward the longer wavelength limit, with adenine and uracil (or thymine) absorbing

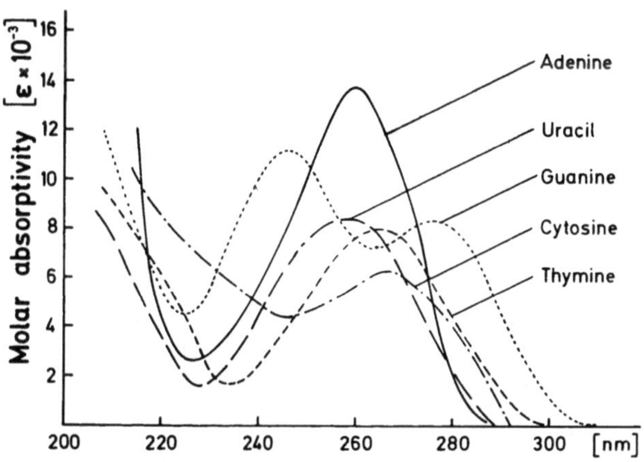

Fig. 1.4. UV absorption curves of the purine and pyrimidine bases at neutral pH

toward the shorter wavelength limit. The second band does not appear in all bases distinctly. It is a shoulder at 237 nm in cytosine and at 256 nm in guanine, and finally the third band comprises the maxima below 200 nm. The pH of the solution has a marked effect on the absorption bands, especially in the cases of cytosine and adenine. Self-consistent field (SCF) calculations with configuration interaction reproduce fairly well the positions and intensities of the observed spectra (PULLMAN and PULLMAN, 1969). Most of the intensity of these bands arises from $\pi \rightarrow \pi^*$ transitions which are polarized in the plane of the base. In addition, the nonbonding electron pairs of the ring nitrogens and carbonyl oxygens give rise to weak $n \rightarrow \pi^*$ transitions which are polarized perpendicular to the base plane. Absorption which could be definitively assigned to $n \rightarrow \pi^*$ transitions has not been reported in aqueous solutions. However, in single crystals, some weak absorption bands have been attributed to $n \rightarrow \pi^*$ transitions in adenine and uracil. The UV absorption in the wavelength range 220-290 nm of the nucleic acid bases is useful for their isolation and characterization. In this respect, the phenomenon of hypochromicity in DNA should be mentioned. Hypochromism is a decrease in the absorption of light by a chromophore. Usually, the peak UV absorption at 260 nm of DNA is observed to be lower than that of the absorption spectra of its constituent bases randomly oriented. This effect arises from the coulombic interactions between the partially aligned transition moments of the stacked bases (RHODES, 1961). The absorption at 260 nm of DNA solutions rises by 20-60% as the DNA undergoes a helix-coil transition, since the amount of stacking is decreased in the single-stranded DNA. The absorption curves of Figure 1.4 show that the increase of absorption at 260 nm is proportional to the content of A-T pairs in the DNA (see II.4).

1.4 Interactions

Following the discovery of the double helical structure of DNA (WATSON and CRICK, 1953), it was commonly believed that specific hydrogen bonds between the purine and pyrimidine bases were solely responsible for the stability of the helix. However, later investigations have shown that base stacking is another major type of base-base interaction. It is now widely held that, while the specific hydrogen bonds between complementary bases are essential for the pairing and, thus, also for the replication and transcription properties of DNA, the vertical stacking of the bases is also responsible for the stability of the DNA helix (SINANOGLU, 1968).

The sugar-phosphate backbone provides a further factor of stability. Because of the negatively charged phosphates, the double-stranded DNA is only stabilized in the presence of counterions, which reduce the electrostatic repulsion between the phosphate groups (see I.1.4.4). However, next to these ionic bonds, the limited number of preferred conformations of the polynucleotide backbone imposes restrictions and promotes the stacked helical arrangement of adjacent bases (SUNDARALINGAM, 1975). Thus, the double helical structure of DNA is largely the result of the conformational energy of the sugar-phosphate backbone which allows for the stacking of the bases. In this way a hydrophobic core is created where hydrogen bond formation between complementary bases is favored. The short-range Van der Waals-London forces are responsible for the hydrogen-bonding interactions and the stacking interactions, as well as for the additional sugar-base and sugar-sugar interactions. Several related reviews have been written and we will note those of VOET and RICH (1970) and SOBELL (1972) for the hydrogen bonding between the bases, those of BUGG (1972) and TS'O (1974) for the base stacking, and that of SUNDARALINGAM (1975) for the role of the polynucleotide backbone.

1.4.1 In-Plane Interactions : The Hydrogen Bonds

A feature of the nucleic acid bases is that only the complementary bases in the Watson-Crick sense (A-T, A-U, G-C) co-crystallize and form hydrogen-bonded complexes. Indeed, despite the fact that several mixtures have been tried, and putting aside the interpurine or interpyrimidine complexes, the base-pair complexes found in the solid state are formed exclusively between derivatives of adenine and uracil (or thymine) or guanine and cytosine. However, a variability in the hydrogen-bonding schemes has been found. HOOGSTEEN (1959) discovered that methylated derivatives of adenine and thymine co-crystallize as a hydrogen-bonded complex with a configuration different from the Watson-Crick one. Instead of being linked to N1 of adenine, the thymine molecule is linked to N7 (Fig. 1.5a: Hoogsteen configuration). This type of hydrogen bonding is quite general for associations between derivatives of

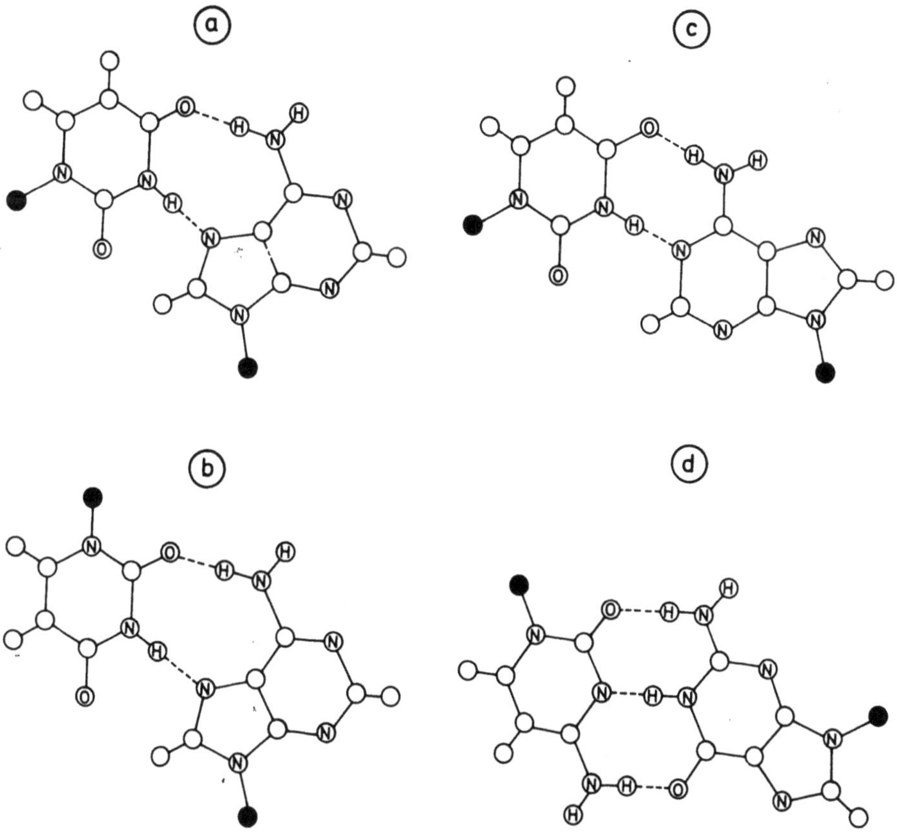

Fig. 1.5. Base-pairing schemes: (a) adenine-uracil in a Hoogsteen pair involving N7 of adenine; (b) adenine-uracil in a reversed Hoogsteen pair (related to *a* by rotation of uracil about N3-C6). The pairing involves C2=O2 of uracil instead of C4=O4 of uracil; (c) adenine-uracil in a Watson-Crick pair; (d) guanine-cytosine in a Watson-Crick pair

adenine and uracil; in some cases, they may be of the "reversed Hoogsteen" type (Fig. 1.5b). However, recently it was found that the sodium salt of adenylyl-3', 5'-uridine phosphate (SEEMAN et al., 1976) crystallizes with the Watson-Crick hydrogen-bonding scheme between uracil and adenine (Fig. 1.5c). The crystalline complex ethidium: 5-iodouridylyl-3',5'-adenosine phosphate also presents the Watson-Crick base pairing between the adenine and uracil residues (TSAI et al., 1975). Derivatives of guanine and cytosine co-crystallize only according to the Watson-Crick scheme (see Fig. 1.5d).

Because of the strong competition between the solute and the solvent for hydrogen-bonding interaction, the formation of hydrogen-bonded pairs in solution can only be observed in nonaqueous solvents like chloroform or carbon tetrachloride which show weaker hydrogen-bonding capacity than water, or none at all. Additionally, in aqueous solution the association between purine and pyrimidine bases is nonspecific and results in the vertical stacking of the bases. However, in nonaqueous solvents, IR and NMR studies have shown that the strongest H-bonds form between adenine and uracil derivatives, as well as guanine and cytosine: stronger than those between derivatives of the same base or between "false" pairs.

In 1962, DE VOE and TINOCO showed that Van der Waals-London forces play a role in the interaction between the horizontally associated bases of DNA. This has been demonstrated more significantly by PULLMAN and has collaborators on the basis of quantum mechanical calculations (PULLMAN and PULLMAN, 1969). The total interaction energy is the sum of the energies of the electrostatic interactions, the polarization interactions, and the dispersion or London interactions. The electrostatic interactions were calculated in the monopole approximation, i.e., each charge on one molecule has Coulomb interactions with each charge on the other molecule. The polarization interactions occur between the dipole moment of one molecule and the electric field induced in that molecule by the charge distribution on the other molecule. Like the London interaction energy, the polarization interaction energy varies with the inverse sixth power of the distance. According to these calculations, the electrostatic component of the Van der Waals-London forces predominated in hydrogen bonding. Comparison of the pairing schemes shows that the Hoogsteen arrangement is the most stable one in the case of the A-U pairing, while it is the Watson-Crick arrangement for the G-C pairing. Calculations also reveal that the interaction energy of A-T and G-C pairs is greater than the energies of self-association of their two constituents. In addition, the four "false" pairs (A-C, G-T, C-T, A-G) have interaction energies smaller than the energies of self-association of one of their constituents or than the mean of the energies of the two constituents.

1.4.2 Vertical Associations : The Stacking of the Bases

Besides hydrogen bonding, a major type of base-base interaction in aqueous solution and in the solid state is the vertical stacking of the bases. In the solid state, the planes of the bases are usually parallel with an interplanar spacing of 3.4 ± 0.2 Å. An outstanding feature of the stacking is the consistency with which specific patterns appear in widely different crystalline environments. Furthermore, the stacking interactions in fibres of polynucleotides show a number of striking similarities to those found in crystals of their constituents. As in the crystal structures of adenine nucleosides and nucleotides, the major interactions between adjacent bases involve contact between an amino group and an imidazole ring in the double-helical structure of polyadenylic acid. Intrastrand stacking of adenine residues in B-DNA also involves the amino group of adenine. In general, only partial base overlap is observed in most of the known structures, with polar bonds of one base interacting with the polarizable ring of an adjacent base. Thus it appears that base stacking is governed to a large extent by polarization forces. The importance of dispersion forces in that kind of vertical association has indeed been suggested by theoretical calculations (DE VOE and TINOCO, 1962; PULLMAN and PULLMAN, 1969).

The purines do not stack in nonaqueous solvents like chloroform, dimethylsulphoxide, and dimethylformamide, while they associate considerably in aqueous solution by vertical stacking. This has been taken to mean that hydrophobic solvent effects are mainly responsible for the base stacking in aqueous solution. However, the nonaqueous solvents possess polar substituents which are able to associate directly with the bases. Such specific solvent-base interactions compete with the stacking forces, thus disrupting the stacking in nonaqueous solvents. Hydrophobic solvent factors probably help promote nucleic acid derivatives stacking, but it is likely that other forces are also involved. Several studies have demonstrated that the extent of association of purine and pyrimidine derivatives in aqueous solution is directly related to the polarizabilities of the bases (TS'O, 1974), suggesting again the possible importance of polarization forces in the base-stacking interactions. Quantum mechanical calculations have been performed to evaluate the relative contribution of the two types of interaction, horizontal and vertical, to the overall stability of the double-helical structure of DNA. The intermolecular forces were calculated in the "monopole" approximation (PULLMAN and PULLMAN, 1969), in continuation of earlier work by DE VOE and TINOCO (1962) on the "dipole" approximation. The calculations show that the two types of interaction are of comparable magnitude and contribute equivalently to the overall stability. Further, the electrostatic term is predominant for the in-plane interactions, whereas it is the dispersion term which contributes most to the stacking interaction.

1.4.3 The Role of the Sugar-Phosphate Backbone

The discussion in I.1.2.2 has shown that, despite the various rotatable single bonds, the number of allowed conformations of a nucleotide is small. Furthermore, the rotations about the single bonds are not only restricted but also correlated among themselves (SUNDARALINGAM, 1969, 1973). The recognition of the strong difference in the conformation properties of nucleosides and nucleotides led SUNDARALINGAM (1973) to the concept of the "rigid" nucleotide. Thus, in constrast to the considerable flexibility of nucleosides with respect to the glycosyl torsion angle and the C4'-C5' bond, 5'-nucleotides and dinucleotides have exclusively been found in the anti conformation with the gauche$^+$ rotamer. This is further exemplified by the fact that all known polynucleotides occur in the anti-gauche$^+$ conformation. The two nucleotide units (Fig. 1.3) together with the rotational flexibility of the P-O bonds facilitate considerably the understanding of helical and folded structures of polynucleotides.

The "rigid" nucleotide unit is to the polynucleotide structure what the planar peptide bond is to the α-helix of proteins. The less preferred conformations may be important for the formation of loops and bends in polynucleotides and contribute to the randomness of coiled polynucleotides. The evidence for the "rigid" nucleotide concept, mainly gained from solid state data, is now amply corroborated by NMR studies and semiempirical or quantum-mechanical calculations.

Semiempirical calculations show that the preferred conformation of the phosphodiester group is such that the polynucleotide will form a helix permitting the stacking of the bases (YATHINDRA and SUNDARALINGAM, 1975). Moreover, the appropriate stacking distance of 3.0 to 4.0 Å, all other parameters optimized, will only be obtained when the glycosyl torsion angle falls in the usual narrow anti or syn ranges observed in the solid state for the monomers (OLSON, 1976). The torsion energy and the short-range interactions, responsible for the preferred conformations of the sugar-phosphate backbone, are thus the source of the driving force yielding the helical and stacked arrangement of the bases in polynucleotides. The hydrogen bonding between base pairs in double helices provides additional stability. In this respect, the fact that Watson-Crick pairing between adenine and uracil has exclusively been observed in dinucleoside phosphate crystals not only stresses the importance of the sugar-phosphate backbone for this type of bonding in polynucleotide structure but suggests also that the observed hydrogen-bonding scheme might depend on the sugar-phosphate conformation.

1.4.4 The Stability of the DNA Helix

The stability of the helical structure of DNA is strongly influenced by several factors (BLOOMFIELD et al., 1974). These include the base composition, the molecular weight, the salt concentration, the pH of the solution, and the nature of the solvent. The "melting curve" or the profile of the UV absorbance at 260 nm against temperature yields a "melting temperature" which is taken to be a measure of the helix stability (see I.1.2.3 and I.1.3.2). The melting temperature increases linearly with the fractional G-C composition showing that the "G-C rich type" DNA has a greater thermal stability than the "A-T rich type". It can be added that below a molecular weight of 2.5 10^6, the stability of the DNA helix decreases with decreasing molecular weight.

The concentration of salts or protons may shift the equilibrium helix $\rightleftarrows$ coil[1] and, accordingly, the melting temperature, depending on the preferred form for binding. Since the charge density and therefore the electrostatic potential of the helix is larger than that of the coil, the helix attracts counterions more ef-

[1] The word "coil" is used here for the denatured form of the DNA in contrast with its native helical form. It does not imply that the single strands are truly random coils (see I.1.2.3).

fectively. Thus an increase in the concentration of counterions stabilizes the helix and raises the melting temperature. On the other hand, the bases protonate or deprotonate more readily in the coil than inside the helix. Therefore, below pH 5 and above pH 9, the helix-coil equilibrium is shifted toward the coil. The dependence of the melting temperature on salt concentration varies with the pH. At high pH, because the coil is more ionized than the helix, its negative charge density becomes larger than that of the helix and it therefore binds more counterions. Thus at high pH values, higher salt concentration destabilizes the helix. However, at acid pH, since the coil binds more protons, the difference between its charge density and that of the helix is increased and the equilibrium is displaced toward the helix. Organic solvents and several anions ($CCl_3CO^-_2$, CNS^-, ClO^-_4) destabilize the DNA helix. Correlations are found between the solubilizing power of these agents on the bases and their denaturing effectiveness. Since this solubilizing action occurs more effectively in the coil form, these agents shift the helix-coil equilibrium toward the coil.

1.4.5 Interactions with Aromatic Molecules

A large number of aromatic molecules influence the biological functions of nucleic acids by binding to them. These include pyrimidine and purine bases; aromatic amino acids and related compounds; polycyclic hydrocarbons; the aminoacridine dyes; and some antibiotics like actinomycin D. These molecules are inactivating, carcinogenic or mutagenic agents. Moreover, the aminoacridine dyes, among others, absorb light in the visible region and give rise to the photodynamic effect, i.e., in the presence of dyes the absorption of visible light by the cells leads to their destruction (see I.7.3). Consequently, the interactions of such molecules with DNA have been extensively studied. Although analogous types of binding have been proposed for molecules like actinomycin D and ethidium bromide, most of the results have been obtained from aminoacridine dyes. We shall select their binding properties as a model. Like most of the other compounds, aminoacridine dyes possess a planar aromatic system with three rings. In addition, some are also positively charged at neutral pH like the aminoacridine dyes. Detailed information about this subject may be found in the reviews of PEACOCKE (1973) and LOCHMANN and MICHELER (1973).

Two modes of binding are usually considered. In the strong binding complexes, up to one dye per 4-5 nucleotides is bound while, in the weak binding complexes, a one-to-one dye per phosphate stoichiometry is present. The ionic strength influences the extent of binding particularly in the weak binding mode and less effectively in the strong binding one. This indicates that electrostatic forces are predominant in the weak binding and that they play a role in the strong binding. In the strong binding complex, the dye molecules intercalate between two base-pairs in the DNA whereas, in the weak binding complex, the dye molecules aggregate on the sugar-phosphate backbone. The intercalation model, originally proposed by LERMAN (1961), relies on viscosity and sedimentation studies, which have shown that the contour length of DNA is increased upon binding. Indeed, according to this model, the intercalation of the dye molecule in the DNA helix is made possible by a local untwisting of the double helix and the subsequent extension of the distance between two base-pairs to twice their distance in the absence of bound dye. In the interspace so achieved, the dye molecule lies parallel to the planes of two successive base-pairs with its positively charged nitrogen atom near the helix axis. In this position, maximum overlap occurs between the π-electrons of the intercalated dye and the neighboring base-pairs.

Furthermore, the positively charged nitrogen atom is equidistant from the two sugar-phosphate chains. This attraction between the intercalated cations and the DNA phosphate contributes to the stabilization of the complex. Van der Waals-London forces are mainly responsible for the stability of the strong binding complexes between acridine dyes and DNA. The limited extent of strong binding shows that it is an anti-cooperative process which implies that, after intercalation of a dye molecule, further binding to neighboring sites is excluded. The detailed structure of intercalation complexes has been recently determined by X-ray diffraction studies on the co-crystals ethidium: 5-iodouridylyl-3'-, 5'-adenosine phosphate (TSAI et al., 1975), and 9-aminoacridine: adenylyl-3', 5'-uridine phosphate (SEEMAN et al., 1975).

In these complexes, the observed unusual conformation of the sugar-phosphate chains was related to the unwinding of the double helix fragment in order to accommodate the ethidium or acridine molecule in accord with the intercalation model.

On the other hand, cooperative and electrostatic interactions are the major sources of stabilization of the weak complexes. In weak binding, the dye molecules are externally bound to the DNA helix by electrostatic forces between the DNA phosphates and the ring nitrogens of the aminoacridine cations. The aggregated dye molecules are oriented parallel to the base planes of DNA and stacked on top of each other in the same direction as the helix axis. External weak binding is thus a cooperative process.

Besides double-helical DNA some single-stranded macromolecules bind dyes strongly. The intercalation model cannot immediately explain the strong binding to single-stranded polymers. A modified model has been proposed by PRITCHARD et al. (1966), in which the acridine cation lies between successive nucleotide bases on the same polynucleotide chain in a plane approximately parallel to the base plane but at an angle such that the positive ring nitrogen of the acridine is close to the polymer phosphate group. In that case, the acridine is eccentric and interacts only with adjacent bases of one DNA chain.

Chapter 2

Interaction of Ionizing Radiation with Matter

2.1 Introduction

Observed chemical or biological changes induced by ionizing radiation result from chains of events. Following PLATZMAN (1958, 1967) and HART and PLATZMAN (1961) one may distinguish more or less distinct, consecutive, temporal stages in this response of a medium to irradiation. The first and physical stage is that of dissipation of radiant energy, followed by the physicochemical stage in which the primary products undergo secondary reactions. The third stage is the chemical stage. With condensed matter this stage may be subdivided into a track period, in which spatial inhomogeneities play an important role, and a subsequent uniform period. Finally, in organisms, there follows a biological stage.

In this section we are concerned primarily with the first two stages. For reasons more fundamental than brevity of presentation the discussion must of necessity be quite unsatisfactory. Firstly, any description of the radiation in the interior of an irradiated material must inevitably be incomplete. Secondly, even for simple molecules much remains to be clarified and understood. This holds for isolated molecules and more so for molecules in irradiated liquids or solids. Clearly, the important task of identifying the initial effects of ionizing radiations on biological molecules, in particular the very large molecules of the nucleic acids, must still proceed with inadequate aid from physics. What is known for simpler molecules in the gas phase has still to serve as an indispensable part of the background for a qualitative understanding of what may take place in condensed materials of a more complex constitution. This is the justification for including, in the following discussion, material which reflects knowledge pertaining to situations which otherwise would appear to be of little relevance to the present book. It should be stressed, however, that theoretical insight and experimental information pertaining to specific molecular species and/or physical conditions should be applied to different molecules or conditions only with great reservations.

2.2 Ionizing Radiation

Ionizing radiation is any radiation consisting of either directly or indirectly ionizing particles. Directly ionizing particles are charged particles (electrons, protons, α-particles, "heavy" ions, etc.) having sufficient kinetic energy to produce ionization by collisions with atoms or molecules. Indirectly ionizing particles are uncharged particles (photons, neutrons, etc.) which can liberate or generate directly ionizing particles. These definitions are essentially as recommended by ICRU, the International Commission on Radiation Units and Measurements (1971).

The high-energy photons of ordinary X- or γ-radiations act on the irradiated material almost exclusively through the fast electrons (secondary radiation) to which they give rise (photoelectric effect, Compton effect, electron-positron pair production). These radiations, together with incident fast electrons, constitute, therefore, the "electronic" radiations. Fast neutrons impart energy to atomic nuclei of the medium to produce - in hydrogenous materials - recoil protons. The question of understanding the action of ionizing radiations becomes, therefore, mainly one of understanding how charged particles act.

Protons and heavier ions act partly through secondary electrons, including occasionally fast ones (δ-rays). Electrons, therefore, play a particularly significant role in radiation research.

2.3 Degradation Spectra and Linear Energy Transfer (LET)

A prerequisite to a complete understanding of the effects induced in an irradiated material would be a description of the nature of the radiation(s) present at any point. This description, in practice, can be but an approximate one. One important approach is that of determining the so-called degradation spectrum for each kind of ionizing particle present. Closely related is the simpler one of estimating the so-called LET distribution, which gives a rough but useful idea of the spatial distribution of the discrete energy transfers from the ionizing particles to the irradiated medium.

The degradation spectrum $y(T)$ is defined so that, for electrons, $y(T)dT$ is the aggregate distance traversed by all electrons (primaries and/or secondaries) in spanning the kinetic-energy range from $T + dt$ to T. The theory of electron-degradation spectra which was developed by SPENCER and FANO (1954) included the occasional production of fast secondaries. Reference to subsequent theoretical as well as experimental work may be found in the papers by EMERSON et al. (1973) and DOUTHAT (1975). A typical aspect of the degradation spectrum of an initially fast electron is a rapid rise in the electron flux at T-values below a few hundred electron volts. While the Spencer-Fano theory is not applicable in this energy region, the numerous low-energy electrons are expected to be important to the initiation of chemical effects.

The essentially continuous slowing-down of a heavy charged particle renders its degradation spectrum essentially equal to the reciprocal of the (average) stopping power $-dT/dl$. The concomitant electron degradation spectrum can only be estimated, due to inadequate knowledge of the initial energy distribution of the secondary electrons. The high-energy part of this distribution, the δ-ray spectrum, can be calculated, however, and the corresponding electrons may contribute decisively to the effects of irradiation. This is true for energetic heavy ions in particular, as discussed by BRUSTAD (1962) and by BUTTS and KATZ (1967). For recent progress in the determination of the energy distribution of secondary electrons the paper by KIM (1975) should be consulted.

The linear energy transfer (LET) has been defined by the International Commision of Radiation Units and Measurements (1970) as

$$L_\Delta = (dE/dl),$$

which stands for the mean energy transferred per unit track length, excluding transfers in a single process of an amount of energy exceeding some specified value Δ. If one chooses Δ = 200 eV, for example, then in a rough sense transfers giving rise to δ-rays are excluded. Such rays are better treated separately. The symbol L_∞ is used for the value of LET with all energy transfers included, in which case LET amounts to the stopping power. During the slowing-down of an ionizing particle, and for particles of different energies, there will be a variation in the rate of energy transfer. The resulting distribution of L-values is expressed alternatively as a "track-length LET distribution" or as an "absorbed-dose LET distribution", with the associated "track-average LET" and "absorbed-dose average LET" respectively. These quantities have been discussed in detail by the International Commission on Radiation Units and Measurements (1970) in a report which also contains useful tables and diagrams for energy losses, stopping powers, ranges, etc.

L divided by the value of the average energy lost by the particle per event gives the average density of energy-transfer events along the particle track. It should be borne in mind, however, that the actual distribution is a stochastic one. For an initially energetic electron moving in a condensed medium of nucleic-acid composition the average separation of adjacent events will be of the order of 100 nm as long as the electron energy stays above 100 keV, and roughly 10 nm at 1 KeV. The diameter of the double helix, in comparison, is about 2 nm. At these energies, therefore, and for electrons taking more or less the shortest path through the helix, the probability is quite small that two or more energy transfers take place within it. This probability increases as the electron slows down further. At a few hundred electron volts it will be close to unity. Below some 100 eV the probability of more than one ionization decreases rapidly and excitations become predominant.

2.4 Primary Processes and Primary Products

The adjective "primary" is used variously in the literature on irradiation effects, with a restriction implied varying from the strictly "initial" to the somewhat ambiguous "early". Taken to be the germane initial events, however, the primary processes are collision processes with energy transfer from the (fast) primary or secondary charged particles to the internal structure of molecules of the irradiated medium. The corresponding primary products are the activated (excited or ionized) molecules formed in such collisions. The initial (physical stage) distribution of the energy deposited by the radiation is represented, therefore, by molecules activated to a great variety of states and having a nonuniform spatial distribution. In condensed media several molecules may be activated in a single primary process.

These primary products, supplemented by the slow fraction of electrons set free in direct-ionization processes, are the precursors of essentially all observed effects of irradiation. They are unstable and, usually, with lifetimes much too short to permit experimental observation of the primary products themselves. About 1960 remarkable progress started in the discovery and identification of transient species produced in pulse radiolysis and detected by optical, ESR, and electrical techniques as discussed, for example, by MATHESON and DORFMAN (1969). The excited molecules, radicals, or ions observed are as a rule not the primary products of irradiation, however. This holds true even in the field of study opened up by the more recent stroboscopic pulse-radiolysis system described by BRONSKILL et al. (1970), in which signals are observed with a time resolution as short as 20 ps ($2 \cdot 10^{-11}$ s) following the initial processes. An approximate time scale for events in radiation chemistry has been presented by SWALLOW (1973).

2.5 Activation Spectra

The relative frequency of production of the different, normally activated states is rather independent of the energy and charge of the impinging particles as long as the latter are swiftly-moving. For electrons this holds for kinetic energies exceeding some 100 electron volts. The approximation discussed below fails progressively for slower electrons. The slow electrons, the effects of which may be quite important, will be considered in a later section.

In the overwhelming majority of collisions of a fast charged particle it loses but a very small fraction of its kinetic energy. The primary products produced by it are formed, therefore, predominantly in such "distant" or "glancing" collisions. The varying electric field acting on the electronic system of an isolated molecule during the passage of the particle turns out to be approximately equivalent to that of a beam of light covering the vast, continuous spectral range from the visible region to the extreme ultraviolet ($h\nu > 100$ eV), and with the number of photons per unit frequency interval (or, photon-energy interval) inversely proportional to the frequency (or, to the photon energy). This fact forms the basis of the valuable optical approximation according to which the energy spectrum of primary events is obtainable from the optical absorption spectrum simply by applying to the latter the conversion factor $1/E$, where E is the transition energy at the respective spectral points and which equals the energy transferred from the impinging particle. For molecules, because transitions are to quasi-continuous excited states or to the ionization continuum, we introduce the differential optical oscillator strength $df(E)/dE$ and arrive at the activation spectrum in the optical approximation given by the expression

$$(1/E)df(E)/dE.$$

The first sufficiently complete activation spectrum deduced from various experimental data by PLATZMAN (1962a), that of CH_4, revealed that, on the average, each individual "package" of energy delivered by a fast, charged particle is quite large. Normal primary processes involve an activation of valence electrons of some 10-30 eV. This statement reflects what appears to be a common trait of the otherwise mutually different action spectra of ordinary organic molecules as well as the water molecule. A direct experimental approach is that of determining electron energy-loss spectra (cf., e.g., LASSETTRE and WHITE, 1974; see also Fig. 2.1). The light from synchrotrons, on the other hand, covers the spectral region required for a determination based on the optical approximation (cf. PERLMAN et al., 1974).

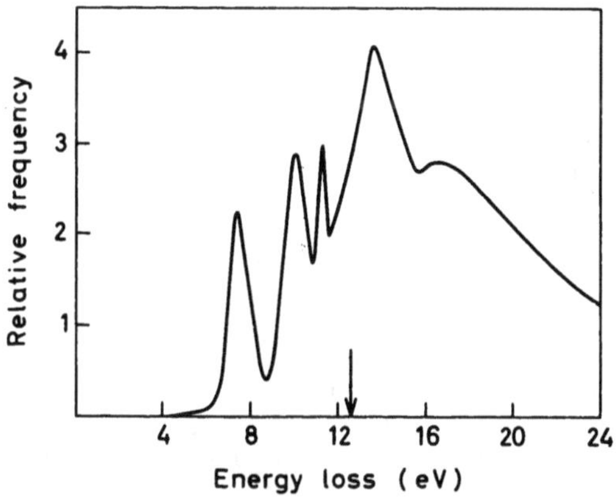

Fig. 2.1. Small scattering-angle, low-resolution, electronimpact spectrum of water molecules for 500 eV electrons. *Arrow* indicates the first ionization potential of H_2O. (Adapted from LASSETTRE and WHITE, 1974)

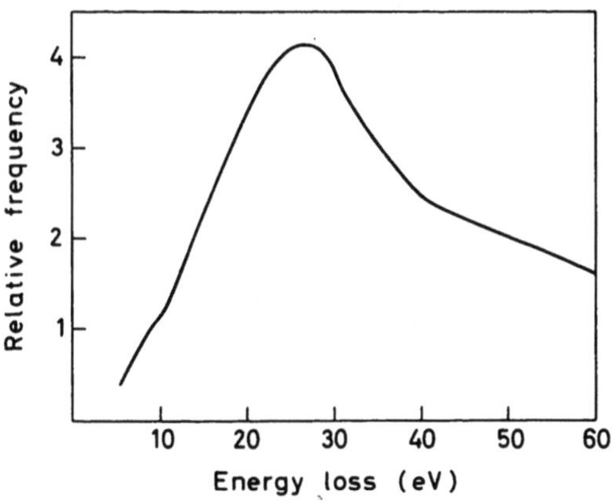

Fig. 2.2. Small scattering-angle energy-loss spectrum for 150 keV electrons passing through a 200 nm thick film of nucleic acid. (Adapted from JOHNSON and RYMER, 1967)

In macromolecules like the nucleic acids effects might show up which are normally associated with condensed media. In the latter a single-act collective excitation of valence electrons of a few or many molecules is possible due to coulombic coupling. Theories of collective excitation have been worked out for idealized models. PLATZMAN (1967) has discussed the expected differences between the spectra of primary activations of isolated molecules and molecules in the condensed phase. Experimentally, condensed-phase activation spectra have been obtained, for a variety of materials, in the form of approximate first-collision energy-loss spectra of fast electrons in thin films. The materials include nucleic acid (Fig. 2.2), studied by JOHNSON and RYMER (1967). A common trait of these spectra is an observed, broad maximum at 20-25 eV, which is usually interpreted as the manifestation of collective excitations. Nevertheless, the effect of this phenomenon on the activation spectra appears to be moderate. The spatial and temporal correlations between the activations may be extremely important in condensed-phase radiolysis, however (cf. SANTAR and BEDNAR, 1969).

2.6 Superexcited States

Molecular activation spectra indicate that fast, charged particles induce transitions predominantly to high-lying states. Direct ionization takes place if an electron is ejected during the excitation act. It was emphasized by PLATZMAN (1962a, b, 1967) that primary activations are frequently to intermediate, so-called superexcited states, also above the first ionization between (indirect) ionization and atomic rearrangements, the latter leading to neutral-product formation. Originally, PLATZMAN (1962a) arrived at the conclusion that the predominant primary products of irradiation are such superexcited states of molecules, and that the observed ionization is mainly indirect. PERSON (1972), who treated also noncompetitive neutral-product formation from superexcited states, estimated that, for various examples studied in the low-energy region of superexcitation, the fraction of ionization from competitive preionization was less than 40% in most cases, and usually in the range of 10-20%.

The nature of superexcited states of molecules is not well known as yet. In a condensed medium the consequences of superexcitation and of collective excitation could be quite similar, however. Eventual localization of the initially delocalized energy in the latter case might result in the production of a molecule in a superexcited state.

Dissociation of a superexcited molecule may lead to chemically very potent, energy-rich products in the form of either electronically excited or high-velocity fragments. Quantitative results of the radiolysis of liquid hydrocarbons were explained in this way first by HATANO et al. (1968), who utilized the optical approximation in their calculation of the yield of "hot" hydrogen atoms assumed to result from decomposition of superexcited hydrocarbon molecules. A review on superexcitation was presented some years ago by MAKAROV and POLAK (1970).

2.7 Inner-Shell Ionizations

The role of multiple ionization in radiation action has been discussed by PLATZMAN (1952) and by ORE (1967). Direct double or multiple ejection of electrons in a primary process plays an insignificant role under ordinary experimental conditions. Indirect multiple ionization, initiated by the ejection in a primary process of an inner-shell electron, is of some interest, however.

Consider a nucleic-acid molecule in which an electron has been expelled from the K-shell of a carbon, nitrogen, or oxygen atom, or from the L-shell of a phosphorus atom (the yield of K-shell ionizations in phosphorus is normally negligible). The vacancy produced represents localized energy of an amount of about 280 eV, 400 eV, 530 eV, and 140 eV, respectively. The electron ejected in the primary process will be followed in some 10^{-15} sec by another one emitted in an Auger transition. This Auger electron will have an energy in the region where the ionization cross-section is particularly large. It will produce, therefore, several nearby ionizations. Again, additional electrons may be given off by the molecule during the adjustment processes. Clearly, an inner-shell ionization creates an exceptionally high degree of disturbance within and, in condensed media, in the immediate neighborhood of the molecule. The energy deposited exceeds the vacancy energy by an amount equal to the kinetic energy of the electron expelled in the primary act. It may give rise to some 5-10 ionizations altogether. The possible consequences are not known for nucleic acids. Results obtained on different molecules bring out certain pertinent points, however.

An inner-shell vacancy in the heavy atom of CH_3I, C_2H_5I, and $Pb(CH_3)_4$, respectively, was shown by CARLSON and WHITE (1966) to result - in the gase phase - in a "coulombic explosion" completely fragmenting the molecule into positive, high-speed atomic ions (see III.2.2.4). More relevant in the present context are the observations by NATH et al. (1970) on various cobalt complexes in the solid phase. Chemically very stable chelates were among those molecules which were found to fragment. In one case fragmentation was enhanced by the dispersal of the complex in an inert matrix. In some large and highly conjugated systems no fragmentation seemed to take place. Rapid dispersal of charge through the conjugated system, combined with the large size, could render the coulombic repulsion inadequate to fragment the molecule in the first stage of development. The next step, involving charge neutralization comprising the molecule and its immediate neighborhood, apparently did not lead to fragmentation either. Clearly, one should be careful not to talk in too general terms of what are the immediate, physicochemical consequences of inner-shell ionization in molecules.

A prescription for the prediction of absolute yields of inner-shell ionizations is due to DURUP and PLATZMAN (1961). In the case of nucleic acids a crude estimate leads to a predicted G-value (defined in I.2.10) somewhat in excess of 0.03 for irradiation with electrons or photons in the MeV energy region (cf. ORE, 1967). Taking into account the resulting multiple ionization, one arrives at an expected fraction of about 5% of the totality of ionizations in the material being associated with inner-shell ionizations. Even if rare, these kinds of primary process might give rise to special chemical effects requiring particularly great amounts of energy or high degrees of local disturbance.

2.8 Slow Particles

The initial energy distribution of slow secondary electrons is not well known. For gases the size distribution of the corresponding small ion clusters reflects this spectrum in a rough sense. Various experimental findings, critically examined by ORE and LARSEN (1964) and by ORE (1971), suggest that, for example, some 60% or more of the secondaries produced by an energetic electron will be unable to ionize on their own. The molecular activation spectra tend to support this conclusion.

An expected accumulation of electrons in the low-energy part of the degradation spectrum adds significance to the role played by slow electrons. According to predictions summarized by SANTAR (1971), for molecules like $(CH_2)_n$, about two out of every three excitations are produced by electrons of energy below about 100 eV. Ionizations, on the other hand, are produced, naturally, predominantly by fast electrons, which typically are responsible for some 80-90% of the ionizations. For fast electrons the frequency of ionizations exceeds that of excitations by a factor of two or more. The presence of the numerous slow electrons tends to make the average ratio of ionizations to excitations equal to about one, however.

For slow electrons the optical selection rules are relaxed. That is, they induce both optically allowed and optically forbidden transitions. At energies below the respective thresholds, however, an electron will be unable to cause ionization ("subionization electron") or electronic excitation ("subexcitation electron"). Collisions leading to vibrational and rotational excitation will still be possible, and various types of electron attachment processes may occur, as discussed by CHRISTOPHOROU and COMPTON (1967).

As previously mentioned, there appears to be no drastic difference between gas-phase and condensed-phase activation spectra - detailed structure being neglected. For a condensed medium the small, perturbed, spatial region associated with the most frequent depositions of energy up to about 100 eV is called a "spur". Spur reactions have played an important role in discussions on radiation chemistry. [Overlapping spurs, as well as the perturbed regions associated with secondaries of moderate energy, are the so-called "blobs", each representing some 100-500 eV (cf. MOZUMDER and MAGEE, 1966)]. It should be emphasized, however, that little is known about either the primary, activated states or the ionization (charge-separation) mechanisms in condensed media. For recent discussions on ionization in organic liquids by electronic radiations the reader is referred to the paper by ZADOR et al. (1975) and references given therein.

For positive particles (protons, α-particles, etc.) the phenomenon of electron capture sets in when they are slowed down to velocities comparable to those of valence electrons in atoms and molecules. At still lower velocities nuclear collisions are expected to predominate. According to SNYDER and NEUFELD (1957), for protons in a material like nucleic acid, nuclear collisions should account for most of the energy losses at kinetic energies below about 1.5 eV. While neither of these two types of processes plays a significant role in the total absorption of energy from an initially high-energy proton, nuclear collisions are of special interest because the energy imparted to an atom in such a collision is likely to cause disruption of the molecule. References to papers dealing with effects ascribed to nuclear collisions in biological macromolecules, including DNA, are to be found in the book by DERTINGER and JUNG (1969). Nuclear collisions are not the only processes which can lead to the release or displacement of an atom. In solids such effects are observed to follow even from nonionizing events, the mechanism probably being energy conversion in a radiationless electron-hole (exciton) recombination.

2.9 Direct and Indirect Effects. Energy Transfer

In discussions on radiation effects at the molecular level one distinguishes between so-called direct effects and indirect effects. Somewhat different interpretations of these concepts are to be found in the literature, however. One definition is that of DERTINGER and JUNG (1969): A direct effect is one which is produced in the same molecule as that in which the primary process took place. In liquid, e.g., aqueous solution, indirect effects are caused by diffusible radical products of the radiolysis of water. Indirect radiation action can occur in dry and solid media as well. Thus, irradiation of organic compounds gives rise to free hydrogen atoms which may react with undamaged molecules.

In a complex molecule - in particular in the very complex molecules of the nucleic acids - a multitude of different primary processes and ensuing intramolecular processes are possible. Our knowledge as to the nature and probabilities of these various processes is quite incomplete and qualitative although a variety of final effects is known.

Various mechanisms have to be considered in attempts to account for how changes - observed or unobserved - in one subunit of a molecule can lead to an effect in a different subunit. Phenomena of this kind are often and rather noncommittally said to be due to "intramolecular energy transfer". They may involve, however, charge transfer also (electron transfer or exchange, proton-position shifts, etc.). If strong coupling exists between the molecular subunits then electronic excitation energy will be more or less completely delocalized. In that case "energy transfer", described in terms of exciton migration, refers to the tansition from a nonstationary state to one of localized energy, represented by the final effect. If, on the other hand, the coupling is very weak, one may regard the electronic excitation as temporarily completely localized on one subunit and interpret the observed effect in terms of "true" transfer of electronic excitation energy, with one subunit being the energy donor and a second being the acceptor.

In the last-mentioned case, as well as for molecules separated in favorable cases by as much as 10 nm, excitation transfers is explained in terms of the so-called Förster mechanism (FÖRSTER, 1948): The electrostatic interaction between the two separate electronic systems supplies the coupling which permits a simultaneous de-excitation of the electronically excited donor and excitation of the acceptor. Efficient transfer requires both transitions to be optically allowed ones (dipole-dipole transfer) and, in addition, a high degree of overlap of the emission band of the donor and the absorption band of the acceptor (resonance transfer). Transfer is favored, furthermore, by a long lifetime of the excited donor with respect to competing de-excitation processes (internal conversion, radiative transitions, etc.). Again, the probability of the transfer process per unit time is inversely proportional to the sixth power of the distance separating the two systems. For example, the Förster mechanism permits energy transfer between separate π-electron systems separated by groups containing σ-electrons, which require considerably higher energy for their excitation and which represent, therefore, insulators for π-electron excitation, allowing no electron transfer inbetween.

The review articles by FÖRSTER (1968) and VOLTZ (1968) are but two out of a great number of reviews of electronic excitation-energy transfer. DEE and BAUR (1974) have presented an extensive theoretical treatment of charge and excitation migration in DNA chains.

9.10 Radiation Yield. G-Value

The yield of a specified type of product (or, effect) of irradiation is commonly presented in the form of a pure number, the so-called G-value. If the product in question is a molecular species either formed or showing a specified type of structural change or damage the G-value equals the total number of these molecules divided by the absorbed radiant energy expressed in units of 100 eV. In the large molecules of the nucleic acids a specified effect may show up in more than one part of the same molecule. In such a case, and in related ones, the G-value presented may be the total frequency of occurrence of this effect per 100 eV absorbed. Usually, the absorbed radiant energy refers to that absorbed in the medium as a whole. However for media of mixed composition it is sometimes more meaningful to let it refer to the energy absorbed in a specified part or component of the medium.

Chapter 3

Structure of Radicals from Nucleic Acid Constituents

3.1 Introduction

Absorption of ionizing radiation in organic material gives rise to abrupt changes in the electronic structure of the molecules which may lead to the dislocation of electrons from their regular state of orbital occupancy in pairs. The resulting molecular species with unpaired electrons are generally referred to as free radicals. An evaluation of their relevance in radiation damage processes, in principle, involves relating both their chemical structure and concentration at various stages in the chain of reactions to the nature of the stable chemical and biological alterations detected at later stages of damage. For fundamental reasons, this problem is difficult to solve. Being forced to indirect argumentation one may lean on two main indications. First, the concentrations of radiation-induced radicals as measured by ESR-spectroscopy (I.5) are found to lie within the range of concentrations observed for stable chemically altered products. Moreover, the observed stable chemical alterations can, in general, be successfully explained on the basis of free radicals as precursors, the structure of the latter being in many cases identical with those found in ESR-studies in the solid state (as presented in I.3 and 4; see also II.1). Both observations leave little doubt about the role of free radicals being of primary importance as detectable intermediates in the pathway of reactions leading to chemical and, finally, biological damage in nucleic acids.

Free radicals are conveniently studied by Electron Spin Resonance (ESR) spectroscopy by way of their paramagnetic properties. To some extent, the attention given to radical species in radiation damage studies rests on the availability of this method which provides detailed information about both the concentration and structural features of free radicals. The major limitation of ESR spectroscopy arises from its applicability to species with lifetimes above μs or preferentially ms. One solution to this problem is to study radicals in the solid state, thereby increasing their lifetimes. It is this aspect of ESR research which will be presented in the remainder of Section I of this book.

Since it was first applied to irradiated DNA by SHIELDS and GORDY (1959), ESR spectroscopy has provided a wealth of data on free radicals in nucleic acids and constituents. Major contributions to the structural knowledge have derived from single crystal studies, which were initiated by PRUDEN et al. (1965) in their work on thymidine. This aspect of ESR research will be discussed in full detail in this chapter. Earlier reviews such as those by WYARD and ELLIOTT (1973), HERAK (1973) and MÜLLER and HÜTTERMANN (1973) have covered the literature on single crystal studies until about 1972. Since then, considerable progress has been made in unravelling pathways of radical formation by including investigations of low-temperature precursors and implementing new techniques like ENDOR spectroscopy. Supplemented by a large number of related studies in noncrystalline solids and crystalline powders, a rather comprehensive picture of the basic solid-state radiation response of the class of molecules represented by the heterocyclic bases and the sugar-phosphate group is beginning to develop.

The various radicals studied can be conveniently separated into two main groups, base radicals and species on the sugar-phosphate moiety. Among the base radicals, the neutral secondary species are found to be distinct from the parent molecules by the net loss or gain of one hydrogen atom. These will be referred to as hydrogen-abstraction or -addition radicals, regardless of the actual reaction involved. This terminology is used since details of the formation reactions are known for only a few radicals.

Whereas mainly secondary radicals contribute to observations made at 300 K, base radicals resulting from loss or addition of an electron are often detected upon low temperature irradiation. Again, the denotation of these species as molecular cations or anions used in this text needs not necessarily reflect the actual state of charge of the radical. For example, adenine crystallizes as a di-cation in dihydrochloride crystals. The electron adduct species in this case would have the same charge as the cation; nevertheless, following the nomenclature pervading the literature, this radical is called anion to indicate that an extra electron has entered the π-system.

Radicals in the sugar-phosphate backbone are found to be restricted to the sugar part. So far, only neutral species have been observed which all may be classified as hydrogen-abstraction radicals.

The current use of the nomenclature to assign the chemical modification of the radicals in the literature is extremely confusing. Frequently, a radical can be found to bear several different names none of which complies with organic chemistry rules. In this chapter, we avoid the use of both incorrect short names and the correct but often lengthy chemical names but introduce instead numbers for each radical and present its structural formula.

3.2 Base Radicals

3.2.1 Hydrogen Addition to Purines

A common feature of purine bases, nucleosides (-tides) irradiated at room temperature is the appearance of a triplet ESR pattern of intensities 1:2:1 and about 40 G component spacing. Frequently, especially in the adenine base, the inner line of the triplet is superimposed by a broad singlet of about 20 G linewidth. Whereas the nature of the radical giving rise to the latter signal has been the subject of a long and continuing debate, the triplet has been assigned unequivocally to a hydrogen-addition radical to the purine base.

Investigations of guanine derivatives include single crystals of guanine$\cdot$HCl$(H_2O)_2$ (ALEXANDER and GORDY, 1967) and deoxyguanosine $\cdot$HCl (DERTINGER, 1967). In both compounds, the hydrogen-addition radical appears to be the only radical produced at room temperature. It is formed by saturation of the double bond linking nitrogen N7 with carbon C8 (3.1)[1]. In guanine$\cdot$HCl$(H_2O)_2$, interaction with the proton at N7 is observed. The spectral parameters of both radicals are listed in Table 3.1.

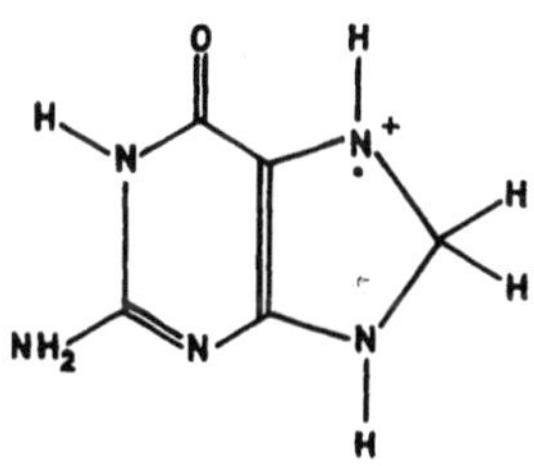

Formula (3.1)

The precursors of the hydrogen-addition radicals are not well established. There is no report on a single crystal study below room temperature. Investigations in alkaline and neutral glasses between 77 K and 200 K (LION and VAN DE VORST, 1971a, b; SEVILLA and MOHAN, 1974) indicate that long-wavelength UV light ($\lambda \geq$ 320 nm) induces protonation of the guanine anion to form the hydrogen-addition radical (cf. I.6).

Adenine derivatives have been studied in more detail. Single crystal investigations include the bases adenine$\cdot$HCl, adenine$\cdot$(HCl)$_2$, and adenine$\cdot 1/2H_2SO_4 \cdot H_2O$ (HÜTTERMANN, 1975; ZEHNER et al., 1977; WESTHOF et al., 1977), 9-methyl- and 9-ethyl-adenine (SCHMIDT and SNIPES, 1969; FLOSSMANN et al., 1974b; ZEHNER et al., 1976a), the nucleosides adenosine, adenosine$\cdot$HCl and deoxyadenosine$\cdot H_2O$ (DERTINGER, 1967; LICHTER and GORDY, 1968; ALEXANDER and FRANKLIN, 1971; ZEHNER et al., 1976a, 1977) as well as one nucleotide, 5'-AMP (HÜTTERMANN, 1969). All components, most prominently the nucleosides (-tides) as well as the alkyl-substituted bases, show the triplet ESR pattern.

Whereas the principal nature of the corresponding radical as being a hydrogen-addition species at the base moiety has been unequivocally accepted, there has been a discussion about the radical site on the adenine ring. Indeed, two positions are possible which can be expected to produce nearly identical ESR spectra. One is analogous to the guanine case and involves saturation of the N7-C8 double-bond (3.2) whereas the equivalent position requires spin-density at N3 (3.3).

---

[1] Numbers in parenthesis refer to formulas.

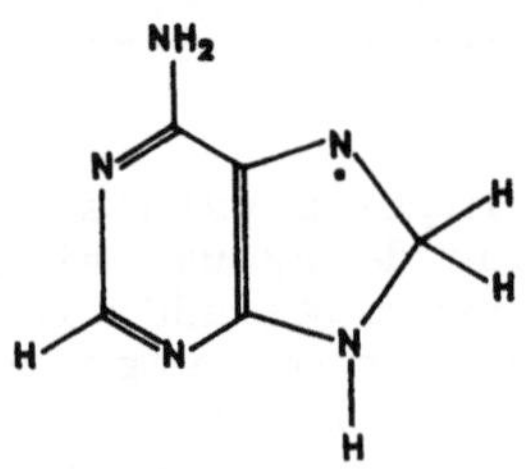

Formula (3.2)

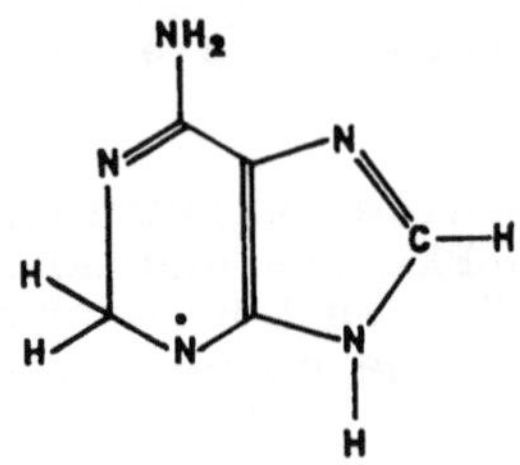

(3.3)

Several workers (SCHMIDT and BORG, 1971; HÜTTERMANN, 1975) have pointed out that a distinction between the two positions on the basis of analysis of ESR data alone is hardly possible; the molecular geometry places the two ESR-active groups in antiparallel positions making them indistinguishable. SCHMIDT and BORG (1971) studied several polycrystalline N9-substituted adenine derivatives which were selectively deuterated at position C8 and/or in the hydrogen-bonding network. They concluded that the N7-C8 double bond was the preferred reaction site and that (3.3) was formed only in minor quantities. ZEHNER and co-workers extended this line of research to single crystal studies and in addition used heat and illumination to differentiate between the two radicals. They were able to show that in most adenine derivatives investigated both addition at C2 and C8 can occur. The C2-addition radical (3.3), however, appears to need a specific environment in order to be stabilized. This explains its concentration varying to a large degree depending on the crystal studied. Another observation was that the magnitude of the β-methylene proton couplings were quite sensitive to the protonation state of adenine in the C2-addition radical whereas it remained nearly constant for radical (3.2) in all crystals investigated. Table 3.1 summarizes the spectral parameters of hydrogen-addition radicals in purine components.

Table 3.1. Spectral parameters of hydrogen-addition radicals in purine derivatives derived from single-crystals

Compound	Radical (formula number)	Hyperfine couplings (G)	g-tensor	Reference
Guanine·HCl$(H_2O)_2$	(3.1)	$(8\ H)_2$: 36.0 ± 0.5	2.0023	ALEXANDER and GORDY (1967)
		7 N : 21.0 ± 0.5 1.5 1.5		
		7 H :− 9.0 ± 0.5 1.0 −15.0		
		9 H :− 2.9 − 0.9 − 5.0		

Table 3.1. Continued

Compound	Radical (formula numbers)	Hyperfine couplings (G)	g-tensor	Reference
Deoxyguanosine•HCl	(3.1)	$(8\ H)_2$: 35.0 ± 1.0		DERTINGER (1967)
Adenosine-5'-mono-phosphate	(3.2) or (3.3)	$(H)_2$: 40.0		HÜTTERMANN (1969)
Deoxyadenosine•H_2O	(3.2)	$(8\ H)_2$: 40.0 ± 0.5		DERTINGER (1967)
	(3.3)	$(2\ H)_2$: 43.7 ± 0.5 3 N : 19.0 ± 0.5 1 N : 8.7 ± 0.5	2.0030 2.0050	LICHTER and GORDY (1968)
9-CH_3-adenine	(3.2)	$(8\ H)_2$: 38.0 ± 0.5		ZEHNER et al. (1976a)
	(3.3)	$(2\ H)_2$: 44.0 ± 0.5		WESTHOF et al. (1977)
9-CH_3-adenine•2HBr	(3.2)	$(8\ H)_2$: 44.0 ± 2.0		WESTHOF et al. (1977)
9-C_2H_5-adenine	(3.2)	$(8\ H)_2$: 39.5 7 N : 27.5 9 N : 6.6		FLOSSMANN et al. (1974b)
Adenine•HCl Adenine•$(HCl)_2$	(3.2)	$(8\ H)_2$: 41.5 ± 2.5 7 N : 26.0	2.0025 2.0050	HÜTTERMANN (1975)
Adenine•HCl•1/2H_2O	(3.2) (3.3)	$(8\ H)_2$: 42.0 ± 0.5 $(2\ H)_2$: 40.0 ± 0.5 3 N : 25.0		ZEHNER et al. (1977)
Adenine•$(HCl)_2$	(3.2) (3.3)	$(8\ H)_2$: 41.0 ± 0.5 $(2\ H)_2$: 45.0 ± 0.5		ZEHNER et al.(1977)
Adenine•1/2 H_2SO_4•H_2O	(3.3)	$(2\ H)_2$: 40.0 ± 0.5		WESTHOF et al. (1977)
Deoxyadenosine•H_2O	(3.2)	$(8\ H)_2$: 38.0 ± 0.5 7 N : 25.0 9 N : 8.6		ZEHNER et al. (1976a)
	(3.3)	2 H : 34.0 ± 2.0 2 H' : 52.0 ± 2.0 3 N : 20.0 1 N : 9.5		
Adenosine	(3.2)	$(8\ H)_2$: 39.0 ± 0.5		ZEHNER et al. (1977)

Table 3.1. Continued

Compound	Radical (formula numbers)	Hyperfine couplings (G)	g-tensor	Reference
Adenosine	(3.3)	$(2\ H)_2$: 44.0 ± 0.5 3 N : 20.5 1 N : 9.5		ZEHNER et al. (1977)
Adenosine•HCl	(3.2)	$(8\ H)_2$: 39.5 ± 0.5	2.0024 2.0038 2.0038	DERTINGER (1967)
	(3.2) (3.3)	$(8\ H)_2$: 40.0 ± 0.5 $(2\ H)_2$: 39.0 ± 0.5 3 N : 23.0 1 N : 6.0		ZEHNER et al. (1977)

5 H : hydrogen bonded to carbon C5
$(6H)_2$: equivalent hydrogens bonded to carbon C6
6H, 6H' : inequivalent hydrogens bonded to carbon C6
7 N : Nitrogen nucleus at position 7; in plane-coupling is not listed if it is zero

3.2.2 Hydrogen Addition to Pyrimidines

Radical formation in pyrimidine bases, base analogs, nucleosides and nucleotides has been studied extensively. As a major route of damage, hydrogen addition was found to occur at either position of the 5,6-double bond in the unsaturated base leading to the production of 5-yl or 6-yl radicals, respectively; other addition radicals on the bases involve the carbonyl groups.

3.2.2.1 5-yl Radicals

Hydrogen addition to carbon C6 and concurrent formation of the 5-yl radical was first observed in thymidine (PRUDEN et al., 1965). The unpaired electron in (3.4) interacts with the methyl protons and the methylene group at C6 by way of hyperconjugation to produce on octet ESR pattern of about 140 G total spread. This spectral extent exceeds that of most of the radicals in nucleic acid constituents and has thus provided a "finger-print" for the identification of thymine damage in overlapped spectra obtained from irradiated nucleic acids. As a consequence, thymine derivatives have attracted considerable interest. Other single crystal studies of (3.4) include ESR investigations of thymine (HÜTTERMANN, 1970; DULČIĆ and HERAK, 1971) and 1-CH_3-thymine (SNIPES and SCHMIDT, 1968) as well as an ENDOR study of thymidine at 4.2 K (BOX et al., 1974). The spectral parameters reported are summarized in Table 3.2, in which the data of the corresponding radical produced in dihydrothymine by hydrogen abstraction (HERAK, 1970c) are included for comparison.

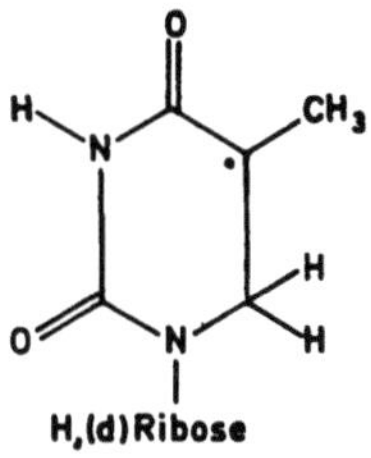

Formula (3.4)

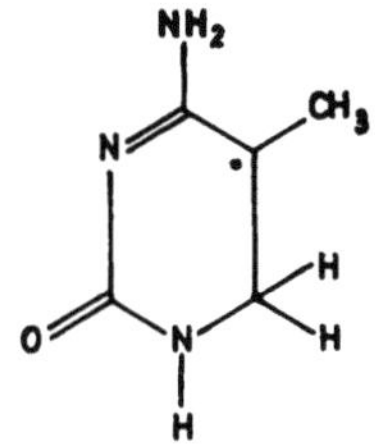

(3.5)

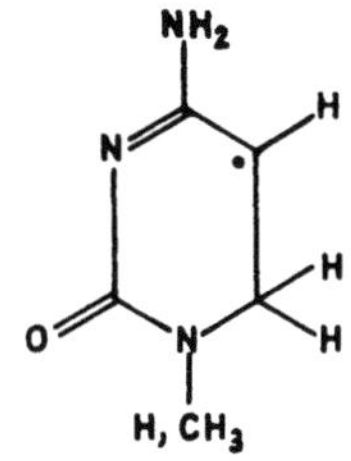

(3.6)

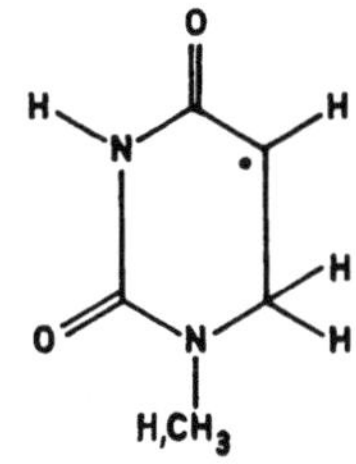

(3.7)

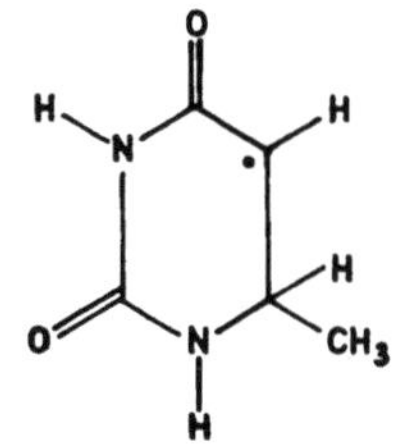

(3.8)

The spin-density on C5 in radical (3.4) is approximately 0.70 to 0.75. The methyl group is found to be freely rotating in the temperature range of 77 K to 300 K; thus, all protons have equivalent coupling of about 20 G to 22 G. At 4.2 K, the methyl rotation is hindered and transitions between different rotational sublevels are observed. The coupling of the two methylene protons is found to be inequivalent in thymine whereas it is seen to be nearly equivalent in the nucleoside, when measured with the high precision of ENDOR.

Table 3.2. Spectral parameters of 5-yl radicals: (3.4) in thymine derivatives

Compound	Methyl coupling (G)	Methylene coupling (G)	g-tensor	Temperature (K)	Reference
Thymine·H_2O	21.7	H : 36.19 H' : 43.75		77	HÜTTERMANN (1970)
	21.7	H : 34.21 H' : 40.79		300	
Thymine·H_2O	$A_{\parallel}$: 21.6 $A_{\perp}$: 19.3	H : $A_{\parallel}$: 35.8 $A_{\perp}$: 32.8 H' : $A_{\parallel}$: 44.4 $A_{\perp}$: 40.6	2.0044 2.0036 2.0028	300	DULČIĆ and HERAK (1971)
Thymine	$A_{\parallel}$: 21.6 $A_{\perp}$: 19.3	H : $A_{\parallel}$: 36.8 $A_{\perp}$: 33.5 H' : $A_{\parallel}$: 43.2 $A_{\perp}$: 40.6	2.0032 2.0042 2.0028	300	
1-CH_3-thymine	19.9	H : 36.8 H' : 41.0		300	SNIPES and SCHMIDT (1968)
	19.9	H : 37.6 H' : 43.1		140	
Thymidine	20.5 ± 0.5	40.5 ± 0.5	2.0024 2.0030 2.0042	300	PRUDEN et al. (1965)

Table 3.2. Continued

Compound	Methyl coupling (G)	Methylene coupling (G)	g-tensor	Temperature (K)	Reference
Thymidine	Methyl-rotation blocked; couplings measured for rotational sublevels; refer to original article for details	H : 43.8 40.4 39.7 H' : 41.9 38.2 37.7		4.2	BOX et al. (1974)
5,6-dihydro-thymine	$A_{\parallel}$: 22.0 $A_{\perp}$: 19.5	$A_{\parallel}$: 39.0 $A_{\perp}$: 36.5		300	HERAK (1970c)

Footnote: See Table 3.1

In cytosine derivatives 5-yl hydrogen-addition radicals have been found to occur in crystals of cytosine·H_2O (FLOSSMANN et al., 1976), the two protonated bases cytosine·HCl and 1-methyl-cytosine·HBr (WESTHOF et al., 1975; WESTHOF et al., 1977), 1-methyl-cytosine (RUSTGI and BOX, 1974; BERNHARD and FARLEY, 1976) as well as of 5-methyl-cytosine (HÜTTERMANN et al., 1971). Whereas in the latter compound, (3.5) is the only hydrogen-addition radical, the 5-yl species (3.6) in the two other cytosine derivatives competes in intensity with other addition products and is sometimes present as a minor product only. The parameters of both radical (3.5) and (3.6) are listed in Table 3.3 which also includes data on the 5-yl radical (3.7) formed in 1-methyluracil·HBr (FLOSSMANN et al., 1975b) and of (3.8) in dihydro-6-methyl-uracil (BERGENE and HENRIKSEN, 1971). The latter components are the only uracil-derivatives which stabilize 5-yl radicals at room temperature.

Before turning to the discussion of 6-yl radicals it is worth mentioning that conversions can take place between 5-yl and 6-yl radical positions. This has been observed first in dihydrothymine (BERNHARD and SNIPES, 1967) in which the corresponding 5-yl or 6-yl hydrogen-addition radicals of thymine are formed by hydrogen abstraction from the respective positions of the 5,6-bond. In that component, the 6-yl radical converts into 5-yl upon heating. The activation energy of the reaction is 23.1 kcal/mol. These findings suggest that in pyrimidines methylated at the 5 or 6 position the more stable radical has its spin-density located on the carbon carrying the substituent. The observation of BERGENE and HENRIKSEN (1971) that in 5,6-dihydro-6-methyl-uracil the 6-yl radical is the most stable species at room temperature, would seem to support the hypothesis. Also, thymine or 5-methylcytosine show a strong preference for stabilizing 5-yl radicals.

In the case of unsubstituted pyrimidines, FLOSSMANN et al. (1976a) find that C5 → C6 addition radical conversions can be induced by irradiation of cytosine and uracil derivatives with visible light (λ > 400 nm).

Table 3.3. Spectral parameters of 5-yl radicals: cytosine and uracil derivatives (ENDOR data presented if available)

Compound	Radical (formula numbers)	Hyperfine coupling (G)		Temperature (K)	Reference
Cytosine•HCl	(3.6)	5 H	: - 7.2 - 26.7 - 17.8	300	WESTHOF et al. (1975)
		$(6\ H)_2$	: 48.8		
$1\text{-}CH_3$-cytosine	(3.5)	5 H	: - 8.16 - 16.70 - 28.45		RUSTGI and BOX (1974)
		6 H	: 50.64 46.68 45.51		
		6 H'	: 53.49 50.28 49.75		
$5\text{-}CH_3$-cytosine	(3.5)	$H(CH_3)$	: 19.5	130	HÜTTERMANN et al. (1971)
		6 H	: 37.6		
		6 H'	: 46.9		
$1\text{-}CH_3$-uracil•HBr	(3.7)	5 H	: - 8.5 - 26.5 - 15.0	300	FLOSSMANN et al. (1975b)
		$(6\ H)_2$	: 47.5		
5,6-dihydro-6-CH_3-uracil	(3.8)	5 H	: (-) 13.3 (-) 23.4 (-) 29.5	295	BERGENE and HENRIKSEN (1971)
		6 H	: 41.4 45.3 49.4		
Cytosine	$(3.11) \xrightarrow{h\nu} (3.6)$	5 H	: - 8.5 - 26.5 - 15.0	300	FLOSSMANN et al. (1976)
		6 H	: 49.0		
		6 H'	: 53.0		
Uracil	(3.6)	5 H	: - 8.1 - 28.0 - 18.0	300	ZEHNER et al. (1976b)
		6 H	: 45.0		
		6 H'	: 51.0		
$1\text{-}CH_3$-cytosine•HBr	(3.6)	$(6\ H)_2$	: 48.0	300	WESTHOF et al. (1977)

Table 3.3. Continued

Compound	Radical (formula numbers)	Hyperfine coupling (G)	Temperature (K)	Reference
Deoxycytidine·HCl	(3.6)	5 H : - 8.5 - 26.5 - 15.0		FLOSSMANN et al. (1976)
		$(6\ H)_2$: 50.0	300	

Footnote: See Table 3.1

3.2.2.2 6-yl Radicals

Hydrogen addition to C5 of the 5,6-double bond of pyrimidines appears to be the preferred route of base damage either if both C5 and C6 carry a hydrogen atom or if a nonhydrogen substituent is bonded to C6. The latter situation is encountered in orotic acid (6-carboxyl-uracil) which stabilizes the 6-yl species (3.9) at room temperature as the only radical involving the 5,6-double bond (HÜTTERMANN et al., 1970).

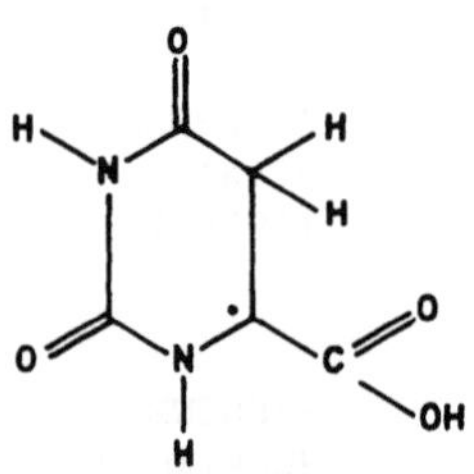

Formula (3.9)

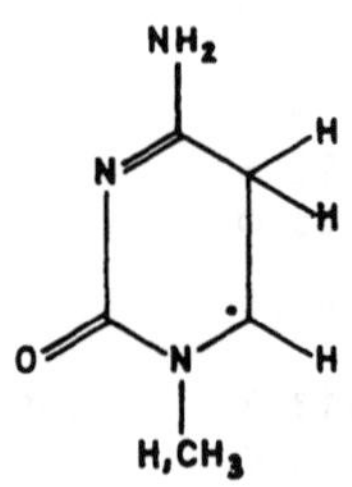

(3.10)

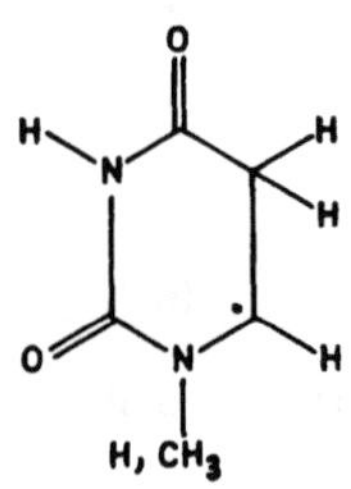

(3.11)

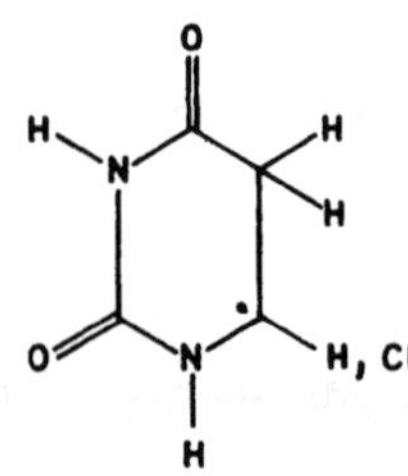

(3.12)

Single-crystal studies of 6-yl species in unsubstituted pyrimidines include the bases cytosine·H_2O (COOK et al., 1967; FLOSSMANN et al., 1976), 1-methyl-cytosine and 1-methyl-cytosine·HBr (ELLIOTT, 1971; RUSTGI and BOX, 1974; BERNHARD and FARLEY, 1976; WESTHOF et al., 1977), uracil (ZEHNER et al., 1976b), 1-methyl-uracil and 1-methyl-uracil·HBr (FLOSSMANN et al., 1973; HERAK, 1973; HERAK and McDOWELL, 1974a; FLOSSMANN et al., 1975b; BERNHARD and FARLEY, 1976) and the nucleoside deoxycytidine·HCl (FLOSSMANN et al., 1976). The corresponding radical structures are (3.10) and (3.11). Their spectral data are listed in Table 3.4 which also includes the parameters of (3.12) obtained from 5,6-dihydrouracil (HERAK, 1970b) and 5,6-dihydro-6-methyl-uracil (BERGENE and HENRIKSEN, 1971).

Table 3.4. Spectral data of 6-yl radicals (ENDOR data presented if available)

Compound	Radical (formula numbers)	Hyperfine coupling (G)	g-tensor	Temperature (K)	Reference
Orotic acid	(3.9)	5 H : 22.8 5 H' : 34.5	2.0023 2.0066 2.0031	300	HÜTTERMANN et al. (1970)
Cytosine·H_2O	(3.10)	$(5\ H)_2$: 37.03 6 H : - 17.81 - 12.82 - 28.13		300	COOK et al. (1967)
	(3.10)	$(5\ H)_2$: 37.0 6 H : - 8.1 - 30.0 - 17.7		300	FLOSSMANN et al. (1976)
1-CH_3-cytosine	(3.10)	5 H : 33.26 29.48 29.34 5 H' : 47.72 43.91 43.52 6 H : - 6.83 - 16.54 - 28.88	2.0038 2.0029 2.0028		RUSTGI and BOX (1974)
1-CH_3-uracil	(3.11)	5 H : 37.16 33.87 34.31 5 H' : 37.80 34.31 34.75 6 H : - 8.81 - 18.53 - 30.68			HERAK and McDOWELL (1974a)
5,6-dihydro-uracil	(3.12)	5 H : 24.0 5 H' : 42.0 6 H : - 6.0 - 21.5 - 31.0	approx.		HERAK (1970b)
5,6-dihydro-6-CH_3-uracil	(3.12)	5 H : 60.2 61.4 66.6 5 H' : 0.0 6 CH_3 : 17.9 18.2 21.2			BERGENE and HENRIKSEN (1971)

Table 3.4. Continued

Compound	Radical (formula numbers)	Hyperfine coupling (G)	g-tensor	Temperature (K)	Reference
Uracil	(3.11)	$(5\ H)_2$: 35.5		300	ZEHNER et al. (1976b)
		6 H : - 8.0			
		- 29.5			
		- 18.0			
$1\text{-}CH_3$-uracil• HBr	(3.11)	$(5\ H)_2$: 33.0		300	FLOSSMANN et al. (1975b)
		6 H : - 8.1			
		- 30.0			
		- 17.7			
Deoxycytidine •HCl	(3.10)	5 H : 13.0 ± 3.0		300	FLOSSMANN et al (1976a)
		5 H' : 47.0 ± 3.0			
		6 H : - 6.5			
		- 30.0			
		- 17.7			

Footnote: See Table 3.1

3.2.2.3 > $\dot{C}$-OH Radicals

Outside the 5,6- or the 7,8-double bonds of pyrimidines and purines, respectively, only the exocyclic keto-oxygens are expected to be the site of hydrogen addition. These > $\dot{C}$-OH species are indeed observed in a number of pyrimidines but not, for example, in guanine or in less common purines containing carbonyl groups. The first report on hydrogen-addition to a keto-oxygen has been for cytosine•H_2O (DERTINGER, 1967a). The same species was studied in more detail recently by WESTHOF et al. (1975) in cytosine•HCl and cytosine•H_2O. In both investigations, Q-band frequencies were employed to separate the anisotropic doublet pattern of about 10 G maximum splitting from lines of other radicals. The two groups of authors give different assignments of the coupling proton. DERTINGER (1967a) identified it as the primary > $\dot{C}$-OH proton in (3.13) whereas WESTHOF et al. (1975) ascribed the coupling to the proton on C6 in structure (3.14). INDO (Intermediate Neglect of Differential Overlaps) calculations were used in the latter work to show that the primary proton was not expected to exhibit observable hyperfine splitting. Experiments with deuterated crystals ascertained structure (3.14) as the correct assignment. A > $\dot{C}$-OH radical at O4, structure (3.16) was also found in crystals of uracil (ZEHNER et al., 1976b). Hydrogen-addition to a carboxyl-group to form > $\dot{C}(OH)_2$ radicals (3.15) has been reported for orotic acid (HÜTTERMANN et al., 1970).

Other > $\dot{C}$-OH species are observed in alloxan and 5-halogen-substituted uracil derivatives. These compounds are discussed in I.7.

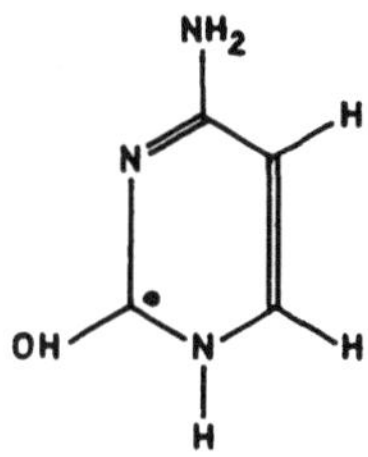

Formula (3.13)

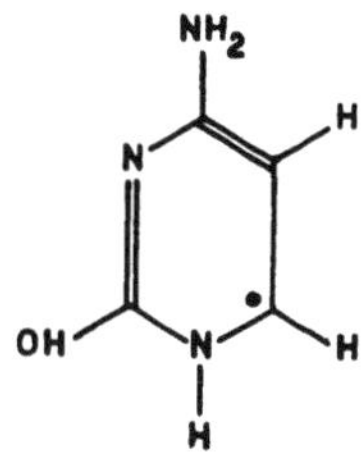

(3.14)

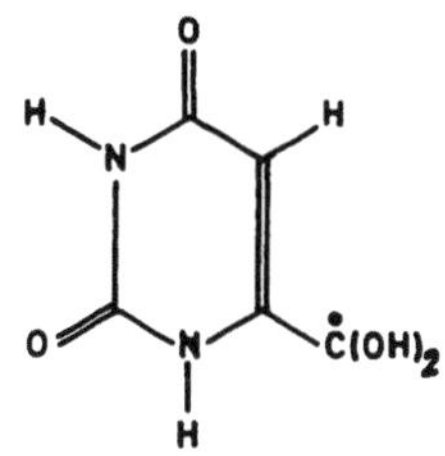

(3.15)

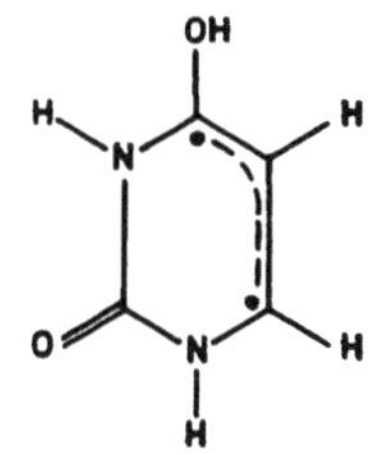

(3.16)

3.2.3 Hydrogen-Abstraction Radicals

3.2.3.1 Abstraction from Pyrimidine Rings

The source of the hydrogen that adds to pyrimidine or purine bases may be any molecule or molecular fragment containing hydrogen. The fact that hydrogen-addition radicals are found in pure and homogeneous systems like anhydrous crystals of thymine (DULČIĆ and HERAK, 1972) and uracil (ZEHNER et al., 1976b) proves that the bases themselves can be hydrogen donors. Direct evidence for the radiation abstraction of hydrogen atoms from the pyrimidine ring comes from single-crystal studies. The first analysis of a hydrogen-abstraction radical was reported for cytosine at room temperature (COOK et al., 1967). A re-investigation showed, however, that this assignment has to be challenged. No abstraction radicals were found at room temperature but at 77 K instead (FLOSSMANN et al., 1976). The site of abstraction is N1 (3.17). At room temperature, this type of abstraction radical has been observed in anhydrous thymine, thymine$\cdot H_2O$ (DULČIĆ and HERAK, 1972) uracil (ZEHNER et al., 1976b) and orotic acid (HORAN and SNIPES, 1970). Similar species are found in serveral halo-uracil bases (see I.7).

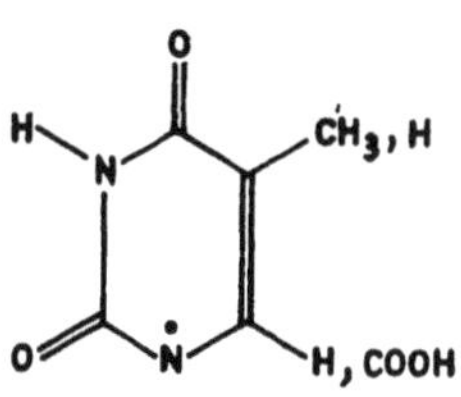

Formula (3.17)

The ESR spectra of all these radicals are characterized by an approximate axial symmetry in the coupling of N1 ($A_{\parallel} \cong$ 15 G, $A_{\perp} \cong$ 0 G) and a relatively large coupling of the C5-substituent.

No positive evidence for the hydrogen-atom abstraction from the N-H fragments in purines exists.

3.2.3.2 Abstraction from Substituent Groups to Pyrimidines and Purines

That hydrogen can be abstracted from substituted alkyl groups in pyrimidines and purines has been proved in a number of single-crystal studies (SCHMIDT and SNIPES, 1969; HÜTTERMANN, 1970; HÜTTERMANN et al., 1971; HERAK and SCHOFFA, 1971; HERAK and DULČIĆ, 1973; FLOSSMANN et al., 1973; HERAK and McDOWELL, 1974b; FLOSSMANN et al., 1974b; SCHMIDT, 1975). In irradiated pyrimidines methylated on C5, radicals of form (3.18) are produced. The spin density on the site of the abstraction is about 0.60. The remaining fraction resides mostly on C6. Radicals (3.19) are formed in irradiated N-methylated pyrimidines. In these radicals, about 80% of the unpaired electron is located on the site of the hydrogen abstraction. Except for methylene hydrogens, no resolved proton splitting can be detected by ESR. The values for the methylene α-hydrogen couplings are shown in Table 3.5.

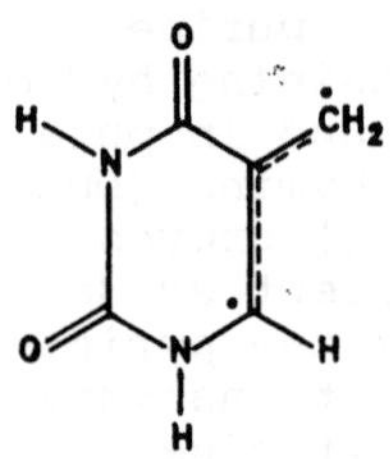

Formula (3.18)

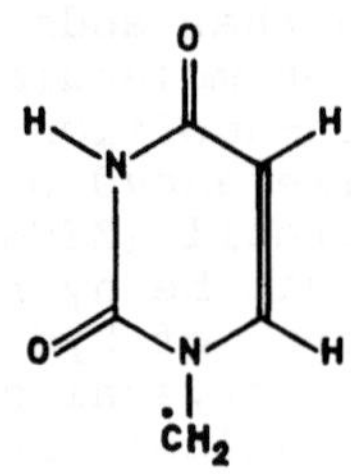

(3.19)

Table 3.5. Methylene α-hydrogen couplings in H-abstraction radicals from methyl groups

Compound	Coupling parameters (G) Principal values			Isotropic	Reference
Thymine	8.1	15.6	22.4	15.0	HÜTTERMANN (1970)
5-Methyl-cytosine	8.6	16.3	24.9	16.6	HÜTTERMANN et al. (1971)
6-Aza thymine				14.2	HERAK and SCHOFFA (1971)
Thymidine	7.7	15.2	23.7	15.5	HERAK and McDOWELL (1974)
9-Methyl-adenine	9.5	19.5	31.2	19.9	SCHMIDT and SNIPES (1969)
1-Methyl-uracil	9.8	20.0	30.2	20.0	HERAK and DULČIĆ (1973)
	8.5	18.2	30.5	19.1	FLOSSMANN et al. (1973)

The hydrogen, if abstracted as an atom may produce further damage in the neighbourhood of the site where it is liberated. It may either abstract hydrogen atoms from the aliphatic portions of the system or add to the conjugated parts, such as the pyrimidine and purine bases. If the site of the reaction is not far from the site of the primary abstraction (10 Å), such a situation can be stabilized in a solid and observed by ESR. The damage sites, each possessing an electronic spin S = 1/2, interact with each other and form the resulting spin triplet state, which can be analysed in a single crystal. Several sorts of radical pairs have been identified in 1-methyl uracil (DULČIĆ and HERAK, 1973a; FLOSSMANN et al., 1973; FLOSSMANN et al., 1975a) and in thymine (DULČIĆ and HERAK, 1973b; BERGENE and MELØ , 1973). Radical pairs might be constituted of two hydrogen-abstraction radicals or of a H-abstraction and a H-addition radical (Fig. 3.1). In the former

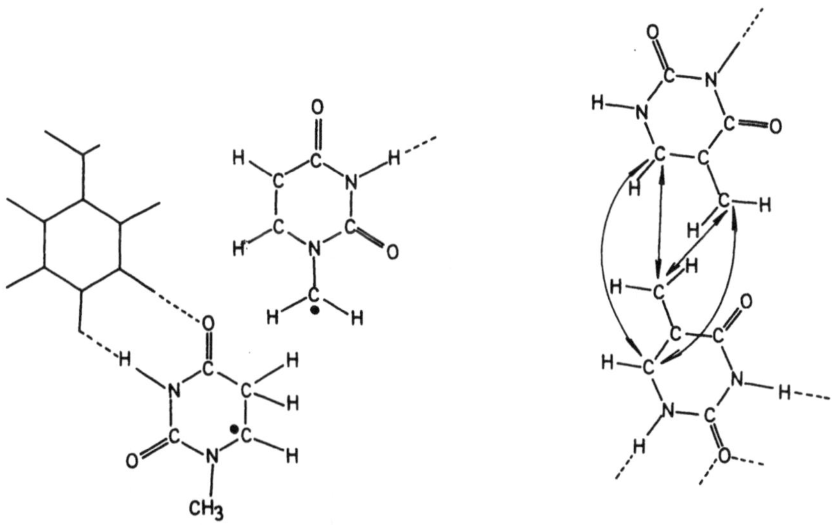

(a) (b)

Fig. 3.1a and b. Examples of geometrical arrangements of radical pairs showing pair of hydrogen addition - and abstraction radicals in 1-methyl uracil (a) and pair of abstraction radicals together with possible reaction sites of abstracted hydrogens in thymine (b)

case, the hydrogen atom produced by homolytic cleavage of a C-H bond abstracts another H atom under formation of H_2. The latter process involves a short-distance transfer and a subsequent addition reaction. In this case there is a conservation of charge, mass and spin within the radical pair. The structure and constitution of radical pairs thus may provide information on the mechanisms and pathways of damaging processes.

3.2.4 Molecular Ion Radicals

Ion radicals are postulated to be unstable precursors of the secondary radicals which are normally observed by ESR at room temperature. That hydrogen-addition radicals are formed from anions was first directly proved in thymine by HOLROYD and GLASS (1968). Most ion radicals studied previously in nucleic acid bases were produced in aqueous glasses. Anions and cations were produced in a nonradiolytic way (cf. I.6).

Recently, there has been considerable progress made in studying primary species in single crystals of nucleic acid constituents. The cytosine cation radical has been analysed in detail at 77 K by this method (HERAK and GALOGAŽA, 1969b; FLOSSMANN et al., 1976). The identification of this radical is based on the comparison between the observed spectral parameters and the expected ones for various anticipated radicals. The hyperfine ESR pattern consists of two ring nitrogen couplings (N1 and N3) and two α-hydrogen couplings (H1 and H5), one of which is exchangeable upon deuteration (H1). The spin densities derived from the coupling parameters agree fairly well with the values calculated from the molecular-orbital theory by BAUDET et al. (1962).

Anions characterized by anisotropic splitting due to the C6 α-hydrogen were tentatively assigned in 5-methyl-cytosine (HÜTTERMANN et al., 1971) and have since then been identified in crystals of 1-methyluracil·HBr (FLOSSMANN et al., 1975b), cytosine·HCl and cytosine·H_2O (WESTHOF et al., 1975; FLOSSMANN et al., 1976) and thymidine (BOX and BUDZINSKI, 1975). The latter group of authors also found electron adduct radicals in two nucleotides at 4.2 K, uridine-5'-phosphate and cytidine-3'-phosphate (BOX et al., 1975). Thus it appears that anions can be stabilized at the base moiety in several pyrimidine derivatives at low temperature.

The information derived from single crystal studies about oxidation products in pyrimidines is less abundant. Besides cytosine, cations, or rather deprotonated neutral oxidized species, have been observed in barbituric acid (see I.7) and in 5,6-dihydro-6-methyluracil (BOX and BUDZINSKI, 1975a). In all three components, N1 is the site of primary spin-density and of deprotonation. This makes the deprotonated radicals structurally identical to those described previously as resulting from hydrogen abstraction from N1 (see I.3.2.3). It may be argued that the oxidation products observed at low temperatures are precursors of the latter, secondary species. The site of damage is shifted when an alkyl-substituent or, more important, a sugar- or sugar-phosphate group is attached to N1. In nucleosides (-tides), two radicals have been identified that are due to oxidation. One is a neutral radical resulting from loss of a hydrogen from O5' and the other involves H abstraction from C5'. These are discussed in I.3.3.

Identification of ion radicals in single crystals of purines has proved difficult owing to the lack of resolution in the spectra of these components at low temperatures. For some time, the average g-factor derived from singlet spectra in powders or glasses appeared to be the only parameters characterizing adenine and guanine anions and cations (LACROIX et al., 1967; SEVILLA and MOHAN, 1974). Recently, HÜTTERMANN (1975) studied adenine hydrochloride crystals. Both dose yield curves and reaction kinetics were used to show the existence of an anion precursor of the hydrogen-addition radical. In the same compound, the presence of an anion was also established at 4.2 K (BOX and BUDZINSKI, 1976).

The spectral parameters of ion radicals in purine and pyrimidine bases are listed in Table 3.6.

Table 3.6. Spectral parameters of ion radicals of pyrimidine and purine bases

Compound	Radical	Hyperfine coupling (G)	g-tensor	Reference
Cytosine•H_2O	cation	5 H : - 5.7 - 22.6 - 16.4	2.0045 2.0030 2.0015	HERAK and GALOGAŽA (1969b)
		3 N : 2.2 1.8 9.3		
		1 N : 3.8 3.1 16.3		
	cation	5 H : - 7.1 - 22.0 - 11.2		FLOSSMANN et al. (1976)
		3 N : 4.0 0.0		
		1 N : 13.1 0.0		
	anion	6 H : - 6.0 - 22.5 - 15.0		FLOSSMANN et al. (1976)
Cytosine•HCl	anion	6 H : - 9.2 - 23.8 - 16.3		WESTHOF et al. (1975)
Uridine-5'-phosphate	anion	6 H : - 5.80 - 21.11 - 11.76		BOX et al. (1975)
Cytidine-3'-phosphate	anion	6 H : - 5.24 - 21.16 - 12.14		BOX et al. (1975)
Thymidine	anion	6 H : - 4.64 - 19.82 - 10.68		BOX and BUDZINSKI (1975)
1-CH_3-uracil•HBr	anion	6 H : - 7.5 - 22.5 - 14.0		FLOSSMANN et al. (1975b)
Adenine•$(HCl)_2$	anion	8 H : - 3.8 - 14.5 - 7.9		BOX and BUDZINSKI (1976)

Footnote: See Table 3.1

3.3 Sugar-Phosphate Backbone Radicals

It is known from various chemical and physical investigations, particularly from the sedimentation analysis of irradiated nucleic acids, that irradiation introduces damage in the sugar-phosphate backbone. The most evident consequence of the absorption of radiation energy in this region is the breakage of the sugar-phosphate bond (see III.2). The ESR technique has not been as successful in analysing this damage as in studying radicals produced in the bases. The reason for this lies in the complexity of the system which gives rise to complex ESR spectra. Another difficulty arises from the fact that there are only some nucleosides and few nucleotides which can be prepared in the form of single crystals of suitable dimensions for the ESR investigation. From the studies carried out so far, it appears that C1' and C5' are the main sites of secondary radicals in the sugar moiety.

The first report on a radical formed in the sugar moiety was for 3'-cytidylic acid at room temperature (BERNHARD and SNIPES, 1968). A dominant quartet ESR pattern was observed, brought about by an α-hydrogen and a β-proton coupling. It was suggested that this radical is produced by breakage of the C4'-O1' bond leaving the unpaired spin on C4'. Although this kind of mechanism (discussed in more detail in II.3) accounts for the formation of a C4'-centered radical in 2'-deoxy-D-ribose (HÜTTERMANN and MÜLLER, 1969a), it appears to be invalid for 3'-cytidylic acid. The precursor to this radical has recently been identified by BERNHARD et al. (1976a) in studies at 77 K as due to hydrogen abstraction from C5' as in (3.20). The knowledge that (3.20) converts into the room temperature species was found to limit the possible radical structures of the latter. As a probable candidate, deprotonation of C5' in (3.20) has been considered.

Hydrogen abstraction from C5' to yield radical (3.20) was first proposed by ALEXANDER and FRANKLIN (1971) for deoxyadenosine. Subsequently, COMBS et al. (1973) used theoretical calculations to support formation of (3.20) as the preferred species in both deoxyadenosine and 3'-cytidylic acid. Most recently, hydrogen abstraction from C5' has been observed in 5-chloro- and 5-bromodeoxyuridine (HÜTTERMANN et al., 1977b).

Another damage residing at C5' is the scission of the sugar-phosphate bond. Radicals formed this way have been found in a crystal of deoxycytidine 5'-monophosphate (KRILOV and HERAK, 1974). At 77 K, the conformation of the sugar remains essentially unaltered by irradiation, and couplings of two equivalent α-hydrogens attached to C5' and a β-hydrogen to C4' yield (3.21). At room temperature, the ESR pattern converts into a spectrum consisting of three α-hydrogen couplings.

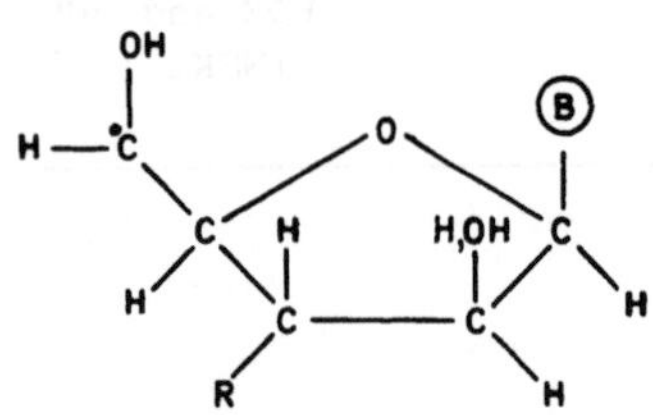

Formula (3.20)

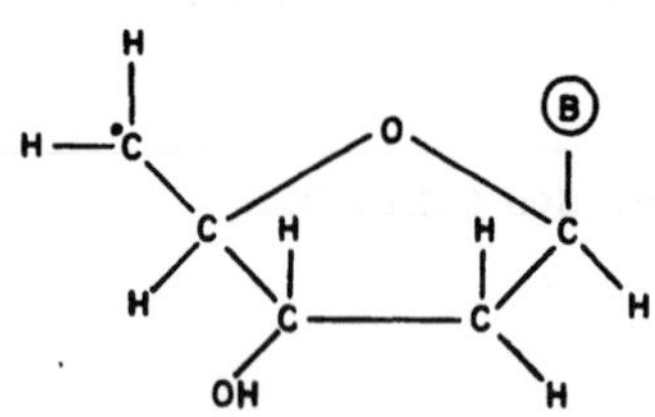

(3.21)

A hypothesis has been put forward (HERAK et al., 1976) that the damaged sugar (3.21) decompose upon warming, yielding radical (3.22).

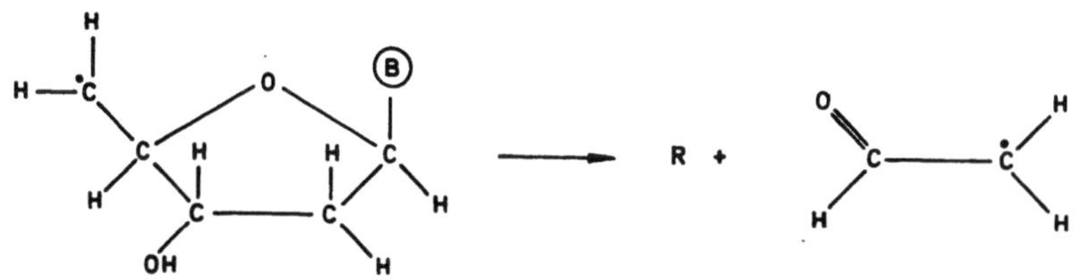

Formula (3.22)

The ESR spectrum belonging to radical (3.22) consists of three typical α-proton couplings whose isotropic components are 16.0, 16.0 and 8.8 G, respectively.

Essentially the same spectral characteristics were observed for a radical in irradiated cytidine (HAMPTON and ALEXANDER, 1973). It was suggested that the coupling hydrogens were those at C4' and C5'. Many radical structures and configurations were considered, but no satisfactory fit between the structure and the observed spectral characteristics was found. It is probable that in cytidine, radical (3.22) accounts for the observed spectra. If so, these radicals must be formed by abstraction of the hydroxyl groups from C5' and subsequent decomposition of the furanose ring, similarly to the process outlined for deoxycytidine 5'-monophosphate. Quite recently, BERNHARD et al. (1976b) detected a radical with similar spectral parameters in still another cytosine derivative, 3'-cytidylic acid. It is conceivable that all three species may eventually be linked together by a common mechanism.

Several authors have reported the radiation abstraction of the hydrogen atom from C1' (3.23). In polynucleotides (HERAK and

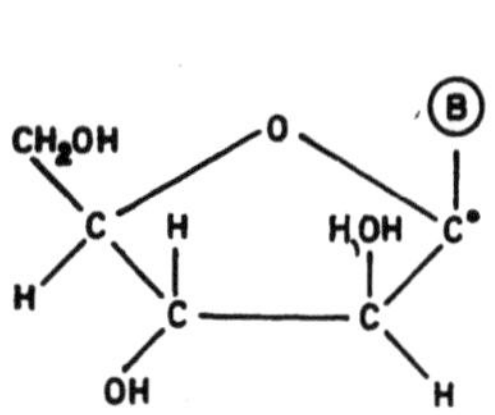

Formula (3.23)

GORDY, 1966b), an isotropic doublet of 18 G splitting was attributed to that radical. The splitting was brought about by the hydrogen on C2'. In cytidine, GORDY (1969) assigned a 1:1:1 triplet to the same radical. The radical was supposed to be in the σ-state and the splitting was believed to originate from N1. Recently, HERAK et al. (1976) observed a pattern consisting of two pronounced predominantly isotropic proton couplings (approx. 19.6 G) in an irradiated single crystal of deoxycytidine 5'-monophosphate. The couplings were assigned to C2' hydrogens in the radical analogous to (3.23). In some cases radicals (3.23) are supposed to be formed as precursors of the observed secondary radicals. However, the appearance and transformation of this species and its eventual role in further evolution of the damage is still to be investigated.

The compilation of radicals in the sugar-phosphate backbone is completed with the description of a well characterized site of radical formation. It is the 5'-oxygen position from which a hydrogen is abstracted to yield radical (3.24) which is stable

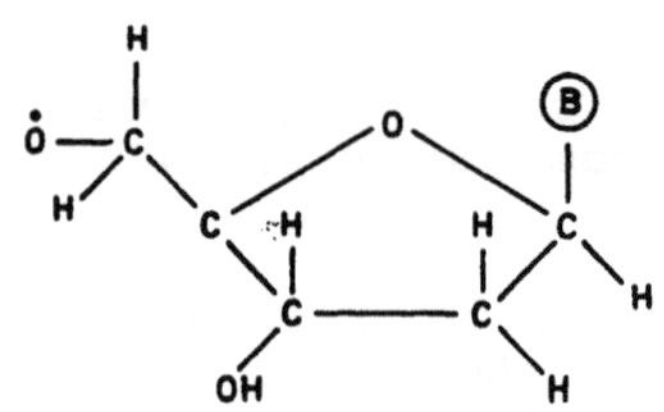

Formula (3.24)

at low temperatures. These radicals exhibit a distinct quartet spectrum resulting from interaction with the β-hydrogens at C5'; their splittings range from 60 to 85 G. Thus, the outer lines of the quartet are positioned outside the region of lines of other radicals allowing for an unequivocal assignment of the radical. Another distinct feature of (3.24) is an axially symmetric g-tensor reaching an unusually high value of about 2.06 to 2.08 along the C5'-O5' bond. This radical, which was first observed in thymidine by Box and co-workers (BOX and BUDZINSKI, 1975), has since then been found as a major product at 77 K in a furanoside derivative of uracil (BERGENE and VAUGHAN, 1976), in adenosine·HCl, in both 5-bromo- and 5-chlorodeoxyuridine, and as a minor species in 3'-cytidylic acid (BERNHARD et al., 1977).

The spectral parameters of sugar-phosphate backbone radicals are compiled in Table 3.7.

3.4 Molecular Orbital Calculations

Theoretical calculations of the coupling constants of an assumed radical structure have proved to be of considerable help in invalidating or substantiating the radical identification. This can be achieved most simply by Hückel molecular orbital calculations. In recent years, however, more sophisticated semi-empirical calculations have been preferred. Because of its wide applicability, the INDO method (POPLE and BEVERIDGE, 1970) has been mostly used. Some examples pertinent to radicals in nucleic acid constituents will be presented here. For details of the calculations, we refer to the original literature (WESTHOF and FLOSSMANN, 1975; ADAMS and BOX, 1975; WESTHOF and VAN ROOTEN, 1976).

Table 3.7. Spectral parameters of sugar-phosphate backbone radicals

Radical (formula number)	Compound	Hyperfine coupling (G)	g-tensor	Temperature (K)	Reference
(3.20)	Deoxyadenosine	5'H : - 16.4	2.0029	300	ALEXANDER and FRANKLIN (1971)
		- 22.6	2.0052		
		- 5.0	2.0075		
		4'H : 7.0			
	3'CMP	5'H : - 32.0	min:2.0024	77	BERNHARD et al. (1976a)
		- 22.0	max:2.0034		
		- 14.2			
		5'OH : 26.6			
		19.3			
		16.5			
		4'H : 7.5			
		4.6			
		1.5			
	5-Chloro- and 5-Bromodeoxyuridine	5'H : - 11.05	2.0034	200-240	HÜTTERMANN et al. (1977b)
		- 17.78	2.0050		
		- 33.15	2.0044		
		5'OH : 12.8			
		7.6			
		5.5			
		4'H : 20.8			
		18.7			
		17.3			
(3.21)		$5'(H)_2$: - 14.0		77	KRILOV and HERAK (1974)
		- 20.4			
		- 29.7			
		4'H : 36.0			
(3.22)	Cytidine	$H(\alpha_1)$: - 23.66	2.0031	300	HAMPTON and ALEXANDER (1973)
		- 15.52	2.0025		
		- 7.38	2.0021		
		$H(\alpha_2)$: - 24.58			
		- 15.48			
		- 7.24			
		$H(\alpha_3)$: - 13.28			
		- 9.63			
		- 4.85			
	3'CMP	$H(\alpha_1)$: - 21.02	2.0048	300	BERNHARD et al. (1976b)
		- 14.17	2.0038		
		- 4.71	2.0028		
		$H(\alpha_2)$: - 21.87			
		- 14.17			
		- 6.49			

Table 3.7. Continued

Radical (formula number)	Compound	Hyperfine coupling (G)	g-tensor	Temperature (K)	Reference
(3.22)	3'CMP	$H(\alpha_3)$: - 14.13 - 10.38 - 2.46			
	5'dCMP	$H(\alpha_1)$: - 25.63 - 15.83 - 6.43	∿ 2.0023		HERAK et al. (1976)
		$H(\alpha_2)$: - 24.63 - 16.10 - 6.94			
		$H(\alpha_3)$: - 12.51 - 9.44 - 4.43			
(3.23)	5'dCMP	2'H : 21.79 18.87 18.22	∿ 2.0023		HERAK et al. (1976)
		2'H' : 22.93 18.24 17.59			
(3.24)	Thymidine	5'H : 85.22 78.72 77.37	2.0831 2.0052 1.9999	4.2	BOX and BUDZINSKI (1975)
		5'H' : 76.02 69.59 68.17			
	Uracil-β-D-arabinofuranoside	5'H : 92.0 89.0 89.0	2.0591 2.0071 2.0029	77	BERGENE and VAUGHAN (1976)
		5'H' : 49.0 47.0 46.0			
	Adenosine•HCl	5'H : 98.8 91.5 90.2	2.0798 2.0055 2.0005	77	
		5'H' : 52.5 46.1 45.3			
(3.24)	5-Chlorodeoxyuridine	5'H : 89.2 (89.9) 81.4 (84.7) 79.2 (82.1)	2.0929 (2.0922) 2.0048 (2.0090) 1.9975 (1.9959	77	BERNHARD et al. (1977)
	(5-Bromodeoxyuridine)	5'H' : 62.2 (61.9) 58.2 (55.2) 55.2 (54.7)			

Footnote: See Table 3.1

In Section 3.2.1, it was remarked that magnitude of the methylene proton couplings in the C2-addition radical of adenine (3.3) is affected by the protonation state of the molecule. The results of the corresponding calculations are presented in Table 3.8. The quantitative agreement with the experimental values is certainly not perfect. However, the relative trends in the methylene proton couplings and in the nitrogen couplings are well reproduced.

Table 3.8. Coupling constants (in G) of the different H-addition radicals of the adenine base in various protonation states as calculated by the INDO method (ZEHNER et al., 1976)

Protonation State		Radical	H(C2)	H(C8)	N'	N"
Free base	INDO	C2-Addition Formula (3.3)	67.8	- 4.0	25.6	18.2
		C8-Addition Formula (3.2)	- 2.7	47.1	31.2	1.5
	Exper.	C2-Addition	42.0	-5.0/-8.0	20.7	9.2
		C8-Addition	--	38.0	27.0	6.4
N1-protonated	INDO	C2-Addition	49.4	- 2.1	29.2	3.5
		C8-Addition	- 2.8	49.9	29.9	3.0
	Exper.	C2 Addition	40.0	-4.0/-5.0	23.0	6.0
		C8-Addition	--	41.0	--	--
N1- and N7-protonated	INDO	C2-Addition	54.2	0.4	29.3	5.0
		C8-Addition	- 0.9	51.1	24.6	7.5
	Exper.	C2-Addition	45.0	--	25.0	--
		C8-Addition	--	41.0	--	--

For the protons the isotropic part is given and for the nitrogens the value which should be observed with the magnetic field perpendicular to the ring. For the C2-addition radical, Formula (3.3), N' and N" are N3 and N1, for the C8-addition radical, Formula (3.2), N' and N" are N7 and N9, respectively. The experimental values are typical values taken from Table 3.1

Table 3.9 shows the results of the calculations for the 5- and 6-yl radicals of some pyrimidine compounds. As a rule, the methylene β-proton couplings are considerably smaller in the 6-yl than in the 5-yl species for any crystal stabilizing both radicals. The average sum of the methylene couplings in 6-yl radicals varies between 66.0 and 70.0 G. Though smaller, this value is comparable to that of 5-yl radicals in 5-methylated uracil and cytosine derivatives. However, in cytosine and uracil derivatives with an unsubstituted C5 = C6 double bond, the sum of the methylene proton couplings in 5-yl radicals is close to 100.0 G. All other parameters kept constant, it can be seen from Table 3.9 that INDO calculations reproduce these large couplings of the 5-yl radicals only when they are protonated at N3 in cytosine and at O4 in uracil. It is important to remark that this protonation is not necessary in the case of thymine (WESTHOF and VAN ROOTEN, 1976).

Table 3.9. Coupling constants (in G) of the different H-addition radicals of uracil, cytosine, and thymine as calculated by the INDO method (FLOSSMANN et al., 1976; WESTHOF and VAN ROOTEN, 1976)

Molecule		Radical	H(C5)	H(C6)
Uracil	INDO	6-yl	38.1	-13.8
		5-yl	-11.6	36.1
		5-yl(O4-H^+)	-11.1	51.9
	Exper.	6-yl	35.5	-18.5
		5-yl	-18.0	48.0
Cytosine	INDO	6-yl	46.7	-17.2
		5-yl	-12.6	40.2
		5-yl(N3-H^+)	-14.3	48.1
	Exper.	6-yl	37.0	18.6
		5-yl	-16.7	51.0
Thymine	INDO	6-yl	-39.2	-17.6
		5-yl	22.1	35.0
	Exper.	6-yl	44.0	-17.4
		5-yl	19.3/21.3	37.0/40.5

The experimental values are taken from Tables 3.3 and 3.4. The 6-yl radical of thymine was observed in single crystals of dihydrothymine (HENRIKSEN and SNIPES, 1970b).

Table 3.10. Coupling constants (in G) as calculated by the INDO method for the cation radical of the thymine molecule (a) and its N1-deprotonated form (b)

Radical	N1 a	N1 b	H(C6)	$H(CH_3)$	H(N1)	
Cation	10.3	6.4	2.8	13.1	-11.6	
H-abstraction from N1	10.1	8.0	4.8	11.7	--	
Exper. (Glass)	4.0-5.3	3.3-4.0	< 4	20.4	--	(SEVILLA, 1971)
(single crystal)	6.1-6.4	3.5-3.8	--	19.1	--	(DULČIĆ and HERAK, 1972)

	H(C6)	$H'(\dot{C}H_2)$	$H''(\dot{C}H_2)$	N1 iso/ N3 iso	
H-abstraction from methyl group	-12.1	-16.1	-16.0	8.8	
IDEM deprotonated at N1	- 3.8	-15.6	-15.5	2.3	
Exper. (Flow system)	- 7.79	-16.06	-15.38	2.2	(NETA, 1972)
(single crystal)	-11.3	-15.4	-15.4	--	(HÜTTERMANN, 1970)

For the methyl protons, the values given are those for a freely rotating methyl group. The results for the radical resulting from hydrogen abstraction from the methyl group and its N1-deprotonated form are also given. The ratio of the isotropic couplings of the nitrogen N1 and N3 is added.

Two examples of calculations for abstraction radicals are given in Table 3.10. These are the radical resulting from hydrogen abstraction from the nitrogen N1 of thymine and that resulting from hydrogen abstraction from the methyl group of thymine. For comparison, the cation radical of the thymine molecule is presented. It can be observed that it is not easy to distinguish the cation radical from the N1-abstraction radical. Only the observation of a proton bonded to N1 would be conclusive. Some imbalance in unpaired electron density, too large nitrogen and too small methyl protons, has been obtained (ADAMS and BOX, 1975). On the other hand, the effect of deprotonation is easily detected in the case of the abstraction radical from the methyl group. When deprotonated at N1, this abstraction radical yields a smaller coupling for H(C6) than for the two $H(CH_2)$ protons. This difference has been experimentally observed.

3.5 Radical Formation Mechanisms

Little is known about quantitative relations for the conversion of molecules activated in the process of energy absorption into radicals. Various qualitative schemes have been considered. In solid material of homogeneous composition of organic molecules MH containing hydrogen, the following reactions appear to be of importance:

1. The process of ionization itself may involve the formation of two charged radicals

$$MH \rightsquigarrow [MH^{\cdot}]^{+} + e^{-} \qquad [3.1]$$

both of which can react further if the electron escapes recombination with the parent cation.

2. The fate of the cation radical $[MH^{\cdot}]^{+}$ may be deprotonation

$$[MH^{\cdot}]^{+} \xrightarrow{-H^{+}} M^{\cdot} \qquad [3.2]$$

the result of which is equivalent to that obtained in the more familiar ion-molecule reaction scheme

$$[MH^{\cdot}]^{+} + MH \longrightarrow MH^{+}_{2} + M^{\cdot} \qquad [3.2a]$$

3. The electron liberated in the process of ionization can be captured by neutral molecules to form anion radicals

$$e^{-} + MH \longrightarrow [MH^{\cdot}]^{-} \qquad [3.3]$$

which may subsequently protonate

$$[MH^{\cdot}]^{-} + MH \longrightarrow MH^{\cdot}_{2} + M^{-} \qquad [3.3a]$$

to form a radical and an ion with paired electrons.

4. A possible pathway from a highly excited molecule MH* to free radicals is the homolytic fission via predissociation into two neutral radicals

$$MH^* \longrightarrow M^{\cdot} + H^{\cdot} \qquad [3.4]$$

One of these should react further or possess sufficient kinetic energy to escape recombination within the cage of the surrounding molecules. This condition favours dissociation into atomic hydrogen $H^{\cdot}$ and a larger molecular radical $M^{\cdot}$.

5. Hydrogen atoms are expected to diffuse through the matrix and attach to other molecules, preferentially at unsaturated bonds, to form addition radicals

$$H^{\cdot} + MH \longrightarrow MH^{\cdot}{}_2 \qquad [3.5]$$

In addition, hydrogen atoms can abstract hydrogen from saturated parts of the molecule

$$H^{\cdot} + MH \longrightarrow H_2 + M^{\cdot} \qquad [3.5a]$$

to yield radicals $M^{\cdot}$ and molecular hydrogen.

These reactions describe pathways from primary charged radicals and excited molecules to two groups of neutral secondary species, $M^{\cdot}$ and $MH^{\cdot}{}_2$ which have a net difference of plus or minus one hydrogen atom in their sum formula as compared to the parent molecule MH. All secondary radicals described in the previous paragraphs satisfy the latter condition. Thus it appears that both ionization and excitation as primary events may lead to the same two groups of secondary radicals.

It would be satisfying to use the detailed structural knowledge obtained from single crystal investigations to discriminate between the two reaction pathways and to give an estimate of their relative efficiencies. A consistent scheme, however, has not yet evolved. Single crystal studies have aimed until now mainly at mapping the various radical structures possible, rather than at following in detail the reactions of their formation and obtaining estimates of the radical stöchiometry. Among the few examples for which the latter goal has been partially achieved are the halodeoxyuridines 5-chloro- and 5-bromodeoxyuridine (cf. I.7). Apart from these, a more or less coherent body of data is available which indicate that a variety of parameters seem to govern the type of reaction actually occurring and its site at the molecule.

Primary oxidation and reduction products have been observed for a variety of bases, nucleosides and -tides at temperatures between 4.2 and 77 K. In all components, the electron adduct radical resides at the base moiety showing main interaction with the C6-bonded hydrogen in pyrimidines and the proton at C8 in purines. Oxidation in the pure bases leaves the main fraction of unpaired electron density at N1 in pyrimidines. Information about a purine base cation does not exist so far. The site of oxidation is shifted to the terminal CH_2OH group of the sugar in a number of nucleosides (-tides), where a $R\dot{O}$ radical is produced.

Although the observation of primary species resulting from ionization would seem to imply that they should be precursors of the secondary hydrogen-addition and -abstraction radicals, direct evidence for such reaction pathways has remained scarce in single crystal studies. In view of the results obtained in aqueous glasses (see I.6) it appears quite certain that hydrogen-addition radicals, for example, can be formed by protonation of the anion under appropriate conditions. The actual parameters governing this reaction, however, are not yet well understood. Among others, the bonding scheme in crystals, the availability of protons in the vicinity of a primary defect site and structural features of the molecule itself may be considered influential.

Progress towards elucidating some of these factors has been achieved in the case of the bases uracil, cytosine and adenine and some of their N-alkylated derivatives. In these compounds, anion radicals and subsequent protonation were observed only if the crystals contained molecules like H_2O, HCl or HBr which either change the protonation state of the bases, or influence the hydrogen-bonding scheme in the crystals. For uracil and cytosine, protonation was found to take place selectively at C6. These findings indicate that addition of atomic hydrogen resulting from C-H bond cleavage upon excitation should take place preferentially at position C5 in uracil and cytosine. The relative stability of 5-yl and 6-yl radicals against treatment of light and heat in crystals with and without cocrystallized polar molecules supports this conclusion; 5-yl radicals are found to be stable only in a polar environment. A similar distinction appears to prevail also in adenine derivatives for the C2- and C8-addition radicals but a detailed evaluation of the different parameters has not yet been worked out.

Little is known about the formation mechanisms of the > $\dot{C}$OH type hydrogen-addition radicals in pyrimidines. Comparison with radiation-chemical findings would seem to indicate that they are formed by protonation of an anion precursor. If so, the parameters directing protonation to C6 or to a carbonyl-function are not understood.

Details of the formation of hydrogen-abstraction radicals and their participation in the radical stöchiometry are again nuclear. In pure bases, abstraction from N1 in pyrimidines is well established. These radicals may result from deprotonation of a cation precursor. Likewise, abstraction radicals in alkyl groups in N-alkylated bases could be formed by this process, but probably also result from C-H bond cleavage upon excitation. Experimental indications for the latter process can be taken from the presence of pairs of abstraction radicals and from the fact that abstraction radicals are found in crystals like 1-methyl-uracil which do not stabilize ion radicals at low temperatures.

Abstraction radicals in the sugar-phosphate backbone may also be produced directly by C-H bond cleavage as a consequence of excitation. Other possible pathways include hydrogen abstraction by small, diffusible radicals like hydrogen atoms or by other radicals formed in the vicinity which are capable of hydrogen abstraction. Evidence for the latter reaction has been obtained in two halodeoxyuridines in which a primary oxidation product,

$R\dot{O}$, was found to abstract hydrogen from a nearby site, C5', in a neighbouring molecule (cf. I.7).

Summing up the present state of knowledge about radiation-induced free radicals in nucleic acid constitutents as viewed from single crystal investigations, the mapping of possible radical structures has reached a satisfactory level of completeness, especially in the case of purine and pyrimidine bases. Further work needs to be devoted to a more detailed elucidation of radical structures in the sugar-phosphate domain. Beyond that, there is an urgent need for information about both structural and quantitative features of radical reaction pathways in order to develop schemes of primary radiation-damage mechanisms which may be applicable to an understanding of the chemical and biological alterations of irradiated nucleic acids.

Chapter 4

Structure of Radicals from Nucleic Acids

4.1 Frozen Solution and Dry State

4.1.1 Radiolysis of DNA in Frozen Solution

Experiments on irradiated DNA in which ESR methods are used to study free radicals have been carried out on DNA in both the dry state and in aqueous solution. In the latter case, it is necessary to solidify the solutions by freezing in order that the radicals should be stable. Irradiation of such frozen solutions or low-temperature glasses is considered in this section. Results for DNA in the dry state are discussed in I.4.1.2.

Radiolysis of materials in frozen solution is necessarily more complicated than in the dry state. Usually, irradiation of the solvent produces radicals, which may then react with the solute, in addition to radicals produced by direct radiation of the solute. It is to be expected that the concentration of solute and pH value will be important parameters.

The ESR spectrum of pure ice after radiolysis at 77 K can be accounted for entirely by the anisotropy and hyperfine splitting of the $OH^{\cdot}$ radical (DIBDIN, 1967; BRIVATI et al., 1967). This species is unstable above 110 K (SIEGEL et al., 1961) but is not affected by UV bleaching ($\lambda > 300$ nm) at 77 K (LENHERR and ORMEROD, 1968). Hydrogen atoms are found at lower temperatures but are not stable above 50 K in neutral ice (FLOURNOY et al., 1962).

SANNER (1965) irradiated neutral aqueous solutions of DNA at 77 K. The ESR spectra were the superposition of the separate spectra of the DNA and the water radicals. A similar finding was reported by LENHERR and ORMEROD (1968), who also found that the $OH^{\cdot}$ radicals did not react with the DNA when the samples were warmed to 118 K. The spectrum remaining at this temperature, at which $OH^{\cdot}$ radicals are unstable, was essentially that of the DNA itself, which showed a single line similar to that obtained after irradiation of dry DNA at the same temperature. Figure 4.1 shows the spectra of pure ice and of 10% DNA solution both immediately after irradiation at 77 K and after annealing at 118 K. Corresponding spectra for D_2O solutions are also shown. ABAGYAN and BUTYAGIN (1969) also obtained ESR spectra from neutral DNA solutions γ-irradiated at 77 K. Their spectra were similar to that of irradiated ice but with several additional lines attributable to the presence of DNA. Thus their DNA spectrum is different from the single line reported by LENHERR and ORMEROD (1968).

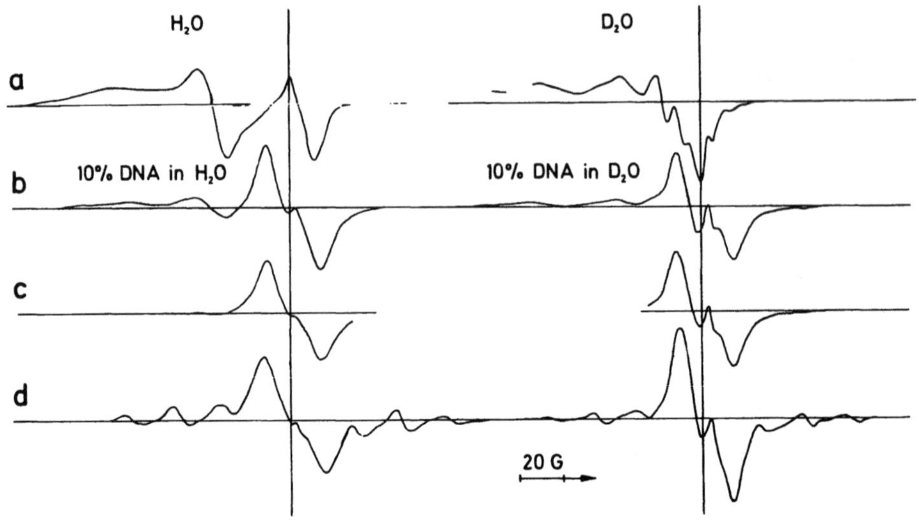

Fig. 4.1. ESR spectra of DNA in H_2O and D_2O at 77 K. *a*: pure ice; *b*: 10% DNA immediately after irradiation; *c*: after annealing at 118 K; *d*: after annealing at 195 K (LENHERR and ORMEROD, 1968)

All workers who have reported ESR spectra of irradiated frozen DNA solutions also report the appearance of an eight-line or octet spectrum when the irradiated samples are warmed. This spectrum, which is also found in irradiated solid DNA, as well as in thymine and thymidine, is discussed in more detail in I.3 and I.4.1.2. It arises from an unpaired electron on the thymine ring at C5 due to a hydrogen atom added at C6. The resulting free radical is often referred to as the 5-thymyl radical as abbreviation of the correct chemical name 5,6-dihydro-5-thymyl radical [see Formula (3.4)]. LENHERR and ORMEROD (1968) showed that it could also be obtained by UV bleaching ($\lambda > 300$ nm) the irradiated DNA solutions at 77 K, a procedure which thus appears to be qualitatively similar to thermal annealing, at least in neutral glasses. However, the concentration of 5-thymyl radicals after UV bleaching was much less than in the thermally annealed samples. UV photolysis of the neutral solutions without prior radiolysis gave no ESR signal (LENHERR and ORMEROD, 1968).

Similar experiments were carried out on thymidine and thymidine-5'-monophosphate (5'-dTMP). In contrast to DNA a doublet spectrum was observed at 77 K for both these substances (LENHERR and ORMEROD, 1968). Figure 4.2 shows the corresponding spectra. The doublets can be seen clearly in Figure 4.2b after annealing at 118 K. Similar spectra were obtained by ABAGYAN and BUTYAGIN (1969) for thymidine in both aqueous and D_2O glasses. The fact that the doublet was unaffected by the solvent led them to propose that a radical anion had been formed on the thymine ring. LENHERR and ORMEROD (1968) had already reached a similar conclusion from experiments with glasses containing electron scavengers e.g., 1% cystamine. However, they reported differences in behaviour, in the presence of electron scavengers, between DNA on the one hand and thymidine and 5'-dTMP on the other.

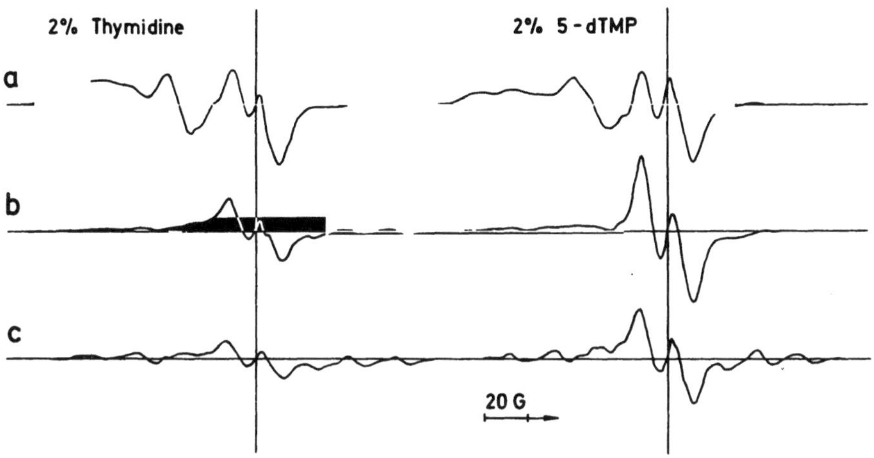

Fig. 4.2. ESR spectra of thymidine and thymidine-5'-monophosphate (5'-dTMP) in aqueous glasses. *a*: immediately after irradiation at 77 K; *b*: after annealing at 118 K; *c*: after annealing at 195 K (LENHERR and ORMEROD, 1968)

In aqueous glasses containing thymidine or 5'-dTMP, electron scavengers reduce the yield of anion radicals formed at 77 K i.e., the scavenger and the thymine base appear to compete for electrons. In DNA glasses, scavengers apparently have little effect upon the number of radicals initially formed, but, after annealing, the intensity of the 5-thymyl octet is very much reduced. Here the scavenger appears to interfere with the secondary processes which occur on warming.

Neutral glasses containing thymidine and 5'-dTMP, in common with DNA, give the 5-thymyl octet spectrum on warming (Fig. 4.2c). LENHERR and ORMEROD (1968) proposed that the radical anion, initially formed at 77 K, is converted to a secondary radical (e.g., 5-thymyl) by subsequent protonation. This process, which is equivalent to the addition of a hydrogen atom in two stages, was originally proposed in connection with the observation of the 5-thymyl octet in irradiated solid DNA (ORMEROD, 1965). It is discussed in more detail in I.4.1.2.

The ESR spectra of aqueous glasses containing DNA itself, after irradiation at 77 K, have not been interpreted directly in terms of the anionic radical species. However, if the mechanism, proposed by LENHERR and ORMEROD (1968) is applicable to DNA, the source of the proton is of some interest. Unfortunately the results of experiments so far reported do not allow this question to be answered unequivocally. Thus LENHERR and ORMEROD (1968) found no trace of a deuterated 5-thymyl radical after annealing a previously irradiated 10% solution of DNA in D_2O, and concluded that the proton did not come from the water. On the other hand, ABAGYAN and BUTYAGIN (1969) reported the conversion of the doublet to a sextet pattern after annealing a previously irradiated frozen solution of thymidine in D_2O. The sextet spectrum arises from the 5-deutero thymyl radical $-\dot{C}(CH_3) - CHD -$. This species was also reported by LION and VAN DE VORST (1973b) in annealed D_2O glasses containing 5'-dTMP.

In summary, there are discrepancies between the results for DNA glasses and glasses containing thymidine and 5'-dTMP. Measurement of radical yields, which are discussed in detail in I.5, have shown that a considerable indirect effect takes place when aqueous DNA glasses are irradiated. As mentioned earlier, $OH^{\cdot}$ radicals formed in the water have been shown not to attack the solute. It must, therefore, be presumed that electrons, formed by ionization of the solvent, attack the DNA at low temperature. This in turn implies that the secondary DNA radicals which are eventually formed have ionic precursors.

No results for irradiation of DNA in alkaline glasses appear to have been published. This is probably because the situation in such glasses is much more complicated than in the neutral glasses discussed above. For example, NAZHAT and WEISS (1970) found that frozen solutions of NaOH form true glasses only above 5 M NaOH and, unlike neutral glasses, do not show ESR spectra of $OH^{\cdot}$ radicals after irradiation. Irradiated frozen alkaline solutions exhibit a blue colour due to trapped electrons and the radical species O^- is also formed (SCHULTE-FROHLINDE and EIBEN, 1962, 1963). If oxygen is present the species O_2^- is formed on annealing (MOORTHY et al., 1972). Figure 4.3a shows the spectrum of 6 M NaOH γ-irradiated at 77 K.

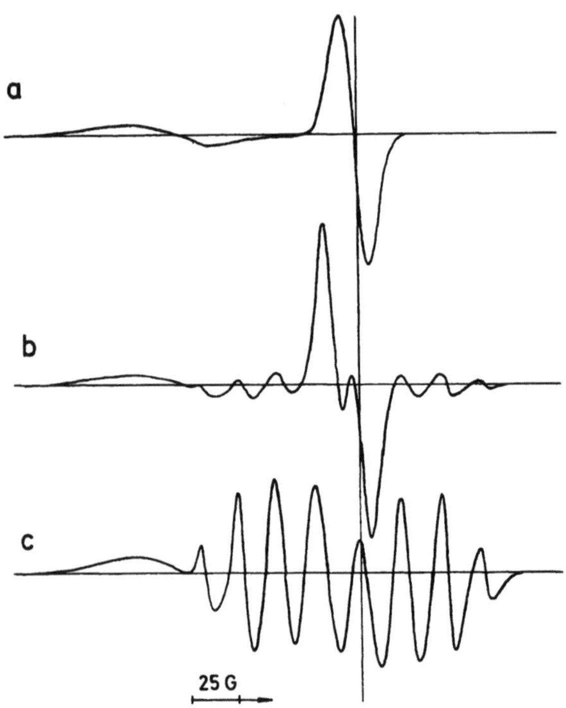

Fig. 4.3. ESR spectra of 6M NaOH glasses irradiated at 77 K. *a*: 6M NaOH glass alone; *b*: with 0.08M thymidine; *c*: sample of *b* after UV bleaching (λ > 310 nm) (LENHERR and ORMEROD, 1970)

Alkaline (NaOH) glasses containing thymine have been studied by NAZHAT and WEISS (1970) and by SRINIVASAN et al. (1969, 1970). Both groups of workers observed the 5-thymyl octet after annealing the irradiated glass. NAZHAT and WEISS (1970) found a deviation from the expected intensity distribution of the lines comprising the octet and suggested that an O^- adduct of thymine might have been formed as well as the 5-thymyl radical. The second group of workers postulated the thymine anion to be the precursor of the 5-thymyl radical partly because the blue colour of the trapped electron disappeared when thymine was present and partly because the formation of 5-thymyl radicals was suppressed in the presence of electron scavengers. Neither groups reported observing the spectrum of thymine anion immediately after irradiation at 77 K. On the other hand, KAALHUS and JOHANSEN (1973), who also studied alkaline thymine glasses, claimed to have observed a doublet at 77 K as well as the single line of the trapped electron. The doublet, which is shown in Figure 4.4a was attributed to thymine anion. Photobleaching the irradiated glass at 580 nm caused the trapped electron signal to disappear and the intensity of the doublet to increase. This suggests that photobleaching mobilizes the previously trapped electrons which can then attach to the thymine molecules to form thymine anions.

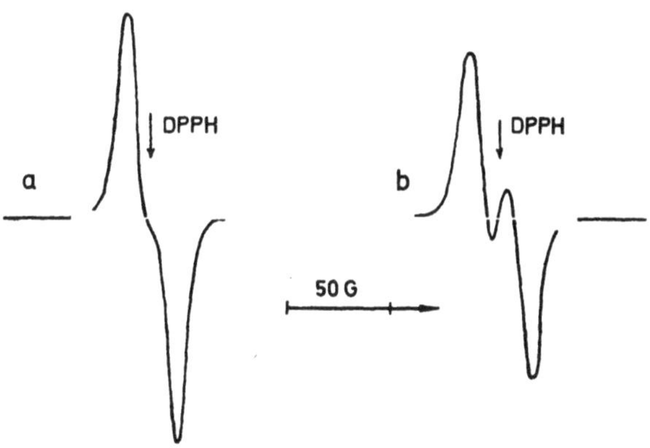

Fig. 4.4. ESR spectra of thymine doublet radicals in low temperature glasses. *a*: 6M NaOH glass; *b*: 5M H_2SO_4 glass (KAALHUS and JOHANSEN, 1973)

The behaviour of alkaline glasses containing thymidine is somewhat different. Both LENHERR and ORMEROD (1970) and NAZHAT and WEISS (1970) found a considerable yield of 5-thymyl radicals immediately after irradiation at 77 K. The spectrum reported by LENHERR and ORMEROD (1970) is shown in Figure 4.3b. It may be compared with that of Figure 4.2b for thymine irradiated in a neutral glass which shows negligible yield of 5-thymyl radicals. The intensity of the octet spectrum could be further increased by subsequent annealing or UV bleaching (Fig. 4.3c). Only LENHERR and ORMEROD (1970) reported observing the anion doublet at 77 K. This shows clearly in Figure 4.3b. NAZHAT and WEISS (1970) found evidence for a radical located on the deoxyribose ring.

LION and VAN DE VORST (1973b) have irradiated 5'-dTMP in alkaline glasses and found only a trace of the octet spectrum at 77 K. Its intensity increased rapidly when the samples were warmed. The anion doublet spectrum was also observed at low temperature.

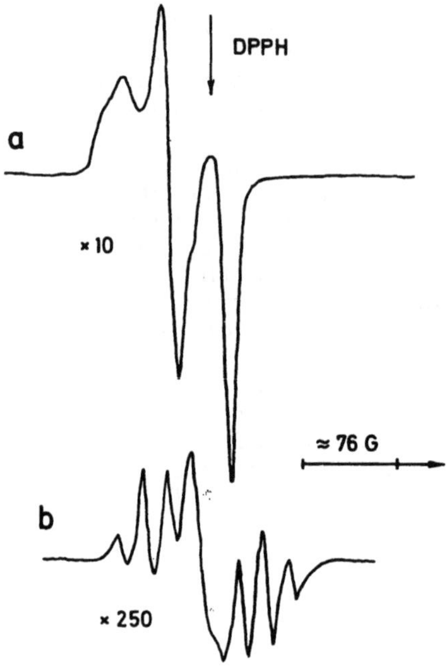

Fig. 4.5. ESR spectra of a DNA glass. *a*: after irradiation and observation at 77 K; *b*: after subsequent annealing to 195 K; the hydrogen atom signal has been omitted (HOLMES and WEISS, 1968)

Acidic glasses are also more complicated than neutral glasses. Radiolysis of sulphuric acid glasses is believed to produce radical species such as SO_2^-, SO_3^-, and SO_4^- (HENRIKSEN and JONES, 1971), the spectra of which may overlap those of other solute radicals. Radiolysis of acid glasses at 77 K also produces trapped hydrogen atoms.

HOLMES and WEISS (1968) irradiated an aqueous glass containing DNA and dihydrogen phosphate anion ($H_2PO_4^-$). The spectrum (Fig. 4.5a) showed a DNA signal together with the widely spaced doublet due to hydrogen atoms. No interpretation of the DNA signal was made. On warming the glass the hydrogen atom doublet disappeared while the 5-thymyl octet spectrum appeared (Fig. 4.5b). Power-saturation studies showed the presence of a second DNA radical giving a singlet spectrum and which was less stable than the 5-thymyl radical; it was thought to be located on the deoxyribose ring. Similar glasses containing either thymine or 5'-dTMP gave the 5-thymyl octet after irradiating and annealing. It was suggested that in these glasses hydrogen atoms react with the solute to give an enhanced yield of 5-thymyl radicals compared with neutral glasses in which hydrogen atoms are not stabilized at 77 K.

Quantitative studies by HENRIKSEN and JONES (1971) on frozen sulphuric acid glasses containing thymine showed that the 5-thymyl radical is not formed by simple addition of a hydrogen atom. In their annealing experiments they found that 5-thymyl radicals are formed at a temperature higher than that at which the hydrogen atom spectrum disappears. Also, the number of 5-thymyl radicals formed was higher than the number of hydrogen atoms initially stabilized. They supposed the thymine anion to be the precursor of the 5-thymyl radical, although in these acid glasses no ESR signal attributable to a thymine anion was actually observed. KAALHUS and JOHANSEN (1973) studied the same system and obtained a doublet spectrum at low temperature by subtracting the background due to the sulphuric acid radicals. This doublet is shown in Figure 4.4b. A similar doublet was obtained from acid glasses containing thymidine and 5'-dTMP. LION and VAN DE VORST (1973b) also obtained a doublet from 5'-dTMP in sulphuric acid glasses. They attributed their doublet spectrum to the radical anion on the thymine base. However, KAALHUS and JOHANSEN (1973) after studying the kinetics of the formation of hydrogen in sulphuric acid glasses concluded that a hydrogen atom adduct of thymine (4-ketyl radical) was responsible for the doublet observed after irradiation at 77 K.

SEVILLA (1971) was able to show the existence of thymine cation radicals after UV photolysis of thymine in phosphoric acid glasses. These glasses also showed spectra from hydrogen atoms and 5-thymyl radicals at 77 K. These latter species were absent if an electron scavenger (e.g., N_2O) was present during photolysis. This shows that in this case the electron is the precursor of both trapped hydrogen atoms and 5-thymyl radicals. However, the production of radical cations by radiolysis of acid glasses containing either thymine or other DNA constituents at low temperature has not been observed.

4.1.2 Radicals from Solid Nucleic Acids

The literature on the radiolysis of solid DNA at both room and low temperature contains many conflicting results. This is partly because of the complexity of the DNA molecule itself, but there is also the difficulty of obtaining samples which are sufficiently pure. This led to early reports of ESR signals in unirradiated DNA (BLYUMENFELD, 1959) which were later shown by WALSH et al. (1961) to be due to transition metal ions. If DNA samples are prepared and purified with sufficient care, no ESR signals should be detected.

The first ESR spectra for nucleic acids irradiated at room temperature all showed single asymmetric lines (SHIELDS and GORDY, 1959; SHEN et al., 1959; BOAG and MÜLLER, 1959). A spectrum showing two satellite lines accompanying the central singlet was published by DORLET et al. (1962). MÜLLER (1963) reported more structure, but no definite identification of a radical could be made from any of these results.

The first identification of a radical structure in irradiated DNA was made independently by EHRENBERG et al. (1963) and by SALOVEY et al. (1963). They found that part of the DNA spectrum is an octet which is very similar to the octet spectrum from irradiated polycrystalline thymine and thymidine (Fig. 4.6). The work was extended to UV photolysis of DNA at 77 K by EISINGER and SHULMAN (1963) who again found the same octet.

SALOVEY et al. (1963) assigned the octet spectrum to the 5-thymyl radical $-CH_2-\dot{C}(CH_3)-$ formed at the 5-6 bond of the thymine ring.

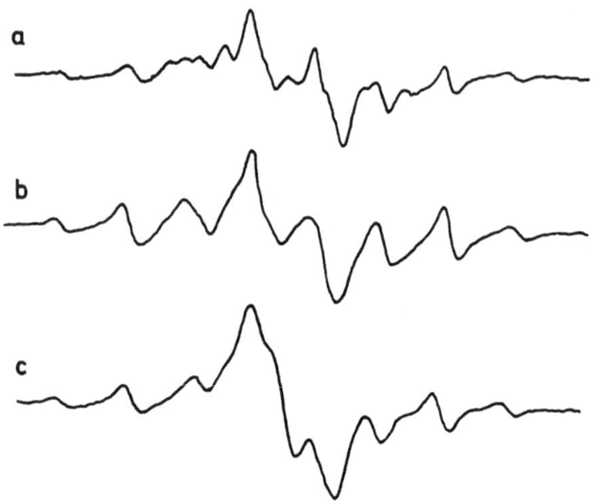

Fig. 4.6. ESR spectra of *a*: thymine; *b*: thymidine and *c*: DNA after γ-irradiation in the dry state at room temperature; the overall width of the spectra is about 135 G (EHRENBERG et al., 1963)

Its structure was subsequently confirmed both by studies on thymine which had been deuterated at the methyl position and on DNA into which the same deutero-methyl-thymine had been incorporated by growing a thymine-requiring mutant of *E. coli* in solutions containing deutero-methyl-thymine (PERSHAN et al., 1964). In both cases spectra consistent with a radical structure $-CH_2-\dot{C}(CD_3)-$ were obtained. In both normal and deuterated DNA the 5-thymyl radical is evidently formed by addition of a hydrogen atom. In a further experiment, PERSHAN et al. (1964) found that if normal dry DNA was exposed to D_2O instead of H_2O vapour before UV irradiation, the deutero-thymyl radical $-CHD-\dot{C}(CH_3)-$ was produced. The extra hydrogen therefore comes either from an exchangeable hydrogen or directly from the water. The signal from dry DNA was found to be the same whether the DNA was dissolved in H_2O or D_2O before being dried. The presence of two sources of hydrogen was inferred, one being dominant in moist and the other in dry DNA (PERSHAN et al., 1965).

The relative contribution made by the 5-thymyl radical to the total DNA spectrum has been reported by several groups of workers. Thus EISINGER and SHULMAN (1963) found the 5-thymyl radical to be the principal radical in DNA UV-irradiated at low temperature whereas GORDY et al. (1965) claimed that the predominant spectrum for DNA γ-irradiated at room temperature was a single line. MÜLLER (1964a) estimated that the relative contribution of 5-thymyl radicals was always less than 20% of the initial radical yield. This was supported by ORMEROD (1965) who estimated that 15% of DNA radicals at room temperature are 5-thymyl and concluded that in DNA radicals are not formed preferentially on the thymine base. COOK and WYARD (1966a) found about 50% thymyl radicals in several DNA samples at room temperature.

In view of the above results it is pertinent to discuss two factors which have been found to influence the appearance of the 5-thymyl spectrum in irradiated DNA. These are the temperature of irradiation together with any subsequent annealing process and the moisture content.

SALOVEY et al. (1963) found an increase in the relative strength of the octet spectrum on warming to room temperature for samples irradiated at ∿ 200 K. ORMEROD (1965) observed an increase in the fraction of 5-thymyl radicals from 6% to 15% as DNA irradiated at 77 K was warmed to room temperature. SANNER (1965) recorded the ESR spectra of DNA irradiated at 77 K after successive heat treatments. The thymine radical spectrum appeared between 200 K and room temperature. This experiment was repeated by COOK and WYARD (1966b) who observed three distinct changes in the spectrum as the sample was warmed. These occurred at 100 K, 210 K (where the 5-thymyl spectrum mainly appeared), and 260 K. The spectra are shown in Figure 4.7. VAN DE VORST (1967) has studied the changes in concentration of 5-thymyl radicals after successive annealing of a DNA sample irradiated at 100 K. He found a linear increase with annealing temperature up to 250 K. Above this temperature the number of thymine radicals in the sample remained constant.

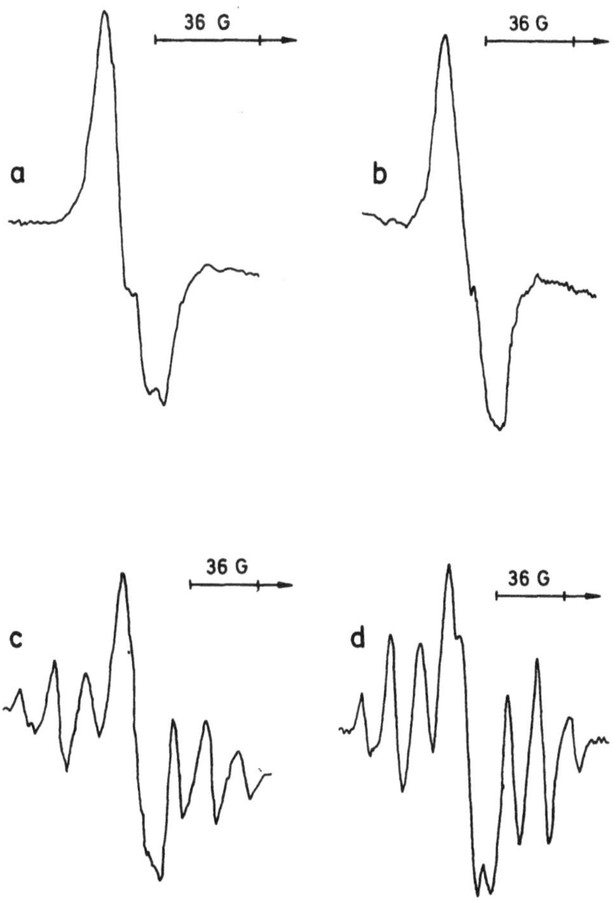

Fig. 4.7. ESR spectra of dry DNA γ-irradiated at 77 K. *a*: immediately after irradiation; *b*: after annealing for 5 minutes at 100 K; *c*: after 5 min at 210 K, *d*: final spectrum recorded at room temperature. Spectra *a*, *b* and *c* were recorded at 77 K (COOK and WYARD, 1966b)

The effect of moisture content upon the production of the thymine radical in DNA UV irradiated at 77 K has been investigated by PERSHAN et al. (1964, 1965). They found that if DNA is equilibrated in moist air at different relative humidities before UV irradiation and measurement at 77 K, the strength of the thymine radical spectrum increases with moisture content. The effect was shown not to be due to changes in the secondary structure of the DNA. The mechanism of thymine radical production with UV may be different from that with ionizing radiation. However, SANNER (1965) has compared spectra from dry DNA with those from a 20% aqueous solution of DNA after successive heat treatments following X-irradiation at 77 K. He found that the thymine radical spectrum appeared at a lower temperature in the frozen DNA solution than in the dry DNA and suggested that in DNA the presence of water facilitated secondary radical reactions which occur on warming. VAN DE VORST (1967) has also carried out annealing experiments on DNA equilibrated in moist air (50% relative humidity). He found a well-marked thymine radical spectrum after X or γ-irradiation and measurement at 100 K as well as at higher temperatures.

Production of the thymine radical in DNA thus appears to be facilitated by an increased temperature of irradiation or of annealing after irradiation at a lower temperature, and by an increased moisture content. It has commonly been observed that irradiation at a particular temperature produces the same spectrum as warming up to that temperature after irradiation at a lower temperature. If the amount of moisture present is excessive, the thymine radicals may not be observed, particularly at room temperature and above, because the stability of the DNA radicals is known to be affected by moisture content. Radical stability is discussed further in II.5. The DNA samples used by EHRENBERG et al. (1963) were treated in such a way as to remove oxygen rather than moisture. This probably accounts for their success in observing the thymine radical. Samples used by earlier workers were carefully dried and, therefore, gave only a single line spectrum.

In addition to the octet pattern other hyperfine lines have been found in ESR spectra of irradiated DNA. Although in many cases tentative assignments to particular types of radical structure have been made, these have never been confirmed. GORDY et al. (1965) have suggested that the spectrum obtained by DORLET et al. (1962) is due to the addition of a hydrogen atom to the guanine ring. GORDY et al. (1965) also reported a triplet spectrum which they compared to the spectrum of cytidine observed earlier by SHIELDS and GORDY (1959). VAN DE VORST and VILLEE (1964) found a resemblance between a spectrum from DNA at 150 K after irradiation at 100 K and the spectrum of cytosine obtained under similar conditions. On warming to room temperature the DNA spectrum became similar to the superposition of spectra from guanine and cytosine. A later report by VAN DE VORST and KRSMANOVIC-SIMIC (1966) showed that at higher doses the DNA spectrum is better approximated by an admixture of thymine and adenine spectra as well as those of guanine and cytosine. The thymine spectrum is also in evidence at lower doses if the molecular weight of the DNA is lowered before irradiation by ultrasonic degradation (VAN DE VORST et al., 1965a). Depurination of DNA before irradiation was reported to yield a cytosine-like spectrum at room temperature for both normal DNA of high molecular weight (VAN DE VORST and RICHIR, 1965) and degraded DNA (VAN DE VORST et al., 1966). Subtracting a theoretical octet spectrum from the spectrum of DNA after UV irradiation at 77K, PERSHAN et al. (1964) obtained five other spectra under differing conditions; these have not been identified, but one is reported to be similar to the spectrum of polycrystalline cytidine.

MÜLLER (1967) has pointed out that the reported resemblance of DNA spectra to spectra of, for example, cytosine and cytidine is probably coincidental on the grounds that the pure bases and the nucleosides are not components of DNA and do not necessarily give the same spectra as the corresponding nucleotides which are components of DNA. It is therefore appropriate to compare the spectrum of irradiated DNA with that of mixtures of the four nucleotides.

An early report by ALEXANDER et al. (1961) showed that the spectrum of DNA obtained after irradiation and observation at 77 K was quite unlike the spectrum obtained by adding together the curves obtained from the four separate nucleotides; from this result it was concluded that energy transfer, or spin migration, occurs prior to radical formation so that certain parts of the DNA molecule may be damaged preferentially. It was originally suggested by SHIELDS and GORDY (1959) that radiation-induced odd electrons may migrate along the polymer chain of DNA and that the spectrum of DNA would consequently be different from the sum of its constituents.

ORMEROD (1965) has proposed an electronic mechanism for radical formation in irradiated DNA. He found that at least part of the DNA spectrum at 77 K is due to trapped electrons. Some of these electrons are presumed to be trapped either on the thymine base or on other sites which transfer an electron to thymine on warming. The suggested reactions are those discussed in I.3:

$$T + e^- \longrightarrow T^-$$

at 77 K followed by either

$$T^- + RH \longrightarrow \dot{T}H + R^-$$

or

$$T^- + RH^+ \longrightarrow \dot{T}H + R^\cdot$$

on warming, where T represents the thymine ring, $\dot{T}H$ the 5-thymyl radical and RH some other part of the DNA molecule. This scheme is based partly on the fact that hydrogen atoms are not observed in DNA after irradiation at 77 K and partly on the fact that the presence of electron scavengers reduces or completely suppresses the yield of thymine radicals on warming. Ormerod's scheme has been disputed by HERAK and GALOGAŽA (1969a) who suggested that an anion formed by the first step might release a hydrogen atom which could then react directly with another unionized base to form a hydrogen addition radical. In both schemes, however, the first step is the formation of a base ion; energy transfer or spin migration then corresponds to the formation of an ion preferentially on one or more DNA bases.

PIHL and SANNER (1966) reported a resemblance between the spectrum of various DNA samples, after irradiation at 77 K and annealing at room temperature, and those of freeze-dried mixtures of nucleotides, provided that the base compositions were similar. Their results did not support the view that a migration occurs to any great extent in DNA. A similar conclusion was reached by PERSHAN et al. (1964, 1965) and by VAN DE VORST (1967). More recently, however, GREGOLI and BERTINCHAMPS (1972a) have reported preferential radical formation on pyrimidines rather than purines and on cytosine rather than thymine in several samples of DNA and various mixtures of nucleotides after freeze drying and irradiating. Subsequently, GREGOLI and BERTINCHAMPS (1972b) were able to match their DNA spectra with suitable linear combinations of thymine and cytosine radical spectra. Despite the conflicting evidence regarding radical or spin migration, close agreement was obtained between the experimental DNA spectra and the computer-reconstructed spectra. These results are discussed in more detail in I.7.2.

ESR has also been used to investigate the radiolysis of certain naturally occurring complexes of DNA and protein. Two types of nucleoprotein have been studied. These are fish sperm heads which contain about 65% DNA tightly bound to the protein salmine and *E. coli* bacteriophages of the T-series in which the DNA is surrounded by a protein coat from which it is easily separated. The different binding between DNA and protein has been found to influence the radiological behaviour.

The spectrum of salmon sperm heads after γ-irradiation and observation at 77 K was shown by ALEXANDER et al. (1961) to bear a much closer resemblance to the spectrum of a sample of isolated DNA than to that of an artificial mixture of DNA and salmine. It was concluded that energy transfer from the salmine to the DNA had taken place in the nucleoprotein. At room temperature all three samples showed different spectra. The difference between the spectra of the sperm heads and the isolated DNA was ascribed to differences in the structure and packing of the DNA which in turn influences the secondary ractions occurring on warming. Further evidence of energy transfer from protein to DNA is shown by the behaviour of sperm heads in the presence of electron scavengers (ORMEROD, 1965) which closely paralles that of isolated DNA. Thus both sperm heads and DNA were found to give a single line spectrum at 77 K whether or not electron scavengers were present. On warming to room temperature both were found to give the thymine octet spectrum which was either reduced or absent in the presence of electron scavengers. VAN DE VORST et al. (1965b) have irradiated nucleohistone and, in contrast to the above findings, have reported a localization of the radical on the protein rather than on the DNA. Thus at room temperature they found a resemblance between the spectrum of the nucleohistone and the isolated histone.

Experiments with sperm heads containing cysteamine (ORMEROD and ALEXANDER, 1962, 1963) showed that the spectrum at 77 K was unaffected if the amount of cysteamine was 5% or less. On warming, the spectrum of a sulphur radical appeared. These results were interpreted as showing that cysteamine protects against radiation by a repair process rather than by transfer of energy from DNA to cysteamine.

Early results for irradiated whole bacteriophage (ZIMMER and MÜLLER, 1965) showed a pronounced variability of the spectra even in apparently identical experiments. It was reported later, however, that under most conditions spectra of bacteriophage can be duplicated by simple superposition of the spectra of their components (MÜLLER, 1967) although an exception was found for T2-phage at 77 K. In this case an asymmetric doublet was found which could not be resolved into the spectra of the protein and DNA irradiated separately (MÜLLER, 1963). Nevertheless, energy transfer has not been reported in bacteriophage. This reflects the much looser association which is found in bacteriophage between DNA and protein than is found in sperm heads.

4.2 Oriented Fibers

The preceding chapter has shown results obtained from unoriented samples of nucleic acids. The observed ESR spectra, often called "powder" spectra, are envelopes over spectra from all possible orientations of the free radical vs. the magnetic field of the spectrometer. If the individual spectrum is almost isotropic, as is the case with the hydrogen-addition radical [Formula (3.4)] on thymine (TH), information on spectral parameters may easily be available also from a powder spectrum, as indeed the early identification of the TH radicals has shown. On the other hand, if there is considerable anisotropy in the hyperfine couplings, as one can expect for hydrogen nuclei in α-position or central nitrogen nuclei, the overlapping of spectra from all directions in a powder spectrum will smear out many of the finer characteristics of the individual spectra, and thus make the spectral interpretation more difficult. As in the case of nucleic acid constituents, a single crystal would be the ideal material to use in order to obtain more information. Unfortunately it is not possible to prepare single crystals from DNA.

Instead, other methods of orientation have been in use which take advantage of the tendency to fiber formation of DNA. In these fibers the DNA molecules are oriented approximately parallel to the fiber axis. A few examples of methods which have been used to give oriented DNA in small amounts are the following:

The DNA fibers used by WILKINS et al. (1951) in their X-ray diffraction work were prepared by withdrawing a glass point from the mass of a sticky DNA gel. More recently fibers obtained with this "dry spinning" method were used in ESR and semiconductivity work (SNART, 1968). Attempts have been made to increase the capacity of the method by using a rotating take-up device (WYCKOFF, 1955; LEDNEV and TICHONENKO, 1966; SNART, 1968).

Thin films of oriented DNA for optical studies have been prepared by shearing a DNA gel between two (quartz) plates or by stroking the gel with a spatula; with these techniques great difficulties were experienced in the preparation of uniformly oriented samples (SEEDS, 1953).

VOROBLIEV and SHAPOT (1953) prepared fibers of DNA or nucleoprotein for mechanochemical investigations by extruding a solution of the polymer through a capillary into an aqueous alcohol saline solution.

SPITKOVSKII (1955) oriented nucleoprotein for mechanochemical studies by letting a drop of nucleoprotein solution slowly fall through a precipitating saline solution, whereby a tail of oriented material was obtained.

KURUCSEV and ZDYSIEWICS (1971) have described three methods for the orientation of DNA molecules in films suitable for optical studies. The first involves stretching of pure isotropic DNA films, the second stretching of a substrate poly(vinyl alcohol) film with DNA films deposited on its surface, the third is stretching of mixed DNA-poly(vinyl alcohol) films.

4.2.1 The Wet Spinning Method

To meet the needs of oriented DNA in larger amounts for a wide range of physicochemical studies, a new orientation method based on wet spinning was developed (RUPPRECHT, 1963, 1966, 1970a, b). The principles of the method are as follows:

DNA solution is continuously extruded through a spinneret into a spinning bath (aqueous ethyl alcohol containing alkali-metal chloride). During precipitation, the DNA fibers are stretched as they pass down a column and via a stationary V-shaped fiber-guide they are converged into a bundle which is wound onto a rotating teflon-coated cylinder. While rotating, this cylinder also performs a slow axial motion back and forth within a preset range and a deposit of parallel DNA fibers is built up to the desired width and thickness. The cylinder, with deposit, is subsequently bathed in an aqueous alcohol solution of predetermined alkali-metal chloride concentration and is finally dried slowly over silica gel or a saturated electrolyte solution. During this drying process the alcohol and part of the water are absorbed by the drying agent and the DNA deposit coalesces into a film of oriented DNA which has the appearance of a piece of adhesive tape. Figure 4.8 shows a front view of the wet spinning apparatus.

Birefringence, UV dichroism and X-ray diffraction have all indicated that the molecular orientation of the DNA films obtained with the wet spinning method is excellent. More specifically WILKINS and ARNOTT (1966), from an X-ray diffraction test of wet-spun DNA films, have reported a maximum detectable disorientation of $\pm 4^{\circ}$ corresponding approximately to a standard deviation of about 2°.

Special equipment for folding (Fig. 4.9) and compressing DNA films into larger samples has been developed to enable the preparation of oriented samples of practically any desired dimension (RUPPRECHT, 1970a). Studies with such samples were initiated in the fields of semiconductivity, ESR, NMR, X-ray diffraction, optics, neutron scattering and mechanochemistry. The wet spinning method has also been modified for preparation of oriented dye-DNA complexes (RUPPRECHT et al., 1969).

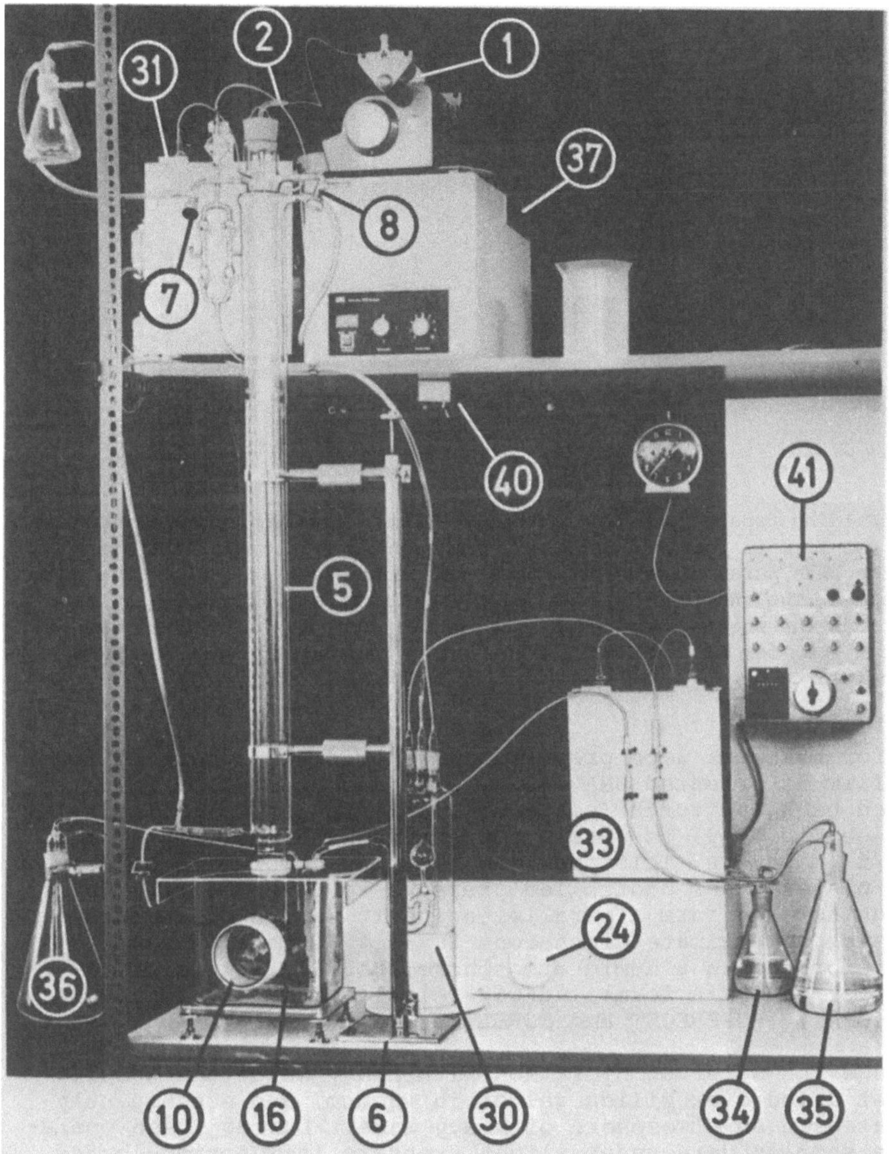

Fig. 4.8. Front view of the wet spinning apparatus. *1*: Precision pump with DNA solution; *2*: Glass capillary through which the DNA solution is conveyed to the glass spinneret at lower end of capillary inside the glass column; *5*: thermostated glass column with spinning bath (70-80% aqueous alcohol containing alkali-metal chloride) where the DNA fibers from spinneret are precipitated under stretching, a process giving molecular orientation along the fibers; *10*: teflon-coated rotating glass cylinder on which the DNA fibers are continuously wound. The cylinder is submerged in the spinning bath in a glass vessel *16*

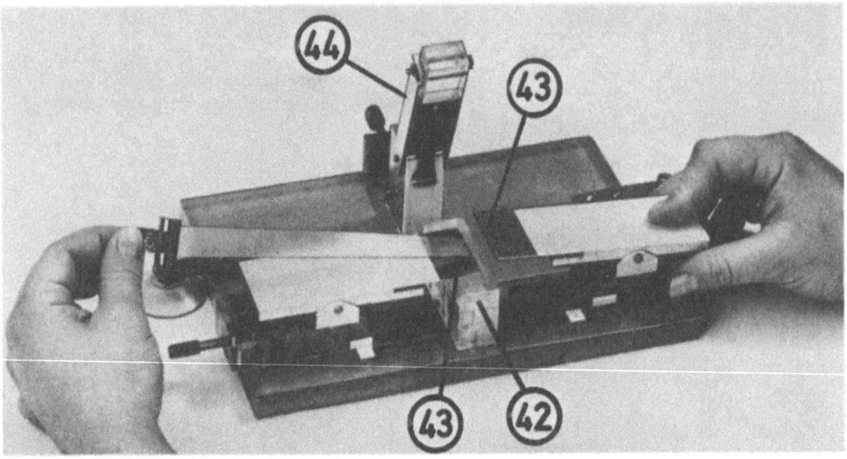

Fig. 4.9. Folding apparatus for oriented DNA films. *42*: sample table onto which the DNA film is fastened initially with adhesive tape; *43*: stainless steel sheets with parallel free end edges over which, one at a time, the DNA film is folded. The sheet holders are displaceable on slades. In their innermost positions the sheets overlap with an amount equal to the desired sample width (about 2 mm for ESR samples); *44*: pressure arm which is pressed down to sharpen fold

Samples for EPR work were prepared in a humid atmosphere as follows: A film of oriented DNA, 10 mm wide and about 0.05 mm thick, was folded back and forth (perpendicularly to the molecular orientation, see Fig. 4.9) to form a concertina-like pack containing about 40 film layers. This pack was slightly compressed between parallel plates under controlled pressure to make the film layers stick together and form a parallel-epipedic sample of oriented DNA with the approximate dimensions 2 × 2 × 10 mm. The samples were equilibrated in a humid atmosphere, usually 75% relative humidity for NaDNA (A form) and 66% relative humidity for LiDNA (B or C forms) (RUPPRECHT and FORSLIND, 1970).

To exchange the water of hydration with heavy water the samples were first dried over silica gel or in vacuum, and subsequently equilibrated in an atmosphere of heavy water. During these operations the samples were under slight pressure in a spring-loaded holder (RUPPRECHT, 1970a).

In order to attempt an exchange of protons attached to ring carbons on the DNA bases the ESR samples were soaked for months at 40°C in a solution containing 80% C_2H_5OD and 20% D_2O. The choice of temperature for this process was influenced by mechanochemical work on oriented DNA (RUPPRECHT, 1970c) indicating risk of denaturation when using higher temperatures.

4.2.2 ESR Results

The first investigation of free radicals in oriented DNA samples was performed on fibers drawn from a solution and wound on to a spectrosil quartz plate (ELLIOTT and WYARD, 1965). Despite the rather crude method of orientation some anisotropies were observed in the ESR spectra obtained after X-irradiation of dry samples at room temperature. The suggested interpretation was that the radicals were π-electron systems located on the DNA bases.

In 1967 the first results were reported from DNA samples prepared by the wet spinning method (EHRENBERG et al., 1967). The samples containing approx. 30% of water, were γ-irradiated at 77 K. The first derivative spectra showed little resolution and were about 24 G wide (peak-to-peak in the derivative spectra). The amplitude was, however, strongly dependent on the angle between the DNA helices and the field. Upon annealing of the samples at ∿ 200 K for 5 min, considerable amounts of the $\dot{T}H$ radicals were observed. The primary ESR spectrum was tentatively assigned to radicals with the unparied electrons in π-orbitals on the DNA bases.

Further work in this laboratory indicated that the amplitude variation observed in the wet-spun samples corresponded to anisotropies in unresolved hyperfine splittings in the radicals, since the total intensity of the signal estimated as the double integral of the first derivative spectrum remained constant when the sample was rotated in the magnetic field. Our work then concentrated on increasing the spectral resolution. This was done by improvements in several ways: new experimental techniques made it possible to reduce the microwave power of the spectrometer and thus minimize saturation effects which were clearly present in the early spectra. Samples were prepared which were equilibrated over D_2O instead of H_2O. A numerical resolution enhancement procedure (HEDBERG and EHRENBERG, 1968) was applied to the experimental spectra. The outcome of all these efforts was ESR spectra with a partial resolution, which in addition to the observed anisotropies have enough information for the proposal of detailed radical models (GRÄSLUND et al., 1971). Figure 4.10 shows the ESR spectra of wet-spun NaDNA from calf thymus containing 30% D_2O, γ-irradiated at 77 K. The proposed radicals were two kinds of ionic radicals with delocalized π-electron spin densities on the DNA bases. From the deduced hyperfine couplings in the radical components, spin concentrations were estimated and compared with calculated spin densities. This comparison made it probable that the radical mixture was composed of anionic radicals on thymine (or cytosine) and cationic radicals on guanine (or cytosine). The quantitative ratio between the anionic and cationic component was estimated to be in the range 3:1 to 1:1. Spectra simulations based on the proposed radical models are also indicated in the figure. It should be observed that the sample orientation, when the magnetic field H is parallel to the fiber (helix) direction Z, is unique in the sense that H is perpendicular to all base planes (assuming the Watson-Crick model for the molecular structure of DNA). On the other hand, when H is perpendicular to Z, H lies in the base planes, and all possible directions in these planes are represented. This had to be taken into account in the spectra simulations.

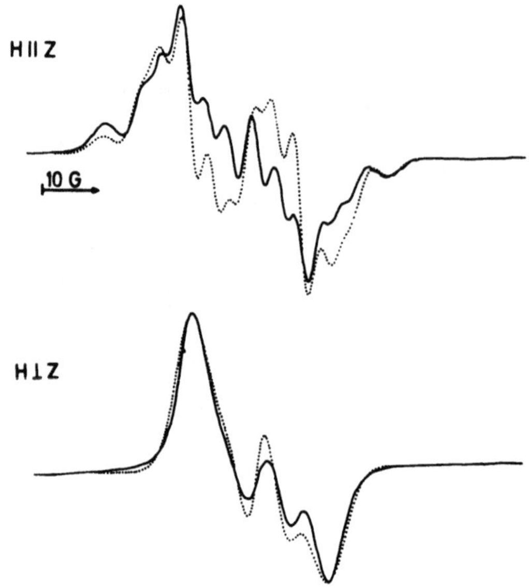

Fig. 4.10. (———): ESR spectra at 77 K of an oriented calf thymus NaDNA sample containing 30% D_2O after γ-irradiation with a dose 450 krad at 77 K. The spectra were recorded with the magnetic field of the spectrometer parallel to and perpendicular to the DNA fiber direction (H ∥ Z and H⊥Z, respectively). The microwave power was 3 μW; (.....) simulated spectra composed of the sum of two component spectra with different line widths and hyperfine coupling schemes. The parameters used were as in GRÄSLUND et al. (1971)

The proposed schemes of hyperfine couplings for the two radical components could be to some extent verified by a partial slow deuterium exchange effect in the spectra. Equilibration of the sample over D_2O at room temperature gave exchange only of the easily exchangeable protons in the DNA (crystal water, OH, NH and NH_2 groups) and had only a line-sharpening effect in the EPR spectra. However, from other studies on DNA (FRITZSCHE, 1967; SHELTON and CLARK, Jr., 1968) it was known that prolonged exposure to D_2O at an elevated temperature could cause exchange of a carbon-bound hydrogen (most likely C8-H of a purine). This hydrogen atom should also be involved in hyperfine coupling according to our radical models (in the guanine cationic radicals).

The sample was kept at 310 K in a deuterated ethanol-water medium as described in I.4.2.1 (wet spinning) for three months prior to irradiation. Following this treatment the ESR spectra were clearly changed. Deuterium has a magnetic moment which is 0.15 that of hydrogen, and a spin quantum number of 1. A hydrogen doublet is thus expected to collapse to a comparatively narrow (often unresolved) triplet spectrum. A closer analysis of the changed ESR spectra showed that the effect could be explained and simulated as a 10% deuterium exchange in the C8-H position of guanine.

In our recent investigations of the primary radicals in oriented DNA we have studied samples of nondeuterated and fully deuterated

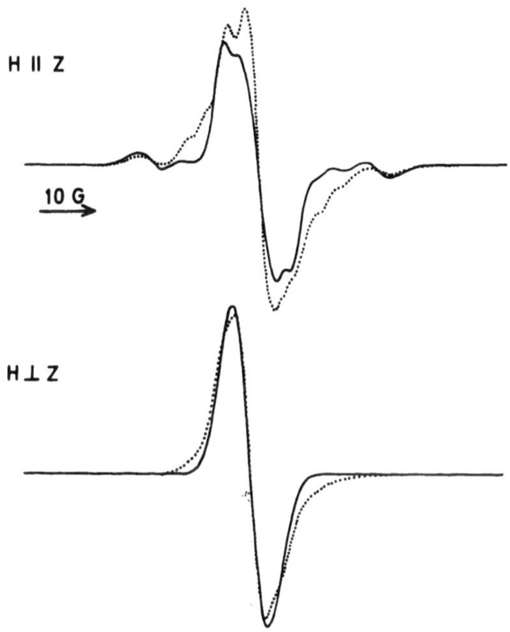

Fig. 4.11. (———) ESR spectra at 95 K of an oriented algal LiDNA sample containing 30% ^2H_2O after γ-irradiation with a dose 450 krad at 77 K. The spectra were recorded with the magnetic field of the spectrometer parallel to and perpendicular to the DNA fiber direction H ∥ Z and H⊥Z, respectively). The microwave power was 3 μW; (.....) simulated spectra composed of the sum of two component spectra with different line widths and hyperfine coupling schemes, which are based on the same radical models as in Figure 4.10 but with all hydrogen couplings changed to deuterium couplings. The parameter used were as in GRÄSLUND et al. (1976)

DNA from algae (GRÄSLUND et al., 1976). The ESR spectra from the nondeuterated sample were found to be similar to those from calf thymus DNA, whereas the spectra from the fully deuterated sample were completely different (Fig. 4.11). However, the radical models proposed earlier with all hydrogen couplings changed to the corresponding deuterium couplings were found to give a satisfactory description of the spectra also from the fully deuterated sample. Spectra simulations illustrating this are included in Figure 4.11. Quantitatively the simulated spectra account for ∼ 75% of the total spin concentration, and from the simulations we estimate the amount of spins corresponding to the anionic radical component to be about twice that of the cationic radical component. Thus the experimental spectra seem to be composed of 50% anionic radicals (on thymine) 25% cationic radicals (probably on guanine), and 25% unknown radicals, X˙.

The presence of ionic radicals in DNA after γ-irradiation at low temperatures was not unexpected. Earlier work on thymidine solutions, described in I.4.1 had shown that anionic thymine radicals are probable precursors to the ṪH radicals. Anionic radicals on thymine in DNA at 77 K is therefore completely in line with the appearance of ṪH radicals upon annealing of the same samples. This radical conversion was studied more in detail in a recent ESR-kinetic study on the wet-spun DNA samples (GRÄSLUND et al., 1975a). As expected, there was a close quantitative correspondence between the heat-induced decay of the anionic radicals and the appearance of the ṪH radicals. The overall conversion rate was found to be independent of radiation dose, but strongly dependent on the water content of the samples (less water, slower conversion rate). Kinetically the proton addition reaction at different elevated temperatures could be described as a first order process with a whole spectrum of rate constants and activation energies. The ṪH radicals are relatively stable only in a relativly narrow temperature interval. In samples containing 30% water, they appear in significant amounts after annealing a few minutes at temperatures between 170 and 200 K. The whole DNA spectrum disappears within 1 or 2 min at room temperature.

For the proposed cationic radicals, supposedly associated with guanine, no such further reaction has been observed. A charge neutralization would in this case probably result in a proton abstraction from the cationic radical. The hypothetic neutral radical might have its ESR spectrum hidden under the primary ionic radical spectra, or its lifetime may be too short for observation.

Thus the identification of the anionic thymine radicals from the DNA spectrum at 77 K seems to be based on firm ground, because of their quantitative conversion to easily recognized secondary radicals (the ṪH radicals). Unfortunately no such corroboration has been obtained for the cationic radicals; despite a rather precise characterization of their ESR spectra, in agreement with deuteration studies, their assignment to a particular base in DNA must yet be regarded with some caution.

Ionic radicals of the nucleic acid constituent bases have lately been the object of many EPR studies, most of them, however, in frozen solutions (see I.4.1). The ionic radical spectra from DNA discussed here are in general agreement with these constituent studies; however, the powder spectra from many of the ionic constituent radicals are not very distinct from each other, and therefore cannot be used to discriminate between different possibilities of assignment of the component radical spectra in DNA at 77 K.

The physical and biological implications of primary ionic radicals in irradiated DNA must now be considered. First it should be mentioned that the word "primary" is not quite correct, since these radicals are definitely localized at specific preferred positions, which must be the result of some initial transfer of excitation energy or radiation product to these locations. They represent an intermediate step from initial ionizations to chemical radiation effects. They are probably formed also under more biological circumstances, but at higher temperatures and higher water content are so reactive that they might be observed only with an in situ irradiation equipment.

It is not known what electron-paired products ionic radicals may give rise to in DNA, but it is a reasonable hypothesis that they initiate base damage. In a recent pulse radiolysis study on solutions of various nucleosides, the formation of guanine cationic radicals was observed (WILLSON et al., 1974). A possible model for the protecting action of cysteamine and promethazine was suggested, in which these agents restored the initially formed cationic radicals to the normal nucleosides. The detailed biological consequences of DNA base damage are of course a complicated function of the specific nature of the biological system under study. These aspects will be considered in the biological section of this book, Section III.

Chapter 5

Radical Yields

5.1 General Aspects

In this chapter the results of numerous measurements of radical yields (G-values) on both DNA and its constituents will be presented and discussed. The radical yield is defined as the number of free spins produced by each 100 eV of energy absorbed by the sample during radiolysis. The sample may be a single substance, a mixture, a solution or complex of more than one substance. Measurement of the yield comes next in importance to the identification of the radical itself. Although the latter task is often difficult, or even impossible to accomplish, the radical yield can nearly always be measured. Nevertheless, absolute measurements on the same substance from different laboratories often show a wide spread of values, due partly to a number of systematic errors that may arise, and partly to various factors related to the sample itself. The sources of error have been considered by WYARD and COOK (1969) who concluded that relative measurements of different samples on the same ESR spectrometer should have errors not exceeding 20%, while errors in absolute determinations should not exceed 40%.

It is well known that when biological systems are irradiated with ionizing radiation they undergo change which may lead to malfunction or even total loss of viability. The only changes which can be detected in the ESR spectrometer are the growth or decay of free radicals. It is important to know whether or not the production of these free radicals is in any way related to the process responsible for causing biological damage. If the latter can be quantified, e.g., by the number of cells destroyed or molecules of a particular type produced or changed, then a corresponding yield can be defined and measured, at least in principle. This yield for biological damage can then be compared with the corresponding yield for radical production in the same system. If the two values are similar it would seem reasonable to infer a relation between radical production and biological damage.

The radical yield, particularly for single substances, has often been interpreted as a measure of radiation resistance. However, as pointed out by SHIELDS and GORDY (1959), if the efficiency for production of detectable long-lived radicals is low, bonds may still be broken and new chemical species formed. With this proviso, it is of interest to compare radical yields for related substances to see how the radiation resistance is modified by changes in the structure of the undamaged molecule. This will be done later in this chapter for the bases, nucleosides and nucleotides of DNA. An allied study is the effect of additives upon radical yield. An additive which increases the radical yield of the parent substance is said to act as a sensitizer. Conversely, when the yield is decreased, the additive protects the parent substance. In these cases there is a transfer of energy from the sensitizer to the parent substance or from the parent substance to the protector.

The definition of radical yield supposes that the radical concentration rises linearly with dose. If the radical concentration is plotted against dose for a particular substance a curve of the type shown in Figure 5.1 is usually produced. Such a curve is typical of most irradiated organic substances. It appears that within experimental limits such dose-effect curves are exponential and can be described by the equation (MÜLLER, 1967)

$$C = C_\infty \ (1 - e^{-D/D_o}) \qquad (5.1)$$

where C is the radical concentration and D is the radiation dose. At very high doses the radical concentration reaches a limiting or saturation value C_∞. D_o is the dose at which C is 37% of C_∞ and is sometimes called the characteristic dose.

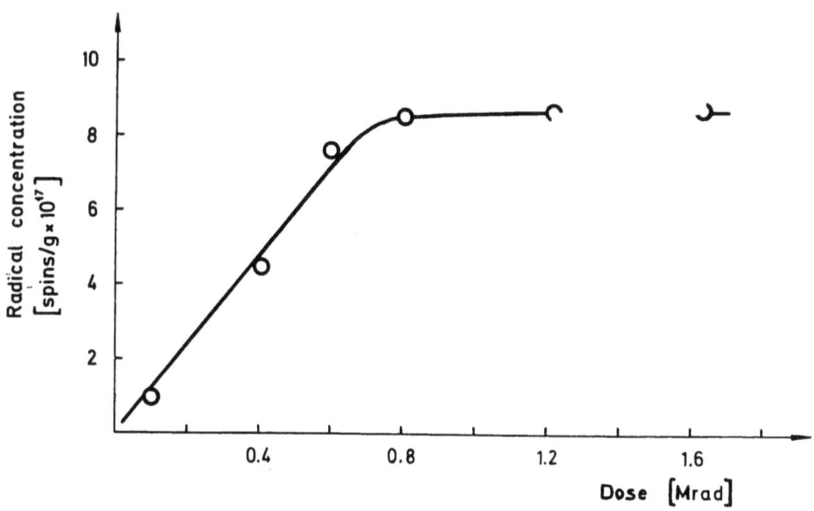

Fig. 5.1. Radical yield or dose-effect curve for a sample of DNA (WYARD and COOK, 1969)

When the dose is small, the curve is approximately linear. The yield can properly be defined as the limiting value of the slope of the dose-effect curve as the dose becomes very small. This quantity is sometimes called the initial yield and is given by $G = C_\infty/D_o$. The quantities C_∞ and D_o, which are constants for the particular substance, define the shape of the dose-effect curve. It follows that if a noninteracting mixture of two substances with different values of D_o and C_∞ is irradiated the dose-effect curve will be a composite of two exponentials and will not therefore be described by the simple equation given above. Nevertheless, a total initial G-value can still be defined which is the sum of the initial G-values of each component. Sometimes when two or more radical species are produced by irradiation of a single substance it is possible to measure separate radical concentrations and G-values. This can be done only when the overlapping of the spectra is favourable, usually because of the differing g-factors.

The practical importance of the dose-effect curve and the proper definition of the radical yield shows up in a historical context. Many early reports of G-values for both DNA and its constituents were later shown to be too low because the dose saturation effect had been neglected. These early radical yields were, in effect, C_∞ divided by a dose of several Mrad. They have accordingly been omitted from this chapter.

Although the exponential dose-effect curve has been found to apply to almost all organic substances investigated, there is still no satisfactory explanation as to why saturation should occur. COOK and WYARD (1966a) have estimated, for example, that at saturation there is one radical for every 15,000, 28,000, 8000 and 34,000 molecules of cytosine, thymine, uracil and guanine respectively. Therefore, saturation occurs long before all the undamaged molecules in the crystal lattice have been converted into free radicals.

The number of radicals which can be trapped in organic solids averages about 10^{18} spins per g. This means that radicals are not stabilized closer together than about 100 Å in most substances. It might seem that saturation arises by interaction of radicals that are closer to each other than a certain minimum distance. Furthermore, any interaction between nearest neighbours would be expected to be independent of the type of radiation. In amino acids it has been found that γ- and α-irradiation lead to different values of C_∞ (MÜLLER et al., 1963) although the ESR spectra show that both types of radiation produce the same radical species. Nucleic acid constituents and amino acids both show an exponential dose-effect curve and it would seem probable that the same mechanism of saturation would be present in both types of substance. Thus an explanation depending upon interaction between the radicals is unsatisfactory.

A similar objection can be raised to diffusion-controlled recombination of the radicals in that the diffusion coefficient would be expected to be independent of the type of radiation. In addition, the saturation concentration would depend upon the dose rate and, following irradiation, a second-order decay of radicals would take place. That these effects have not generally been observed shows this type of mechanism to be unimportant.

ROTBLAT and SIMMONS (1963) supposed that radiation destroys the radicals as well as producing them and thereby derived an exponential dose-effect curve. The rate of formation of radicals dN/dD, where N is the number of radicals and D is the dose, is just the initial yield which is a constant. Let this be denoted by n. The rate of destruction is taken to be proportional to N and can be written as αN where α is the rate of destruction or recombination. The net rate of radical formation is then given by

$$\frac{dN}{dD} = n - \alpha N \tag{5.2}$$

with solution

$$N = \frac{n}{\alpha} (1 - e^{-\alpha D}) \tag{5.3}$$

so that the saturation population is n/α. This equation is clearly of the same form as Equation (5.1) and shows that $1/\alpha$ corresponds to the characteristic dose D_o. This model, while being in agreement with experimental dose-effect curves, does not explain how the radicals might be destroyed.

The possibility that otherwise stable radicals are destroyed by reaction with hydrogen atoms has been discussed by MÜLLER (1967). Such atoms are produced by irradiation of most organic compounds and are sufficiently mobile for them not to be stabilized even at 77 K. Loss of a stable radical could result either by simple addition or by abstraction of another hydrogen atom to form molecular hydrogen. The probability of such an event on a given path length of the hydrogen atom is directly proportional to the concentration of stable radicals and so provides a possible physical explanation of Equation (5.2) and consequently of the exponential dose-effect curve. MÜLLER (1967) has pointed out that on this model the characteristic dose D_o should be of the same order of magnitude in different substances, assuming that radicals are produced mainly by hydrogen addition, and that the rate constant for radical destruction by reaction with hydrogen is fairly constant. Experimental observations show that values of D_o extend from several hundred krad for purine bases to several Mrad for a component such as thymidine. Values of D_o and C_∞ have generally not been given the same prominence in the literature as the yields. Nevertheless, destruction of radicals has been demonstrated in crystalline alanine (SNIPES and HORAN, 1967) at room temperature and in several other amino acids and dihydrothymine (HORAN and SNIPES, 1969) at 77 K.

More recently RIESZ et al. (1970) have studied the radiation-induced transfer of tritium to nonexchangeable sites in salmon sperm DNA, ribonuclease and poly-DL-alanine. The samples were first treated with tritiated water, then dried and γ-irradiated. Afterwords they were dissolved in water and lyophilized to constant specific activity. All three samples showed a linear increase of specific activity with dose well into the region of radical saturation. From these experiments it was concluded that saturation in ribonuclease and poly-DL-alanine is due principally to radiation-induced destruction of radicals. The same applied to at least an appreciable fraction of the radicals in the DNA sample.

5.2 Yields in Constituents

5.2.1 Solid Bases, Nucleosides and Nucleotides

The yields that have been published for the nucleic acid constituents after irradiation at room temperature are given in Table 5.1. The nucleosides and nucleotides are in all cases deoxynucleosides and deoxynucleotides i.e., constituents of DNA. It is immediately apparent that different workers have obtained very different yields for the same substance and that conclusions based upon small differences in yield must be treated with care. In this respect it helps to compare results from the same laboratory rather than relying on the absolute accuracy of the determinations.

Table 5.1. Radical yields in constituents of DNA at room temperature

Substance	(1)	(2)	(3)	(4)	(5)
Adenine	0.1	0.1		0.04	
Deoxyadenosine		1.4			
5'-dAMP		2			
Guanine		0.8		0.05	
Deoxyguanosine		0.9			
5'-dGMP		3			
Cytosine	0.4	0.4		0.1	
Deoxycytidine		1.0[a]			
5'-dCMP		5			
Uracil	0.1	0.8		0.8	
Deoxyuridine					
5'-dUMP					
Thymine	< 0.1	0.1	0.4	0.1	
Thymidine		0.4	0.33	0.4	
5'-dTMP		2		2	
Deoxyribose		4			
Deoxyribose-5'-phosphate					9.4

[a] deoxycytidine-HCl

(1) HENRIKSEN (1963); (2) KÖHNLEIN and MÜLLER (1964); MÜLLER (1964); MÜLLER and KÖHNLEIN (1964); (3) ORMEROD (1965); (4) COOK and WYARD (1966a); (5) VAN DE VORST (1970)

If the yields for any one base and its derivatives are compared, it can be seen that there is a progressive increase from the pure base through the nucleoside to the nucleotide. The yields of the nucleotides are comparable with those of the sugars and of ribose-5'-phosphate. A G-value of 4.6 has been reported by LACROIX and VAN DE VORST (1964) for the latter compound. Qualitative study of the spectra from irradiated nucleosides and nucleotides indicates that the free radical is located primarily on the base.

To the extent that the yield of base radicals is increased when the base is combined with either the sugar or sugar-phosphate group, the bases can be said to have been sensitized. Energy is thus transferred to the bases during irradiation of the nucleosides and nucleotides, although the mechanism involved appears to be unknown.

5.2.2 Influence of Environmental Parameters

5.2.2.1 Temperature

Radical yields have also been reported after irradiation of the constituents at 77 K. The yields can be determined directly by measuring the radical concentration at 77 K after irradiation at this temperature; this gives the low temperature yield. If the sample is allowed to warm up, the radical concentration can then be measured at room temperature to give a second value of yield. This is the yield after warming, which will nearly always be less than the low temperature yield because the radical concentration nearly always falls when samples are warmed. Again, the yield after warming is usually different from the room temperature yield obtained after both irradiation and measurement at room temperature. Table 5.2 gives the low temperature yields and the yields after warming in brackets. PIHL and SANNER (1966) are the only

Table 5.2. Radical yields in constituents of DNA at low temperature (77 K) and on warming (in brackets)

Substance	(1)	(2)	(3)	(4)	(5)
Adenine	0.3 (0.1)	0.14 (0.03)		0.2 (0.04)	
Deoxyadenosine		1.0 (0.5)		0.2 (0.1)	
5'-dAMP				1.3 (0.6)	
Guanine	0.6 (0.1)	1.3 (0.3)		0.6 (0.2)	
Deoxyguanosine		1.8 (0.4)		1.2 (0.2)	
5'-dGMP				2.8 (1.2)	
Cytosine	0.7 (0.2)	0.8 (0.3)		0.9 (0.05)	
Deoxycytidine		1.0 (0.4)		1.2 (0.5)	
5'-dCMP				2.1 (0.8)	
Uracil	0.6 (0.2)	1.2 (0.6)		0.8 (0.2)	
Deoxyuridine		0.9 (0.4)		0.3 (0.1)	
5'-dUMP				1.7 (0.6)	
Thymine	0.1 (<0.1)	1.5 (0.7)	0.47 (0.20)	0.13 (0.11)	
Thymidine		0.4 (0.4)	0.17 (0.085)	0.4 (0.3)	
5'-dTMP				4.3 (2.1)	
Deoxyribose		2 (1.0)		3.1 (2.5)	
Deoxyribose-5'-phosphate					8.5

(1) HENRIKSEN (1963); (2) KÖHNLEIN and MÜLLER (1964); MÜLLER (1964); MÜLLER and KÖHNLEIN (1964); (3) ORMEROD (1965); (4) PIHL and SANNER (1966); VAN DE VORST (1970)

workers to have investigated both ribose and deoxyribose derivatives. They observed no consistent difference between the yields for the two sets of derivatives. Only the results for the deoxyribose derivatives are given in the table.

In common with the room temperature yields, those at low temperature usually show a similar progressive increase in passing from the pure base to the nucleotide. The low temperature yields are usually larger than their room temperature counterparts, although in a few cases, e.g., the sugars, the situation appears to be reversed. It can be argued that comparison of yields at different temperatures is not very meaningful because different radical species are involved. It was mentioned in I.4 that when DNA is warmed to room temperature after irradiation at low temperature, the spectrum produced is the same as after irradiation at room temperature. A similar result holds for the constituents of DNA. This implies that the same radicals are present in the two cases and that in consequence a more meaningful comparison would be between the room temperature yield and the yield after warming. The data show that the room temperature yields are of the order of two or three times the yields after warming for most of the constituents (thymine appears to be an exception). Evidently irradiation at low temperature provides a measure of radiation protection, but the sensitization of the bases by the sugar and phosphate groups is present at low temperature, as well as at room temperature.

5.2.2.2 Crystallinity

HENRIKSEN (1966) has reported results which indicate that the physical dimensions of the crystals in a sample may influence the yield of radiation-induced radicals. After irradiation at room temperature, he found a 10% increase in yield for freeze-dried glycine compared with a single crystal sample. The difference was larger after low temperature irradiation when the yield for freeze-dried glycine was 24% larger than in a polycrystalline sample. Similar results were obtained for glycyl-glycine. The enzymes trypsin and lysozyme were also studied at room temperature. A progressive increase in yield with diminishing particle size was found. Freeze-dried samples showed the largest yields. Although the variations in yield were never large, and in some cases were smaller than the expected error for measurements of relative yield, they were consistent. The results support the idea that the yield depends upon the availability of suitable radical trapping sites which might in turn be associated with lattice imperfections and disorder. It is likely that the concentration of imperfections is larger on the surface compared with the interior of the crystal. Similar experiments have not been reported for DNA constituents but it seems that a similar effect might well be found.

5.2.2.3 Glasses

Very few measurements of yields of DNA constituents in irradiated aqueous glasses have been reported. It is necessary to measure the yield of the solute, i.e., number of solute radicals formed for 100 eV absorbed by the solute, as well as the yield for the whole frozen solution, over a range of solute concentration.

Table 5.3. Radical yields in frozen aqueous solutions of thymidine-5'-monophosphate irradiated at 77 K (VAN DE VORST, 1974)

Conc. of 5'-dTMP mol dm^{-3}	G (Solution)	G (Solute)	G ($OH^{\bullet}$)
0.1	1.4	24	1.5
0.2	1.8	17	2.0
0.35	2.1	14	2.4
0.5	2.3	14	2.7
0.65	2.5	9	3.2
0.8	2.2	7	3.2

VAN DE VORST (1974) has studied yields in irradiated aqueous glasses of thymidine-5'-monophosphate (5'-dTMP). The results are given in Table 5.3. It will be noticed that the yield of solute radicals G (solute) shows an inverse variation with the concentration of 5'-dTMP. Also, the presence of an indirect effect is indicated because G (solute) is always much greater than 1.8, this being the G-value obtained for freeze-dried 5'-dTMP irradiated at 77 K. The yield of $OH^{\bullet}$ radicals was found to vary directly with the concentration of 5'-dTMP, as can be seen from the fourth column of Table 5.3, but the stability of these radicals was found to be independent of the solute concentration. From this it was concluded that $OH^{\bullet}$ radicals do not react with 5'-dTMP.

Quantitative data for both thermal annealing and UV bleaching of frozen irradiated solutions of thymidine and 5'-dTMP have been published by LENHERR and ORMEROD (1968); their results are summarized in Table 5.4. The second column gives the total number of radicals left after annealing or bleaching as a percentage of the number present immediately after radiolysis at 77 K. The third and fourth columns give the estimated number of 5-thymyl radicals formed as a percentage of the total initial radicals at 77 K and of the number remaining after annealing or bleaching respectively. The quantity G ($\dot{T}H$) is the number of 5-thymyl radicals eventually produced per 100 eV of energy initially absorbed by the solution. Annealing took place at 195 K. UV bleaching (λ > 300 nm) was always followed by heat treatment at 118 K. This composite treatment provides a better basis for comparing the two processes. Thus in both cases $OH^{\bullet}$ radicals are destroyed and do not contribute to the final spectrum from which the concentration of thymyl radicals was estimated. UV bleaching alone at 77 K has been shown not to affect $OH^{\bullet}$ radicals. Table 5.4 shows that in the cases considered 5-thymyl radicals form the whole of the final radical population. The efficiency for production of 5-thymyl radicals is evidently less after UV bleaching than after thermal annealing.

Table 5.4. Effect of thermal annealing and UV bleaching upon frozen irradiated solutions of thymidine and thymidine-5'-monophosphate (LENHERR and ORMEROD, 1968)

Sample	% Radicals remaining	% 5-thymyl of Initial radicals	% 5-thymyl of Final radicals	G ($\dot{T}H$)
Thermal annealing:-				
2% thymidine in H_2O	16	18	100	0.10
2% 5'-dTMP in H_2O	20	20	96	0.12
UV Bleaching:-				
2% thymidine in H_2O	14	14	100	0.078
2% 5'-dTMP in H_2O	7.5	7.5	100	0.045

In conclusion it can be said that there now seems to be fairly general agreement about the importance of indirect effects in the radiolysis of frozen aqueous solutions. However, GREGOLI et al. (1974), after studying the radiolysis of frozen solutions of deoxyadenosine-5'-monophosphate (5'-dAMP) concluded that indirect action does not take place when nucleotides are irradiated in the frozen state, since for both 5'-dAMP and 5'-dTMP (GREGOLI et al., 1976) they reported a linear dependence of the radical yield upon the solute concentration.

5.3 Yields in Nucleic Acids

5.3.1 Influence of Environmental Parameters

Before giving the results that have been reported for radical yields in nucleic acids, it is first desirable to discuss briefly a few factors which have been found to influence the yield.

5.3.1.1 Effect of Moisture Content

It was found by SHIELDS and GORDY (1959) that the ESR signals in DNA are stable for several days under vacuum or in dry air at room temperature, but decay rapidly in moist air. Similar findings were reported by BOAG and MÜLLER (1959). EISINGER and SHULMAN (1963) found that the radicals in γ-irradiated DNA were stable in air at 50°C in their laboratory during the winter when the relative humidity was low (∿ 10%), but not at room temperature in summer when the humidity was relatively high.

In view of these results it might be expected that measured values of radical yield would depend upon the moisture content. Radical decay would take place both during irradiation and in the interval between irradiation and recording of the ESR spectra.

The relation between radical yield and moisture content has not been investigated directly. However, VAN DE VORST et al. (1960)

reported that the slope of the dose-effect curve increased with the dose rate for DNA γ-irradiated in air at room temperature. This effect is easily understood since the shorter the time of irradiation for a given dose, the less the radical decay and the higher the measured radical yield. It was also found that samples which had been given the same dose at different rates showed different radical decay curves after irradiation. This showed that at least two different radical species with different lifetimes had been produced. Thus the higher the dose rate the higher the proportion of short-lived radicals and the steeper the decay curve. A later study of the stability of DNA radicals by VAN DE VORST (1964) showed that the dose-rate effect is absent in DNA samples irradiated in vacuo. In this case the results were in qualitative agreement with those for irradiation under dry oxygen. These studies confirmed the relation between the stability of the DNA radicals and the moisture content and also showed that oxygen itself has a negligeable effect on both the spectra and the stability of the radicals as long as water is absent. Of the radicals produced by irradiation of DNA, COOK and WYARD (1966b) have shown that the least stable is the thymine radical.

5.3.1.2 Effect of Molecular Weight

VAN DE VORST et al. (1965a) reported that the radical yield of irradiated DNA shows an inverse dependence upon the molecular weight. Thus at room temperature a sample of chicken erythrocyte DNA of molecular weight 6×10^6 gave a yield of 1. After ultrasonic degradation to molecular weight 2.7×10^5 the yield was 1.6. Similarly, at low temperature, samples of salmon sperm DNA of molecular weights 1.9×10^6 and 4.1×10^5 gave yields of 1.8 and 2.4 respectively. It was estimated that a ten-fold reduction in molecular weight gives a 50% increase in radical yield. All the samples were irradiated in vacuo and the results, which are self-consistent, are unlikely to be artefacts arising from the presence of variable amounts of moisture. A possible explanation of the effect is that when bonds are broken during irradiation, diffusion of groups away from the parent molecule is more difficult when the molecular weight of the latter is high. Recombination is then more likely to occur leading to a lower radical yield.

In related experiments it was shown that if DNA is denatured by warming in aqueous solution so as to separate the two strands normally intertwined by hydrogen bonding, there is no change in the radical yield, whatever the temperature of irradiation (KRSMANOVIC-SIMIC et al., 1966). As was mentioned in I.4 the spectra are also unaffected by this treatment. A similar invariance of yield with denaturation was also found in poly A-U (RAJALAKSHMI et al., 1965).

5.3.1.3 Yields at Room Temperature

The results of radical yield measurements on DNA at room temperature are given in Table 5.5. As far as it has been possible to ascertain, all the G-values were obtained at doses sufficiently low to give an approximation to the initial yield well within experimental error. In most cases doses were much less than 1 Mrad. MÜLLER (1963) has reported a value of 205 krad for the characteristic dose D_o of T2 phage DNA.

Table 5.5. Radical yields in DNA at room temperature. Molecular weights, where known, are given in the second column

Reference	Source of DNA	G-value
MÜLLER (1962)	Calf thymus	0.6
EHRENBERG et al. (1963)	Calf thymus	0.4
HENRIKSEN (1963)	-	5.7
MÜLLER (1963)	*E.coli* phage T2	3.3-12.5
VAN DE VORST (1964)	Herring sperm	0.3
ORMEROD (1965)	Salmon sperm	2.0
VAN DE VORST et al. (1965b)	Calf thymus (3×10^6)	0.5
VAN DE VORST et al. (1965a)	Chicken erythrocyte (6×10^6)	1.0
	" " (2.7×10^5)	1.6
VAN DE VORST et al. (1966)	Salmon sperm (1.9×10^6)	0.6
	" " (4.0×10^5)	0.9
COOK and WYARD (1966b)	Calf thymus I	1.4
	II	1.4
	III	0.8
	Salmon sperm I	2.0
	II	0.8
	Herring sperm	1.5

All the samples appear to have been irradiated in vacuo. Thus any variation in the yields would not be due to instability of the radicals which has been shown to arise when excessive moisture is present. For the majority of the samples the molecular weights are unknown. The exceptions are those samples used by VAN DE VORST and co-workers to investigate the effect of varying the molecular weight. These values are given in the table alongside the source of DNA. In all other cases it may be assumed that natural (i.e., not degraded) DNA was used. The molecular weights of samples of natural DNA should all lie within an order of magnitude so that the upper limit for any variation in yield on this account should be 50%.

Despite these considerations, inspection of the table shows that there is a difference of a factor of more than 40 between the smallest and the largest G-value. However, it is noticeable that the variation is much smaller among results from any one laboratory. For example, those reported by COOK and WYARD (1966b) vary by a factor of no more than 2.5. As these authors have pointed out (WYARD and COOK, 1969) this variation can be attributed to experimental and systematic errors from known sources, including those associated with the sample such as impurity and moisture. Otherwise the reasons for the large differences in radical yields at room temperature are unknown.

5.3.1.4 Effect of Temperature

Radical yields have also been determined in DNA following irradiation at low temperature. Table 5.6 gives the low temperature yields and the corresponding yields after warming, in brackets, whenever these have been determined.

Table 5.6. Radical yields in DNA at low temperature (77 K) and on warming (in brackets)

Reference	Source of DNA	G-value
HENRIKSEN (1963)	-	3.0
MÜLLER (1963)	*E. coli* phage T2	(3.3-12.5)
PATTEN and GORDY (1964)	-	1.0[a]
ORMEROD (1965)	Salmon sperm	1.3 (0.8)
SANNER (1965)	Thymus	2.8
VAN DE VORST et al. (1965a)	Salmon sperm (1.9×10^6) " " (4.1×10^5)	1.8[b] 2.4[b]
PIHL and SANNER (1966)	Calf thymus	3.5 (1.5)
	Salmon sperm	3.2 (1.6)
	B. cereus T	4.1 (2.4)
	B. subtilis 168	5.1 (3.0)
	E. coli B	4.0 (2.5)
	Ps. aeruginosa	3.0 (2.2)
LENHERR and ORMEROD (1968)	Salmon sperm	3.1
VAN DE VORST (1970)	Salmon sperm (1.9×10^6) " " (4×10^5)	2.2 3.0
	Chicken erythrocyte (6×10^6) " " (2.7×10^5)	3.0 4.6
	Calf thymus (3×10^6)	1.5
GRÄSLUND et al. (1971)	Calf thymus (oriented)	1.6

[a] at 4.2 K. [b] irradiation at 100 K, ESR at 150 K.

Molecular weights, where known, are given in the second column.

It is evident from Table 5.6 that there is less variation among the yields at low temperature than at room temperature. The smallest and largest yields now differ by a factor of about 10 and are, on the whole, larger than at room temperature. The yields on warming, with one exception, are less than those at low temperature.

ORMEROD (1965) has compared the yield on warming salmon sperm DNA (0.8) with the room temperature yield (2.0) for DNA from the same source and concluded that the formation of primary radicals is reduced by lowering the temperature i.e., there is a protection effect. However, many of the yields on warming reported by PIHL and SANNER (1966), for example, are greater than the majority of room temperature yields in Table 5.5. It may thus be necessary to confine comparison of this sort to results from the same laboratory which, as noted earlier, show less fluctuation. PIHL and SANNER, unfortunately, have not reported room temperature yields for their samples. It should also be pointed out that MÜLLER (1963) found that for T2-phage DNA the yields after warming and at room temperature were the same.

5.3.1.5 Glasses

Yield measurements on frozen irradiated neutral solution of DNA have indicated the presence of an indirect effect. SANNER (1965), after studying aqueous DNA at various concentrations, found that the measured radical yields for the whole solution, i.e., G (solution) were always higher than the expected values. The latter were calculated from the yields of the components, namely, G = 2.8 ± 0.2 for dry DNA irradiated at 77 K and G (ice) = 1.0 ± 0.1, also at 77 K; his results are given in Table 5.7. Despite the difficulties of accurately measuring the radical yield of dry DNA, comparison of the yields in the second and third columns of the table shows the production of "excess ESR centres" by interaction between the solute and solvent.

Table 5.7. Radical yields in frozen aqueous solutions of DNA irradiated at 77 K (SANNER, 1965)

Conc. of DNA (%)	G (Solution) measured	expected	G (Solute)
9	1.5 ± 0.1	1.2	6.5
18	1.9 ± 0.1	1.3	6.0
36	2.5 ± 0.1	1.6	5.2

The presence of an indirect effect can also be inferred by comparing the radical yield of the solute, G (solute), with the yield of dry DNA. As mentioned in I.5.2.2.3, G (solute) is the number of solute radicals formed when the solute absorbs 100 eV. It can be estimated in the following way. Suppose f is the fraction of solute in the solution by weight. Let the whole solution absorb energy 100 eV. Assuming that the absorbed energy is partitioned by weight fraction, the energy absorbed by the solvent is 100 (1-f) eV. If a second assumption is made, namely that the yield of water radicals in the solution is unaffected by the presence of solute, and is equal to the yield in pure ice, then the number of water radicals is (1-f) G (ice). The total number of radicals in the solution is just G (solution), so that the number of radicals associated with the solute is the difference G (solution) - (1-f) G (ice). From the first assumption the energy absorbed by the solute is 100 f eV. This gives a value of G (solute):-

$$G\ (\text{solute}) = \frac{1}{f}\ [G\ (\text{solution}) - (1-f)\ G\ (\text{ice})] \qquad (5.4)$$

The fourth column of Table 5.7 gives values of G (solute) calculated from the corresponding G (solution) and the previous value of G (ice) using this expression. Again, the high values of G (solute) compared with the value for dry DNA (2.8 ± 0.2) show the presence of excess ESR centres.

The validity of the above expression for G (solute) depends upon the assumption that the yield of water radicals in the solution is the same as in ice. LENHERR and ORMEROD (1968) estimated that within experimental error the yield of $OH^{\cdot}$ radicals was the same in DNA solutions as in pure ice. Their value for G (ice) was 0.51 ± 0.03. They also measured yields of DNA radicals in frozen solution. Their results, together with calculated values of G (solute), based upon their value for G (ice), are given in Table 5.8. The value of G (solute) are greater than the measured yield of dry DNA [Eq. (5.1)] as given in Table 5.6. They also found that the expected values of G (solution) calculated from the yields of ice and dry DNA were lower than the measured values of G (solution). Their results are thus in agreement with those of SANNER (1965) in showing the presence of an indirect effect in irradiated frozen DNA solutions. LENHERR and ORMEROD (1968) also found that there is no reaction between the water ($OH^{\cdot}$) radicals and the DNA. Since the yield of the former is unaffected by the presence of DNA it follows that the excess radicals associated with the indirect effect are located on the solute.

Table 5.8. Radical yields in frozen solutions (H_2O and D_2O) of DNA at 77 K (LENHERR and ORMEROD, 1968)

Conc. of DNA (%)	G (Solution) measured	G (Solute)
2 (H_2O)	0.96 ± 0.03	24 ± 3
10 (H_2O)	0.85 ± 0.04	4.3 ± 0.8
10 (D_2O)	0.75 ± 0.04	3.2 ± 0.7

The data for 10% DNA in D_2O solution in Table 5.8 shows that the yields in D_2O solution are expected to be lower than those in the corresponding aqueous solution.

If the results of SANNER (1965) and LENHERR and ORMEROD (1968) are compared (Tables 5.7 and 5.8), they will be found to disagree over the variation of the yield of DNA radicals with respect to solute concentration. Thus SANNER (1965) found the yield of DNA radicals to be substantially independent of DNA concentration and inferred from this that only energy absorbed within a small envelope of water surrounding each solute molecule would be transferred to the latter. LENHERR and ORMEROD (1968), on the other hand, found very different values of G (solute) for solutions of 2 and 10% DNA, which indicated an inverse dependence of solute yield upon concentration. Such a dependence was found by VAN DE VORST (1974) for frozen aqueous solutions of 5'-dTMP as discussed in I.5.2.2.3. The reason for this discrepancy is not clear and it will probably be resolved only after further investigation in the laboratory.

5.3.2 Comparison with Yields of Constituents

There appears to be greater variation in the yields reported for DNA than for the constituents. The largest and smallest yields for DNA differ by a factor of 40 as mentioned earlier, while for the constituents a four-fold variation in values from different laboratories is not uncommon. However Table 5.1 shows that even yields for the bases can show surprising differences. Guanine, for example, is reported to have yields of 0.05 and 0.8. As already mentioned, early results, in which saturation of the dose was neglected, have been omitted from all the tables in this chapter. In addition, various factors which affect the yield of radicals in DNA, such as moisture content and molecular weight, would not apply to the isolated bases.

Low temperature yields for DNA show less variation than at room temperature and a similar situation exists for yields in the constituents. Thymine is exceptional in that low temperature yields as diverse as 1.5 and 0.1 have been reported (Table 5.2).

The increase in yield in passing from base, through nucleoside to the nucleotide has been noted already. As far as can be ascertained from the data of Table 5.5, room temperature yields of DNA tend to be somewhat less than the mean of the yields of the four nucleotides. As was discussed in I.5.2.1, the sugar and sugar-phosphate groups can be regarded as having a sensitizing effect upon the base in the nucleosides and nucleotides respectively. An alternative approach is to regard the base as protecting the remainder of the molecule. This effect can be seen quantitatively by comparing the experimental yields of the nucleosides and nucleotides with the theoretical yields calculated from the known yields of the component base and either the sugar or sugar-phosphate group. Calculated yields, reported by VAN DE VORST (1970), are given in Table 5.9, together with the experimental yields for comparison. The latter have already been given in Tables 5.1 and 5.2. The calculated yields for the nucleotides at room temperature use the yields for the bases reported by KÖHNLEIN and MÜLLER (1964), given in Table 5.1. Those at low temperature use the base yields of PIHL and SANNER (1966) given in Table 5.2. The same yields of the bases were used to calculate yields for the nucleotides. The yields of 2-deoxy-D-ribose-5'-phosphate, as determined by VAN DE VORST (1970), were 9.4 and 8.5 at room and low temperature respectively. From Table 5.9 it can be seen that all the experimental yields for the nucleosides and nucleotides are less than the corresponding calculated values. This indicates the presence of a radioprotective effect. It is possible to define the efficiency of protection as the difference between the calculated and experimental G-values expressed as a percentage of the calculated G-value. However, because this quantity depends upon the difference between two G-values, it is subject to very wide fluctuation, and no consistent conclusions can be drawn.

Table 5.9. Experimental and calculated radical yields in nucleosides and nucleotides (VAN DE VORST, 1970)

Substance	G-value at 300 K		G-value at 77 K	
	Experimental	Calculated	Experimental	Calculated
Deoxyadenosine	1.4	2.19	0.2	1.93
5'-dAMP	2	5.95	1.3	5.3
Deoxyguanosine	0.9	2.46	1.2	1.99
5'-dGMP	3	6.0	2.8	5.4
Deoxycytidine	1.0	2.55	1.2	2.06
5'-dCMP	5	6.5	2.1	6.0
Thymidine	0.4	2.27	0.4	1.67
5'-dTMP	2	6.1	4.3	5.8

DUCHESNE and VAN DE VORST (1965) calculated the expected yield for DNA at room temperature from the corresponding yields of the four nucleotides, assuming that the latter were present in equal proportions and without interaction between them. Their calculated yield of 2.8 is higher than most of the experimental yields in Table 5.5 but comparable with the yields on warming reported by PIHL and SANNER (1966), given in Table 5.6. These latter authors concluded from their data on nucleotides (Table 5.2) that the number of radicals stabilized in equimolar mixtures of dAMP and dTMP should be similar to that in corresponding mixtures of dGMP and dCMP, at least at low temperature. This was confirmed by experiments on freeze-dried mixtures of nucleotides which showed that both the low temperature yield and the yield on warming decreased only slightly with increasing guanine-cytosine content. However, the results for the yields of DNA were not significantly different from those of the mixtures of nucleotides.

In later calculations, VAN DE VORST (1970), starting from the yields of the isolated bases and of deoxyribose phosphate, also found that the theoretical yields of mixtures of nucleotides should be almost independent of the base composition. However his values, namely 6.1 and 5.6 at room and low temperature respectively, are noticeably higher than most of the experimental G-values of DNA itself.

In summary it can be said that nucleosides and nucleotides show a radioprotective effect. The sugar moiety, which in isolation shows a high yield, is protected when combined with a base to form a nucleoside, and the observed radical is located on the base. In the nucleotides, similar protection of the sugar-phosphate moiety by the base occurs. Yields of nucleotides are higher than those of nucleosides, but yields of DNA are of the same order as the yields of nucleotides. Calculations by DUCHESNE and VAN DE VORST (1965) and by VAN DE VORST (1970) suggest that a radioprotective effect is present in DNA in that the yields of the latter are less than that expected from a noninteracting mixture of nucleotides. Experimental work by PIHL and SANNER (1966) on DNA and mixtures of nucleotides, on the other hand, suggests that yields from both systems are very similar. Yields of both DNA and mixtures of nucleotides are not greatly dependent upon the base composition. In DNA the yield has been found to vary inversely with the molecular weight - an effect which has been considered as a form of autoprotection (VAN DE VORST et al., 1965a). Thus, together with the large observed spread in G-values of DNA, for reasons which are still largely unknown, makes comparison with the constituents rather difficult and unsatisfactory. It is expected that further progress in this direction will depend upon a better understanding of the experimental factors which so far have impaired the reproducibility of the yields of irradiated DNA.

Chapter 6

Radiomimetic Radical Production

6.1 Introduction

The investigation of radiation-induced free radical formation in DNA and its constituents by means of ESR spectroscopy has yielded considerable information on the response of this important macromolecule to ionizing radiation. However, many details of the mechanisms ranging from the formation of primary radicals to the appearance of more stable secondary species, have been under debate and some of them remain still unexplained. It is tempting to elucidate specific routes of damage selectively by inducing the corresponding radical species using nonradiolytic techniques. In this way, their ESR spectra are frequently undisturbed by overlap of spectra of other radicals. This "radiomimetic" radical formation not only serves to test the validity of radiation damage pathway models but also allows confident extraction of spectral parameters of the radicals under study. The latter information is, in turn, useful in the interpretation of spectra from irradiated samples, especially when polycrystalline powders or noncrystalline solids are concerned.

The radiomimetic approach was used for the first time to study the interaction of solid nucleic acid constituents with H-atoms and, some time later, with OH-radicals. Later the fate of anion radicals of nucleic acid bases and base analogues was studied in some detail. More recently, the radicals resulting from deposition of energies of specific magnitude transferred to purine and pyrimidine derivatives via excited inert gases have been investigated. This chapter summarizes the results of these studies.

6.2 Interaction with H-Atoms and OH-Radicals

As was shown in I.3, hydrogen-addition radicals are found in all pyrimidine and purine constituents of nucleic acids. It is also known from the structure of radical pairs, for example, that hydrogen atoms released upon irradiation are capable of abstracting further hydrogens from the aliphatic parts of the molecules. Hydrogen atoms therefore could play an important role in the overall radiation damage of nucleic acids.

6.2.1 Addition and Abstraction Reactions with Nucleic Acid Bases and Sugars

The first experiments with hydrogen atoms and nucleic acid constituents were reported on all the nucleic acid bases by HERAK and GORDY (1965), on thymine by HELLER and COLE (1965) and on thymine and thymidine by HOLMES et al. (1966). It was shown that thermal hydrogen atoms readily react with pyrimidine and purine bases and form stable radicals in powdered samples. The radicals are of the hydrogen-addition type, as discussed in I.3.1. Figure 6.1 presents the ESR spectra of powdered thymine exposed to H- or D-atoms. One immediately recognizes the pattern belonging to the hydrogen addition radical [Formula (3.4)]. That the radical

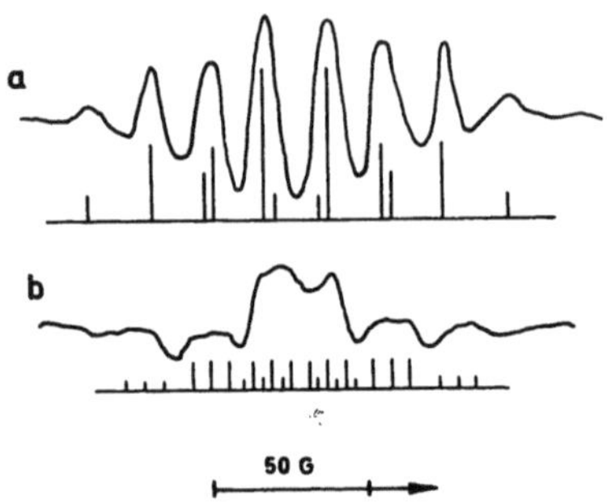

Fig. 6.1. ESR spectra (second derivative) of *a* thymine bombarded with H-atoms and *b* with D-atoms (after HERAK and GORDY, 1965)

was formed by addition of an H-atom from the discharge was proved in the experiment with deuterium atoms; if the D_2 gas was discharged instead of H_2, the addition of the deuterium atom on C6 of the pyrimidine ring (Fig. 6.1b) was observed. In a similar way it was proved that thermal hydrogen atoms add on C5 in uracil and cytosine, yielding radical Formulas (3.11) and (3.10).

When the purine bases, guanine and adenine, were exposed to H-atoms, broad 1:2:1 ESR triplet spectra were observed. From these spectra and the analogous ones recorded after D-bombardment, it was found that the radicals thus formed were also of the hydrogen-addition type [Formulas (3.1) and (3.2)].

The sugars, ribose and deoxyribose, were found to be less reactive with thermal hydrogen atoms. Radicals were observed both in pure ribose, in deoxyribose in the pyranose form and in deoxyribose phosphate where the sugar is in the furanose form, but the radical concentration was found to be much lower than in the bases under identical experimental conditions (HOLMES et al., 1967; HOLROYD and GLASS, 1969). The radicals have not been identified inspite of the well-resolved ESR hyperfine pattern. It is obvious, however, that the radicals must be of the hydrogen-atom-abstraction type.

Nucleosides and nucleotides exhibited much more expressed base radicals, although sugar radicals were also present in almost all samples studied. The apparent lower production rate and lower saturation concentration of sugar radicals is not completely understood; however, it should be brought about by the kinetic factors (HOLMES et al., 1967; HENRIKSEN, 1969).

6.2.2 Reactions with Nucleic Acids

It is expected that both DNA and RNA react with H-atoms in a similar manner as do the isolated constituents. The first experiments to prove this were, however, only partly successful (HELLER and COLE, 1965; HERAK and GORDY, 1966a). The ESR spectra characterizing the hydrogen-addition radicals of the bombarded RNA were easily observed, while DNA did not show any signs of H-addition radicals. Later it was shown by HOLROYD and GLASS (1969) that the lack of hydrogen-addition radicals in DNA could be attributed to the protective action of the protein in the impure preparation of DNA. Relatively pure DNA preparations exhibited a well-resolved hyperfine pattern. As low as 3% of the histone present significantly reduced the hydrogen-addition-radical lines.

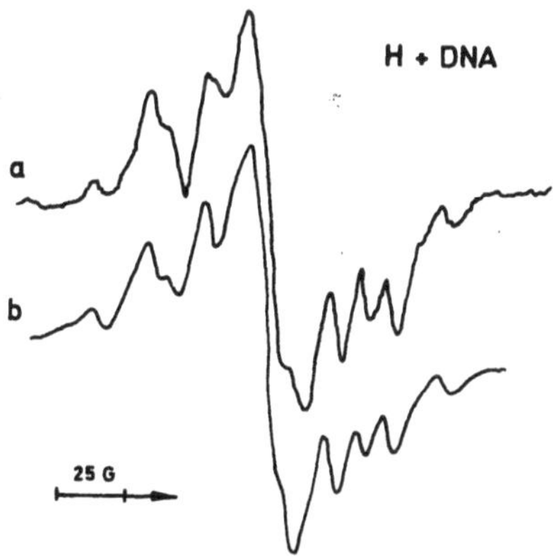

Fig. 6.2. The ESR spectrum of *a* salmon sperm DNA bombarded with H-atoms and *b* the computed curve with the assumption that it is composed of 18% deoxyribose radicals, 10% cytosine H-addition radicals, 23% thymine H-addition radicals (after HOLROYD and GLASS, 1969)

In RNA, both purine and pyrimidine H-addition radicals were present (HERAK and GORDY, 1966a). The exact ratio was not determined. In DNA (Fig. 6.2), HOLROYD and GLASS (1969) identified the addition of hydrogen atoms to all four bases (adenine and guanine were not distinguished). With the help of the computer simulation of the spectra, they found 82% of the radicals to be on the bases and 18% on the sugars. Among the base radicals, 23% was on thymine, 10% on cytosine and 49% on the purine bases. It is interesting to note that the above ratios rather closely reflect the base composition of the investigated DNA (29% thymine, 21% cytosine, 50% purines). This would indicate that all four bases are approximately equally capable of accepting hydrogen atoms.

The hydrogen-bombardment technique has given impetus to the elucidation of radiation damage pathways of nucleic acids and their constituents, especially the bases. At the time of the first experiments, radiation-induced hydrogen-addition radicals were known only in thymine compounds. After the addition of H-atoms to all nucleic acid bases had been performed nonradiolytically, these radicals were soon found in other irradiated purine and pyrimidine nucleic acid constituents, both in frozen solutions and in the crystalline form. Thus, it was established that one possible pathway leading to the hydrogen-addition type of radicals at the nucleic acid bases was the addition of thermal hydrogen atoms.

6.3 Reaction with Electrons

About twelve years ago, ORMEROD (1965) suggested the existence of an anionic stage in the formation by ionizing radiation of hydrogen adduct radicals at the C6-position of thymine. It was therefore of interest to study anion radicals of DNA bases, separate from other possible radicals produced during radiolysis, and their conversion into H-addition radicals. One approach to this problem consists of studying the reaction of nucleic acid components with electrons generated by photolysis of ferrocyanide ions in alkaline glasses at 77 K (AYSCOUGH et al., 1965), according to Equation (6.1).

$$Fe(CN)_6^{4-} + h\nu_{254nm} \longrightarrow Fe(CN)_6^{3-} + e^- \qquad (6.1)$$

These electrons are characterized in ESR by a singlet 14 G wide (g = 2.0006) and spectrophotometrically by an intense absorption at 585 nm. They have been shown to have the properties of hydrated electrons formed during water radiolysis (KEVAN, 1969). On bleaching of the glass with light corresponding to their absorption band or on warming the sample to temperatures higher than 173 K, the electrons become mobile and may react with solute molecules. Recently, it was found that 12 M LiCl is also compatible with the ferrocyanide technique for generating electrons (SEVILLA et al., 1972a).

6.3.1 ESR Characteristics of the Pyrimidine and Purine Anion Radicals

6.3.1.1 Pyrimidine Derivatives

HOLROYD and GLASS (1968) reported the first ESR spectra of the anions of some pyrimidine derivatives: they found 24 G wide singlets in an NaOH glass at 80 K. Only in the case of uracil anion at 190 K, did they observe a well-resolved doublet with a hyperfine splitting equal to 16 G. In a great number of pyrimidine compounds, the best doublet resolution was attained by SEVILLA and his collaborators (SEVILLA, 1971; SEVILLA and VAN PAEMEL, 1972; SEVILLA et al., 1972a, b, 1974), employing D_2O in place of H_2O in alkaline and neutral matrices. The set of their results is presented in Table 6.1. Some representative ESR spectra are given in Figure 6.3.

In general, the anions are observed after photo-bleaching of the trapped electrons. In the case of 5-halouracils, filtered light is used as several of the anion radicals formed dehalogenate when exposed to unfiltered light (SEVILLA et al., 1974).

From the theoretical calculations (BAUDET et al., 1962; MANTIONE and PULLMAN, 1964), a spin-density value of 0.5 is expected at position C6 in complete agreement with the large doublet splitting measured in all investigated pyrimidine derivatives. The quartet of 15.4 G observed with 6-methyluracil confirms in addition that the C6 position is that of high spin density. As to the halopyrimidines, the g-factors found for the halopyrimidines at 113 K are in the range 2.0033 ± 0.0003 (fluorouracil) (SEVILLA et al., 1974) to 2.0022 ± 0.0004 (iodouracil) (SEVILLA and VAN PAEMEL,

Table 6.1. Hyperfine splitting constants of pyrimidine anions

Pyrimidine derivatives	Matrix	T Observation K	Description of the signal	Hyperfine splitting G	Figure	Reference
Thymine	8 M NaOD/D_2O	113	doublet	15.5 ± 0.5		1, 2
	12 M LiCl/H_2O	110	doublet	14.0		4
Thymidine	8 M NaOD/D_2O	113	doublet	13.7		2
	12 M LiCl/H_2O	110	doublet	14.0		4
Cytosine	8 M NaOD/D_2O	113	doublet	16.9	6.3a	2
Cytidine	8 M NaOD/D_2O	113	doublet	13.4		2
Uracil	8 M NaOD/D_2O	113	doublet	15.2		2
Uridine	8 M NaOD/D_2O	113	doublet	14.2		2
5-methylcytosine	8 M NaOD/D_2O	110	doublet	15.0		3
1,3-dimethyluracil	8 M NaOD/D_2O	113	doublet	12.3		2
6-methyluracil	8 M NaOD/D_2O	113	quartet, 1:3:3:1	15.4	6.3b	2
5-fluorouracil	8 M NaOD/D_2O	77	doublet	12.0		3
	12 M LiCl/D_2O or H_2O	77-170	doublet split by a second doublet	13 and 5.3	6.3c	5
5-bromouracil[a]	12 M LiCl/D_2O or H_2O	77	doublet	13.0		5
5-chlorouracil	8 M NaOD/D_2O	77	doublet	14.0		2
	12 M LiCl/D_2O or H_2O	77-165	doublet	15.0		5
5-iodouracil	8 M NaOD/D_2O	113	doublet	12.0		2
5-bromo-6-methyluracil	12 M LiCl/D_2O	77	quartet,1:3:3:1	16.8	6.3d	5

[a] At 77 K, in 8 M NaOD, the bromouracil anion is unstable whereas the fluorouracil and chlorouracil anions are found to be very stable (SEVILLA et al., 1974a).

References: (1) SEVILLA, 1971; (2) SEVILLA and VAN PAEMEL, 1972; (3) SEVILLA et al., 1972b; (4) SEVILLA et al., 1972a; (5) SEVILLA et al., 1974.

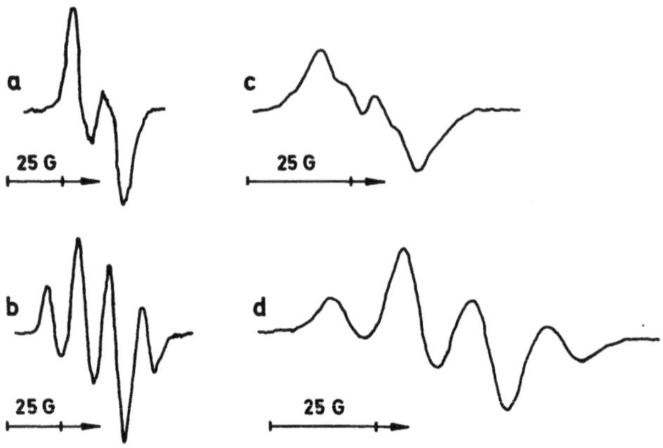

Fig. 6.3. ESR spectra of *a* Cytosine anion observed at 113 K in 8 M NaOD/D_2O (after SEVILLA and VAN PAEMEL, 1972); *b* 6-methyluracil anion observed at 113 K in 8 M NaOD/D_2O (after SEVILLA and VAN PAEMEL, 1972); *c* 5-fluorouracil anion observed at 140 K in 12 M LiCl/H_2O (after SEVILLA et al., 1974); *d* 5-bromo-6-methyluracil anion observed at 77 K in 12 M LiCl/D_2O (after SEVILLA et al., 1974)

1972). For other pyrimidine anions, a g-factor of 2.0030 ± 0.0004 is found (SEVILLA and VAN PAEMEL, 1972). The small difference in g-factor for the halopyrimidines from the pyrimidine anions suggests very small density on the halogen. This experimental result agrees with the theoretical predictions which give a spin-density smaller than 3.10^{-3} on the halogens (SEVILLA and VAN PAEMEL, 1972).

6.3.1.2 Purine Derivatives

Until now, only the adenine and guanine anion radicals were produced by photoionization of $K_4Fe(CN)_6$ and subsequent photobleaching (λ > 430 nm) of the trapped electrons (SEVILLA and MOHAN, 1974). In 12 M LiCl/D_2O glass at 77 K, their ESR-spectra are poorly resolved singlets, 9 G wide in the two cases, with corresponding g-factors of 2.0034 ± 0.0003.

6.3.2 Conversion of the Anion Radicals

6.3.2.1 Pyrimidine Base Anions as Precursors of H-Addition Radicals

HOLROYD and GLASS (1968) have investigated the protonation reaction of thymine derivative anions in alkaline glasses. They showed that anions of thymidine and thymidine-5'-monophosphate easily undergo proton transfer from surrounding water molecules to the C6-position of the thymine ring, either by photobleaching or by thermal annealing (Fig. 6.4a). As to the base thymine itself, it was anomalous in its reactivity during red light illumination: the eight-line spectrum characteristic of the "thymyl" radical (3.4) was not observed. However, when the bleached sample was warmed to 180 K, the spectrum appeared and remained when recooled to 80 K. The same transformation also took place upon warming the sample containing the anion without previous bleaching in 8 M NaOH/H_2O (SEVILLA, 1971) and in 12 M LiCl/H_2O (SEVILLA et al., 1972a).

On the contrary, uracil and deoxycytidine anions have not been found to protonate in alkaline glasses (HOLROYD and GLASS, 1968).

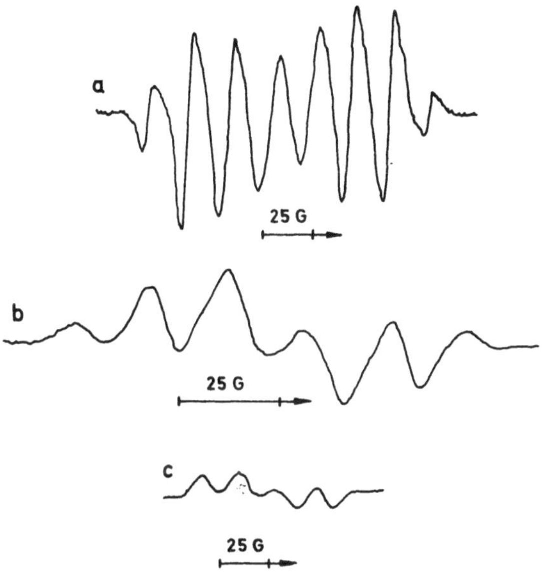

Fig. 6.4. ESR spectra of *a* thymidine, after its reaction with electrons (8 M $NaOH/H_2O$) and thermal annealing to 180 K. Observed at 77 K (after HOLROYD and GLASS, 1968); *b* 5-bromo-6-methyluracil, after its reaction with electrons (12 M $LiCl/H_2O$) and thermal annealing to 165 K. Observed at 77 K (after SEVILLA et al., 1974); *c* 5-bromouracil, after its reaction with electrons (12 M $LiCl/H_2O$) and thermal annealing to 140 K. Observed at 77 K (after SEVILLA et al., 1974)

6.3.2.2 Dehalogenation Radicals in 5-Halouracils

The results for the halouracils in 12 M LiCl (SEVILLA et al., 1974) suggest dehalogenation radicals for 5-bromo- and 5-bromo-6-methyluracil. In the latter case, warming the anion to 140 K or irradiating with unfiltered light from an incandescent lamp produced a 24 G singlet attributed to a radical produced by debromination. Further warming of this radical to 160 K resulted in the spectrum shown in Figure 6.4b. The spectrum consists of six lines in the approximate ratio of intensities 1:3:4:4:3:1. The only radical consistent with the observed splittings is a radical with one β proton at the 5 position and a radical site at position 6: it is the 5-hydroxy-6-methyl-5,6-dihydrouracil-6-yl radical, resulting from the hydration of the debromination radical. In 5-bromouracil, the debromination of the anion leads to one broadened spectrum due to the uracil-5-yl-radical. Upon warming this species in $LiCl/H_2O$, a quartet spectrum (Fig. 6.4c) is found. The interpretation of this spectrum agrees with the hydration of the uracilyl radical to produce the 5-hydroxy-5,6-dihydrouracilyl radical.

6.3.2.3 Protonation Reactions of Purine Anions

LION and VAN DE VORST (1971a) have shown that the free radicals induced in adenine, guanine, their nucleosides and their nucleotides by photoionization of ferrocyanide ions (8 M $NaOH/H_2O$) with immediate subsequent thermal annealing (173 K) were essentially H-adduct radicals. The spectrum obtained at 183 K with guanine nucleosides is illustrated in Figure 6.5. However, this approach does not lend itself to a demonstration of the intermediate anion stage. This was isolated in 12 M $LiCl/H_2O$ by SEVILLA and MOHAN (1974) who then warmed the anion radicals of adenine and guanine

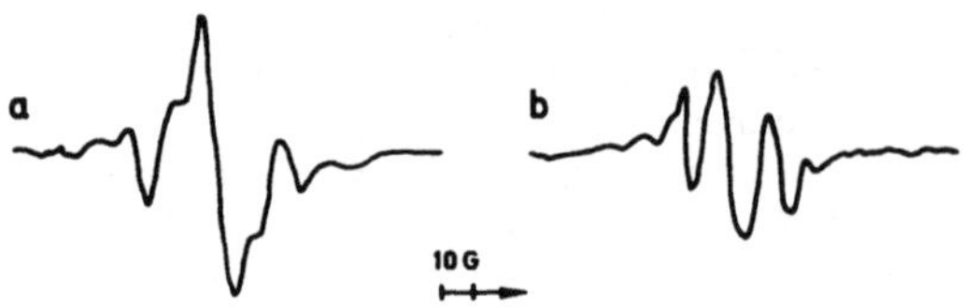

Fig. 6.5. ESR spectra of *a* guanosine, and *b* deoxyguanosine, after their reaction with electrons (8 M NaOH/H_2O) and thermal annealing to 173 K; observed at 173 K (after LION and VAN DE VORST, 1971a)

to 170 K. No evidence for a protonation reaction was found. For the guanine anion, irradiation with UV light (λ > 320 nm) led to the ESR triplet associated with an H-addition radical at a carbon site [Formula (3.1)]. On the contrary, in 8 M NaOH, after warming up to 190 K, protonation was noted for adenine (1:2:1 triplet of ΔH 41 G) but not for guanine. This last result is in conflict with those previously reported (LION and VAN DE VORST, 1971a). However, it seems established that, under appropriate conditions, protonation reactions can be induced at carbon sites in the purine anion radicals.

In summary, the above findings demonstrate that pyrimidine and purine derivatives in general, and more specifically the four DNA bases, react with electrons to form anion radicals. In the case of adenine, guanine and thymine, these anions become transformed into H-adduct radicals after protonation. No protonation reaction has been observed with cytosine and uracil. The tendency towards protonation can be ordered as follows: thymine ≥ adenine ≥ guanine (SEVILLA and MOHAN, 1974). In neutral and alkaline aqueous glasses, the protonation is achieved by proton transfer from the surrounding water molecules.

6.4 Interaction with Excited Inert Gases

In order to obtain a more clear-cut picture of the early radical processes, the paramagnetic species resulting from energy transfer from inert gas atoms and molecules excited to metastable states in a glow discharge to powdered nucleic acids and constituents were investigated by ESR spectroscopy in vacuo (WEIBEZAHN and DERTINGER, 1973a, b). Helium, argon, and nitrogen were used as bombarding gases. After passage through the radiofrequency discharge region, the unstable excited states (singlet states) decay in approx. 10^{-8} s leaving the gases in metastable excited states. The energies of the lowest metastable states transferred to the samples are (BROCKLEHURST, 1968): 19.81 eV for He, 11.56 eV for Ar, and 6.17 eV for N_2 (vibrational state). Bombardment with all gases yielded qualitatively identical spectra. Subsequently only the spectra obtained after exposure to excited Ar are shown.

6.4.1 Purine and Pyrimidine Derivatives

From Figure 6.6 it can be seen that there is no significant difference in the ESR spectra of (a) Ar-bombarded adenine, (b) deoxyadenosine and (c) deoxyadenosine-5'-monophosphate, indicating that paramagnetic base damage only is involved. This is in sharp contrast to the γ-irradiated adenine compounds which exhibit different spectra (Fig. 6.6d-f). The central parts of the latter spectra have been interpreted as due to ion radicals and deoxyribose radicals (DERTINGER and HARTIG, 1972). Obviously, they have no feature in common with the central parts of the spectra of the Ar-bombarded adenine compounds (Fig. 6.6a-c) which are essentially single lines, whereas a doublet character is predominant in the spectra after γ-irradiation. Whether irradiated or bombarded, adenine and its derivatives show hydrogen addition radicals. This is indicated by the presence of a 1:2:1 triplet of approximately 40 G coupling which has been attributed to H-adducts at the position C2 and C8 of the adenine ring (cf. I.3). The outer lines of this triplet can be seen in each of the spectra displayed in the Figure 6.6, the relative line intensities depending somewhat on the compound; with guanine, deoxyguanosine, and deoxyguanosine-5'-monophosphate similar observations are made.

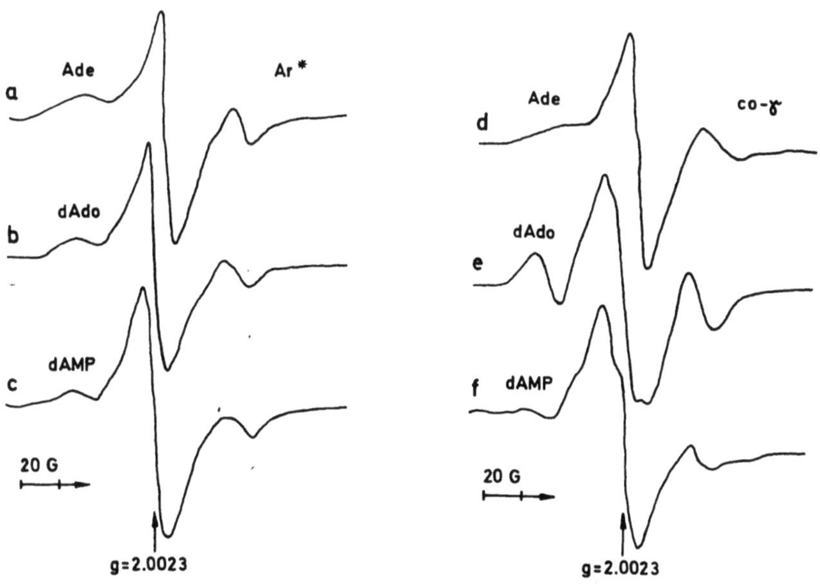

Fig. 6.6. First derivative of the ESR spectra of *a* adenine, *b* deoxyadenosine, *c* deoxyadenosine-5'-monophosphate after exposure to discharge-excited argon; *d* adenine, *e* deoxyadenosine, and *f* deoxyadenosine-5'-monophosphate after gamma-irradiation (after WEIBEZAHN and DERTINGER, 1973b)

From these results it is quite evident that the two different ways of energy transfer operating under bombardment with excited gases on one hand and under ionizing irradiation on the other hand, lead to different pathways of radical formation. Whereas for the irradiated purines ion radicals are assumed to be the primary radicals their fragmentation giving rise to the production of atomic hydrogen (DERTINGER and HARTIG, 1972), the pathway of radical processes following bombardment with excited gases must be derived from the characteristic ESR singlets in the corresponding spectra. The singlet resonance may be due to a σ-radical at the base moiety formed by rupture of C-H bond. In this event atomic hydrogen is liberated, giving rise to H-adduct formation (triplet resonance). To ascertain this assumption we will look for further results obtained with bombardment and irradiation of thymine, thymidine, and thymidine-5'-monophosphate.

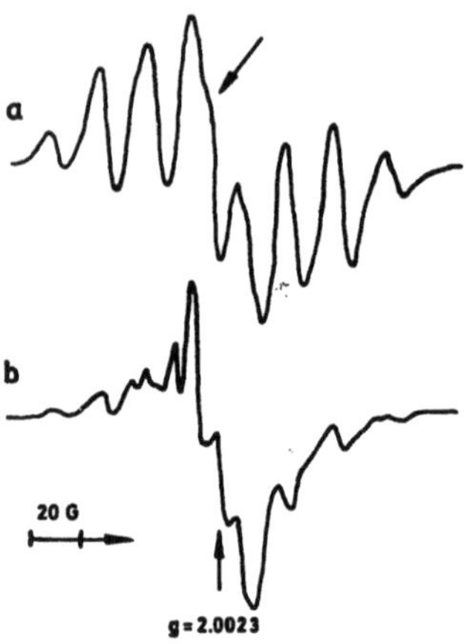

Fig. 6.7. First derivative of thymine ESR spectra (g_E = 2.0023): *a* exposure to discharge-excited argon; *b* gamma-irradiation (after WEIBEZAHN and DERTINGER, 1973a)

In Figure 6.7 spectra of thymine are given (a) after Ar-bombardment and (b) after exposure to γ-radiation in vacuo. The pronounced difference between these spectra may be explained by considering the underlying radicals. Obviously, an octet spectrum interpreted as being a hydrogen addition radical to the position C6 of the thymine ring [Formula (3.4)] is the most prominent structure in Figure 6.7a. In contrast, this octet pattern makes a relative small contribution to the spectrum of irradiated thymine in Figure 6.7b. Earlier ESR analysis revealed that the characteristics of this spectrum are due to a hydrogen abstraction radical from the methyl group [Formula (3.18)] (HÜTTERMANN, 1970; HARTIG and DERTINGER, 1971). However, there is another signal contributing to the central portion of Figure 6.7a, besides the octet. As indicated by the arrow, a distortion of the central part of this spectrum is observed being due to a single line. During exposure, a small increase in amplitude of this line is detected which, however, reverts to the amount displayed when the discharge is switched off. This single line is more pronounced in the corresponding thymidine spectrum, although the octet component is again the dominant structure there.

We may conclude that in Ar (He, N_2) bombarded thymine, thymidine, and dTMP, extensive hydrogen addition to the position C6 of the thymine moiety takes place. We consider the source of hydrogen liberation as being the radical underlying the single line. This assumption is supported by the fact that upon ionizing irradiation the C6-adduct cannot only be formed by hydrogen additon to thymine but also through protonation of the anion radical. HERAK (1970a) has shown that thymidine single crystals grown from D_2O and irradiated give rise to an ESR spectrum indicating that deuterium adds to the C6-position. Thus, in this case protonation of the anion is the mechanism of C6-adduct formation, the proton coming from an easily exchangeable site (hydrogen bonds). The same experiment for thymidine bombardment with Ar (He, N_2) samples does not show any deuterium exchange.

Therefore, the events following Ar-bombardment appear to be characterized by (1) the absence of ionization and consequently the lack of protonation and exchange reactions, and (2) the liberation of hydrogen from a nonexchangeable site with subsequent reaction of this species to yield the C6-adduct. Furthermore, the results permit the conclusion that the singlet-line is the precursor radical in the sense that it is formed by the rupture of a C-H bond of the thymine moiety. The radical structure compatible with our results is a C6-centered σ-type radical formed by dissociation of the C6-H bond. This hypothesis is supported by an INDO MO calculation (WESTHOF et al., 1974) showing that the C6-σ-radical should indeed not give rise to observable couplings. Also, it has a binding energy of -3530.32 Kcal/mol, i.e., this radical is relatively stable.

6.4.2 DNA

The spectra of Ar-bombarded and irradiated DNA are presented in Figure 6.8. The spectrum obtained after Ar-bombardment (a) is characterized by a dominating singlet-like resonance and is completely different from the spectrum of irradiated DNA (b); not only is there a large difference in the central region but also in the low- and high-field directions where a strong contribution of the eight-line spectrum of the 5,6-dihydro-5-thymyl radical is observed after irradiation. Apparently these lines are only weakly developed after Ar-bombardment.

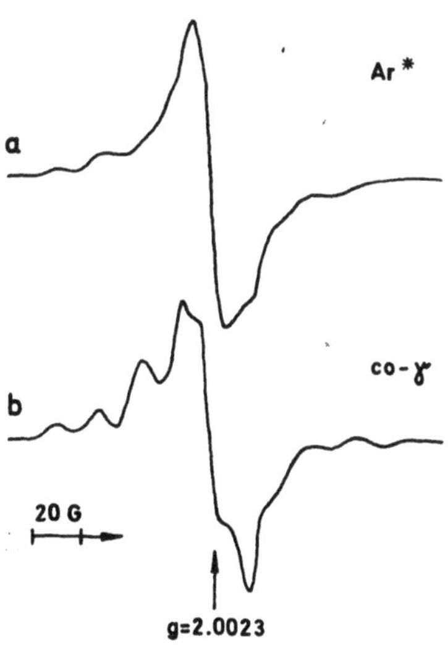

Fig. 6.8. First derivative of the DNA ESR spectra: *a* exposure to discharge-excited argon; *b* gamma-irradiation (after WEIBEZAHN and DERTINGER, 1973b)

6.4.3 Mechanisms of Radical Formation

The profound differences in mechanisms of radical formation after Ar-bombardment and irradiation should be explained in terms of the distinct modes of energy transfer in both cases. As discussed in I.2, ionization is the predominant result of high energy irradiation. So-called super-excited states may also be formed on irradiation. These are believed to dissipate excitation energy by internal conversion and vibrational modes leading ultimately to bond dissociation. Radicals resulting from this effect should be observable in the ESR spectra of the irradiated compounds. However, the absence of such observations agrees with existing hints that most radiation-induced DNA base radicals can be related mainly to ion radicals as precursors (HERAK and GALOGAZA, 1969; HARTIG and DERTINGER, 1971; SEVILLA, 1971; DERTINGER and HARTIG, 1972).

On the other hand, such radicals resulting from dissociation are thought to be formed on bombardment with excited gases. In this case, radiationless transfer of excitation energy to the target molecules takes place, leaving them collectively in a highly excited state. The decay of these states may give rise to extensive fragmentation processes without ionization of the molecules. The argument is in agreement with the observation that the primary radicals formed by bombardment with excited gases are σ-type species due to bond rupture and not due to ionization, although e.g., the energy of the Ar and He metastables exceeds the ionization potential of thymine. Although the ESR results for the purines do not allow as rigorous conclusions as in the thymine case, there are indications that the same mechanisms should be operating. This is based on (1) the occurrence of single lines in both cases being indicative of the same radical type, (2) the observation of hydrogen addition radicals, (3) the absence of ion radicals, (4) the independence of the radical types from the presence of a sugar or sugar-phosphate moiety.

Finally we consider the action of the excited inert gases on DNA. The corresponding ESR spectrum is a slightly asymmetric singlet-like resonance with indications of shoulders thus being compatible with a superposition of several distinct single lines at slightly different g-factors probably due to different types of σ-radicals (i.e, purine or pyrimidine types).

In summary, these results indicate that dissociative effects due to superexcitation should be negligible in the case of ionizing irradiation. Conversely, ionization does not contribute to radical formation on bombardment with excited gases. The value of using excited gases as radiomimetics for energy transfer can be seen as offering the opportunity to rule out certain hypothetical pathways of radiation action in the dry state. From this point of view, these studies are likely to contribute to the elucidation of the primary radical processes after ionizing irradiation. On the other hand, they represent a simple method for producing certain types of base damage in DNA useful for biological experiments (see LÜCKE-HUHLE and JUNG, 1973, 1974).

Chapter 7

Transfer Phenomena

7.1 Introduction

A brief introduction into the concept of "energy transfer" as a term used in describing a variety of more or less well-defined mechanisms by which radiation-induced damage is transferred between different sites within a system has been given in I.2. Energy transfer, which may actually involve charge or mass migration besides transfer of electronic excitation energy, is generally taken to characterize processes by which a given defect is shifted within one molecule. Intermolecular transfer, on the other hand, is usually termed an indirect effect. In both cases, it is implied that the nature of the defect prior to the transfer differs from that of the species after completion of the process. In the context of ESR research, the term spin transfer is frequently found in the literature to indicate by whatever mechanism the site of one radical, usually called primary, has transformed to another one with a concurrent change in the radical structure. Thus, spin transfer leads to a distribution of secondary radicals over molecules or, more generally, subunits of a system which differs from random.

It has been pointed out in I.2 that the distinction between inter- and intramolecular transfer loses its applicability in the case of large and complex molecules of heterogeneous composition like nucleic acids. Consequently, the term transfer is used here in its broadest sense as a process by which the yield of specific radicals is influenced. There are many factors which bring about a change in the radical distribution. For example, the temperature of irradiation or the chemical or crystalline environment of the molecule under study may influence the pathway of radical reactions. In the present discussion we shall concentrate on three different aspects of transfer which are not necessarily directly related but may serve to add to the overall picture of radiation-induced radical reactions occurring in nucleic acids and their constituents.

In the first section, an intramolecular factor altering the radical distribution is discussed. It is the introduction of a non-hydrogen substituent at position C5 of uracil or thymine which gives rise to base analogues not occurring normally in nucleic acids. An example of radiobiological significance is the replacement of the thymine methyl group by halogen atoms which brings about an increased radiosensitivity in vivo as is described in III.3. In addition, a number of barbituric acid derivatives with different C5-substituents are included in the discussion.

Another aspect of energy transfer deals with the elucidation of the radical distribution in γ-irradiated solid DNA and the role of transfer of electrons between purine and pyrimidine bases at the stage of anion radicals. The discussion of this problem is somewhat related to the third transfer topic which is the photosensitization of nucleic acids by dyes in frozen aqueous solution since the latter also involves electron transfer from the dye to the nucleic acid bases.

7.2 Substituent Effects

An analogue of a nucleic acid base formed by introduction of a different substituent at a specific position usually experiences a two-fold difference in its radiation response, as compared to the unaltered component. Both its total radical yield may be changed and the distribution of radical structures altered by shifting their relative proportions or by introducing new radical species. The latter structural aspect will be emphasized here in description of radical structures in base analogues employing mainly single-crystal results.

7.2.1 Barbituric acid and Derivatives

Barbituric acid (7.1)[1] is a pyrimidine derivative (pyrimidinetrione) related to the nucleic acid base uracil by replacing the C6-hydrogen with a keto-oxygen and by saturating the 5,6-double bond via addition of a hydrogen to C5. Its radical formation has been investigated in some detail by ESR and ENDOR methods between 4.3 K and room temperature (BERNHARD and SNIPES, 1966; MELØ et al., 1973; BOX and BUDZINSKI, 1973; HÜTTERMANN et al., 1976c). The primary species observed at 4.2 K are the cation (7.2) and the anion (7.3) both of which undergo neutralization by deprotonation and proton capture, respectively. The protonated anion appears to be stable upon annealing to room temperature whereas the deprotonated cation is thought to be the precursor of the main room temperature radical (7.4) formed by hydrogen abstraction of C5. ENDOR measurements of the weakly coupled protons hint at a strong participation of the water molecules in dihydrate crystals in stabilizing radical (7.4) and in addition indicate a tautomeric shift to radical (7.4a). At 77 K pairs of radicals (7.4) or (7.4a) are observed. The spectral parameters of the three radicals are listed in Table 7.1.

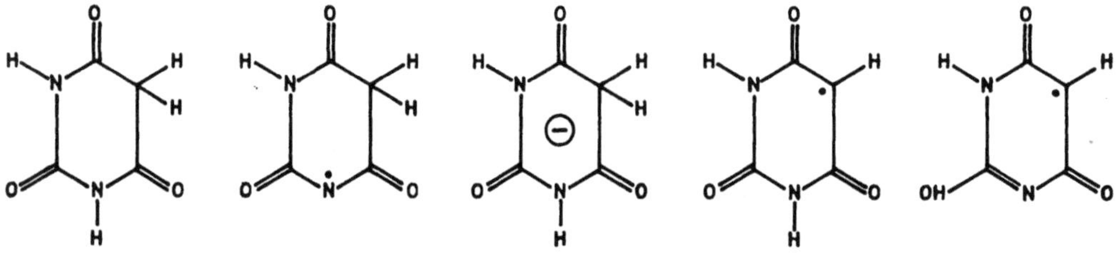

Formula (7.1) (7.2) (7.3) (7.4) (7.4a)

[1] Numbers in parenthesis refer to formulas.

Table 7.1. Main spectral parameters of barbituric acid radicals

Radical (formula numbers)	Matrix	Temperature (K)	Hyperfine Couplings (G)	g-tensor	Reference
(7.2)	Dihydrate crystals	42	^{14}N : $A_{\parallel}$ = 34.9 $A_{\perp}$ ≃ 0	2.0139 2.0074 2.0020	BOX and BUDZINSKI (1973)
	Dideuterate crystals	77	^{14}N : $A_{\parallel}$ = 28.8 $A_{\perp}$ ≃ 0		HÜTTERMANN et al. (1976c)
(7.3)	Dihydrate crystals	4.2	^{1}H : 31.26 28.06 27.71 $^{1}H'$: 34.90 30.12 29.91		BOX and BUDZINSKI (1973)
(7.4)	Dihydrate crystals	300	^{1}H : 11.2 30.0 20.4	2.0044 2.0056 2.0022	BERNHARD and SNIPES (1966)
	Dideuterate crystals	300	^{2}H : 1.4 4.6 3.1		HÜTTERMANN et al. (1976c)
(7.4a)	Dihydrate crystals	300	^{1}H : 9.25 29.20 19.20		

The main secondary species in barbituric acid is thus an abstraction radical. BENSON and co-workers studied a variety of derivatives of barbituric acid with different substituents at C5 to elucidate some of the parameters governing the abstraction (BENSON and SNIPES, 1967, 1969; HAAK and BENSON, 1971; GUTIERREZ and BENSON, 1974a, b). Their findings are summarized in Table 7.2. In compounds with identical substituents at C5, abstraction of OH (alloxan) and of an ethyl group (barbital) is found to occur in addition to H-abstraction in barbituric acid. Of the dissimilar substituents, hydrogen, if present is always the only one abstracted. The same tendency, but less efficient, holds for the hydroxyl group since the other substituent is also partly abstracted. Also, there is the exception of alloxantine where one whole alloxantine group is eliminated by C-C bond cleavage instead of OH-abstraction. Finally, of the two alkyl substituents, the ethyl group appears to be more amenable to elimination than the methyl substituent.

It is probably premature to speculate about details of the abstraction mechanisms on the basis of these data, since precursor radicals have been reported only for the case of ethylmethylbarbituric acid, where a radical formed from hydrogen abstraction

Table 7.2. Abstraction radicals in 5-substituted barbituric acid derivatives

Substituents R_1	R_2	Name of Compound	Abstracted Groups	References
H	H	Barbituric acid	H	BERNHARD and SNIPES (1966)
H	CH_3	5-methyl-barbituric acid	H	GUTIERREZ and BENSON (1974b)
H	CH_2-CH_3	5-ethyl-barbituric acid	H	GUTIERREZ and BENSON (1974a)
CH_2-CH_3	CH_2-CH_3	Barbital	CH_2-CH_3	HAAK and BENSON (1971)
OH	OH	Alloxan	OH	BENSON and SNIPES (1967)
OH	CH_3	Hydroxymethyl-barbituric acid	70% OH 30% CH_3	GUTIERREZ and BENSON (1974b)
OH	CH_2-CH_3	Hydroxyethyl-barbituric acid	65% OH 35% CH_2-CH_3	GUTIERREZ and BENSON (1974b)
OH	Alloxanyl	Alloxantine	Alloxanyl	BENSON and SNIPES (1969)
CH_3	CH_2-CH_3	Ethylmethyl-barbituric acid	60% CH_2-CH_3 after conversion via storage 100% CH_2-CH_3	GUTIERREZ and BENSON (1974b)

within the ethyl group has been observed to convert into the final radical, which results from elimination of the whole group, probably via ethane production. Still, the series of measurements indicates that abstraction of a C5-substituent in barbituric acid derivatives may be quite selective and also that it can involve elimination of rather large substituent groups via C-C bond cleavage.

7.2.2 5-Halogen Substituted Uracil Bases and Nucleosides

There has been considerable interest over the past ten years in the radical formation in 5-halouracil derivatives owing to their radiosensitizing properties when incorporated into nucleic acids instead of thymine. Structural information from ESR studies of single crystals, however, has been evolving only quite recently. This is mainly due to complications arising in the interpretation of spectra because of nuclear quadrupole interactions when hyperfine coupling to chlorine, bromine or iodine is involved. Some tensor data on α- and β-halogen couplings in simpler organic radicals have become available recently (KISPERT and MYERS, 1972; MISHRA et al., 1973a, b; PICONE and ROGERS, 1974a, b). Whereas the chlorine parameters appear to be understood reasonably well, there has been some debate about the correct assignment of α-bromine and -iodine interactions.

Radical formation in halouracil bases has been investigated in single crystals of 5-fluorouracil (NEUMÜLLER, 1976), 5-chlorouracil (OLOFF, 1976), 5-bromouracil (HÜTTERMANN and MÜLLER, 1969b) and in a cocrystal of 1-methylcytosine: 5-fluorouracil (FARLEY, 1974; FARLEY and BERNHARD, 1975). In fluorouracil, (7.5), two radicals are produced at room temperature. Structure (7.6) is the hydrogen adduct at position C6 which shows hyperfine interaction with the α-fluorine nucleus and two inequivalent β-protons. This species

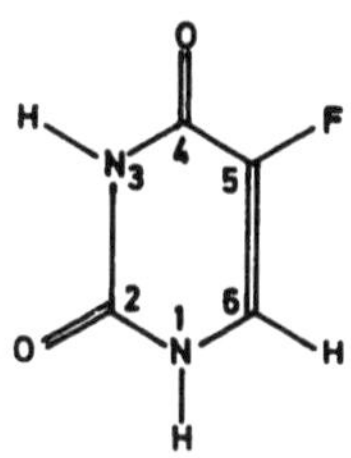
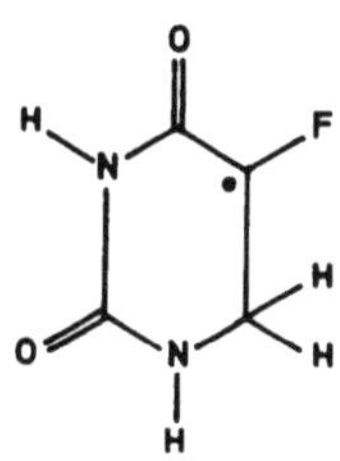
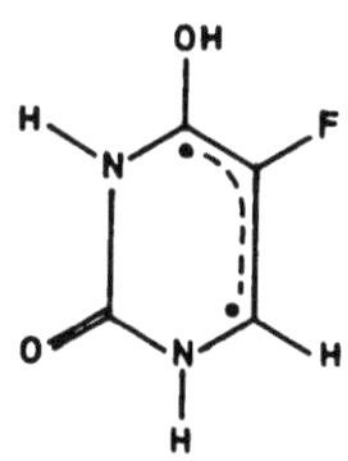
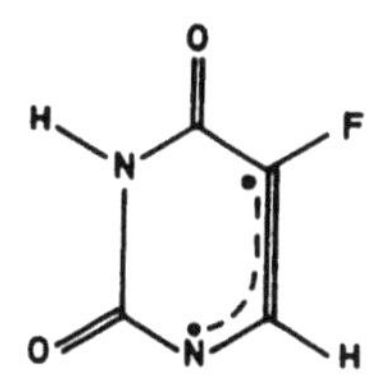

Formula (7.5) (7.6) (7.7) (7.8)

is also observed in the co-crystal at room temperature. The other radical is (7.7), the hydrogen adduct at O4. It is formed in much higher concentration than (7.6). Both radicals are probably the result of protonation of the anion. This is suggested by the results obtained with deuterated crystals in which the added protons are found to be exchanged by deuterons. Protonation at O4 is facilitated by the hydrogen bond between O4 and N3 of the adjacent molecule, thus explaining the high concentration of (7.7). A third radical (7.8), is produced in the co-crystals at 77 K. It involves loss of a hydrogen from N1 and shows characteristic interactions with both ^{14}N and ^{19}F nuclei.

Radical formation in 5-chloro- and 5-bromouracil is not as well understood; 5-chlorouracil (7.9), exhibits four different radicals at room temperature. The hydrogen adduct at C6 (7.10), and radical (7.11) resulting from hydrogen loss at N1 are both well characterized by their spectral parameters. The assignment of (7.12), the hydrogen adduct at O4 is tentative, however, since its lines show extensive overlap with lines of other radicals preventing a thorough identification. Likewise, a spectrum obtained at low doses showing π-type interaction with a chlorine nucleus is assigned to the cation (7.13) only tentatively. Theoretical calculation of the spin-density distribution in the cation partially supports this assignment. The situation is still more complex in

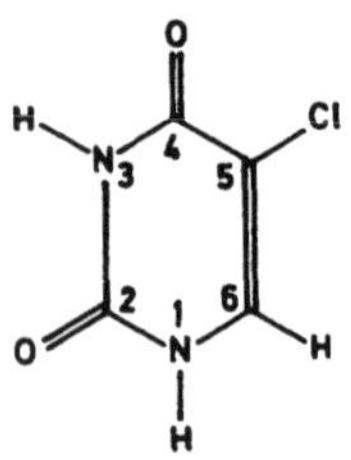
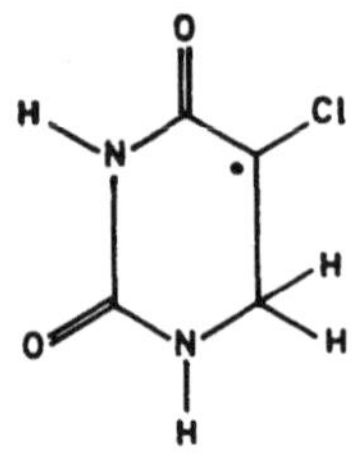
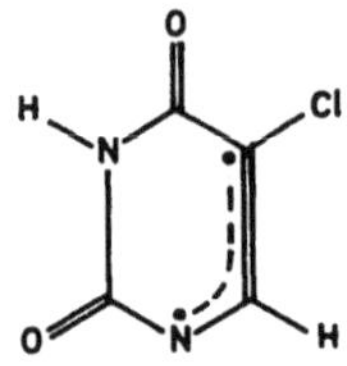
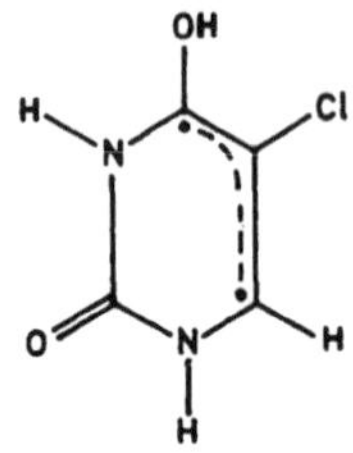
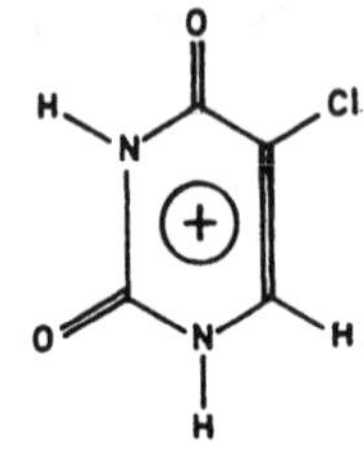

Formula (7.9) (7.10) (7.11) (7.12) (7.13)

5-bromouracil where extremely complicated spectra involving bromine interaction are observed at room temperature. The structural analogue of (7.13) has been proposed to account for these features but a more precise assignment is still to be worked out. Another radical in 5-bromouracil shows features of (7.12) but its complete identification is again still pending.

Although the information is far from being complete, three main radical features seem to emerge from the halouracil base studies. Two of these are hydrogen adducts at C6 and O4, respectively, and one is formed by loss of a hydrogen from N1. The latter radical is suppressed in ribo- and deoxyribosides by the glycosidic bond whereas the hydrogen adducts do prevail in the nucleosides. Results have been reported for the riboside 5-chlorouridine (OLOFF and HÜTTERMANN, 1977; REISS and GORDY, 1971) and the deoxynucleosides 5-chloro-, 5-bromo-, and 5-iododeoxyuridine (NICE and RORKE, 1969a, b; HÜTTERMANN and MÜLLER, 1969b; HÜTTERMANN et al., 1976a, b; HAINDL and HÜTTERMANN, 1978). In all components, the hydrogen adduct at C6 (7.15), is one of the dominant radicals observed at

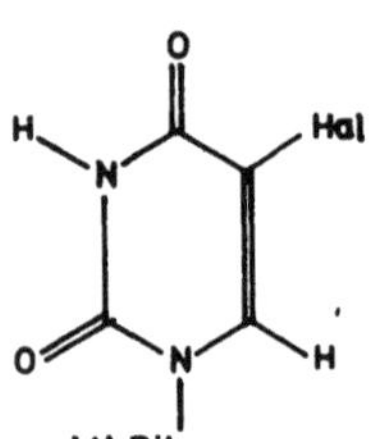

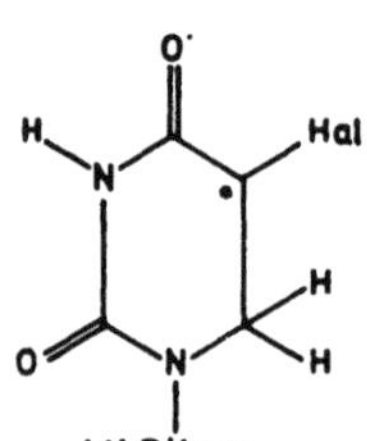

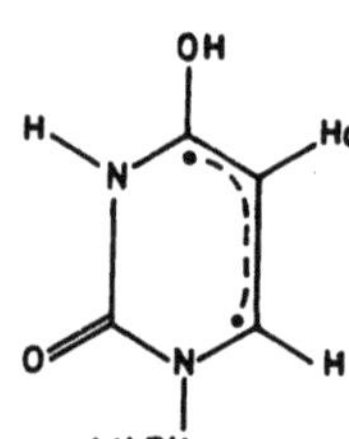

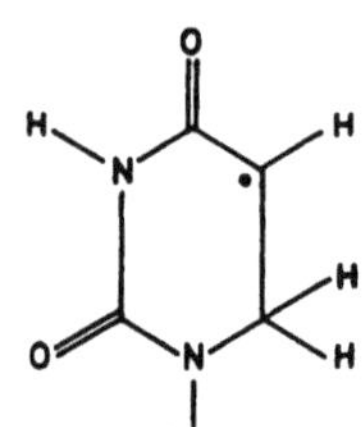

Formula (7.14) (7.15) (7.16) (7.17)

room temperature and competes in intensity with the other hydrogen adduct at O4 (7.16). A controversial exception to this scheme appears to be the result of REISS and GORDY (1971) in 5-chlorouridine who claim that neither of the two radicals are produced but radical (7.17) instead. The latter is formed by dehalogenation under consumption of two hydrogen atoms. In view of the contradictory findings by OLOFF and HÜTTERMANN (1977) with the same compound and the complementary information of the three deoxyribosides, radical (7.17) appears to be a wrong assignment. It is indeed likely that the spectral features in question should rather be attributed to radical (7.15). Provided a rotation about an axis at a skew angle to the principal tensor directions, as is the case for 5-chlorouridine crystals, the coupling to an α-chlorine nucleus may easily be mistaken for an α-proton interaction (cf. HÜTTERMANN et al., 1976a).

Another controversy concerns the assignment of lines not belonging to (7.15), which are observed in 5-bromo- and 5-iododeoxyuridine at room temperature. NICE and RORKE (1969a) attribute the spectral features to a deoxyribose radical resulting from H-abstraction from the C5' position [cf. (3.22)] whereas HÜTTERMANN and MÜLLER (1969) propose structure (7.16). The latter assignment appears to be more appropriate. Radical (3.22) was recently observed in single crystals of 5-chloro- and 5-bromodeoxyuridine at low temperatures (HÜTTERMANN et al., 1977b). Apart from having different spectral features than the lines in question it was found to decay into radical (7.15) well below room temperature (HÜTTERMANN et al., 1977a).

The α-halogen interaction in the hydrogen adduct radical (7.15) was not revealed completely in the original work on 5-bromo- and 5-iododeoxyuridine but has been established recently by theoretical simulation (HÜTTERMANN et al., 1976b; HAINDL and HÜTTERMANN, 1978). These results together with the information about the α-chloro- and α-fluoro-radicals in the other halouracil bases and nucleosides provide for a series of α-halogen interaction parameters in isostructural radicals (7.6), (7.10), and (7.15), respectively, which may facilitate future interpretations of spectra in more complex systems like halouracil-substituted nucleic acids. The parameters, together with those of radicals (7.7), (7.8), (7.11) and (7.16) which also appear well established are listed in Table 7.3.

Table 7.3. Spectral parameters of α-halogen and other radicals in 5-halouracil derivatives

Radical (formula numbers)	Compound	Nucleus	Hyperfine coupling (G)	g-tensor	Reference
(7.6)	5-Fluorouracil	F^{19}	+ 178		NEUMÜLLER
			- 18		(1976)
			- 9		
		6H	47		
			45		
			47		
		6H'	31.5		
			27.		
			25.5		
	$1\text{-}CH_3$-Cytosin:	F^{19}	162.7	2.0016	FARLEY
	5-Fluorouracil		60.2	2.0038	(1974)
			11.8		
		6H	44.5		
			43.2		
			42.0		

Table 7.3. Continued

Radical (formula numbers)	Compound	Nucleus	Hyperfine coupling (G)	g-tensor	Reference
(7.6)	1-CH_3-Cytosin: 5-Fluorouracil	6H'	31.1 31.5 33.7		FARLEY (1974)
(7.7)	5-Fluorouracil	4(OH)	- 4.1 - 5.0 - 1.7	2.0027 2.0088 2.0058	NEUMÜLLER (1976)
		6H	- 10.8 - 14.9 - 4.4		
		F^{19}	+ 11.2 0.0 0.0		
(7.8)	1-CH_3-cytosine: 5-Fluorouracil	1N	14.4 0.0 0.0	2.0011 2.0056 2.0062	FARLEY and BERNHARD (1975)
		F^{19}	156.4 16.4 16.1		
(7.11)	5-Chlorouracil	1N	12.8 0.0 0.0	2.0016 2.0083	OLOFF (1976)
		Cl^{35}	21.4 - 4.3 - 5.7		
(7.12)	5-Chlorouridine	4(OH)	- 0.96 + 1.18 - 0.49	2.0008 2.0076 2.0030	OLOFF and HÜTTERMANN (1977)
		6H	- 11.95 - 17.21 - 5.64		
(7.10)	5-Chlorouracil	Cl^{35}	+ 17.10 - 4.50 - 6.20	2.0018 2.0083 2.0072	OLOFF (1976)
		6H	24.4 28.9 24.4		
		6H'	45.8 46.7 45.8		
(7.15)	5-Chlorouridine	Cl^{35}	+ 18.9 - 5.2 - 5.8	2.0011 2.0075 2.0057	OLOFF and HÜTTERMANN (1977)

Table 7.3. Continued

Radical (formula numbers)	Compound	Nucleus	Hyperfine coupling (G)	g-tensor	Reference
(7.15)	5-Chlorouridine	6H	24.0 28.8 24.0		OLOFF and HÜTTERMANN (1977)
		6H'	44.4 49.5 44.4		
	5-Chlorodeoxy-uridine	Cl^{35}	+ 16.76 - 3.92 - 4.67	2.0021 2.0086 2.0068	HÜTTERMANN et al. (1976a)
		$6(H)_2$	36.89 39.30 36.89		
	5-Bromodeoxy-uridine	Br^{81}	+ 100.0 - 20.0 - 34.0	2.0039 2.0250 2.0219	HAINDL and HÜTTERMANN (1978)
		$6(H)_2$	34.0 38.5 34.0		
	5-Iododeoxy-uridine	I^{127}	+ 90.0 - 50.0 - 40.0	1.987 2.039 2.050	HÜTTERMANN et al. (1976b)
		$6(H)_2$	38.0		

Surprisingly, in none of the halouracil crystals studied any major fraction of radicals involving dehalogenation has been observed. This is in striking contrast to radiation-chemical findings in dilute aqueous solution which are presented in Section II. In the latter systems, halogen elimination upon attack of either solvated electrons or $OH^{\cdot}$ radicals at the halouracil base is found to be the dominant process. Radical reactions in solid halouracil derivatives have been investigated recently to establish the mode of formation of α-halogen radicals (7.15) in 5-chloro- and 5-bromodeoxyuridine (HÜTTERMANN et al., 1977a). Two pathways were found to be operational. One involves protonation of the anion at C6. This is a fast process which has been assessed so far only indirectly by measurements with deuterated crystals. The other pathway starts with oxidation of the deoxyribose at the terminal $C5'H_2OH$ groups where an $RO^{\cdot}$ radical [cf. (3.24)] is formed at low temperatures. This species decays between 90 K and 120 K, probably by intermolecular hydrogen abstraction, into a C5'-centered second deoxyribose radical [cf. (3.20)]. The latter transfers a hydrogen from a nonexchangeable position to the 5,6-double bond of the base upon decay between 220 K and 260 K to form the α-halogen radical (7.15). Both pathways exhibit nearly identical reaction rates in the two iso-structural crystal systems. In 5-chlorodeoxyuridine they account for practically all radicals observed at room temperature whereas in 5-bromodeoxyuridine the spectra exhibit sizable concentrations of radical (7.16) in addition to the α-halogen radical (7.15).

While the protonation of the anion to form (7.15) may be a ubiquitous event in all solid halouracils, the oxidative pathway starting with RȮ at the deoxyribose is certainly restricted to specific conformational and hydrogen-bonding conditions in the crystal. This is evidenced, for example, by results obtained from 5-chlorouridine in which neither of the two deoxyribose radicals is observed at low temperatures.

It is yet premature to relate the present knowledge about radicals and reactions in solid halo-uracil derivatives to the radiosensitizing mechanisms evoked by these components in nucleic acids. The results, however, suggest that the dehalogenation model that has been derived from aqueous solution studies should be viewed with some reservation regarding its ubiquitons applicability.

7.3 Electron Transfer in γ-Irradiated DNA

Although the energy transfer processes that take place in DNA excited by UV light have been extensively investigated with consistent results, little is known about energy migration in DNA submitted to γ-irradiation. The primary steps of the radiative processes resulting from such high-energy radiation are essentially different from those caused by UV light. Ionizing radiations dissipate energy in a medium mainly by ionization. In γ-irradiated DNA, the primary cause of biological effects is thus the formation of positive and negative ions. These species, randomly produced in the DNA molecule, are chemically unstable and, under suitable experimental conditions, can undergo a series of free radical reactions ending with the formation of stable radiation products.

The problem of the final distribution of γ-radiation damage in DNA can be investigated by different techniques, according to the stage at which the complex phenomenon is observed. On of the most direct approaches would be to perform a complete radiochemical analysis of the γ-irradiated DNA, with the aim of characterizing qualitatively and quantitatively all the products formed in the DNA and determining their specific locations in the different constituents. However, the majority of such investigations were performed with derivatives γ-irradiated in aqueous solution, so that most of the radiation products identified were produced by indirect action, i.e., through the attack of reactive intermediates released by the radiolysis of the water itself. The only attempts to approach this problem in conditions of predominantly direct effects were made by ESR. This work is described in more detail since it enabled the existence of transfer phenomena in γ-irradiated DNA to be determined and to outline the corresponding mechanisms.

7.3.1 Preliminary ESR Experiments on γ-Irradiated DNA

Irradiation of three different, commercially available, samples of DNA, extracted from calf thymus, salmon sperm and herring sperm, respectively, produces ESR spectra, which, recorded under identical experimental conditions, display only slight differences in shape. Since the DNAs examined do not all have the same chemical constitution (molecular weight, sequence of nucleotides, nucleotide molar ratios), the dissimilarities could, in principle, be attributed to the total effect of these factors on the corresponding radical populations. On the other hand, when the three samples are simply dissolved in water, freeze dried and sealed under vacuum, their ESR spectra are rigorously identical to each other. Varying the pH of the DNA solutions prior to freeze drying induces significant changes in the ESR response, but, once again, at each pH value, the three DNA samples display identical ESR patterns. Interpretation of these spectra in terms of the different radical species involved is, for the moment, impossible, but in all the DNA spectra examined, the outermost resonances of the 5-6 dihydro-5-thymyl radical (3.4) can be recognized. This species is not, however, the principal product: The central part of each of the different DNA spectra is obviously due to the overlapping of several patterns and remains unresolved. These preliminary results allow to draw the following conclusions:

1. The final radical population induced by γ-rays in the DNAs examined does not depend on their relative nucleotide composition;

2. some changes in the method of preparing the DNA samples may induce significant changes in the resulting ESR spectrum. These spectral changes would seem to be due to different concentrations of the same radical species rather than to the formation of different radical structures.

The above conclusions are practically all that can be deduced from a first qualitative analysis of the complex and unresolved spectra of DNA γ-irradiated at room temperature in the dry state. To continue with the analysis, the problem must be approached from another angle.

7.3.2 Mechanical Mixtures and Molecular Complexes of DNA Nucleotides

The first point to be investigated is whether radicals are formed at random in γ-irradiated DNA. If so, the final radical population of the DNA should correspond to that of a mixture of nucleotides in the same molar ratio. This can easily be checked by comparing the ESR spectra of the above-mentioned DNAs, "straight from the bottle", with those of finely mixed mechanical mixtures containing the four nucleotides in corresponding molar ratios. The result is, that the ESR spectra of the mechanical mixtures are found to be substantially different from those of the three DNAs examined as is shown in Figure 7.1. In particular, typical patterns arising from H-addition radicals in the two purine constituents, dGMP and dAMP, which are easily detectable in all nucleotide mixtures (marked A and G), are found to be missing

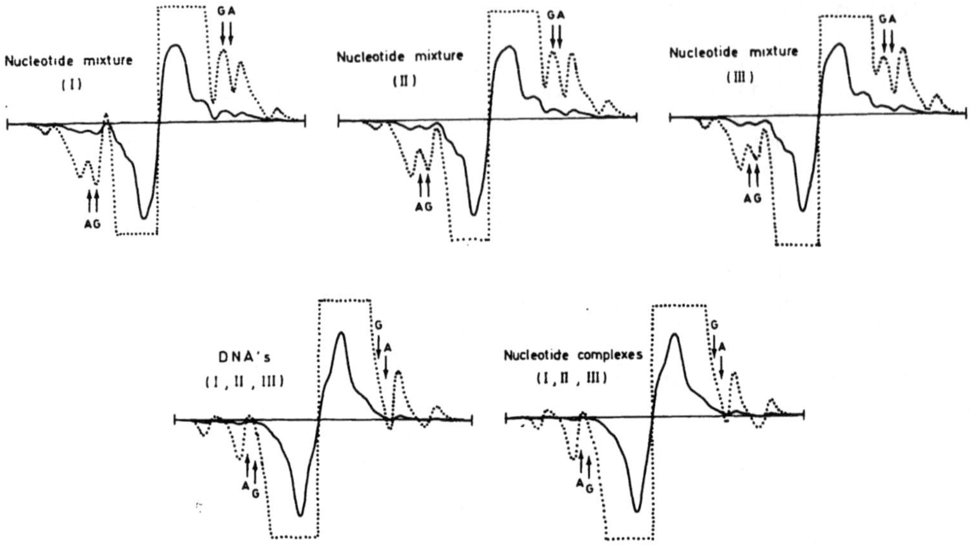

Fig. 7.1. Comparison between the ESR spectra obtained after γ-irradiation of: three different DNA samples previously freeze dried (identical spectra); three different mechanical mixtures containing the four DNA nucleotides in the same molar ratios as in the preceding DNAs; three different molecular complexes of the four DNA nucleotides obtained by freeze drying the above mixtures (identical spectra).
All samples were sealed under a vacuum of 10^{-5} mmHg. Irradiations and measurements at 25°C. Dose 2 Mrads. Arrows in the spectra indicate the relative positions of the triplet lateral satellites arising from H-adduct radicals on 5'-dGMP and 5'-dAMP respectively. DNA (I) is from herring sperm. Nucleotide molar ratio: A (27.8) : G (22.2) : C (20.7) : T (27.5). DNA (II) is from salmon sperm. Nucleotide molar ratio: A (28.7) : G (20.4) : C (19.8) : T (29.3). DNA (III) is from calf thymus. Nucleotide molar ratio: A (29.0) : G (19.5) : C (20.5) : T (31.2)

completely in the DNA spectra. If the above mixtures of nucleotides are now dissolved in water, freeze dried, structural changes at a molecular level are induced. Whereas, in the mechanical mixtures, the molecules of the different nucleotides do not interact, the nucleotides in the freeze dried samples form large random molecular complexes, stuck together by the strong base-base couplings exerted vertically by stacking forces (TS'O, 1974). Unlike mechanical mixtures, but in the same way as in DNA, the structure of these aggregates is now such that energy can be transferred between the different subunits. The ESR results obtained by irradiating these molecular aggregates are of the greatest interest. Despite their different nucleotide composition, these complexes now have ESR spectra which are strictly identical to each other. Moreover, their common spectrum is also identical to the common spectrum of the three DNA samples freeze dried from water. This series of observations is illustrated in Figure 7.1.

The absence of purine-like signals becomes clear-cut in the following stacking complexes which contain one purine (pu) and one pyrimidine (py) nucleotide in equimolar proportions: (dGMP·dCMP); (dGMP·dTMP); (dAMP·dCMP); (dAMP·dTMP). Although the ESR spectra of the corresponding mechanical mixtures are, in each case, the sum of the spectra of the single constituents irradiated alone, the spectra of the above complexes contain essentially the resonances of the respective pyrimidine constituents (GREGOLI et al., 1970a). Moreover, first moment calculations performed on the spectra of these two-nucleotide systems reveal that the selective formation of pyrimidine radicals in the pu-py complexes, can only be explained quantitatively by a complete pu → py transfer of damage, which takes place at some early stage of the radical formation.

The above results permit definition of the problem. In γ-irradiated systems containing purine and pyrimidine nucleotides kept together by stacking forces, the degradation of the energy occurs in such a way that the primary random ionizations are transformed into stable damage centres, which are preferentially localized in the pyrimidine constituents. Since the mechanism responsible for this transfer phenomenon is as yet unknown, for the moment, the term "spin transfer" will be used.

7.3.3 A Working Hypothesis for Internucleotide Spin-Transfer Processes

To elucidate the mechanism governing the spin-transfer processes previously disclosed in DNA and in stacking complexes of nucleotides, the pathways of the formation of stable radical species detectable with ESR has to be retraced. The first point to be re-examined is the lack of "purine-like" signals in all the purine-containing complexes investigated. These signals are the well-resolved triplets associated with radicals of H-adduct on the respective base groups of dGMP and dAMP. On the other hand, it is known that these H-adduct radicals are secondary species which may arise from the protonation of a radical anion (see I.3). The corresponding reaction can be written as follows:

$$\dot{pu}^{-} + RH \longrightarrow \dot{p}uH + R^{-} \qquad (7.1)$$

In this equation, RH is the proton donor molecule and can represent a base, a sugar or even a molecule of water, which is always present in considerable amounts even in the condensed phase. Since ionizations occur at random in irradiated media, purine radical anions must be formed in both the mechanical mixtures and the stacking complexes of nucleotides. The absence of purine H-adduct radicals in the complexes, and the subsequent increase in the pyrimidine radical yield, must therefore be due to a transfer process taking place at the anionic stage of the radical formation. Considering the scale of electron affinity of the different nucleotides, the pyrimidine constituents are found to be much better electron acceptors than the purines (BERTHOD et al., 1966; MONTENAY-GAREŚTIER and HELENE, 1971). This provides the key to an explanation of the transfer phenomena previously disclosed and allows suggestion of the following multi-step mechanism:

1. Absorption of γ-rays leads to fast electron production.

2. Fast electrons, losing energy along their path by secondary ionizations, are finally captured by some molecular orbital. In DNA and its constituents, the captured electrons occupy the lowest empty molecular orbital (LEMO) of the respective π-electron systems.

3. In DNA and stacking complexes of nucleotides, the successive overlapping of the π-orbitals along the stacks allows electron transfers to occur from one base orbital to another. Since the electron transfer direction is towards a minimization of energy, the primary random anionic population is transformed into a new population, of lower energy, consisting mainly of pyrimidine anions: $pu^{\dot{-}} + py \rightarrow pu + py^{\dot{-}}$.

4. Once the thermodynamic equilibrium has been achieved, the pyrimidine anions start to react and are finally converted into stable radical species, localized on the pyrimidine constituents.

This mechanism, consistent with previously reported qualitative and quantitative results, is put forward as a working hypothesis. Its applicability will be substantiated by a series of experiments, described in the following sections.

7.3.4 Other Examples of Intermolecular Spin-Transfer Processes

The validity of the mechanism already described can be tested with stacking complexes formed by each of the DNA nucleotides with proflavine; these complexes are obtained by freeze drying the corresponding aqueous solutions (GREGOLI et al., 1970b). Proflavine has a much greater electron affinity than any nucleotide (PULLMAN and PULLMAN, 1963) and, according to the transfer model, should in all cases become the final acceptor of the radiation damage. The conditions for detecting a spin-transfer towards proflavine by simple analysis of the resulting ESR spectra are, in this case, particularly favorable. Proflavine irradiated alone shows a characteristic narrow singlet of 12 G width, and has a G-value much lower than the relatively high G-values of the single nucleotides.

The influence of the stacking-interaction on the spin-transfer can be normalized by comparing the ESR spectra of freeze dried equimolar complexes of nucleotides and proflavine with those of the corresponding mechanical mixtures. As expected, the ESR spectra of the mechanical mixtures are identical to those of the single nucleotides irradiated alone, because of the great differences in the respective G-values. The results are completely different with freeze-dried mixtures. Here, the molecules of the DNA nucleotides are allowed to interact by stacking with the molecules of the dye. The ESR spectra of these complexes are identical to each other and consist of an intense singlet of 12 G width, clearly arising from proflavine-located free radicals. This result is illustrated in Figure 7.2. Once again, the intensity of the proflavine-like signals is such that it can only be explained as being due to a complete quantitative spin-transfer

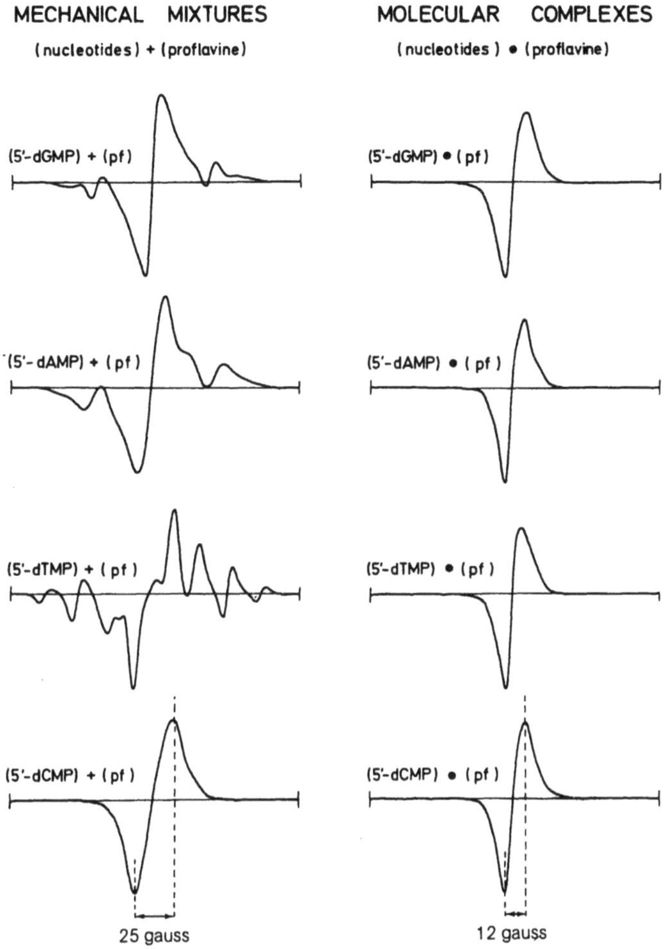

Fig. 7.2. ESR spectra of the mechanical mixtures (*left hand column*) and of the stacking complexes formed by the four DNA nucleotides with proflavine. Each sample contains equimolar amounts of the given nucleotide and of proflavine. Irradiations and measurements at 25°C. Dose 2 Mrad

from the radiosensitive DNA derivatives towards the strongly resistent molecule of the dye.

A progressive increase of the various nucleotide:proflavine molar ratios shows that a complete nucleotide → proflavine spin transfer takes place even at molar ratios as high as 50:1. This observation indicates that the spin transfer does not require a regular alternative sequence of donor and acceptor molecules. It reveals, in addition, that this transfer may be a long-range one, the stacking allowing the electron to tunnel across many bases before it reaches the trap.

A further confirmation of the electron transfer model can be obtained with ESR studies performed on molecular complexes formed by the DNA nucleotides, on the one hand, tryptamine or histamine, on the other (OLAST and BERTINCHAMPS, 1973). The aromatic amines form large stacking complexes with the nucleotides, thus providing the material support for electron transfer phenomena to occur (MONTENAY-GARESTIER and HELENE, 1971). However, the relative scale of electron affinities is now such that electron transfers should be expected to occur from the amines towards the DNA nucleotides. Once more, experimental results confirm the above predictions.

7.3.5 Formal Analysis of the ESR Spectra of Dry DNA

To supply further proof that, in γ-irradiated DNA, the final radical population is localized in the pyrimidine constituents, it should be demonstrated that all the DNA spectra can be interpreted in terms of linear superimpositions of pyrimidine patterns only. This is quite difficult, since there is no way of predicting how many and which pyrimidine radicals, previously identified by single crystal studies, will be present in DNA γ-irradiated in the dry state. Moreover, apart from the well-known radical of H-adduct at C6 of thymine, giving rise to a characteristic octet spread over 140 G, the remaining pyrimidine radicals would give rise to ESR spectra overlapping each other in the central field zone. Any linear combination would then give unresolved patterns. An attempt can, however, be made to unravel some interesting data from a formal computer-assisted analysis of the different, experimentally produced ESR spectra of DNA (GREGOLI and BERTINCHAMPS, 1972a, b).

As mentioned in I.7.3.1, a change in experimental conditions during preparation of the DNA sample will produce considerable differences in the resulting ESR spectrum. Similar investigations performed separately on the two pyrimidine nucleotides dTMP and dCMP could supply significant data for better understanding of the phenomenon.

Experiments with dCMP give disappointing results. Whatever the experimental conditions (irradiation dose, microwave power, pH of the dCMP solution prior to freeze drying, time-decay effect) may be, the ESR spectrum of dCMP remains unchanged and always shows a broad, singlet-type signal of 23 G width. Unlike dCMP, the ESR spectrum of dTMP undergoes significant changes in shape following variations of the experimental parameters. As with DNA, the spectral changes observed for dTMP again appear to be due to changes in the relative proportions of the same radical species. The great variety of dTMP spectra thus produced, lends itself to a computer analysis with a view to isolating the constituent patterns. Three patterns can be unraveled by this technique, one of them being, of course, the octet associated with H-adduct radicals on thymine. As for the remaining patterns, no positive identification of the corresponding radical species is, at present, available, so they will be considered only as formally involved in forming the dTMP spectra.

An attempt can now be made to verify whether the different ESR spectra of DNA can be reconstructed by means of various linear combinations of the three dTMP spectral components and the constant spectrum of dCMP. This can be performed by computer, by combining the four pyrimidine patterns with different relative proportions until a spectrum is obtained which matches a given spectrum of DNA. Since numerous DNA spectra, obtained under different experimental conditions or taken from the literature, were successfully subjected to this test, it may be concluded that the different ESR spectra of DNA γ-irradiated in the dry state are formally constructed by the superimposition of pyrimidine patterns alone.

In summary, it can thus be assumed that, in γ-irradiated DNA, energy-transfer processes do take place. In these processes, the energy carriers are electrons, the transfer pathway is the vertical stacking of the bases and the end points are the pyrimidine moieties. Since hydrogen bonding between complementary bases is not essential for this transfer to occur, energy migration within single-stranded DNA is also predictable. According to this mechanism, intercalation of electron traps within the DNA strands should prevent damage to the DNA itself. This point contains important biological implications. The observation by means of ESR of extensive migrations of radiation damage from DNA to intercalated proflavine was recently confirmed, at a biological level, by experiments on bacteriophage λ and its isolated DNA (MAENHAUT-MICHEL, 1975). Post-irradiation infectivity assays performed on these systems revealed in fact that the intercalated proflavine strongly enhances the survival.

7.4 Photosensitization by Dyes in Frozen Solution

A series of early studies (LEWIS and LIPKIN, 1942; BENNENA et al., 1959) has shown that photoionization of aromatic molecules in rigid glasses is possible with relatively low energy quanta from 3 to 5 eV i.e., about half the energy required in the gas phase. In most cases, the ejection of an electron results from the successive absorption of two photons, the second being absorbed into an intermediate metastable state, thereby raising the excited molecule into the ionization continuum (JOUSSOT-DUBIEN and LESCLAUX, 1970). About 12 years ago, the lowest triplet state of the molecules was assigned to the metastable intermediate (JOUSSOT-DUBIEN and LESCLAUX, 1964; CADOGAN and ALBRECHT, 1965).

The ESR spectrum of aromatic molecules in frozen aqueous media after two-photon ionization consists essentially of the signal from the cation radical of the solute as well as of the lines of the trapped electrons. The latter can induce radical formation in other solute molecules which are not directly affected by the incident photons. This process of photosensitization is known to occur in a great number of biomolecules. The following lines describe the results obtained with nucleic acids and their constituents as sensitized molecules.

7.4.1 Nature of the Photosensitizers

The majority of the experiments has been performed with acridine dyes (Fig. 7.3) and, among them, with proflavine as the sensitizing substrate. The dyes are characterized by an intense absorption in the visible region. They can penetrate into cells and associate to various cell components e.g., nucleic acids and

Fig. 7.3

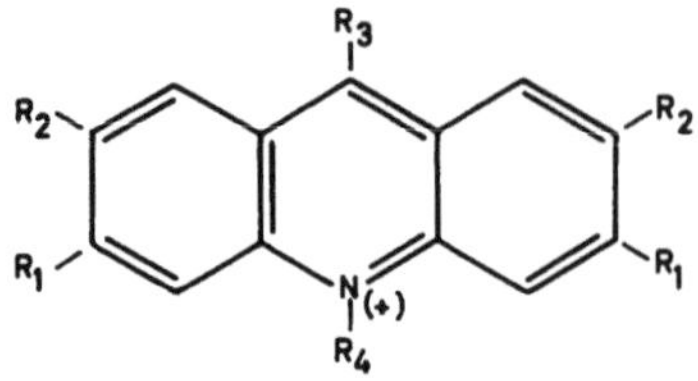

Name	Abbreviation	R_1	R_2	R_3	R_4	λ max (nm)
Proflavine	PF	NH_2	H	H	H	445
Acridine orange	AO	$N(CH_3)_2$	H	H	H	490
Acriflavine	AF	NH_2	H	H	CH_3	447
Acridine yellow	AY	NH_2	CH_3	H	H	413
9-amino acridine	AA	H	H	NH_2	H	398

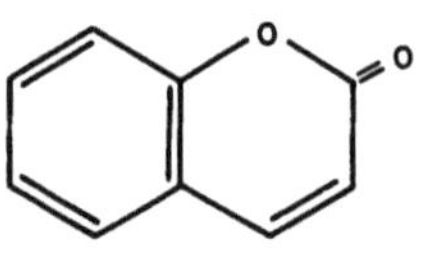

Coumarine

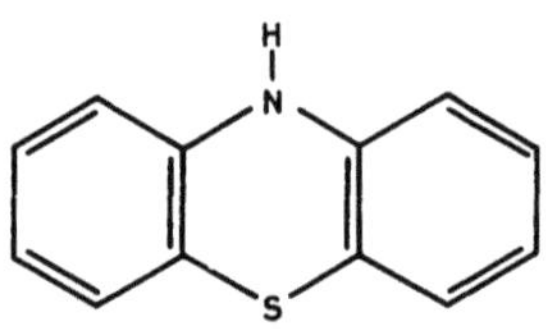

Phenothiazine

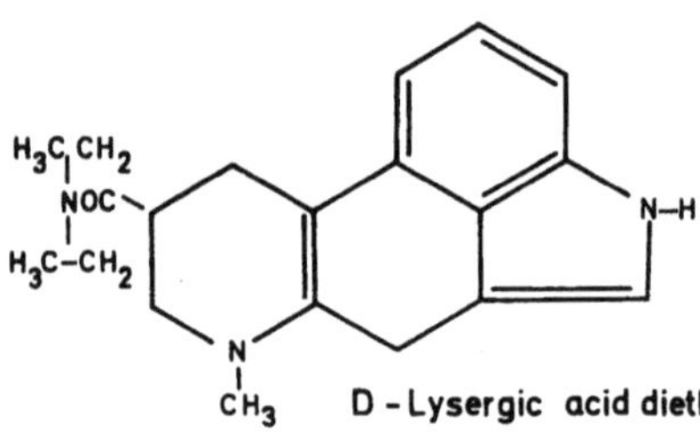

D-Lysergic acid diethylamide (LSD - 25)

proteins. Other sensitizing compounds include coumarine and phenothiazine derivatives, and the drug LSD.

7.4.2 Free Radicals Induced in Nucleotides

7.4.2.1 Description and Interpretation of the ESR Spectra

If frozen solutions of nucleotide (10^{-1} M) - proflavine (5.10^{-4} M) complexes are illuminated at 77 K with $\lambda > 300$ nm in the absence of oxygen, free radicals are induced in the nucleic acid constituents. Figure 7.4 shows the ESR spectra obtained with the four DNA nucleotides (VAN DE VORST and LION, 1971a). In the dAMP-proflavine (PF) system (Fig. 7.4a) a triplet pattern similar to that obtained by the reaction of this compound with trapped electrons (LION and VAN DE VORST, 1970) is observed. The hyperfine

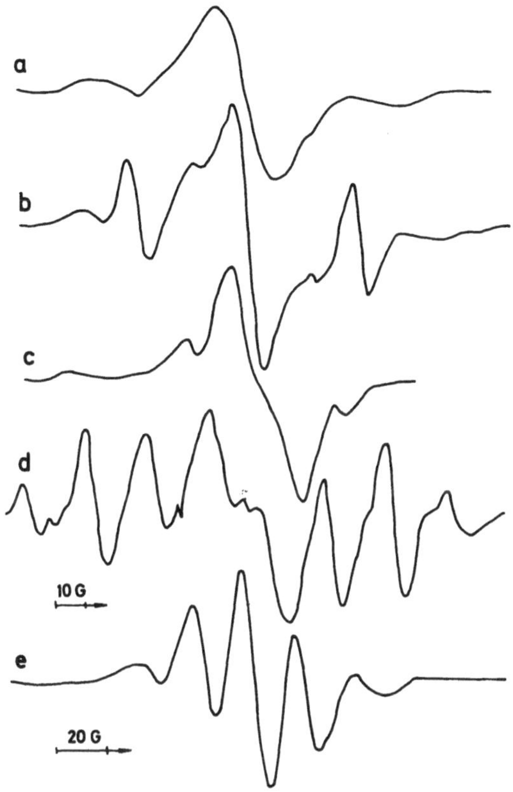

Fig. 7.4. ESR spectra of *a*: dAMP, *b*: dGMP, *c*: dCMP, *d*: dTMP, photosensitized by proflavine in the absence of oxygen. Observed at 77 K; *e*: dTMP, photosensitized by proflavine in the absence of oxygen after warming up 5 min to 225 K. Observed at 77 K. Irradiation at 77 K with $\lambda > 300$ nm. Microwave power 0.8 mW. (VAN DE VORST and LION, 1971a)

splitting of about 40 G is comparable to the values measured for the hydrogen adduct radical to the adenine base in nucleosides (DERTINGER, 1967; LICHTER and GORDY, 1968). Also, the spectrum of dGMP shown in Figure 7.4b resembles closely the one produced upon reaction with electrons in an alkaline glass at 77 K (LION and VAN DE VORST, 1970) or that observed at 300 K of a powder of this nucleotide exposed to hydrogen atoms (HOLROYD and GLASS, 1969). Its assignment to the hydrogen adduct radical of the guanine base is thus straightforward. The spectrum of dCMP is a multiple one (Fig. 7.4c): the central peak exhibits an unresolved doublet, probably associated with a type of radical different from those producing the two lateral bands. The latter are symmetrical with respect to the centre and are not influenced in the same way as the central peak by the incident microwave power. The structure of the radicals giving rise to both patterns are not known. Photosensitization of dTMP (Fig. 7.4d) leads to the well-known eight bands of the hydrogen adduct radical at the thymine base. The spectrum is identical with the one obtained from the reaction of thymidine with hydrogen atoms released by radiolysis of H_2SO_4 at 157 K (ABAGYAN and BUTYAGIN, 1969). Nearly the same spectra are observed when this nucleotide reacts with trapped electrons (HOLROYD and GLASS, 1968).

Generally, all these spectra can be observed up to 260 K. In the case of dTMP, however, a detailed analysis of evolution of the overall shape of the patterns with temperature reveals that, at 225 K, one is left with the signal shown in Figure 7.4e. Similar spectra are observed after photolysis (LION and VAN DE VORST, 1973a) or radiolysis (LION and VAN DE VORST, 1973b) of frozen aqueous solutions containing thymine derivatives. They have been attributed previously to a triplet pattern (VAN DE VORST and LION, 1973) but new experiments, performed in D_2O, have allowed to assign them to a 1:4:6:4:1 quintet with a hyperfine splitting equal to 20 G. Similar patterns were observed upon radiolysis (KAALHUS and JOHANSEN, 1973; GREGOLI et al., 1976). They have also been assigned to a radical at the 5,6-double bond of the thymine base.

Generally, the types of free radical induced in the nucleotides are independent of the structure of the dye used as photosensitizer.

7.4.2.2 Mechanism of Radical Formation

The main feature emerging from the above experiments is the H-adduct radical formation at the base moieties in the illuminated nucleotide-proflavine complexes. The following experimental evidences indicate that successive absorption of two photons into the chromophore which leads to its ionization, mediates radical formation on the nucleotide via transfer of an electron to the constituent base (VAN DE VORST and LION, 1971b):

1. the radical formation rate exhibits a quadratic dependence on light intensity;

2. the transition $\Delta m = \pm 2$ characteristic of the triplet state of the dye is observed during irradiation of substrate-proflavine complexes in the ESR cavity. Furthermore, a direct relation between the formation of the radicals in the substrate and the decrease of the population of the dye triplet state was established;

3. the first stage involves the absorption of a photon corresponding to the normal absorption band of proflavine, i.e., a photon whose wavelength is close to 450 nm and the second photon has been shown to have an energy located in the wavelength range 320 to 365 nm;

4. in the presence of an electron scavenger, the observed spectrum is invariably the same as that characterising the photosensitization of the electron scavenger alone. In the acid phase (H_2SO_4 6M), the signal obtained after irradiation with $\lambda > 300$ nm of proflavine solutions shows always the doublet characteristic of the resonance of $H^{\bullet}$ atoms.

Thus the H-adduct radicals finally detected appear to be the result of a reaction between the nucleotide molecules and the electrons released by photoionization of the sensitizer. According to the experiments performed with X-rays and UV-light in alkaline matrices (see I.6.2), this pathway involves an anionic stage followed by a protonation with a hydrogen from the solvent. However, with the wavelength range used in the experiments described here, this intermediate cannot be isolated; the corresponding absorption bands are situated between 310 and 330 nm for all the nucleotides (LION and VAN DE VORST, 1971a, b).

Similar experimental arguments show that the radicals corresponding to the quintet observed in photosensitized dTMP result of an energy transfer from the excited chromophore to the nucleotide (VAN DE VORST and LION, 1973).

7.4.2.3 Influence of Oxygen

At 77 K, the diffusion of oxygen in frozen aqueous solutions is largely restricted. Consequently, the population of the dye triplet state is not affected significantly by its presence. On the other hand, O_2 can scavenge electrons and give rise to $HO_2^{\bullet}$ radicals according to the following scheme (HOLMES et al., 1970).

$$e^- + O_2 \longrightarrow O_2^- \xrightarrow{H^+} HO_2^{\bullet} \qquad (7.2)$$

$HO_2^{\bullet}$ radicals are effectively identified in the ESR spectrum by a low field peak with a g-factor of about 2.036 (Fig. 7.5a). A consequence of the scavenging of electrons is a decrease of H-adduct radicals in the presence of oxygen. Upon warming to 190 K,

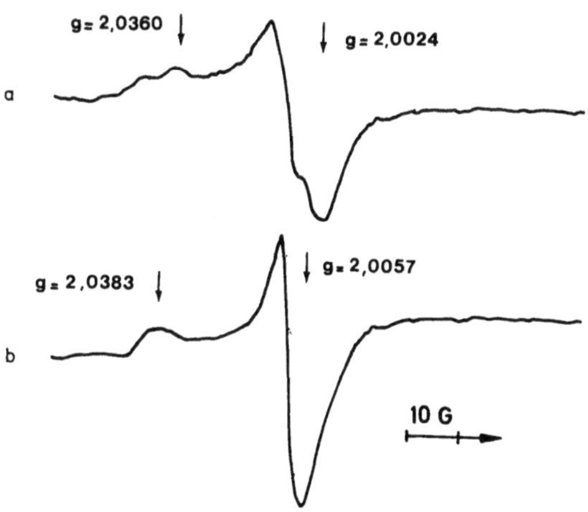

Fig. 7.5. ESR spectra of dCMP photosensitized by proflavine. Irradiation at 77 K with λ > 300 nm, in the presence of oxygen. *a*: Observed at 77 K. Microwave power 200 mW. The peak at g = 2.036 is due to $HO_2^{\bullet}$ radicals (HOLMES et al., 1970); *b*: after warming up 5 min to 190 K. Microwave power 0.8 mW. Observed at 77 K

only peroxide radicals are observed (Fig. 7.5b), probably as the result of a reaction between the substrate radicals and oxygen, which at that temperature can diffuse throughout the matrix. This reaction may also occur to a small extent at 77 K. Similar radical transformations have been observed by HOLMES et al. (1970) in X-irradiated ices containing thymine and oxygen.

7.4.2.4 Influence of the Relative Molar Concentration of Dye and Substrate

At molar concentrations of dye above approximately 1:50 of the substrate, the ESR spectra for all nucleotides consist of a singlet with 17 ± 0.5 G width and a g-factor of 2.0044 ± 0.0001 (VAN

DE VORST and LION, 1976). Its amplitude increases with the concentration of substrate regardless of the dye concentration once it has reached this threshold level. The singlet concentration varies for the different nucleotides. As shown in Table 7.4, deoxyribonucleotides give rise to more intense singlet spectra than the corresponding ribonucleotides. As an exception, dAMP and AMP are characterized by the same intensity of the singlet.

Table 7.4. Influence of nucleotide type on the ESR amplitude in photosensitization[a]

Ribonucleotide 10^{-2} M	ESR Amplitude (arb. units)	Deoxyribonucleotide 10^{-2} M	ESR Amplitude (arb. units)
GMP	96	dGMP	140
AMP	54	dAMP	60
CMP	20	dCMP	120
		dTMP	70

[a] proflavine concentration: 10^{-3} M.

The singlet pattern is most probably associated with a proflavine radical. The increase of its amplitude with the concentration of the nucleotides indicates that these play a role in the formation of the radical. Two mechanisms may be considered: either the electron, released by photoionization, can migrate from one molecule of dye to another and finally be trapped by one of them or is trapped by the substrate, which in turn transfers an electron to the dye. As pointed out by GREGOLI et al. (1970), this transfer could be facilitated by dye-substrate stacking. If so, the analysis of the results noted in Table 7.4 is particularly significant. It reveals that, in general, the deoxyribonucleotides give a more intense transfer than the corresponding ribonucleotides. This is in agreement with the general tendency of stacking with acridine dyes shown by the deoxyribonucleotides (TOMITA, 1967; HRUSKA and DANYLUK, 1968).

7.4.2.5 *Influence of Dye Aggregation*

The population of the triplet state, and consequently the yield of photosensitization is known to be influenced strongly by the formation of aggregates (HELENE et al., 1966). In frozen aqueous solutions, the acridine dyes, more especially acridine orange form stacked aggregates (SCHMIDT and ZELLHOFER, 1974). The dependence of the dye triplet state population on the extent of aggregation can be examined through intensity measurements of the $\Delta m = 2$ transition. It is observed (LION et al., 1975) that insertion of aromatic molecules, like DNA nucleotides, between dye molecules increases the intensity of the corresponding ESR spectrum. This effect, similar to a disaggregation one, can be increased or decreased by adding salts or organic solvents.

7.4.3 Free Radicals Induced in DNA

7.4.3.1 Description and Interpretation of the ESR Spectra

Earlier studies (DELMELLE and DUCHESNE, 1967a; KUBOTA and MIURA, 1967; GRÄSLUND et al., 1969) realized that free radicals could be detected in DNA-proflavine complexes after their illumination with visible light. In the spectra first published the lateral structure of the H-adduct radical at the C6 position of thymine could already be identified. The more recent work of CALBERG-BACQ and VAN DE VORST (1974) has established the similarity be-

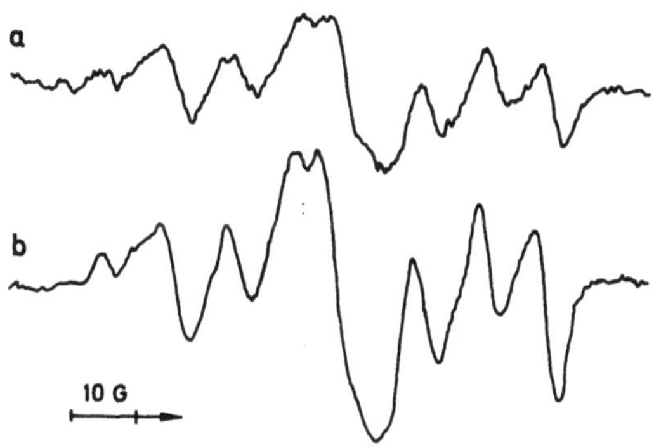

Fig. 7.6. ESR spectra of frozen aqueous solutions of *a*: DNA [1.56 mM (P)] photosensitized by proflavine (2.25×10^{-4} M), $\mu = 0.75$ in $MgCl_2$; *b*: dTMP (10^{-1} M) photosensitized by proflavine (5×10^{-4} M), $\mu = 0.75$ in $MgCl_2$. Irradiation with $\lambda > 300$ nm at 77 K, in the absence of oxygen. Observation at 77 K. Microwave power 1 mW (CALBERG-BACQ and VAN DE VORST, 1974)

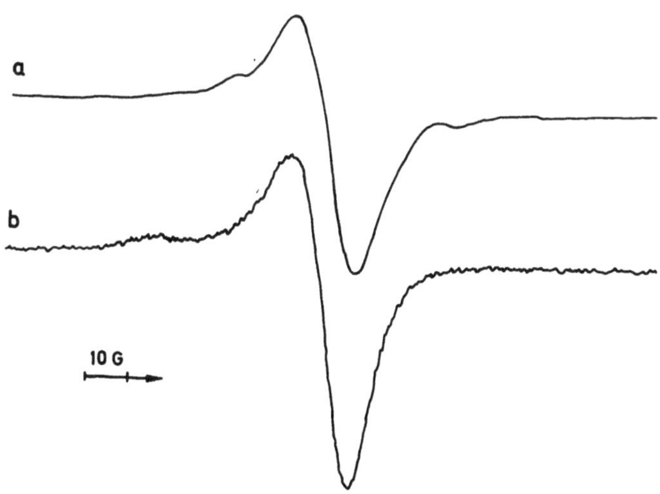

Fig. 7.7. ESR spectra of frozen aqueous solution of DNA [1.56 mM (P)] photosensitized ($\lambda > 300$ nm; T = 77 K) by proflavine (2.25×10^{-4} M) in the presence of O_2, $\mu = 0.375$ in NaCl. Observation at 77 K; Microwave power *a*: 1 mW *b*: 200 mW (CALBERG-BACQ and VAN DE VORST, 1974)

tween the spectra of the two complexes DNA-proflavine and dTMP-proflavine (Fig. 7.6). The presence of O_2 decreases the contribution of H-adduct radicals and another type of free radical is detected. It is characterized by an asymmetrical signal similar to the one generally associated with peroxide radicals (Fig. 7.7).

7.4.3.2 Mechanism of Formation: Anionic Stage

As in the case of nucleotide-proflavine complexes free radicals are observed in the DNA-proflavine complex after absorption of two photons of λ = 445 nm and 320 < λ < 360 nm respectively (CALBERG-BACQ and VAN DE VORST, 1974; GRÄSLUND et al., 1975b). Also, in the presence of the electron scavenger cystamine, only the resonance of this last compound is detected (DELMELLE and DUCHESNE, 1967b). Thus, the free radicals induced in DNA also result from a two-photon ionization of proflavine.

Furthermore, the very recent work of GRÄSLUND et al. (1975b) strengthens the postulate of an anionic stage in the pathway of reactions leading to the final radicals. The analysis of the spectra these authors have obtained by visible irradiation (λ > 395 nm) of oriented wet DNA-proflavine complexes at 77 K yielded the presence of anionic free radicals in thymine and cationic free radicals, probably in guanine. When the exciting light includes short wavelengths below 360 nm, significant amounts of hydrogen-addition radicals on thymine were formed resulting from a protonation of the corresponding anion which absorbs at about 320 nm (HAYON, 1969).

7.4.3.3 Quantitative Influence of Oxygen and Ionic Strength

The presence of oxygen in the photosensitization of DNA in frozen aqueous solutions brings about a clear enhancement of free radical concentration. The effects obtained in N_2 are small but still significant. Another factor influencing the photosensitization efficiency is the ionic strength. The binding of acridine dyes by DNA involves two modes of fixation (PEACOCKE and SKERRETT, 1956): an intercalation of the dye molecule between two pairs of bases and a binding of the dye cations to the phosphate groups external to the DNA double helical structure. The second mode predominates after saturation of the intercalation process. The relative importance of the two types of fixation is greatly dependent upon the ionic strength of the medium (DRUMMOND et al., 1965). CALBERG-BACQ and VAN DE VORST (1974) have investigated this influence and the main conclusions of their work can be summarized as follows:

1. In the absence of O_2, the reaction increases in a range of very low concentration in bound proflavine and reaches a saturation value much before maximum intercalation occurs. The plateau value of the signal intensity increases slightly with the ionic strength of the medium (Fig. 7.8). Therefore, under N_2, the photosensitization of DNA is due to intercalated dye molecules, as the first bound molecules are important and as the reaction is maintained, even slightly increased, at high ionic strength.

2. In air or after saturation with O_2 the number of free radicals increases with the concentration of bound proflavine corresponding to increasing intercalation. Moreover, the maximum intensity is observed in a concentration range which coincides with that of maximal intercalation.

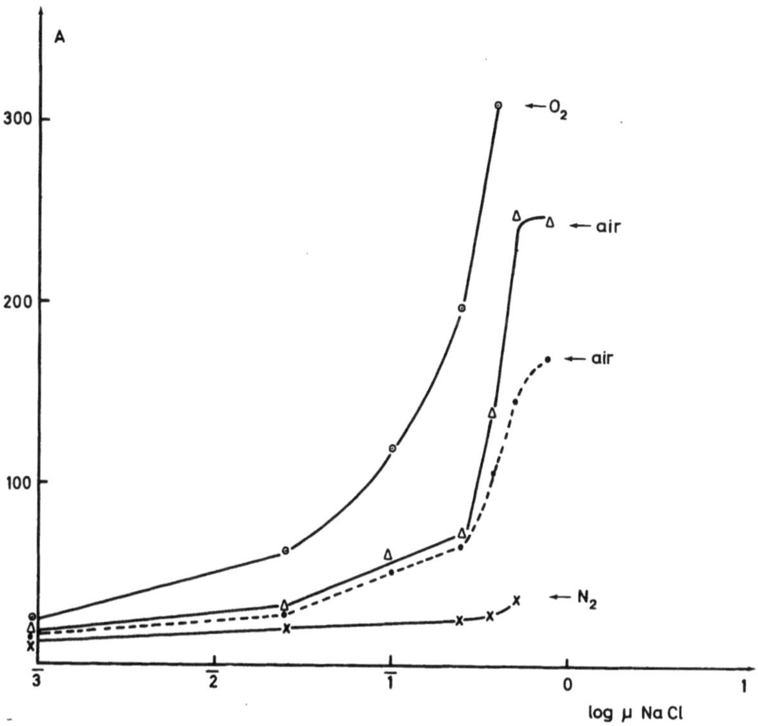

Fig. 7.8. Influence of ionic NaCl strength (log μ) on signal intensity: (A, arbitrary units) obtained with a given concentration of bound proflavine, in O_2, air and N_2, as indicated on the graph: *Solid line:* 15.6×10^{-5} M PF bound by 1.56 mM (P) DNA (i.e., r = 0.1). *Dashed line:* 5×10^{-5} M PF bound by 0.55 mM (P) or 1.56 mM (P) DNA. (CALBERG-BACQ and VAN DE VORST, 1974)

Provided that the proflavine molecules are strongly bound to DNA, they act independently as photosensitizers since the ESR intensity is not influenced by the distance which separates the dye molecules on the DNA helix. Therefore, there would be no recombination between induced free radicals till maximum intercalation of the dye, i.e., no transfer of the effect would take place beyond the base pair next to the dye molecule.

7.4.4 General Conclusions

ESR spectra obtained after photosensitization of DNA or its constituents by acridine dyes are just the same as those resulting from direct radiolysis or photolysis of the same compounds under identical conditions i.e., in frozen aqueous solutions. In each case, the free radicals finally observed are the result of a reaction between these biomolecules and the electrons released by double-photon ionization of the chromophore, X-ray absorption or UV absorption.

The single protonated proflavine (PH^+) seems to be the active species (LION et al., 1975) in the process of photosensitization, which would be favoured by the insertion plane to plane of nucleotide molecules between proflavine molecules. Furthermore, it is well established (VAN DE VORST et al., 1976) that

1. each molecule of proflavine can induce, at the maximum, one free radical in the nucleotides (RH);

2. the chromophore is largely restored in the process, when oxygen is not present;

3. in the presence of oxygen, peroxyde radicals are induced through the matrix;

4. formyl radicals are observed.

Starting from this set of information, the following general scheme can be proposed:

1. Photoionozation of the dye

$$PH^+ \xrightarrow{h\nu_1} {}^1PH^+ \longrightarrow {}^3PH^+ \longrightarrow {}^3P + H^+ \qquad (7.3)$$

$$^3P \xrightarrow{h\nu_2} (P^+)^{\cdot} + e^- \qquad (7.4)$$

2. Formation of anions and of H-adduct radicals in the base

$$RH + e^- \longrightarrow RH^- \qquad (7.5a)$$

$$RH^- + H^+ \longrightarrow (RH^-)H^+ \longrightarrow RH_2^{\cdot} \qquad (7.5b)$$

The protons being liberated by Equation (7.3). In presence of oxygen, the Equation (7.5a) is in competition with the Equation (7.5c) which explains the formation of peroxyde radicals

$$O_2 + e^- \longrightarrow O_2^- \xrightarrow{+H^+} HO_2^{\cdot} \qquad (7.5c)$$

3. Restoration of the dye

$$(P^+)^{\cdot} + H_2O \longrightarrow (P^+) + H^+ + OH^- \longrightarrow PH^+ + OH^{\cdot} \qquad (7.6)$$

favoured by irradication in the absorption band of the radical cation of proflavine ($\lambda \approx 810$ nm).

In presence of oxygen, another reaction could take place which would explain the destruction of the chromophore

$$(P^+)^{\cdot} + HO_2^{\cdot} \longrightarrow POOH$$

Under these conditions, the OH radicals liberated by Equation (7.6) would be the forerunners of the formyl radicals (CANNISTRARO et al., 1975).

References

ABAGYAN, G.V., BUTYAGIN, P.Yu.: O prirode i prevrashcheniya paramagnitni tsentrov v γ-oblichenni rastvora pyrimidinov ocnovani i DNA. Biofizika 14, 785-790 (1969)

ADAMS, S.M., BOX, H.C.: Spin density distribution in primary oxidation and reduction products of irradiated nucleic acid constituents. J. Chem. Phys. 63, 1185-1189 (1975)

ALEXANDER, C., Jr., FRANKLIN, C.E.: ESR study of the sugar radical in irradiated deoxyadenosine. J. Chem. Phys. 54, 1909-1913 (1971)

ALEXANDER, C., Jr., GORDY, W.: Electron spin resonance of an irradiated single crystal of guanine hydrochloride dihydrate. Proc. Natl. Acad. Sci. 58, 1279-1285 (1967)

ALEXANDER, P., LETT, J.T., ORMEROD, M.G.: Energy transfer in irradiated deoxyribonucleic acid and nucleoprotein. Biochim. Biophys. Acta 51, 207-209 (19 (1961)

ARNOTT, S.: The geometry of nucleic acids. Progr. Biophys. Mol. Biol. 21, 265-319 (1970)

ARNOTT, S.: DNA secondary structures. In: Organization and Expression of Chromosomes. Berlin: Dahlem Konferenzen, 1976

AYSCOUGH, P.B., COLLINS, R.G., DAINTON, F.S.: Some elementary processes in radiation and photo-chemistry revealed by electron spin resonance. Nature (London) 205, 965-969 (1965)

BAUDET, J., BERTHIER, G., PULLMAN, B.: Recherches thêoretiques sur la rêpartition des densités de spin dans les molécules d'interêt biochimique. C. R. Acad. Sci. (Paris) 254, 762-767 (1962)

BENNENA, P., HOIJTINK, G.J., LUPINSKI, J.M., OOSTERHOFF, L.J., SELLIER, P., VAN VOORST, J.D.: Photo-ionization of aromatic hydrocarbons in boric acid glasses. Mol. Phys. 2, 431-435 (1959)

BENSON, B., SNIPES, W.: Radiation induced radicals in pyrimidines: ESR of irradiated alloxan. Radiat. Res. 31, 542-543 (1967)

BENSON, B., SNIPES, W.: Free radicals in pyrimidines: E.S.R. of gamma-irradiated single crystals of alloxantin. Intern. J. Radiat. Biol. 15, 583-585 (1969)

BERGENE, R., HENRIKSEN, T.: Free radicals in pyrimidines: a single crystal of dihydro-6-methyluracil irradiated at 295 K. Intern. J. Radiat. Biol. 20, 201-211 (1971)

BERGENE, R., MELØ, T.B.: Formation and trapping of radical pairs in irradiated thymine. Intern. J. Radiat. Biol. 23, 263-270 (1973)

BERGENE, R., VAUGHAN, R.A.: ESR of free radicals in irradiated uracil-β-D-arabinofuranoside. Intern. J. Radiat. Biol. 29, 145-157 (1976)

BERNHARD, W.A., CLOSE, D., CORELLI, J.C., MERCER, K.: ESR of X-irradiated single crystals of 3'-cytidylic acid: hydrogen abstraction from C5' of the sugar moiety. Radiat. Res. 66, 19-32 (1976a)

BERNHARD, W.A., CLOSE, D.M., HÜTTERMANN, J., ZEHNER, H.: The alkoxy radical, $RCH_2\dot{O}$, as a free radical product in X-irradiated single crystals of nucleosides and nucleotides. J. Chem. Phys. 67, 1211-1219 (1977)

BERNHARD, W.A., FARLEY, R.A.: ESR of hydrogen addition radicals in single crystals of 1-methylcytosine and 1-methyluracil. Radiat. Res. 66, 189-198 (1976)

BERHARD, W.A., HÜTTERMANN, J., MÜLLER, A., CLOSE, D.M., FOUSE, G.W.: A stable radical in the ribose moiety of X-irradiated 3'-cytidylic acid and its relation to a similar radical in cytidine: an ESR-ENDOR study. Radiat. Res. 68, 390-413 (1976b)

BERNHARD, W., SNIPES, W.: Electron spin resonance of a gamma-irradiated single crystal of barbituric acid dihydrate. J. Chem. Phys. 44, 2817-2820 (1966)

BERNHARD, W.A., SNIPES, W.: ESR of free radical conversion in gamma-irradiated dihydrothymine. J. Chem. Phys. 46, 2848-2849 (1967)

BERNHARD, W., SNIPES, W.; Electron spin resonance of a gamma-irradiated single crystal of 3'-cytidylic acid. Proc. Natl. Acad. Sci. 59, 1038-1044 (1968)

BERTHOD, H., GIESSNER-PRETTRE, C., PULLMAN, A.: Theoretical study of the electronic properties of the purine and pyrimidine components of the nucleic acids. A semi-empirical Self-Consistent-Field calculation. Theoret. Chim. Acta 5, 53-68 (1966)

BLOOMFIELD, V.A., CROTHERS, D.M., TINOCO, I.,Jr.: Physical Chemistry of Nucleic Acids. New York: Harper and Row, 1974

BLYUMENFELD, L.A.: Anomalnie magnitnie svoystva noklenovie kislot. Biofizika 4, 515-520 (1959)

BOAG, J.W., MÜLLER, A.: Electron spin resonance in irradiated deoxyribonucleic acid. Nature (London) 183, 831-832 (1959)

BOX, H.C.: ESR and ENDOR studies of radiation damage. J. Magn. Res. 14, 323-328 (1974)

BOX, H.C., BUDZINSKI, E.E.: Oxidation and reduction of barbituric acid dihydrate by ionizing radiation at 4.2 K. J. Chem. Phys. 59, 1588-1591 (1973)

BOX, H.C., BUDZINSKI, E.E.: Primary radiation damage in thymidine. J. Chem. Phys. 62, 197-199 (1975)

BOX, H.C., BUDZINSKI, E.E.: Primary radiation products in X-irradiated dihydromethyluracil. J. Chem. Phys. 62, 2006-2007 (1975a)

BOX, H.C., BUDZINSKI, E.E.: Reduction of adenine dihydrochloride by ionizing radiation. J. Chem. Phys. 64, 1593-1595 (1976)

BOX, H.C., BUDZINSKI, E.E., POTTER, W.R.: ENDOR characterization of the free radical in irradiated thymidine. J. Chem. Phys. 61, 1136-1139 (1974)

BOX, H.C., POTTER, W.R., BUDZINSKI, E.E.: The reduction of nucleotides by ionizing radiation: uridine-5'-phosphate and cytidine-5'-phosphate. J. Chem. Phys. 62, 3476-3478 (1975)

BRIVATI, J.A., SYMONS, M.R.C., TINLING, D.J.A., WARDALE, H.W., WILLIAMS, D.O.: Electron spin resonance studies of the hydroxyl radical in γ-irradiated ice. Trans. Faraday Soc. 63, 2112-2116 (1967)

BROCKLEHURST, B.: Fundamental processes of inert-gas sensitization. Radiat. Res. Rev. 1, 223-240 (1968)

BRONSKILL, M.J., TAYLOR, W.B., WOLFF, R.K., HUNT, J.W.: Design and performance of a pulse radiolysis system capable of picosecond time resolution. Rev. Sci. Instrum. 41, 333-340 (1970)

BRUSTAD, T.: Heavy ions and some aspects of their use in molecular and cellular radiobiology. Advan. Biol. Med. Phys. 8, 161-224 (1962)

BUGG, C.E.: Solid-state stacking patterns of purine bases. In: The Purines. Theory and Experiment. pp. 178-200. BERGMANN, E.D., PULLMAN, B. (eds.). Jerusalem: Israel Acad. Sci. Humanities, 1972

BUTTS, J.J., KATŽ, R.: Theory of RBE for heavy ion bombardment of dry enzymes and viruses. Radiat. Res. 30, 855-871 (1967)

CADOGAN, K.D., ALBRECHT, A.C.: Two-photon ionizations in rigid organic solutions and the triplet-state intermediate. J. Chem. Phys. 43, 2550-2552 (1965)

CALBERG-BACQ, C.M., VAN DE VORST, A.: Induction of free radicals in DNA by proflavine and visible light: influence of oxygen and ionic strength. Photochem. Photobiol. 20, 433-439 (1974)

CANNISTRARO, S., LION, Y., VAN DE VORST, A.: Formyl radicals in X-irradiated frozen aqueous solutions of DNA and nucleic acids constituents. Radiat. Res. 64, 581-592 (1975)
CARLSON, T.A., WHITE, R.M.: Measurement of the relative abundances and recoil-energy spectra of fragment ions produced as the initial consequences of X-ray interaction with CH_3I, HI, and DI. J. Chem. Phys. 44, 4510-4520 (1966)
CHARGAFF, E.: Structure and function of nucleic acids as cell constituents. Federation Proc. 10, 654-659 (1951)
CHARGAFF, E., DAVIDSON, J.N.: The Nucleic Acids. New York: Academic Press, 1955
CHRISTOPHOROU, L.G., COMPTON, R.N.: Interaction of low-energy electrons with polyatomic molecules. Health Phys. 13, 1277-1305 (1967)
CLARK, L.B., TINOCO, I., Jr.: Correlations in the ultraviolet spectra of the purine and pyrimidine bases. J. Am. Chem. Soc. 87, 11-15 (1965)
COMBS, L.L., HOLLOMAN, M., YOUNG, J.K.: A semi empirical study of the sugar radicals in irradiated cytidylic acid b and deoxyadenosine. Radiat. Res. 56, 8-20 (1973)
COOK, J.B., ELLIOTT, J.P., WYARD, S.J.: Electron spin resonance of an irradiated single crystal of cytosine. Mol. Phys. 13, 49-64 (1967)
COOK, J.B., WYARD, S.J.: Electron spin resonance studies of radiation damage in deoxyribonucleic acid. Intern. J. Radiat. Biol. 11, 357-365 (1966a)
COOK, J.B., WYARD, S.J.: Electron spin resonance study of thermal effects in irradiated deoxyribonucleic acid. Nature (London) 210, 526-527 (1966b)
COOK, J.B., WYARD, S.J.: Radical yields in some nucleic acid constituents. Intern. J. Radiat. Biol. 11, 225-227 (1966c)
DEE, D., BAUR, M.E.: Charge and excitation migration in DNA chains. J. Chem. Phys. 60, 541-560 (1974)
DELMELLE, M., DUCHESNE, J.: Sur le mécanisme de l'action photomutagénique. C.R. Acad. Sci. (Paris) 264, 138-140 (1967a)
DELMELLE, M., DUCHESNE, J.: Protection de l'acide désoxyribonucléique par la cystamine dans l'effet photomutagène. C.R. Acad. Sci. (Paris) 264, 1651-1652 (1967b)
DERTINGER, H.: Freie Radicale in bestrahlten Einkristallen von Purin-Nukleosiden: Adenosin, Desoxyadenosin und Desoxyguanosin. Z. Naturforsch. 22b, 1261-1266 (1967)
DERTINGER, H.: Der Ursprung des ESR-Dubletts in der Nucleinsäurebase Cytosin: eine Einkristallstudie bei 8 mm Wellenlänge. Z. Naturforsch. 22b, 1266-1269 (1967a)
DERTINGER, H., HARTIG, G.: Mechanisms of radical formation in irradiated purine bases and derivatives: an E.P.R. study using computer technique. Intern. J. Radiat. Biol. 21, 279-292 (1972)
DERTINGER, H., JUNG, H.: Molekulare Strahlenbiologie, pp 76-78. Berlin-Heidelberg-New York: Springer, 1969
DIBDIN, G.H.: ESR of γ-irradiated single crystals of ice at $77^{O}K$. Trans. Faraday Soc. 63, 2098-2111 (1967)
DONOHUE, J.: The hydrogen bond in organic crystals. J. Phys. Chem. 56, 502-510 (1952)
DONOHUE, J., TRUEBLOOD, K.N.: Base pairing in DNA. J. Mol. Biol. 2, 363-371 (1960)
DORLET, C., VAN DE VORST, A., BERTINCHAMPS, A.J.: Electron spin resonance in γ-irradiated deoxyribonucleic acid. Nature (London) 194, 767 (1962)
DOUTHAT, D.A.: Electron degradation spectra in helium. Radiat. Res. 61, 1-20 (1975)
DRUMMOND, D., SIMPSON-GILDEMEISTER, V., PEACOCKE, A.: Interaction of aminoacridines with deoxyribonucleic acid: effects of ionic strength, denaturation and structure. Biopolymers 3, 135-153 (1965)

DUCHESNE, J.: Esquisse des recherches récentes effectuées au laboratoire de spectroscopie des radiofréquences de l'Université de Liège. Arch. Sci. Geneva 10, 257-266 (1957)
DUCHESNE, J., VAN DE VORST, A.: Origine de la radiorésistance des acides nucléiques en phase solide. Bull. Acad. Roy. Belg. Cl. Sci. 51, 778-809 (1965)
DULČIĆ, A., HERAK, J.N.: Crystal structure dependence of radiation-induced radicals in thymine: an ESR study. Radiat. Res. 47, 573-580 (1971)
DULČIĆ, A., HERAK, J.N.: Radiation abstraction of hydrogen atoms in single crystals of thymine. J. Chem. Phys. 57, 2537-2540 (1972)
DULČIĆ, A., HERAK, J.N.: Radical pairs in irradiated single crystals of 1-methyl uracil. Mol. Phys. 26, 605-614 (1973a)
DULČIĆ, A., HERAK, J.N.: Radiation-induced pair-wise radical formation in single crystals of thymine. Biochim. Biophys. Acta 319, 109-115 (1973b)
DURUP, J., PLATZMAN, R.L.: Role of the Auger effect in the displacement of atoms in solids by ionizing radiation. Disc. Faraday Soc. No. 31, 156-166 (1961)
EHRENBERG, A., EHRENBERG, L., LÖFROTH, G.: Thymidine-like electron spin resonance spectra in gamma-irradiated deoxyribonucleic acid. Nature (London) 200, 376-377 (1963)
EHRENBERG, A., RUPPRECHT, A., STRÖM, G.: Electron spin resonance of gamma-irradiated oriented DNA prepared by wet spinning. Science 157, 1317-1319 (1967)
EISINGER, J., SHULMAN, R.G.: UV-induced electron spin resonances in DNA and thymine. Proc. Natl. Acad. Sci. 50, 694-696 (1963)
ELLIOTT, J.P.: An electron spin resonance study of free radiclas in an γ-irradiated single crystal of 1-methyl cytosine. In: Proc. 1. Europ. Biophys. Congr. II, pp. 17-22. Broda, E., Locker, A., Springer-Lederer, H. (eds.). Verlag der Wiener Med. Akademie, 1971
ELLIOTT, J.P., WYARD, S.J.: Electron spin resonance in X-irradiated oriented deoxyribonucleic acid. Nature (London) 208, 483-485 (1965)
EMERSON, L.C., BIRKHOFF, R.D., ANDERSON, V.E., RITCHIE, R.H.: Electron slowing-down spectrum in irradiated silicon. Phys. Rev. B7, 1798-1811 (1973)
FARLEY, A.E.: ESR studies of free radical formation in irradiated single crystals of 1-methylcytosine, 5-fluorouracil, and a molecular complex containing 1-methylcytosine and 5-fluorouracil. Thesis, Univ. Rochester (1974)
FARLEY, R.A., BERNHARD, W.A.: Electron spin resonance of gamma- and X-irradiated nucleic acid base pairs: 1-methylcytosine:5-fluorouracil cocrystals at 77 K. Radiat. Res. 61, 47-62 (1975)
FLOSSMANN, W., HÜTTERMANN, J., MÜLLER, A., WESTHOF, E.: Electron spin resonance of free radicals and radical pairs in irradiated single crystal of 1-methyluracil. Z. Naturforsch. 28c, 523-532 (1973)
FLOSSMANN, W., MÜLLER, A., WESTHOF, E.: Radicals in irradiated monocrystals of the base-pair complex 9-ethyl adenine: 1-methyl uracil. Intern. J. Radiat. Biol. 26, 481-492 (1974a)
FLOSSMANN, W., MÜLLER, A., WESTHOF, E.: Electron spin resonance study of radiation-produced radical pairs in single crystals of 1-methyluracil. Mol. Phys. 29, 703-712 (1975a)
FLOSSMANN, W., WESTHOF, E., MÜLLER, A.: ESR-spectroscopy of radiation-produced radicals in 9-ethyl adenine. Intern. J. Radiation Biol. 25, 437-443 (1974b)
FLOSSMANN, W., WESTHOF, E., MÜLLER, A.: Radical formation in salts of pyrimidines: I. 1-methyluracil·HBr crystals. Intern. J. Radiation Biol. 28, 105-115 (1975b)
FLOSSMANN, W., WESTHOF, E., MÜLLER, A.: Radical formation in salts of pyrimidines: III. Cytosine·H_2O crystals. Intern. J. Radiation Biol. 30, 301-315 (1976)

FLOSSMANN, W., WESTHOF, E., MÜLLER, A.: Light-induced displacement of hydrogen in pyrimidine H-addition radicals. J. Chem. Phys. 64, 1688-1691 (1976a)
FLOURNOY, J.M., BAUM, L.H., SIEGEL, S.: Disappearance of trapped hydrogen atoms in gamma-irradiated ice. J. Chem. Phys. 36, 2229-2230 (1962)
FÖRSTER, TH.: Zwischenmolekulare Energiewanderung und Fluoreszenz. Ann. Physik. 2, 55-75 (1948)
FÖRSTER, TH.: Mechanisms of energy transfer. In: Energetics and Mechanisms in Radiation Biology. pp. 183-201 Philips, G.O. (ed.). London-New York: Academic Press, 1968
FRITZSCHE, H.: Infra red studies of deoxyribonucleic acids, their constituents and analogues. III. Evidence of slow isotopic exchange of hydrogen attached to carbon in DNA. Biochim. Biophys. Acta 149, 173-179 (1967)
GORDY, W.: Free Radicals from biological purines and pyrimidines. Ann. N.Y. Acad. Sci. 158, 100-123 (1969)
GORDY, W., PRUDEN, B., SNIPES, W.: Some radiation effects on DNA and its constituents. Proc. Natl. Acad. Sci. 53, 751-756 (1965)
GRÄSLUND, D.A., EHRENBERG, A., RUPPRECHT, A., STRÖM, G.: Ionic base radicals in γ-irradiated DNA. Biochim. Biophys. Acta 254, 172-186 (1971)
GRÄSLUND, A., EHRENBERG, A., RUPPRECHT, A., STRÖM, G., CRESPI, H.: Ionic base radicals in γ-irradiated oriented nondeuterated and fully deuterated DNA. To be published (1976)
GRÄSLUND, A., EHRENBERG, A., RUPPRECHT, A., TJÄLLDIN, B., STRÖM, G.: ESR kinetics of a free radical conversion in γ-irradiated oreinted DNA. Radiat. Res. 61, 488-503 (1975a)
GRÄSLUND, A., RIGLER, R., EHRENBERG, A.: Light-induced free radicals in DNA-acridine complexes studied by ESR. FEBS Lett. 4, 227-230 (1969)
GRÄSLUND, A., RUPPRECHT, A., STRÖM, G.: Light-induced free radicals in oriented DNA-proflavine complexes. Photochem. Photobiol. 21, 153-157 (1975b)
GREGOLI, S., BERTINCHAMPS, A.: Free radical distribution in gamma-irradiated DNA. Intern. J. Radiat. Biol. 21, 65-73 (1972a)
GREGOLI, S., BERTINCHAMPS, A.: Computer analysis and reconstruction of gamma-irradiated DNA. Intern. J. Radiat. Biol. 21, 75-85 (1972b)
GREGOLI, S., OLAST, M., BERTINCHAMPS, A.: Spin transfer in gamma-irradiated nucleotide complexes. Intern. J. Radiat. Biol. 17, 39-46 (1970a)
GREGOLI, S., OLAST, M., BERTINCHAMPS, A.: Free radicals in γ-irradiated frozen solutions of deoxyadenosine-5'-monophosphate. A computer analysis of temperature-dependent ESR spectra. Radiat. Res. 60, 388-404 (1974)
GREGOLI, S., OLAST, M., BERTINCHAMPS, A.: Free radical formation in deoxythymidine-5'-monophosphate γ-irradiated in frozen solution. A computer-assisted analysis of temperature-dependent ESR spectra. Radiat. Res. 65, 202-219 (1976)
GREGOLI, S., TAVERNA, C., BERTINCHAMPS, A.: Electron transfer in gamma-irradiated complexes between DNA nucleotides and proflavine. Intern. J. Radiat. Biol. 18, 577-585 (1970b)
GUTIERREZ, P., BENSON, B.: Free radicals in pyrimidines: ESR of 5-ethylbarbituric acid and its potassium salt. J. Chem. Phys. 60, 640-644 (1974a)
GUTIERREZ, P., BENSON, B.: Free radicals in pyrimidines: ESR study of selective a-straction of C(5) substituents by X-irradiation. Radiat. Res. 58, 141-153 (1974b)
HAAK, R., BENSON, B.: Free radicals in pyrimidines: ESR of X-irradiated single crystals of barbital (5,5-diethylbarbituric acid). J. Chem. Phys. 55, 3693-3697 (1971)
HAINDL, E., HÜTTERMANN, J.: Magnetic resonance parameters of an α-bromo radical in X-irradiated 5-bromodeoxyuridine crystals. J. Magn. Res. (1978) In press
HAMPTON, D.A., ALEXANDER, C., Jr., ENDOR and ESR studies of irradiated cytidine. J. Chem. Phys. 58, 4891-4898 (1973)

HART, E.J., PLATZMAN, R.L.: Radiation chemistry. In: Mechanisms in Radiobiology. pp. 93-257 Errera, M., Forssberg, A. (eds.). London-New York: Academic Press, 1961

HARTIG, G., DERTINGER, H.: A computer controlled E.P.R. analysis of free radical formation in dry thymine and derivatives after electron irradiation. Intern. J. Radiat. Biol. 20, 577-588 (1971)

HATANO, Y., SHIDA, S., INOKUTI, M.: Hydrogen formation and superexcited states in the radiolysis of liquid olefins. J. Chem. Phys. 48, 940-941. (1968)

HAYON, E.: Optical-absorption spectra of ketyl radicals and radical anions of ionic pyrimidines. J. Chem. Phys. 51, 4881-4892 (1969)

HEDBERG, A., EHRENBERG, A.: Resolution enhancement of ESR spectra from irradiated single crystals of glycine. J. Chem. Phys. 48, 4822-4828 (1968)

HELENE, C., SANTUS, R., PTAK, M.: Etude de l'état triplet de la purine, des nucléosides puriques et des acides aminés aromatiques en solution aqueuse congelée. C.R. Acad. Sci. (Paris) 262, 1349-1352 (1966)

HELLER, H.C., COLE, T.: ESR of H-addition in thymine and DNA. J. Chem. Phys. 54, 1486-1490 (1965)

HENRIKSEN, T.: Electron spin resonance studies on the formation and properties of free radicals in irradiated sulfur-containing substances. Oslo: Norsk Hydro's Institute for Cancer Research, 1963

HENRIKSEN, T.: Effect of crystal dimensions on the yield of radiation-induced radicals in organic substances. Acta Chem. Scand. 20, 2898-2900 (1966)

HENRIKSEN, T.: Effect of temperature on the formation of radicals by hydrogen-atom bombardment. J. Chem. Phys. 50, 4653-4657 (1969)

HENRIKSEN, T., JONES, W.B.G.: ESR studies of the mechanisms for radical formation in irradiated frozen sulfuric acid solutions of thymine. Radiat. Res. 45, 420-433 (1971)

HENRIKSEN, T., SNIPES, W.: Radiation-induced radicals in thymine: ESR studies of single crystals. Radiat. Res. 42, 255-269 (1970a)

HENRIKSEN, T., SNIPES, W.: Electron spin resonance studies of irradiated single crystals of dihydrothymine. J. Chem. Phys. 52, 1997-2000 (1970b)

HERAK, J.N.: E.P.R. study of radiation damage of deuterated single crystal of thymidine. J. Chem. Phys. 52, 6440-6441 (1970a)

HERAK, J.N.: Electron spin resonance of gamma-irradiated single crystal of dihydrouracil. J. Chem. Phys. 53, 576-578 (1970b)

HERAK, J.N.: Radiation-induced conformation changes of the pyrimidine ring in a single crystal of dihydrothymine. Z. Naturforsch. 25a, 1299-1302 (1970c)

HERAK, J.N.: E.P.R. of irradiated single crystals of the nucleic acid constituents. In: Physico-chemical Properties of Nucleic Acids. Duchesne, J. (ed.). London: Academic Press, 1973

HERAK, J.N., DULČIĆ, A.: ESR study of radiation-induced defects in 1-methyl uracil. In: Proc. 17. Collogue Ampère! Hovi, V. (ed.). Amsterdam: North Holland, 1973

HERAK, J.N., GALOGAZA, V.: Radical transformation in irradiated DNA and its constituents. Proc. Natl. Acad. Sci. 64, 8-12 (1969a)

HERAK, J.N., GALOGAZA, V.: Electron spin resonance of gamma-irradiated single crystal of cytosine monohydrate at 77 K. J. Chem. Phys. 50, 3101-3105 (1969b)

HERAK, J.N., GORDY, W.: Free radicals formed by hydrogen atom bombardment of the nucleic-acid bases. Proc. Natl. Acad. Sci. 54, 1287-1292 (1965)

HERAK, J.N., GORDY, W.: Hydrogen addition radicals on RNA and the nucleic-acid bases at 77 K. Proc. Natl. Acad. Sci. 55, 1373-1378 (1966a)

HERAK, J.N., GORDY, W.: ESR study of gamma-irradiated polynucleotides. Proc. Natl. Acad. Sci. 55, 698-704 (1966b)

HERAK, J.N., KRILOV, D., McDOWELL, C.A.: ENDOR study of the stable radicals in an irradiated single crystal of deoxycytidine 5'-phosphate. J. Magn. Res. 23, 1-7 (1976)

HERAK, J.N., McDOWELL, C.A.: ENDOR study of long-range spin interactions in molecular crystals. I. 1-methyl uracil. J. Chem. Phys. 61, 1129-1135 (1974a)
HERAK, J.N., McDOWELL, C.A.: ENDOR study of the minority radicals formed in X-ray-irradiated single crystals of thymidine. J. Magn. Res. 16, 434-438 (1974b)
HERAK, J.N., SCHOFFA, G.: E.P.R. spectroscopy of the radicals formed in a single crystal of 6-azathymine by irradiation. Mol. Phys. 22, 379-383 (1971)
HOLMES, D.E., INGALLS, R.B., MYERS, L.S., Jr.: An ESR study of free radicals formed by reaction of nucleotides and their constituents with hydrogen atoms. Intern. J. Radiat. Biol. 12, 415-426 (1967)
HOLMES, D.E., MYERS, L.S., Jr., INGALLS, R.B.: ESR technique for investigation of reactions of free radicals with components of biological interest: nucleic acid constituents. Nature (London) 209, 1017-1018 (1966)
HOLMES, D.E., NAZHAT, N.B., WEISS, J.J.: Effects of dissolved oxygen on the electron spin resonance signal intensities of trapped hydrogen atoms and some of their reactions in acidic ice matrices. J. Phys. Chem. 74, 1622-1626 (1970)
HOLMES, D.E., WEISS, J.J.: The formation and reactions of hydrogen atoms in gamma-irradiated deoxyribonucleic acid gels. Intern. J. Radiat. Biol. 14, 187-192 (1968)
HOLROYD, R.A., GLASS, J.W.: Radicals formed by electron transfer to pyrimidine derivatives. Intern. J. Radiat. Biol. 14, 445-454 (1968)
HOLROYD, R.A., GLASS, J.W.: The reaction of hydrogen atoms with DNA. Radiat. Res. 39, 758-769 (1969)
HOOGSTEEN, K.: The structure of crystals containing a hydrogen-bonded complex of 1-methylthymine and 9-methyladenine. Acta Cryst. 12, 822-823 (1959)
HORAN, P.K., SNIPES, W.: Free radical destruction by gamma-irradiation at 77°K. Intern. J. Radiat. Biol. 15, 157-161 (1969)
HORAN, P.K., SNIPES, W.: Electron spin resonance of irradiated single crystals of orotic acid. Radiat. Res. 41, 24-38 (1970)
HRUSKA, F.E.: Mapping nucleoside conformation in aqueous solution. A correlation of some furanose structural parameters. In: Conformation of Biological Molecules and Polymers. pp. 345-360. Bergmann, E.D., Pullman, B. (eds.). Jerusalem: Israel Acad. Sci. Humanities, 1973
HRUSKA, F.E., DANYLUK, S.S.: An NMR study of the interaction of purines and pyrimidines derivatives with acridine orange. Biochim. Biophys. Acta 161, 250-252 (1968)
HÜTTERMANN, J.: Erzeugung von Ionenradikalen in bestrahlten Einkristallen von Adenin-Hydrochlorid und Adenosin-5'-Monophosphat. Z. Naturforsch. 24b, 1648-1649 (1969)
HÜTTERMANN, J.: Electron-spin-resosance spectroscopy of radiation-induced free radicals in irradiated single crystals of thymine monohydrate. Intern. J. Radiat. Biol. 17, 249-259 (1970)
HÜTTERMANN, J.: Electron spin resonance of free radicals, V-centers and electrons trapped in irradiated crystals of adenine hydrochlorides. J. Magn. Res. 17, 66-88 (1975)
HÜTTERMANN, J., BERNHARD, W.A., HAINDL, E., SCHMIDT, G.: ESR and ENDOR of an α-chloro radical in X-irradiated 5-chlorodeoxyuridine single crystals. Mol. Phys. 32, 1111-1121 (1976a)
HÜTTERMANN, J., BERNHARD, W.A., HAINDL, E., SCHMIDT, G.: Reactions of radiation-induced free radicals in solid halodeoxyuridines: single crystals of 5-chloro- and 5-bromodeoxyuridine. J. Phys. Chem. 81, 228-232 (1977a)
HÜTTERMANN, J., BERNHARD, W.A., HAINDL, E., SCHMIDT, G.: A C5'-located deoxyribose radical in 5-chloro- and 5-bromodeoxyuridine X-irradiated at low temperatures. Intern. J. Radiat. Biol. (1977b) In press
HÜTTERMANN, J., MÜLLER, A.: Free radicals in gamma-irradiated 2-deoxy-d-ribose: Electron spin resonance of single crystals and polycrystalline material. Radiat. Res. 38, 248-259 (1969a)

HÜTTERMANN, J., MÜLLER, A.: Free radicals in irradiated single crystals of 5-halo-uracil derivatives: 5-bromouracil, 5-bromodeoxyuridine and 5-iodo-deoxyuridine. Intern. J. Radiat. Biol. 15, 297-313 (1969b)
HÜTTERMANN, J., NEILSON, G.W., SYMONS, M.C.R.: Hyperfine coupling to iodine in an α-iodo radical. Mol. Phys. 32, 269 (1976b)
HÜTTERMANN, J., SCHMIDT, G., WEYMANN, D.: ESR and ENDOR of free radicals in X-irradiated barbituric acid. J. Magn. Res. 21, 221-234 (1976c)
HÜTTERMANN, J., WARD, J.F., MYERS, L.S., Jr.: Electron spin resonance of free radicals formed from orotic acid. J. Phys. Chem. 74, 4022-4029 (1970)
HÜTTERMANN, J., WARD, J.F., MYERS, L.S., Jr.: Electron spin resonance studies of free radicals in irradiated single crystals of 5-methylcytosine. Intern. J. Radiat. Phys. Chem. 3, 117-129 (1971)
JOHNSON, C.D., RYMER, T.B.: Existence of collective-excitation energy losses from an electron beam passing through biological materials. Nature (London) 213, 1045-1047 (1967)
JORDAN, D.O.: The Chemistry of the Nucleic Acids. London: Butterworths, 1960
JOUSSOT-DUBIEN, J., LESCLAUX, R.: Photoionisation des composés aromatiques dans l'acide borique. Mise en évidence d'un processus à deux photos. C.R. Acad. Sci. (Paris) 258, 4260-4262 (1964)
JOUSSOT-DUBIEN, J., LESCLAUX, R.: Photochemical effects in polar rigid media. Israel J. Chem. 8, 181-207 (1970)
KAALHUS, O., JOHANSEN, S.A.: Radiation induced radicals in thymine and its derivatives in low temperature glasses. Ann. N.Y. Acad. Sci. 222, 432-445 (1973)
KEVAN, L.: Radiation Chemistry of Frozen Polar Systems in Actions chimiques et biologiques des radiations. Haissinsky, M. (ed.), 13, 57-117 Paris: Masson et Cie., 1969
KIM, Y.-K.: Energy distribution of secondary electrons. Radiat. Res. 61, 21-35 (1975)
KISPERT, L.D., MYERS, F.: ESR study of the chlorofluoroacetamide radical in irradiated dichlorofluoroacetamide single crystals. J. Chem. Phys. 56, 2623-2631 (1972)
KISTENMACHER, T.J., SHIGEMATSU, T.: Structural chemistry of N(7)-substituted purines: 7-methyladenine dihydrochloride. Acta Cryst. B31, 211-217 (1975)
KÖHNLEIN, W., MÜLLER, A.: Quantitative ESR-measurements of radiation-induced radicals in nucleic-acid bases and pentoses. Intern. J. Radiat. Biol. 8, 141-155 (1964)
KRILOV, D., JERAK, J.N.: ESR evidence for the radiation-induced breakage of the sugar-phosphate bonds in nucleotides: single crystal of deoxycytidine 5'monophosphate. Biochim. Biophys. Acta 366, 396-401 (1974)
KRSMANOVIC-SIMIC, D., VAN DE VORST, A., RICHIR, M.: Influence de la dénaturation thermique sur le rendement radiolytique de l'acide désoxyribonucléique en phase solide. Intern. J. Radiat. Biol. 11, 171-178 (1966)
KUBOTA, Y., MIURA, M.: Photochemistry of the DNA-acridine dye complexes. Bull. Chem. Soc. Japan. 40, 2989 (1967)
KURUCSEV, T., ZDYSIEVICZ, J.R.: Simple methods for the orientation of DNA molecules in films suitable for optical studies. Biopolymers 10, 593-599 (1971)
LACROIX, M., DEPIREUX, J., VAN DE VORST, A.: On the nature of free radicals induced by X-rays in puric bases. Proc. Natl. Acad. Sci. 58, 399-401 (1967)
LACROIX, M., VAN DE VORST, A.: Sur le déterminisme de la radiorésistance des nucléotides en phase solide. C.R. Hebd. Séanc. Acad. Sci. (Paris) 259, 2027-2028 (1964)
LASSETTRE, E.N., WHITE, E.R.: Generalized oscillator strengths through the water vapor spectrum to 75 eV excitation energy; electron kinetic energy 500 eV. J. Chem. Phys. 60, 2460-2463 (1974)

LEDNEV, V.V., TICHONENKO, T.I.: X-ray diffraction study of highly polymerized DNA from bacteriophages C_D and T_2. Biofizika II, 209-212 (1966)
LENHERR, A.D., ORMEROD, M.G.: Formation of the thymine radical in irradiated DNA. Biochim. Biophys. Acta 166, 298-310 (1968)
LENHERR, A.D., ORMEROD, M.G.: Electron reactions in γ-irradiated thymidine and cystamine. Nature (London) 225, 546-547 (1970)
LERMAN, L.S.: Structural considerations in the interaction of DNA and acridines. J. Mol. Biol. 3, 18-30 (1961)
LEWIS, G.N., LIPKIN, D.: Reversible photochemical processes in rigid media: the dissociation of organic molecules into radicals and ions. J. Am. Chem. Soc. 64, 2801-2808 (1942)
LICHTER, J.J., GORDY, W.: Electron spin resonance of an irradiated single crystal of deoxyadenosine monohydrate. Proc. Natl. Acad. Sci. 60, 450-455 (1968)
LION, Y., SAUCIN, M., VAN DE VORST, A.: Triplet state population study in relation to the aggregation of proflavine in frozen solutions. Radiat. and Environm. Biophys. 12, 337-344 (1975)
LION, Y., VAN DE VORST, A.: Radicaux libres induits par réaction de dérivés puriques avec des électrons piégés. Intern. J. Radiat. Phys. Chem. 2, 51-58 (1970)
LION, Y., VAN DE VORST, A.: Formation de radicaux anioniques par irradiation X de bases puriques dans des verres alcalins. Intern. J. Radiat. Phys. Chem. 3, 521-525 (1971a)
LION, Y., VAN DE VORST, A.: On the mechanism of formation of free radicals induced in nucleic acid bases by ionizing radiation. Intern. J. Radiat. Phys. Chem. 3, 513-520 (1971b)
LION, Y., VAN DE VORST, A.: Free radicals induced by photolysis of frozen solutions of thymidine-5'-monophosphate: an ESR study. Photochem. Photobiol. 18, 169-173 (1973a)
LION, Y., VAN DE VORST, A.: Radicaux libres induite par radiolyse de solutions gelées de thymidine-5'-monophosphate. Intern. J. Radiat. Biol. 23, 279-284 (1973b)
LOCHMANN, E.R., MICHELER, A.: Binding of organic dyes to nucleic acids. In: Physico-Chemical Properties of Nucleic Acids, 1, pp. 223-267. Duchesne, J. (ed.). New York: Academic Press, 1973
LÜCKE-HUHLE, C., JUNG, H.: Energy requirements for damaging DNA molecules I. Energy transfer from metastable states of excited gases. Intern. J. Radiat. Biol. 24, 221-228 (1973)
LÜCKE-HUHLE, C., JUNG, H.: Energy requirements for damaging DNA molecules II. Energy transfer from metastable states of excited gases to double stranded Φχ 174 RF-DNA. Intern. J. Radiat. Biol. 25, 339-349 (1974)
MAENHAUT-MICHEL, G.: Mechanisms of protection of γ-irradiated bacteriophage λ by proflavine. Intern. J. Radiat. Biol. 27, 425-435 (1975)
MAKAROV, V.I., POLAK, L.S.: Excited and superexcited electronic states of atoms and molecules and their role in radiation chemistry. High Energy Chem. (USSR) 4, 1-18 (1970), translated from Khim. Vys. Energ. 4, 3-23 (1970)
MANTIONE, M.J., PULLMAN, B.: Sur le mécanisme de la photo-dimérisation de la thymine. Biochim. Biophys. Acta 91, 387-398 (1964)
MATHESON, M.S., DORFMAN, L.M.: Pulse Radiolysis. pp. 20-45. Cambridge (US)-London: M.I.T. Press, 1969
MELØ, T.B., BERGENE, R., HENRIKSEN, T.: Early effects of radiation. Radical pairs in a single crystal of barbituric acid. Mol. Phys. 25, 1377-1383 (1973)
MICHELSON, A.M.: The Chemistry of Nucleosides and Nucleotides. New York: Academic Press, 1963
MISHRA, S.P., NEILSON, G.W., SYMONS, M.C.R.: Unstable intermediates. 133. Electron spin resonance studies of 1-monochloro and 1,1-dichloro radicals

from irradiated alkyl chlorides and α-chlorocarbocylic acids. J.C.S. Faraday II, 1425-1437 (1973a)
MISHRA, S.P., NEILSON, G.W., SYMONS, M.C.R.: Unstable intermediates. 137. α-bromalkyl radicals in irradiated organic bromides: an electron spin resonance study. J.C.S. Faraday II, 1165-1174 (1973b)
MONTENAY-GARESTIER, T., HELENE, C.: Reflectance and luminescence studies of molecular complex formation between tryptophan and nucleic acid components in frozen aqueous solutions. Biochemistry 10, 300-305 (1971)
MOORTHY, P.N., RAO, K.N., MISHRA, K.P., SINGH, B.B.: New paramagnetic species formed in γ-irradiated and annealed frozen aqueous alkali hydroxide matrices. Radiochem. Radioanal. Lett. 10, 45-51 (1972)
MOZUMDER, A., MAGEE, J.L.: Model of tracks of ionizing radiations for radical reaction mechanisms. Radiat. Res. 28, 203-214 (1966)
MÜLLER, A.: The efficiency of radical production by X-rays in dry proteins and nucleic acids. Intern. J. Radiat. Biol. 5, 199-200 (1962)
MÜLLER, A.: Observations on X-ray-induced radicals in whole bacteriophage and phage nucleic acid. Intern. J. Radiat. Biol. 6, 137-142 (1963)
MÜLLER, A.: Quantitative ESR-measurements of radiation-induced radicals in nucleotides. Intern. J. Radiat. Biol. 8, 131-140 (1964)
MÜLLER, A.: Spektrographische Untersuchungen mittels paramagnetischer Elektronenresonanz über die Wirkung ionisierender Strahlen auf elementare biologische Objecte. In: Abhandl. Math.-Naturw. Kl., Akademie der Wissenschaften und der Literatur, Mainz: 1964a
MÜLLER, A.: The formation of radicals in nucleic acids, nucleoproteins and their constituents by ionizing radiations. Progr. Biophys. Mol. Biol. 17, 99-147 (1967)
MÜLLER, A., HÜTTERMANN, J.: Structure and formation of free radicals in irradiated single crystals of nucleic acid constituents. Ann. N.Y. Acad. Sci. 222, 411-431 (1973)
MÜLLER, A., KÖHNLEIN, W.: Quantitative ESR-measurements of radiation-induced radicals in nucleosides. Intern. J. Radiat. Biol. 8, 121-130 (1964)
MÜLLER, A., SCHAMBRA, P.E., PIETSCH, E.: Comparative ESR measurements of radical production in amino acids by ^{210}Po alpha- and ^{60}Co gamma-radiation. Intern. J. Radiat. Biol. 7, 587-599 (1963)
NATH, A., KLEIN, M.P., KÜNDIG, W., LICHTENSTEIN, D.: Mössbauer studies of after-effects of Auger ionization following electron capture in cobalt complexes. Radiat. Eff. 2, 211-224 (1970)
NAZHAT, N.B., WEISS, J.J.: Matrix isolation of thymine radicals in γ-irradiated alkaline ice matrixes. Trans. Faraday Soc. 66, 1302-1308 (1970)
NETA, P.: Electron spin resonance study of radicals produced in irradiated aqueous solutions of uracil and related pyrimidines. Radiat. Res. 49, 1-25 (1972)
NETA, P.: ESR radiation chemical studies of halouracils and nitrouracils. In: Fast Processes in Radiation Chemistry and Biology. pp. 235-240. Adams, G.E., Fielden, E.M., Michael, B. (eds.). New York: Wiley, 1973
NEUMÜLLER, W.: ESR- und ENDOR-Untersuchungen zur Struktur freier Radikale in bestrahlten Einkristallen von 5-Fluoruracil. Thesis, Univ. Regensburg, Germany (1976)
NICE, E.G., RORKE, D.: An E.S.R. study of an X-irradiated single crystal of 5-Iododeoxyuridine. Intern. J. Radiat. Biol. 15, 197-205 (1969a)
NICE, E.G., RORKE, D.: E.S.R. study of X-irradiated single crystals of 5-bromodeoxyuridine and 5-iododeoxyuridine. Intern. J. Radiat. Biol. 15, 207-213 (1969b)
OLAST, M., BERTINCHAMPS, A.: Electron transfer in gamma-irradiated complexes between aromatic amino acids and DNA derivatives. Intern. J. Radiat. Biol. 24, 589-594 (1973)
OLOFF, H.: Strahlen-induzierte freie Radikale in Einkristallen von 5-Chloruracil und 5-Chloruridin: ESR- und ENDOR-Untersuchungen. Thesis, Univ. Regensburg, Germany (1976)

OLOFF, H., HÜTTERMANN, J.: Free radicals in solid 5-halouracils: Single crystals of 5-chlorouracil and -uridine. J. Magn. Res. 27, 197-213 (1977)
OLSON, W.K.: The spatial configuration of ordered polynucleotide chains. I. Helix formation and base stacking. Biopolymers 15, 859-878 (1976)
ORE, A.: The role of multiple ionization in radiation action. In: Radiation Research, 1966, 54-67, Silini, G. (ed.). Amsterdam: North-Holland, 1967
ORE, A.: Size distribution of ion clusters along charged particle tracks in gases. Phys. Norvegica 5, 259-265 (1971)
ORE, A., LARSEN, A.: Relative frequences of ion clusters containing various numbers of ion pairs. Radiat. Res. 21, 331-338 (1964)
ORMEROD, M.G.: Free-radical formation in irradiated deoxyribonucleic acid. Intern. J. Radiat. Biol. 9, 291-300 (1965)
ORMEROD, M.G., ALEXANDER, P.: Repair of radiation damage in a nucleoprotein by cysteamine, Nature (London) 193, 290-291 (1962)
ORMEROD, M.G., ALEXANDER, P.: On the mechanism of radiation protection by cysteamine: an investigation by means of electron spin resonance. Radiat. Res. 18, 495-509 (1963)
PATTEN, R.A., GORDY, W.: Electron spin resonance investigations of radiation-induced free radicals in DNA and RNA at low temperatures: effects of water. Nature (London) 201, 361-363 (1964)
PEACOCKE, A.R.: The Interaction of acridines with nucleic acids. In: Acridines. The Chemistry of Heterocyclic Compounds 9, 723-757. Acheson, R.M. (ed.). New York: John Wiley, 1973
PEACOCKE, A., SKERETT, J.: The interaction of aminoacridines with nucleic acids. Trans. Faraday Soc. 52, 261-279 (1956)
PERLMAN, M.L., ROWE, E.M., WATSON, R.E.: Synchrotron radiation-light fantastic. Phys. Today 27, July 30-37 (1974)
PERSHAN, P.S., SHULMAN, R.G., WYLUDA, B.A., EISINGER, J.: Electron spin resonance of irradiated DNA. Physics 1, 163-182 (1964)
PERSHAN, P.S., SHULMAN, R.G., WYLUDA, B.J., EISINGER, J.: Electron spin resonance of irradiated DNA. Science 148, 378-380 (1965)
PERSON, J.C.: Isotope effects on molecular ionization yields. Radiat. Res. 52, 1-16 (1972)
PICONE, R.F., ROGERS, M.T.: ESR study of a bromine-containing radical in γ-irradiated bromomalonamide. J. Chem. Phys. 61, 4814-4819 (1974a)
PICONE, R.F., ROBERS, M.T.: An ESR study of the $\cdot CHICONH_2$ radical in γ-irradiated single crystals of iodoacetamide. J. Magn. Res. 14, 279-285 (1974b)
PIHL, A., SANNER, T.: Radiation sensitivity of cells and base composition of DNA: an electron spin resonance study. Radiat. Res. 28, 96-108 (1966)
PLATZMAN, R.L.: On the primary processes in radiation chemistry and biology. In: Symposium on Radiobiology. pp. 97-116. Nichson, J.J. (ed.). New York: Wiley; London: Chapman and Hall, 1952
PLATZMAN, R.L.: The physical and chemical basis of mechanisms in radiation biology. In: Radiation Biology and Medicine, pp. 15-72. Claus, W.D. (ed.). Reading (U.S.): Addison-Weseley, 1958
PLATZMAN, R.L.: Superexcited states of molecules, and the primary action of ionizing radiation. Vortex 23, 372-385 (1962a)
PLATZMAN, R.L.: Superexcited states of molecules. Radiat. Res. 17, 419-425 (1962b)
PLATZMAN, R.L.: Energy spectrum of primary activations in the action of ionizing radiation. In: Radiation Research 1966, 20-42. Silini, G. (ed.). Amsterdam: North-Holland, 1967
POPLE, J.A., BEVERIDGE, D.L.: Approximate Molecular Orbital Theory. New York: McGraw-Hill, 1970
PRITCHARD, N.J., BLAKE, A., PEACOCKE, A.R.: Modified intercalation model for the interaction of amino acridines and DNA. Nature (London) 212, 1360-1361 (1966)
PRUDEN, B., SNIPES, W., GORDY, W.: Electron spin resonance of an irradiated single crystal of thymidine. Proc. Natl. Acad. Sci. 53, 917-924 (1965)

PULLMANN, B., PULLMAN, A.: Quantum Biochemistry. New York: Wiley-Interscience, 1963
PULLMAN, B., PULLMAN, A.: Quantum-mechanical investigations of the electronic structure of nucleic acids and their constituents. Progr. Nucl. Acid. Res. Mol. Biol. 9, 327-402 (1969)
PULLMAN, A., PULLMAN, B., BERTHIER, G.: Electronic structure of some active compounds in chemotherapy of cancer. C.R. Acad. Sci. (Paris) 243, 380-383 (1956)
RAJALAKSHMI, K.V., VENKATESWARLU, K., VAN DE VORST, A.: Les liasons hydrogêne et la radiorésistance des polynucléotides en phase solide. C.R. Hebd. Séans. Acad. Sci. (Paris) 261, 4879-4881 (1965)
REISS, K.W., GORDY, W.: Electron spin resonance of an irradiated single crystal of 5-chlorouridine. Proc. Natl. Acad. Sci. 68, 2008-2010 (1971)
RHODES, W.: Hypochromism and other spectral properties of helical polynucleotides. J. Am. Chem. Soc. 83, 3609-3617 (1961)
RIESZ, P., SMITHERMAN, T.C., SCHER, C.D.: Hydrogen transfer from exchangeable to carbon-bound sites in γ-irradiated proteins and nucleic acids: The mechanism of radical saturation. Intern. J. Radiat. Biol. 17, 389-393 (1970)
ROTBLAT, J., SIMMONS, J.A.: Dose-response relationship in the yield of radiation-induced free radicals in amino acids. Phys. Med. Biol. 7, 489-497 (1963)
RUPPRECHT, A.: Preparation of oriented DNA in large amounts. Biochem. Biophys. Res. Commun. 12, 163-168 (1963)
RUPPRECHT, A.: Preparation of oriented DNA by wet spinning. Acta Chem. Scand. 20, 494-504 (1966)
RUPPRECHT, A.: A wet spinning apparatus and auxiliary equipment suitable for preparing samples of oriented DNA. Biotechnol. Bioeng. 12, 93-121 (1970a)
RUPPRECHT, A.: A wet spinning method for preparing highly oriented DNA and some physico-chemical studies on the material (Diss.). Roy. Inst. Technol., Stockholm (1970b)
RUPPRECHT, A.: Mechanochemical study of wet spun lithium-DNA fibers. Biopolymers 9, 825-842 (1970c)
RUPPRECHT, A., FORSLIND, B.: Variation of electrolyte content in wet-spun lithium- and sodium-DNA. Biochim. Biophys. Acta 204, 304-316 (1970)
RUPPRECHT, A., RIGLER, R., FORSLIND, B., SWANBECK, G.: Optical properties of oriented dye-LiDNA complexes prepared with the wet spinning method. Europ. J. Biochem. 10, 291-301 (1969)
RUSTGI, S.N., BOX, H.C.: Free radicals in irradiated methyl cytosine. J. Chem. Phys. 60, 3343-3344 (1974)
SALOVEY, R., SHULMAN, R.G., WALSH, W.M.: Electron spin resonance of irradiated DNA. J. Chem. Phys. 39, 839-840 (1963)
SANNER, T.: Transfer of radiation energy to solute molecules in irradiated frozen aqueous solutions. Radiat. Res. 25, 586-600 (1965)
SANTAR, I.: Current state of a priori yield predictions. In: Progress and Problems in Contemporary Radiation Chemistry, Proceedings of the Tenth Czechoslovak Annual Meeting on Radiation Chemistry, Vol. I, 51-90. Teply, J. (ed.). Prague: 1971
SANTAR, I., BEDNAR, J.: Theory of radiation chemical yield - V. Initial structure of the track of a fast electron in a dense medium. Intern. J. Radiat. Phys. Chem. 1, 133-152 (1969)
SCHMIDT, J.: An ESR analysis of a heat-stable radical in γ-irradiated single crystals of 1-methylthymine. J. Chem. Phys. 62, 370-375 (1975)
SCHMIDT, J., BORG, D.C.: Identification of radiation-induced hydrogen-addition radicals in adenine derivatives. Radiat. Res. 46, 36-51 (1971)
SCHMIDT, J., SNIPES, W.: Free radical formation in a gamma-irradiated pyrimidine-purine co-crystal complex. Intern. J. Radiat. Biol. 13, 101-109 (1967)
SCHMIDT, J., SNIPES, W.: Electron spin resonance of gamma-irradiated single crystals of 9-methyladenine. Radiat. Res. 38, 274-284 (1969)

SCHMIDT, H., ZELLHOFER, R.: ESR triplet spectra of acridine dyes and their aggregates. Z. Physik. Chem. Neue Folge 91, 204-218 (1974)

SCHULTE-FROHLINDE, D., EIBEN, K.: Solvatisierte Electronen in eingefrorenen Lösungen. Z. Naturforsch. 17a, 445-446 (1962)

SCHULTE-FROHLINDE, D., EIBEN, K.: O-Radikalionen bei der Radiolyse von alkalischem Eis. Z. Naturforsch. 18a, 99-101 (1963)

SEEDS, W.E.: Polarised ultraviolet microspectrography and molecular structure. Progr. Biophys. 3, 27-46 (1953)

SEEMAN, N.C., DAY, R.O., RICH, A.: Nucleic acid-mutagen interactions: Crystal structure of adenylyl-3',5'-uridine plus 9-aminoacridine. Nature (London) 253, 324-326 (1975)

SEEMAN, N.C., ROSENBERG, J.M., SUDDATH, F.L., KIM, J.J.P, RICH, A.: RNA double-helical fragments of atomic resolution: I. The crystal and molecular structure of sodium adenylyl-3',5'uridine hexahydrate. J. Mol. Biol. 104, 109-144 (1976)

SEVILLA, M.D.: An electron spin resonance study of the photoionization of thymine. The thymine cation and anion radicals. J. Phys. Chem. 75, 626-631 (1971)

SEVILLA, M.D., FAILOR, R., ZORMAN, G.: Radicals formed after electron attachment to 5-halouracils in aqueous glasses. J. Phys. Chem. 78, 696-699 (1974)

SEVILLA, M.D., MOHAN, P.A.: The purine DNA base cation and anion radicals: an ESR study. Intern. J. Radiat. Biol. 25, 635-638 (1974)

SEVILLA, M.D., VAN PAEMEL, C.: An ESR study of pyrimidine, pyrimidine nucleoside and halopyrimidine anions. Photochem. Photobiol. 15, 407-409 (1972)

SEVILLA, M.D., VAN PAEMEL, C., NICHOLS, C.: An electron spin resonance study of several DNA base cation radicals produced by photoionization. J. Phys. Chem. 76, 3571-3577 (1972b)

SEVILLA, M.D., VAN PAEMEL, C., ZORMAN, G.: Reactions of the cation and anion radicals of several DNA bases. J. Phys. Chem. 76, 3577-3582 (1972a)

SHELTON, K.R., CLARK, Jr., J.M.: In vitro tritium labelling of DNA. Biophys. Res. Commun. 33, 850-854 (1968)

SHEN, P.G., BLYUMENFELD, L.A., KALMANSON, A.E., PASYNKII, A.G.: Spektry electronnogo paramagnitnogo resonansa biologicheskikh obektov. III Deistvie ioniziruyushchego izlucheniya na nucleinovye soedineniya. Biofizika 4, 263-274 (1959)

SHIELDS, H., GORDY, W.: Electron-spin resonance studies of radiation damage to nucleic acids and their constituents. Proc. Natl. Acad. Sci. 45, 269-281 (1959)

SIEGEL, S., FLOURNOY, J.M., BAUM, L.H.: Irradiation yields of radicals in gamma-irradiated ice at 4.2^{o} and 77°K. J. Chem. Phys. 34, 1782-1788 (1961)

SINANOGLU, O.: Solvents effects on molecular associations. In: Molecular Associations in Biology. pp. 427-444. Pullman, B. (ed.). New York: Academic Press, 1968

SNART, R.S.: Photoelectric effects of deoxyribonucleic acid. Biopolymers 6, 293-297 (1968)

SNIPES, W., HENRIKSEN, T.: Radiation-induced radicals in thymidine single crystals at 77 K. Intern. J. Radiat. Biol. 17, 367-373 (1970)

SNIPES, W., HORAN, P.K.: Electron spin resonance studies of free radical turnover in gamma-irradiated single crystals of alanine. Radiat. Res. 30, 307-315 (1967)

SNIPES, W., SCHMIDT, J.: Conformation of the hydrogen addition radical in 1-methyl thymine. J. Chem. Phys. 49, 1443-1445 (1968)

SNYDER, W.S., NEUFELD, J.: On the passage of heavy particles through tissue. Radiat. Res. 6, 67-78 (1957)

SOBELL, H.M.: Hydrogen-bonding studies of base-pairing in solution and in the crystalline state. In: The Purines. Theory and Experiment. pp. 124-186. Bergman, E.D., Pullman, B. (eds.). Jerusalem: The Israel Acad. Sci. Humanities, 1972

SPENCER, L.V., FANO, U.: Energy spectrum resulting from electron slowing down. Phys. Rev. 93, 1172-1181 (1954)
SPITKOVSKII, D.M.: Method for obtaining structure in nucleoproteins and for determining their rheological properties. Biokhimiya 20, 566-570 (1955)
SRINIVASAN, V.T., SINGH, B.B., GOPAL-AYANGAR, A.R.: Formation of thymyl radicals in gamma-irradiated frozen aqueous systems: an ESR study. Intern. J. Radiat. Biol. 15, 89-93 (1969)
SRINIVASAN, V.T., SINGH, B.B., GOPAL-AYENGAR, A.R.: Possibility of anion formation in irradiated frozen solutions of nucleic acid bases. Intern. J. Radiat. Biol. 17, 577-585 (1970)
SUNDARALINGAM, M.: Stereochemistry of nucleic acids and their constituents. IV. Allowed and preferred conformations of nucleosides, nucleoside mono-, tri-, tetraphosphates, nucleic acids and polynucleotides. Biopolymers 7, 821-860 (1969)
SUNDARALINGAM, M.: The concenpt of a conformationally "rigid" nucleotide. In: Conformation of Biological Molecules and Polymers. pp. 417-455. Bergman, E.D., Pullman, B. (eds.). Jerusalem: The Israel Acad. Sci. Humanities, 1973
SUNDARALINGAM, M.: Principles governing nucleic acid and polynucleotide conformations. In: Structure and Conformation of Nucleic Acids and Protein-Nucleic Acid Interactions. pp. 487-524. Sundaralingam, M., Rao, S.T. (eds.). Baltimore: Univ. Park Press, 1975
SWALLOW, A.J.: Radiation Chemistry. p. 64. London: Longmans, 1973
TOMITA, G.: Molecular complexes of acridine orange and nucleosides. Biophysik 4, 118-128 (1967)
TRUS, B.L., MARSH, R.E.: The crystal structure of 1-methylcytosine hydrochloride. Acta Cryst. B28, 1834-1840 (1972)
TSAI, C.C., JAIN, S.C., SOBELL, H.M.: X-ray crystallographic visualization of drug-nucleic acid intercalative binding: Structure of an ethidium-dinucleoside monophosphate crystalline complex, ethidium: 5 iodouridylyl (3'-5') adenosine. Proc. Natl. Acad. Sci. 72, 628-632 (1975)
TS'O, P.O.P.: Basic Principles in Nucleic Acid Chemistry 1, 537-562. New-York: Academic Press, 1974
VOE, H. DE, TINOCO, I., Jr.: The stability of helical polynucleotides: base contributions. J. Mol. Biol. 4, 500-517 (1962)
VOET, D., RICH, A.: The crystal structure of purines, pyrimidines and their intermolecular complexes. Progr. Nucl. Acid Res. Mol. Biol. 10, 183-265 (1970)
VOLTZ, R.: Electronic energy transfer in irradiated aromatic materials. Radiat. Res. Rev. 1, 301-360 (1968)
VOROBIEV, V.I., SHAPOT, V.S.: Concerning protein derivatives of high polymers serving as building blocks in protoplasmic structures. Biokhimiya 18, 603-617 (1953)
VORST, A. VAN DE: Stabilité des radicaux libres induits par le rayonnement gamme du ^{60}Co dans l'acide désoxyribonucléique. Intern. J. Radiat. Biol. 8, 111-120 (1964)
VORST, A. VAN DE: Radicaux d'addition et réactions radicalaires dans l'ADN en phase solide. Intern. J. Radiat. Biol. 12, 153-161 (1967)
VORST, A. VAN DE: Effets des rayonnements ionisants sur la matière en phase solide. Mém. Acad. Roy. Belg. Cl. Sci. 39, 1-136 (1970)
VORST, A. VAN DE: Effets indirects dans les solutions aqueeuses de thymidine-5'-monophosphate irradieés par rayons X à 77°K. Intern. J. Radiat. Phys. Chem. 6, 143-155 (1974)
VORST, A. VAN DE, KRSMANOVIC-SIMIC, D.: Sur la localisation des radicaux libres induits par le rayonnement ionisant dans l'acide désoxyribonucléique en phase solide. C.R. Hebd. Séanc. Acad. Sci. (Paris) 262, 2288-2290 (1966)

VORST, A. VAN DE, LION, Y.: Formation de radicaux libres dans les constituants du DNA photosensibilisés par la proflavine. Biochim. Biophys. Acta 238, 417-428 (1971a)
VORST, A. VAN DE, LION, Y.: Mécanisms de la photosensibilisation des constituants du DNA par la proflavine: une étude par résonance parmagnétique électronique. Biochim. Biophys. Acta 246, 421-429 (1971b)
VORST, A. VAN DE, LION, Y.: Radicaux libres caractéristiques de la thymidine-5'-monophosphate photosensibilisée par la proflavine. Biochim. Biophys. Acta 294, 349-355 (1973)
VORST, A. VAN DE, LION, Y.: Electron transfer in photosensitization of biomolecules by acridine dyes. Photochem. Photobiol. 21, 85-89 (1975)
VORST, A. VAN DE, LION, Y., SAUCIN, M.: Photosensibilisation des constituants des acides nucléiques par la proflavine: mécanisme de formation de radicaux d'addition d'hydrogène en solutions aqueuses gelées. Biochim. Biophys. Acta 430, 467-477 (1976)
VORST, A. VAN DE, VAN DER KAA, J.M., DEPIREUX, J., DUCHESNE, J., BERTINCHAMPS, A.J.: Applications de la résonance de spin électronique à l'étude des radicaux libres créés par irradiation de l'acide désoxyribonucléique. Bull. Ampère 9th year, special edition 297-300 (1960)
VORST, A. VAN DE, VILLÉE, F.: Sur la nature des radicaux libres induits par le raxonnement X dans l'acide désoxyribonucléique a l'état solide. Bull. Ampère, 9th year, special edition 259, 928-931 (1964)
VORST, A. VAN DE, VILLÉE, F., DUCHESNE, J.: Influence du poids moléculaire sur le rendement radiolytique de l'acide désoxyribonucléique en phase solide. Intern. J. Radiat. Biol. 9, 269-274 (1965a)
VORST, A. VAN DE, RICHIR, R.: Sur l'identification des radicaux libres induits par le raxonnement X dans l'acide désoxyribonucléique en phase solide. Intern J. Radiat. Biol. 260, 6458-6461 (1965)
VORST, A. VAN DE, RICHIR, M., KRSMANOVIC-SIMIC, D.: Sur la radiorésistance de la nucléohistone. C.R. Hebd. Séanc. Acad. Sci. (Paris) 261, 5682-5685 (1965b)
VORST, A. VAN DE, RICHIR, M., KRSMANOVIC-SIMIC, D.: Sur l'accroissement de la radiosensibilité de l'acide désoxyribonucléique privé de bases puriques. C.R. Hebd. Séanc. Acad. Sci. (Paris) 262, 401-402 (1966)
WALSH, W.M., SHULMAN, R.G., HEIDENREICH, R.D.: Ferromagnetic in nucleic acid samples. Nature (London) 192, 1041-1043 (1961)
WATSON, J.D., CRICK, F.H.C.: Molecular structure of nucleic acids. Nature (London) 171, 737-738 (1953)
WEIBEZAHN, K.F., DERTINGER, H.: E.p.R. of free radicals formed in DNA and its constituents after exposure to discharge-excited inert grases. I. Thymine and thymidine. Intern. J. Radiat. Biol. 23, 271-277 (1973a)
WEIBEZAHN, K.F., DERTINGER, H.: E.p.r. of free radicals formed in DNA and its constituents after exposure to discharge-excited inert grases. II. Purines and DNA. Intern. J. Radiat. Biol. 23, 447-455 (1973b)
WESTHOF, E.: Formation of H adduct radicals in cytosine and its derivatives. Intern. J. Radiat. Biol. 23, 389-400 (1973)
WESTHOF, E., FLOSSMANN, W.: INDO analysis of the imidazole anion and hydrogen addition radicals. J. Am. Chem. Soc. 97, 6622-6623 (1975)
WESTHOF, E., FLOSSMANN, W., MÜLLER, A.: Radical formation in salts of pyrimidines: II. cytosine·HCl crystals. Intern. J. Radiat. Biol. 28, 427-438 (1975)
WESTHOF, E., FLOSSMANN, W., ZEHNER, H., MÜLLER, A.: Formation of H-addition radicals in some pyrimidine and purine derivatives: Radiation Effects in Liquids and Solids. Faraday Gen. Disc. No. 63 (1977)
WESTHOF, W., HÜTTERMANN, J.: ESR of irradiated single crystals of cytosine. Intern. J. Radiat. Biol. 24, 627-630 (1973)
WESTHOF, E., VAN ROOTEN, M.: Z. Naturforsch. 31c, 371 (1976)

WESTHOF, E., WEIBEZAHN, K.F., DERTINGER, H.: INDO MO calculations of the hydrogen abstraction radical formed by bombardment of thymine and derivatives with excited inert gases. Z. Naturforsch. 29c, 303 (1974)
WILKINS, M.H.F., ARNOTT, S.: Personal communication to A. Rupprecht (1966)
WILKINS, M.H.F., GOSLING, R.G., SEEDS, W.E.: Nucleic acid: an extensible molecule? Nature (London) 167, 759-760 (1951)
WILLSON, R.L., WARDMAN, P., ASMUS, K.D.: Interaction of dGMP radical with cysteamine and promethazine as possible model of DNA repair. Nature (London) 252, 323-324 (1974)
WYARD, S.J., COOK, J.B.: Solid State Biophysics. New York-St. Louis-San Francisco-Toronto-London-Sydney: McGraw-Hill, 1969
WYARD, S.J., ELLIOTT, J.: ESR studies of radiation damage in nucleic acids, bases, nucleosides and nucleotides. Ann. N.Y. Acad. Sci. 222, 628-639 (1973)
WYCKOFF, H.W.: X-ray diffraction analysis of the structure of deoxyribonucleic acid (Diss.). MA Inst. Technol. (1955)
YATHINDRA, N., SUNDARALINGAM, M.: The conformational role of the phosphate groups in nucleic acids. Correlation between the sugar pucker and the internucleotide P-O bond conformations. In: Structure and Conformation of Nucleic Acids and Protein-Nucleic Acid Interactions. pp. 649-676. Sundaralingam, M., Rao, S.T. (eds.). Baltimore: Univ. Park Press, 1975
ZADOR, E., WARMAN, J.M., HUMMEL, A.: Ionization in the track of a high-energy electron in hydrocarbon liquids; study of positive and negative ion scavenging using nanosecond pulse radiolysis. J. Chem. Phys. 62, 3897-3906 (1975)
ZEHNER, H., FLOSSMANN, W., WESTHOF, E.: Formation of H-addition radicals in adenine derivatives. Z. Naturforsch. 31c, 225-231 (1976a)
ZEHNER, H., FLOSSMANN, W., WESTHOF, E., MÜLLER, A.: Electron spin resonance of irradiated single crystals of uracil. Mol. Phys. 32, 869-878 (1976b)
ZEHNER, H., WESTHOF, E., FLOSSMANN, W., MÜLLER, A.: Formation of H-addition radicals in adenine derivatives: Part II. Z. Naturforsch. 32c, 1-10 (1977)
ZIMMER, K.G., MÜLLER, A.: New light on radiation biology from electron spin resonance studies. Current Top. Radiat. Res. 1, 1-47 (1965)

Section II
Chemical Aspects

Chapter 1

Primary Events in the Radiolysis of Aqueous Solutions of Nucleic Acids and Related Substances

1.1 Reactive Species in Irradiated Aqueous Systems

When water absorbs ionizing radiation, it decomposes according to the overall equation:

$$H_2O \rightsquigarrow OH,\ H,\ e^-_{aq},\ H_2,\ H_2O_2,\ H_3O^+ \tag{1.1}$$

Hydroxyl radicals, hydrogen atoms and hydrated electrons (e^-_{aq}) are the short-lived species which can react chemically with solutes when these are present. Hydrogen and hydrogen peroxide are the so-called molecular products of water radiolysis and are assumed to arise, at least partly, from recombination processes in local regions of high concentrations of the primary reactive species ("spurs"). The production of the hydronium ion, H_3O^+, is a direct consequence of ionization of the solvent molecule giving H_2O^+ which then undergoes the rapid ion-molecule reaction:

$$H_2O^+ + H_2O \longrightarrow H_3O^+ + OH \tag{1.2}$$

On charge conservation grounds the yield of H_3O^+ must equal that of e^-_{aq}. It is generally held that the formation of both OH and e^-_{aq} is completed within 10^{-11} s; molecular product formation from recombination is completed within 10^{-8} s.

The yields of the products of Equation (1.1) have been measured for many differing radiations (see I.2.11).

In neutral water irradiated with γ-rays or with electrons of energy ∿ 1 MeV the yields of the primary products are: G_{OH} = 2.7, G_H = 0.55, $G_{e^-_{aq}}$ = 2.7, $G_{H_3O^+}$ = 2.7, G_{H_2} = 0.45, $G_{H_2O_2}$ = 0.7. The measured yields are pH-dependent, due to secondary processes involving H_3O^+ and OH^-, and below pH ∿ 3 and above pH ∿ 12 there are slight increases in the extent of water decomposition. The effect of radiation quality on the primary yields is such that the relative extent of molecular product formation compared to the free reactive species increases with increasing density of ionization (the so-called linear energy transfer, LET).

Ionization and excitation of the components of an irradiated medium depend approximately upon the electron density of the components. In dilute aqueous solutions, therefore, essentially all of the energy is absorbed by the water. Hence an important aspect of the radiation chemistry of these systems is the elucidation of the mechanisms of interaction of the oxidising species, OH, and of the reducing species, e^-_{aq} and H, with the added solutes. Some solutes in particular can lead to interconversion of the primary species. Thus, addition of acid converts e^-_{aq} to hydrogen atoms;

$$e^-_{aq} + H_3O^+ \longrightarrow H + H_2O \qquad (1.3)$$

whilst addition of alkali leads to the processes:

$$H + OH^- \rightleftharpoons e^-_{aq} + H_2O, \; pK = 9.7 \qquad (1.4)$$

$$OH + OH^- \rightleftharpoons O^- + H_2O, \; pK = 11.9 \qquad (1.5)$$

In irradiated aqueous solutions containing nitrous oxide, the hydrated electrons are converted to OH radicals:

$$N_2O + e^-_{aq} \xrightarrow{H_2O} N_2 + OH + OH^- \qquad (1.6)$$

Under these conditions the initial radical concentration ratio $[OH]_o/[H]_o$ is ∿ 10 so that it is very nearly a one-radical system. Furthermore, since N_2O does not react with either OH radicals or organic free radicals, N_2O-saturated solutions have been extensively used for studies of reactions of OH radicals with additional solutes.

In solutions containing oxygen, the perhydroxy radical and its ion are formed by reaction with the two reducing species,

$$H + O_2 \longrightarrow HO_2 \qquad (1.7)$$

$$e^-_{aq} + O_2 \longrightarrow O_2^- \qquad (1.8)$$

$$HO_2 \rightleftharpoons O_2^- + H^+ \quad pK = 4.9 \qquad (1.9)$$

and, under alkaline conditions, the basic form of the OH radical can add to oxygen to form the ozonide radical ion:

$$O^- + O_2 \rightleftharpoons O_3^- \qquad (1.10)$$

Additional solutes, such as nucleic acids and related substances, can compete with the above reactions of e^-_{aq}, H, OH and O^- and the extent of such competition can be readily deduced for any particular solution from a knowledge of the specific reaction rate constants.

In general, OH, O^- and H are considered to react with organic solutes either by an addition process (at an unsaturated centre) or by a hydrogen abstraction reaction. The hydrated electron can be regarded as a reactive nucleophile and with organic acceptor molecules produces radical anions which may subsequently protonate. On the other hand, HO_2 and O_2^- are unreactive to most organic compounds and their fate is either disproportionation (to form hydrogen peroxide) or secondary reactions with the organic free radicals. It should be noted that organic free radicals can also react with oxygen to form organic peroxy-radicals:

$$RH + OH \longrightarrow R\cdot + H_2O \xrightarrow{O_2} RO_2\cdot \qquad (1.11)$$

and this can result in specific radiolysis products, such as organic hydroperoxides (RO_2H).

For general aspects of the radiolysis of aqueous systems, readers should consult the books of DRAGANIC and DRAGANIC (1971) and SWALLOW (1973).

1.2 Rates of Reaction with the Primary Species

Values for the reaction rate constants for attack by OH, e^-_{aq} and H have been determined and have yielded information about the relative efficiencies of attack by the primary species, the effects of the ionic states of the solutes and also the influence of the conformation of the polynucleotides upon their reactivities. Details of the methodology have already been reviewed (DORFMAN and ADAMS, 1973; SCHOLES, 1976) and only a very brief resume is necessary here. The technique of pulse radiolysis is of importance in this particular area, since absolute rate constants can be directly obtained. Various types of electron generator have been used to deliver short pulses of high energy electrons (most commonly in the range 10^{-9}-10^{-6} s) and the reaction under investigation is followed either by monitoring decay of the primary reactive species or the formation of the transient product of the reaction. Physical properties monitored can be optical absorption, electron spin resonance absorption, polarographic behaviour, and, if the species is charged, electrical conductivity. Of these four, optical absorption has been used most extensively particularly for reaction with the hydrated electron which has a strong optical absorption around 720 nm. Although some direct measurements of the rates of reaction of OH radicals have been carried out by following the transient builtup, most of the data for OH, as well as for H reactivities towards nucleic acids and related substances, have been obtained by competitive methods.

Table 1.1 gives some selected values for the absolute rate constants of reaction of the primary species with the free bases, nucleosides and nucleotides, when these are irradiated in neutral aqueous solution. It is clear that the rates are high, approaching diffusion-controlled in the case of the reaction of e^-_{aq} with the free bases and nucleosides. The relative rates of reaction of OH and H are as expected by comparison with other chemically similar solutes. The reactivities of the primary species with the purine and pyrimidine bases and with their corresponding nucleosides are in the order e^-_{aq} > OH > H; with the nucleotides, however, the presence of the negatively charged phosphate group lowers the rate constant for reaction with e^-_{aq} and thus the difference between $k_{(e^-_{aq} + \text{nucleotide})}$ and $k_{(OH + \text{nucleoside})}$ is narrowed. The rate constant data of Table 1.1 predict that the site of attack of e^-_{aq} is exclusively on the base components of the various monomeric compounds and that H atoms also prefer to attack the heterocyclic bases rather than the sugar moiety. There is rather little variation in the reactivities of the bases and their nucleosides and nucleotides towards OH radicals; the lower value with sugar is consistent with the conclusion from stationary radiolysis, that only about 25% of the oxidizing species react

Table 1.1. Rate constants for the reaction of the primary species with polynucleotides and their components in neutral solution

Solute	k (× 10^9 M^{-1} s^{-1})					
	e^-_{aq}	Ref	OH	Ref	H	Ref
Bases						
Thy	18	A	5.6	A	0.38	F
Cyt	13.2	B	4.7	A	-	-
Ura	15	C	5.2	A	-	-
Ade	9	B	4.4	A	0.14	F
Pentose	<0.01[a]	D	1.6[a]	E	0.02[c]	F
			1.3[b]	E		
Nucleosides						
Thyd	-	-	4.8	A	0.38	F
Cytd	13.2	B	4.8	A	-	-
Urd	15	C	4.3	C	-	-
Ado	9.2	B	4.0	A	0.23	F
Nucleotides						
TMP	1.5	-	5.3	A	0.42	F
CMP	6.8	-	4.5	A	-	-
dCMP	-	-	4.4	A	-	-
AMP	3.8	A	3.0	A	0.25	F
dAMP	-	-	3.5	A	-	-
dGMP	-	-	5.0	G	-	-
Polynucleotides[d]						
Poly C	-	-	1.2	E	-	-
Poly U	0.75	B	1.25	E	-	-
Poly A	0.25	B	0.9	E	-	-
Poly (U + A)	0.13	B	0.5	E	-	-
DNA	0.14	B	0.4	E	0.08	F
	0.06	A	0.3	A		
			0.6	C		
			0.8	H		

[a] Ribose; [b] Ribose-5-phosphate; [c] Deoxyribose; [d] Calculated rate constant per nucleotide base.

A: SCHOLES et al. (1965); B: SHRAGGE et al. (1971); C: GREENSTOCK et al. (1968, 1969); D: HART et al. (1964); E: MICHAELS and HUNT (1973); F: SCHOLES and SIMIC (1964, 1968); G: WILLSON et al. (1974); H: SCHOLES et al. (1969)

with the pentose moiety (SCHOLES et al., 1960). With regard to the single-stranded polynucleotides (poly U and poly A) and the double-stranded polynucleotides [poly (U + A) and DNA] the reaction rate constants recorded in Table 1.1 have been calculated in terms of the appropriate nucleotide concentrations. All of the known values for DNA are quoted in view of possible sample variation. It can be seen that there is a lowered reactivity of the nucleotide unit in the bound state and that the difference is largest with the solvated electron - the order of reactivity of the polynucleotides now being, OH > e^-_{aq} > H. These observations have been discussed in terms of the collision frequency between a small spherical molecule (e.g., OH, e^-_{aq}) and a rod-like polymer (BRAAMS and EBERT, 1968; SHRAGGE et al., 1971; MICHAELS and HUNT, 1973). In this particular situation, the collision frequency per polymer molecule, γ, is given by the equation:

$$\gamma = \frac{4\,\pi\,ND}{10^3}\;\frac{L}{2\,\ln(L/r)}\quad M^{-1}\,s^{-1}$$

where L is the length of the rod, r is the axial radius of the rod, D is the combined diffusion coefficient and N the Avogadro number. For the reaction of OH with a straight rod polynucleotide of molecular weight > 10^5 [the molecular weights of poly U, poly A and poly (U + A) are taken as 2×10^5, 2×10^5 and 5×10^5 respectively]. MICHAELS and HUNDT (1973) calculated that the collision frequency per base is ∿ $1.2 \times 10^9\ M^{-1}\ s^{-1}$ for a single-stranded polymer and ∿ $0.6 \times 10^9\ M^{-1}\ s^{-1}$ for a double-stranded polymer. Since these values agreed well with the experimentally determined rate constants, per base, for OH radical attack it was concluded that the reactivity is not particularly influenced by the conformation of the polymers, i.e., random coil (poly U), base stacking (poly A) and double-stranded helix with hydrogen bonding [poly (U + A) and DNA]. The lowered values of the rate constants of OH attack are, therefore, attributed to changes in the collision frequency only. On the other hand, application of similar considerations to the reaction with e^-_{aq} led SHRAGGE et al. (1971) to conclude that, in the double-stranded polynucleotides, there is some shielding of base attack.

Determinations of values of the rate constants at different pH values have shown that various ionic states of the solutes can have an important influence on their reactivities; this is particularly evident, as noted above, in the case of reaction of e^-_{aq} where the rate constant falls off as molecules become more negatively charged. Ionic dissociation in both base and sugar moieties can also give rise to such effects and the titration curves are fairly accurately followed (GREENSTOCK et al., 1968; SHRAGGE et al., 1971). Examples of this are the dissociation in the imidazole ring of adenine (pK = 9.8) and the ribose dissociation in adenosine and cytidine (pK ∿ 12). In the case of OH radical attack, it has been found that cytosine and adenine and their derivatives are less reactive in acid solution, presumably as a result of the deactivating effect of the protonated amino group (SCHOLES et al., 1965). With the pyrimidines there is a significant drop in reactivity above pH 12 (MYERS et al., 1968; GREENSTOCK et al., 1969) which may well be associated with the ionization of the OH radical to O^- [Eq. (1.5)] rather than ionization of the pyrimidine molecule.

1.3 Intermediates in the Radiolysis of Aqueous Solutions of Nucleic Acids and Their Components

This section discusses available information on the nature and subsequent reactions of the organic intermediates (radicals and radical ions) formed as a result of attack by OH, H and e^-_{aq}. It should perhaps be stressed that interpretation of the physical data obtained by fast-reaction techniques e.g., optical absorption spectra of transients, is often very dependent upon a knowledge of the nature of the permanent radiolysis products; thus, the information from pulse and stationary techniques is often complementary.

1.3.1 Reaction with Hydroxyl Radicals

1.3.1.1 Pyrimidines

Detailed studies of the products of radiolysis of aqueous pyrimidine solutions clearly indicate that the C5-C6 bond is the major site of attack by the oxidizing primary species. Thus, in the case of uracil, for example, the two initial OH adducts should have the structures:

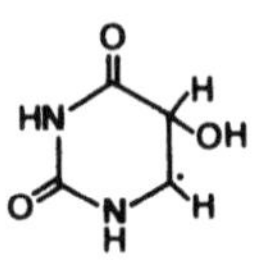

Formula (1)

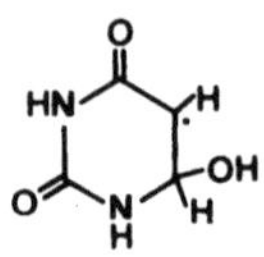

(2)

NETA (1972a), using an in situ steady-state ESR method (in which samples are irradiated inside the cavity of an ESR spectrometer), has established the structures of some of the pyrimidine OH adducts. Solutions were saturated with N_2O to convert e^-_{aq} to OH [EQ. (1.6)]. The irradiations were carried out under alkaline conditions, where the radicals are present in the negatively charged basic forms, thus allowing a higher steady-state concentration since radical-radical reaction rates are slower because of electrostatic repulsion. Addition of OH to both C5 and C6 of uracil was found to take place. It was suggested that the C5 adduct could have Formulas (3) and (4) where enolization between positions 4 and 5 has occurred.

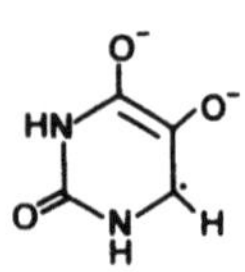

pH = 9.2

Formula (3)

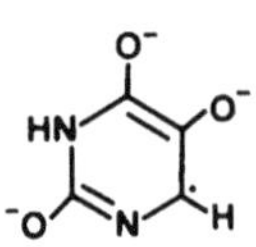

pH = 11.8-13.7

(4)

The C6 adduct in alkaline solution (pH 10.8-12.4) is unstable and undergoes ring fission across the N1-C6 position to yield the radical $H_2NCONHCO\dot{C}HCHO$; a similar ring fission occurs after OH addition to thymine, 6-methyl uracil and iso-orotic acid. It should be noted that OH radicals are in their basic form (O^-) in alkali media [Eq. (1.5)] and can react with the methyl groups of thymine and 6-methyl uracil (MYERS et al., 1965a, b) according to the general process:

$$R\text{-}CH_3 + O^- \longrightarrow R\text{-}\dot{C}H_2 + OH^- \qquad (1.12)$$

Indeed, in the experiments of NETA (1972a) the ESR spectra from thymine and 6-methyl uracil solutions irradiated at pH 13 were compatible with the formation of resonance-stabilized allyl-type radicals produced in Equation (1.12).

The ESR spectra of the OH adducts of various pyrimidines were examined in pulsed solutions containing N_2O (NUCIFORA et al., 1972). From thymine the C6 adduct was positively identified as was the C5 adduct from 6-methyl uracil. The evidence favoured OH addition at C5 in the case of uracil, but no assignment could be made for the reaction with cytosine.

The optical absorption spectra of the pyrimidine-OH adducts have been examined by the pulse radiolysis technique (SCHOLES et al., 1965; DANZIGER et al., 1968; MYERS et al., 1968; SCHOLES, 1968; GREENSTOCK et al., 1969; MYERS and THEARD, 1970; MYERS et al., 1970a, 1970b; FEL et al., 1971; HAYON and SIMIC, 1973; SHRAGGE and HUNT, 1974). Although, in general, the spectra do not readily allow an assignment of the structure of the adducts, they are important for investigations of the kinetics of decay and, thereby, for further elucidation of the subsequent chemical transfor-

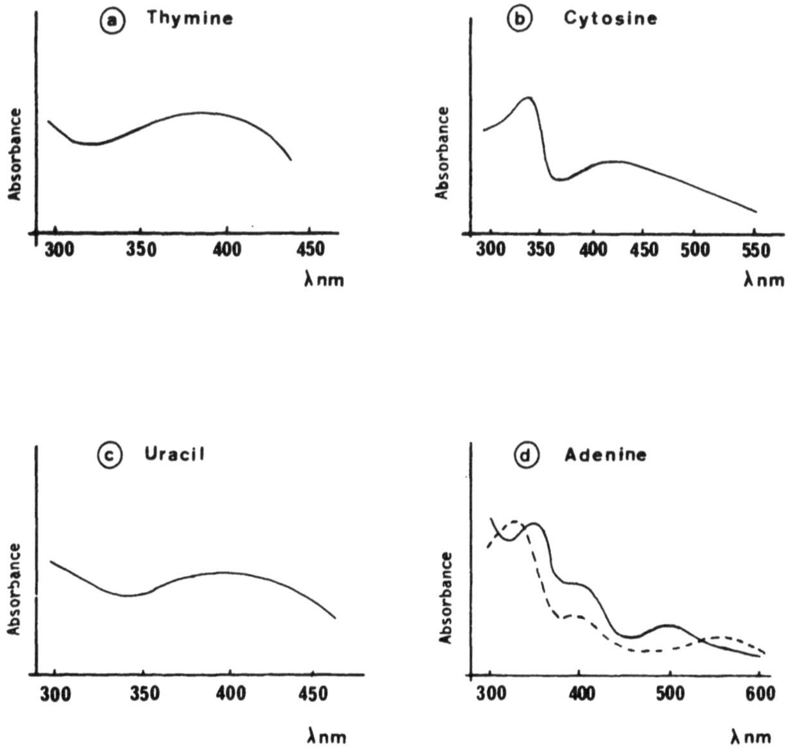

Fig. 1.1a-d. Pulse radiolysis of neutral solutions of nucleic acids bases in pulsed N_2O-saturated solutions: Absorption spectra of the OH adducts (a) thymine, (b) cytosine, (c) uracil, (d) adenine —— immediately after the pulse, ---- after 30 μs

mations; due to the strong optical absorption of the pyrimidine solutions themselves, most measurements have been limited to wavelengths greater than ∿ 300 nm. In pulsed neutral solutions (containing N_2O) the OH adducts of uracil, thymine and cytosine exhibit broad optical absorption bands around 400 nm, cytosine having an additional band at ∿ 350 nm (Fig. 1.1). The decays around the maxima are adequately described by second-order kinetics ($k \approx 10^9\ M^{-1}\ s^{-1}$) indicating radical-radical recombination reactions. There is a change in the transient optical absorption spectra of pulsed pyrimidine solutions as these are made alkaline; there is an overall increase in absorbance as well as some change in shape. HAYON and SIMIC (1973) measured the initial absorption, at a fixed wavelength, as a function of pH and found that the increase followed a titration curve corresponding closely to the pK_a value of the parent pyrimidine. Thus, the pK values of the OH adducts of thymine and uracil were found to be 9.6 and 9.55 respectively, which should be compared with the corresponding pK_a values of 9.9 and 9.5. It was therefore concluded that the changes in spectra were due to tautomerization of the molecule rather than to acid-base properties of the OH-radical adducts to the pyrimidines (i.e., dissociation of the OH group). The transient spectrum of the radical produced from OH addition to 1,3-dimethyluracil was identical in neutral and alkaline solutions tending to support the above conclusion. With cytosine, the transient optical absorption was investigated over the pH range 3-6 and it was found that the pK of the radical (4.5) corresponded to the pK_a (4.6) for the proton dissociation of N_3. More recently, SHRAGGE and HUNT (1974) have reported that, in uracil solutions, in addition to a radical with a pK of 9.5 there is another one of pK ∿ 7. One explanation offered is that the two different forms of the uracil OH-adducts [Formulas (1) and (2)] dissociate with different pKs, or, alternatively, that one of the adducts has two pKs. This is an interesting problem which has yet to be resolved.

On pulsing solutions of thymine or uracil at pH 11 an absorption band is observed at 320 nm (HAYON and SIMIC, 1973). This decays by first-order kinetics and may be associated with ring-opening of the pyrimidine radical as described above. Additional evidence for a base-catalysed rearrangement of the OH adducts of uracil, and also of cytosine, arises from studies of their polarographic behaviour (BANSAL and SELLERS, 1975).

In the pulse radiolysis of N_2O-saturated pyrimidine solutions, a phenomena is observed which is associated with the acidic properties of these molecules and the production of OH^- ions during radiolysis (SCHOLES, 1968; FIELDEN et al., 1970). Because of the reaction of solvated electrons with N_2O [Eq. (1.6)] significant quantities of OH^- ions are formed and these can react with pyrimidines in competition with protons:

$$OH^- + Py \longrightarrow Py^- + H_2O \qquad (1.13)$$

$$OH^- + H_3O^+ \longrightarrow 2H_2O \qquad (1.14)$$

The equilibrium is established through the reverse of Equation (1.13), and, since the ionised form (Py^-) absorbs at a different wavelength, a first-order decay is observed. This occurs over a few μs and is noted only at lower wavelengths (< 300 nm). As expected, pyrimidines which cannot undergo keto-enol tautomerism (e.g., 1,3-dimethyl uracil) do not exhibit this behaviour in pulsed solutions. The rate constants of Equation (1.13), and of its reverse, have been determined for a number of pyrimidines and purines by pulse radiolysis and the values obtained (in the range 1-2 × 10^{10} M^{-1} s^{-1} for reactions with OH^- and in the range 3-4 10^4 s^{-1} for the back reactions) are in good agreement with those measured by conductivity techniques using relaxation methods (EIGEN, 1964).

1.3.1.2 Purines

Evidence for the structures of the primary OH adducts of purines is fragmentary. NUCIFORA et al. (1972), using the ESR coupled pulse-radiolysis technique, investigated N_2O-saturated neutral adenine solutions and observed a single broad line which became narrower as the pH was decreased; no fine structure was observed, but the experiments were taken to be compatible with some addition of OH at the central 4.5 bond as previously proposed (SCHOLES and WEISS, 1952). The optical absorption spectrum in pulsed adenine-N_2O solutions was found to be rather complex (WILLSON, 1966; SCHOLES, 1968). There is an initial first-order process ($t_{1/2}$ ∿ 5 μs) which can lead to an increase or decrease in optical absorption at different wavelengths (see Fig. 1.1). After sufficient time the decay is adequately second order with a rate constant of 7 × 10^8 M^{-1} s^{-1}. Since the rate of the first-order process is decreased on substitution at N9, as in adenosine and adenylic acid, this suggested a chemical process in the imidazole ring. These findings have been recently confirmed by MICHAELS and HUNT (1973) and by VAN HEMMEN (1975). The latter author has also observed first-order decays of the transient spectra in pulsed N_2O-saturated solutions of guanine, hypoxanthine and xanthine, and suggests that the species responsible is an OH adduct of the imidazole ring which then opens. In the case of adenine the ring-opened radical species could be the precursor of 4,6-diamino-5-formamido pyrimidine Formula (5), a known radiolysis product of deoxygenated adenine solutions (HEMS, 1960b), possibly formed according to the scheme (VAN HEMMENS and BLEICHRODT, 1971).

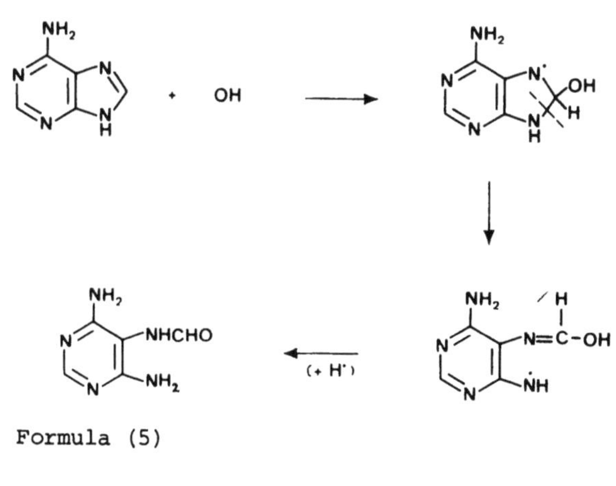

Formula (5)

The parent purine gives neither a first-order decay on pulse radiolysis nor a 5-formamido-pyrimidine as a permanent radiolysis product; VAN HEMMENS (1975) regards this as support for the above involvement of the imidazole ring.

1.3.1.3 Nucleosides, Nucleotides and Polynucleotides

The optical absorption spectra in pulsed N_2O-saturated solutions of pyrimidine and purine nucleosides and nucleotides at wavelengths greater than ∿ 300 nm are very similar to those of the free bases. Optical contribution from the sugar radical is rather small in this region of the spectrum and, in any case, there is predominant OH-attack on the base components. Figure 1.2 gives the absorption spectra of the transients from solutions of the deoxyribonucleotides of adenine, guanine, thymine and cytosine and also from deoxyribose-5-phosphate (SCHOLES et al., 1969). The nucleotide radicals decay more slowly than do base radicals (by a factor of ∿ 10) which is not unexpected in view of the charge on the phosphate groups. Recently, WILLSON et al. (1974) have applied the conductimetric analyzing technique to the pulse radiolysis of solutions of some nucleosides and nucleotides. Solutions (saturated with N_2O) were pulsed under slightly acidic conditions (pH = 4-5) and it was found for guanosine and guanylic acid that there was a large decrease in the conductivity immediately after the radiation pulse, followed by a slow increase over several ms corresponding to the decay kinetics of the optical absorption monitored at 370 nm. No conductivity changes were observed with thymidine cytidine or adenosine. It was concluded that with the guanine compounds the products of the reaction with OH are not adducts but radical-cations formed by the electron-transfer process,

$$G + \cdot OH \longrightarrow G^+ + OH^- \qquad (1.15)$$

where G is guanosine or guanylic acid. The immediate drop in conductivity is attributed to neutralization of the OH^- by solvent protons, thus replacing highly conducting protons by the less conducting organic radical-cations; the subsequent slow increase in conductivity is presumed to arise from the liberation of protons in the decay of the radical-cations. If the explanation is correct one would need to understand why electron transfer is apparently specific to the guanine compounds. An alternative possibility is that OH addition to guanine does, in fact, occur but that the pK of the adduct is such that immediate protonation takes place, i.e., that Equation (1.16) replaces Equation (1.15).

$$G + \cdot OH \longrightarrow \dot{G}OH \xrightarrow{H_2O} (\dot{G}OH)H^+ + OH^- \qquad (1.16)$$

The OH-adduct spectra of poly U and poly A have been compared with that of poly (U + A) (MICHAELS and HUNT, 1973). Assuming that attack by hydroxyl radicals on the double-stranded polynucleotides is random, with equal probability of reaction at a uracil or adenine residue, a resultant theoretical spectrum was calculated from the experimental data for the single-stranded molecules. The results indicate that the transient spectrum from poly (U + A) can, in fact, be obtained by averaging those from poly U and poly A.

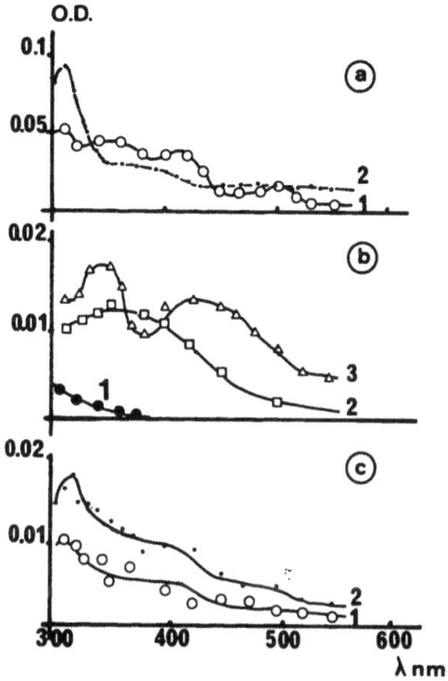

Fig. 1.2a-c. Transient optical absorption spectra in pulsed-N_2O saturated solutions of (a) Purine nucleotides *1*: Deoxyadenylic acid *2*: Deoxyguanylic acid. (b) Pyrimidine nucleotides and deoxyribose phosphate *1*: Deoxyribose-5-phosphate *2*:Thymidylic acid *3*: Cytidylic acid. (c) DNA. (0.01% w/v) spectra taken 15 μs after the pulse. *1*: Single pulse (1.5 krad) *2*: Solution pre-irradiated with 18.5 krad. Curves are computed (see text)

Optical absorption spectra in pulsed DNA-N_2O solutions have been obtained (SCHOLES et al., 1969) the results being shown in Figure 1.2. Given that sugar radicals do not absorb at wavelengths greater than ∿ 400 nm it seems that again one is dealing mainly with transients arising from attack on the base components of the polynucleotide. Both the rate of formation of the transient (due to reaction with OH) as well as the yield, increased on increasing pulsing; this observation is consistent with (1) an enhanced reactivity of the denatured DNA and (2) an increased effective scavenger concentration due to radiation-induced fragmentation of the chains. The spectra of the pulsed polynucleotide solutions were compared with weighted-average spectra of the separate mononucleotides, assuming equal reactivities of mononucleotides towards OH, an AT:GC ratio of 50:42 in the DNA sample used, and a factor due to incomplete scavenging of OH in the polynucleotide solution. There is quite good agreement with experiment (see Fig. 1.2). The transient decay in DNA solutions is quite complex. There is some partial decay over the first few hundreds of microseconds (particularly noticeable at 300 and 550 nm) but a substantial fraction appears to be long-lived, there being little change after a ms. Given the low mobility of these macromolecules in solution and also the fact that the majority of the free radicals (∿ 80%) are formed inside the double helix, it is not surprising that the OH transients from DNA have a long lifetime in solution.

1.3.2 Reaction with Hydrogen Atoms

Since the yield of hydrogen atoms in the radiolysis of water (G_H = 0.55) accounts for only ∿ 10% of the total yield of reactive species, products of hydrogen-atom reactions will only be relatively minor ones in irradiated neutral deoxygenated solutions. However, under acidic conditions where conversion of solvated electrons to hydrogen atoms occurs [Eq. (1.3)], such products will be very substantial. Like OH radicals, hydrogen atoms can react with the pentose moiety by a dehydrogenation process, but here, we are concerned only with the (predominant) addition reactions of H to the heterocyclic bases. Probable sites of attack in pyrimidines and purines are indicated from studies of the ESR spectra of the free radicals from bombardment of the solid materials with externally generated H atoms (HERAK and GORDY, 1966; HOLMES et al., 1967) (see I.6). Thymine and its compounds give an ESR line with a hyperfine structure of eight components, corresponding to the production of the 5-yl radical Formula (6a):

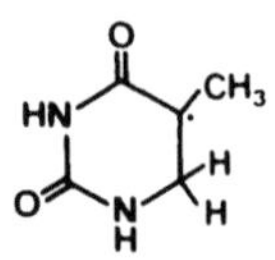

Formula (6a)

[It has been pointed out (NICOLAU et al., 1969) that radical stability and steric factors, rather than electronic factors, may influence the position of attack; C6 is less hindered and the resultant radical is a tertiary one rather than a secondary one.] Cytosine and uracil give similar radicals, although it is not known whether these have 5-yl or 6-yl structures. In the case of adenine and guanine there is evidence for attack at C8, e.g., Formula (6b) from adenine.

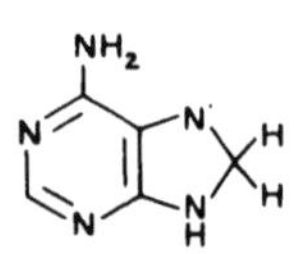

Formula (6b)

SCHMIDT and BORG (1971) report that ∿ 20% of the H-atoms react at C2 in adenine dihydrate and in adenosine. Furthermore, there is a central singlet due to unknown radicals, which suggests that effects of molecular environment, largely determined by crystal structure, may be a factor in the pattern of radical production in such experiments; this necessitates some caution in extrapolating the observations to H atom reactions in an aqueous environment as is indeed clear from the fact that hydroperoxides resulting from attack by H at both C5 and C6 of thymine have been identified in irradiated aerated thymine solutions (see II.2.1).

Pulse-radiolysis experiments confirm that H atoms add to the C5-C6 bond of the pyrimidine molecule. The optical properties and decay characteristics of pyrimidine H-atom adducts can be more conveniently studied in pulsed N_2O-saturated solutions of the corresponding dihydropyrimidine. Here the only radicals produced are those resulting from oxidation (dehydrogenation) at the C5-C6 bond and correspond to the H-atom adducts of the parent pyrimidines e.g., for dihydrouracil Formula (7) oxidation takes place according to the process:

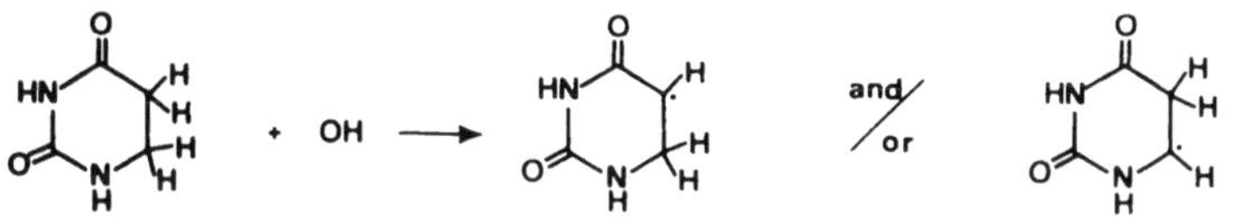

(1.17)

Formula (7)

The optical absorption spectra of the transients produced from dihydrouracil and dihydrothymine (Fig. 1.3) are rather similar to those of the OH adducts of uracil and thymine, as expected;

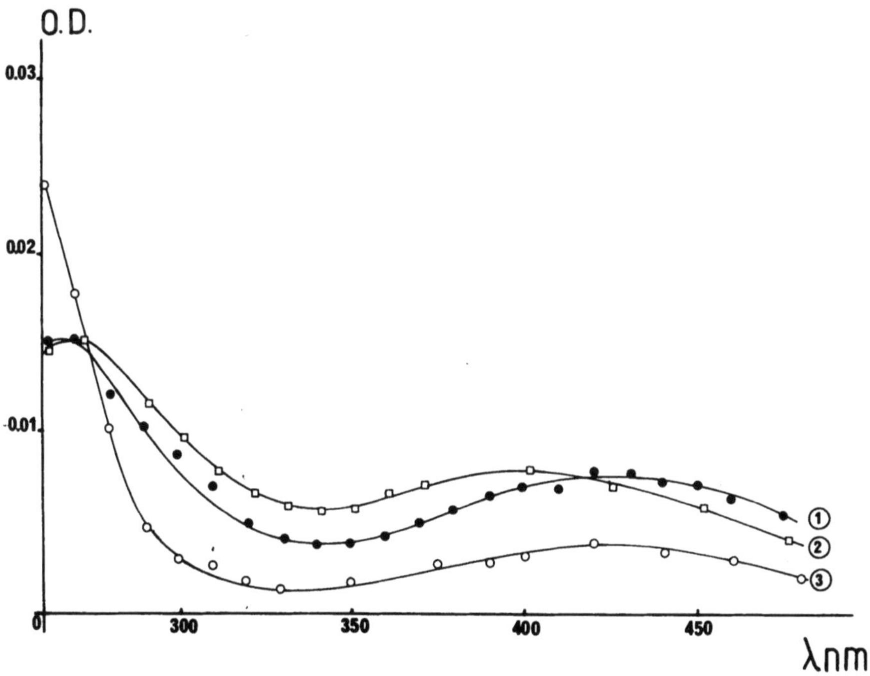

Fig. 1.3. Pulse radiolysis of N_2O-saturated solutions of dihydropyrimidines; optical absorption spectra of uracil and thymine H-atom adducts. *1*: -●-●- uracil adduct immediately after the pulse; *2*: -□-□- thymine adduct immediately after the pulse; *3*: -o-o- uracil adduct after 150 μs

they are, in fact, essentially identical to the spectra attributed to the H adducts in pulsed acidic deaerated solutions of uracil (GREENSTOCK et al., 1969) and of thymine (THEARD et al., 1971). The dihydropyrimidyl radicals decay mainly by second-order kinetic with rate constants around $10^9\ M^{-1}\ s^{-1}$. The parent pyrimidine is one of the products of radical decay and this is seen experimentally in a growth in optical absorption below ∿ 300 nm (Fig. 1.3).

It has been found (PHILLIPS, 1971; BARSZCZ and FIELDEN, 1974; HAYSOM et al., 1975) that the pyrimidine is produced by a first-order process ($t_{1/2} \sim 150\ \mu s$) subsequent to the bimolecular radical interaction. Detailed studies of the effects of pH and of substitution in the dihydropyrimidine ring system on the first-order rate constant and on the yield of parent pyrimidine (AL-YAMOOR et al., 1977) are in agreement with the suggestion of HAYSOM et al. (1972) that electron transfer can take place between dihydropyrimidyl (pyrimidine H-adduct) radicals, e.g., in the interaction of 6-yl radicals [Formula (8)] from dihydrouracil

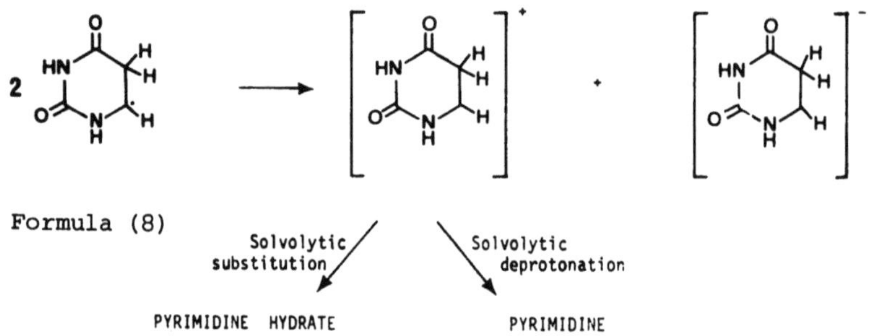

(1.18)

The pulse radiolysis data indicate that the intermediate undergoing the first-order decay to produce pyrimidine is not a classical carbenium ion, which would have much too short a lifetime under these particular solvent conditions.

1.3.3 Reaction with the Solvated Electron

The rate data of Table 1.1 indicate that e^-_{aq} reacts exclusively with the bases in pyrimidine and purine compounds. The electron adducts can be readily observed optically in pulsed deaerated solutions containing an excess of alcohol as an OH and H scavenger (SCHOLES, 1968); t-butanol is probably the best for this purpose since the alcohol radicals formed absorb at much shorter wavelengths than those of the adducts. The electron-adduct spectra have been determined, in this way, for many pyrimidines and purines and their derivatives. (SCHOLES, 1968; HAYON, 1969; GREENSTOCK et al., 1970; THEARD et al., 1971; SHRAGGE and HUNT, 1974; MOORTHY and HAYON, 1975); Figure 1.4 gives those for the nucleic acid bases in neutral aqueous solution. There is a difference between the electron-adduct and H-atom adduct spectra of a given base (cf. Figs. 1.3 and 1.4 for uracil and thymine), indicating a different site of attack of the two reducing species.

The reaction of e^-_{aq} with pyrimidine to form negative ions which may then undergo protonation [Eq. (1.19)] was suggested some time ago (SCHOLES, 1963).

$$Py + e^-_{aq} \longrightarrow (Py\cdot)^- \xrightleftharpoons{H_3O^+} PyH\cdot + H_2O \qquad (1.19)$$

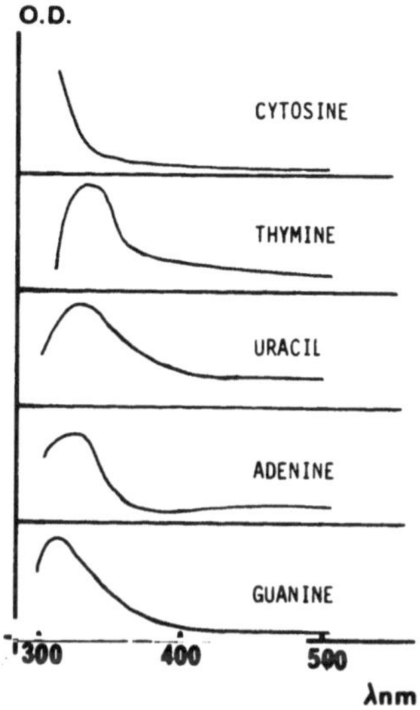

Fig. 1.4. Electron adducts of nucleic acid bases (pH ≈ 6.5)

Pulse-radiolysis studies indeed show pH effects which are indicative of ionic dissociations of the electron adducts, (that of uracil, for example, has $pK_a \sim 7.5$) and HAYON (1969) ascribed such effects to ketyl radical dissociations of the type:

$$> C = O + e^-_{aq} \longrightarrow > \dot{C} - O^- \underset{}{\overset{H_3^+O}{\rightleftharpoons}} > \dot{C} - OH + H_2O \qquad (1.20)$$

Using the ESR coupled pulse technique NUCIFORA et al. (1972) examined the structure of the pyrimidine-electron adducts of uracil and thymine and observed doublets which were related to α-coupling of the proton on C6. It was suggested that e^-_{aq} attacks the carbonyl functions, but, that in the case of the C4 adduct, conjugation occurs. Thus, in the case of uracil, the transient absorbing above 300 nm should have the neutral and ionized forms Formulas (9) and (10) respectively.

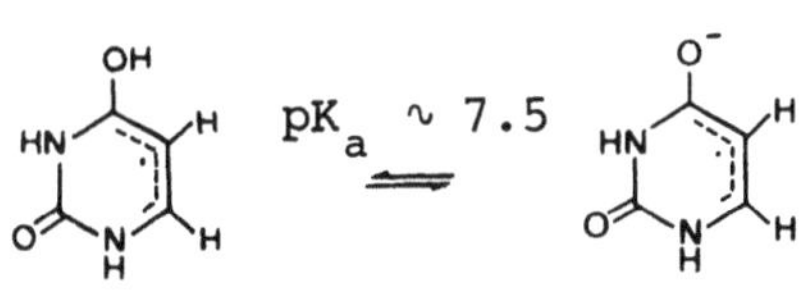

Formula (9) (10)

The bimolecular rate constants for decay of the pyrimidine radical anions are again in the region of 10^9 M^{-1} s^{-1}. Protonation of the adducts is a very fast process, but the reaction with water is quite slow; e.g., for thymine, the measured values for the rate constants of Equations (1.21) and (1.22)

$$(Thy\cdot)^- + H_3O^+ \longrightarrow ThyH\cdot + H_2O \qquad (1.21)$$

$$(Thy\cdot)^- + H_2O \longrightarrow ThyH\cdot + OH^- \qquad (1.22)$$

are $k_{21} = 6.4 \times 10^{10}\ M^{-1}\ s^{-1}$ and $k_{22} = 10^4\ s^{-1}$ (THEARD et al., 1971).

The radiation chemistry of 5-bromouracil [Formula (11)] has been studied in some detail because its incorporation into both mammalian and bacterial cells renders them more radiosensitive. In particular, it has been suggested that 5-bromouracil reacts with e^-_{aq} by dissociative electron capture to give the uracilyl free radical [Formula (12)] and bromide ions (ADAMS, 1967; VOLKERT et al., 1967; ZIMBRICK et al., 1969)

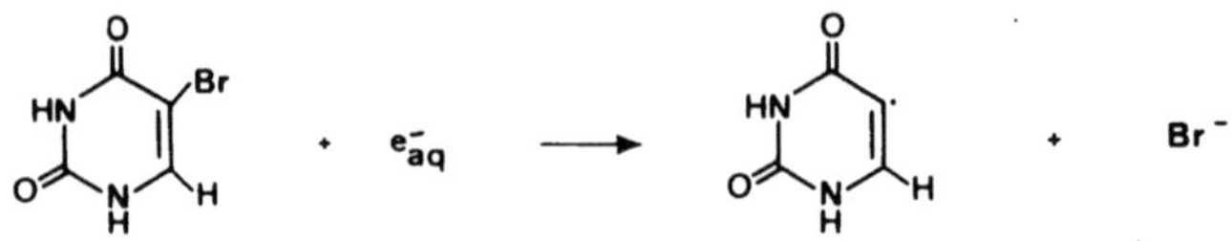

(1.23)

Formula (11) (12)

The uracilyl radical is very reactive and can immediately abstract a hydrogen atom from a convenient source [Eq. (1.24)]

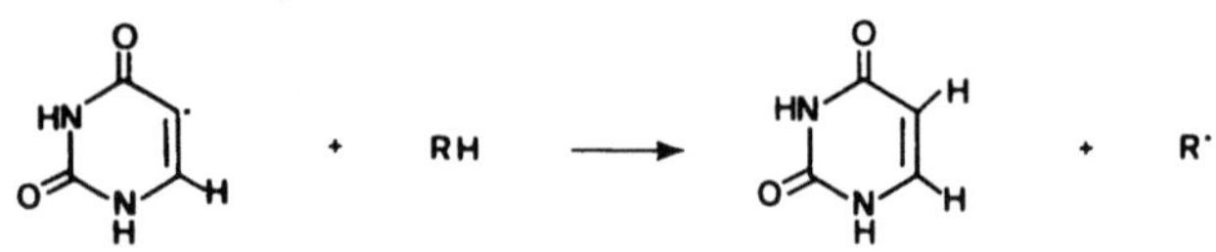

(1.24)

It has been emphasized that the incorporation of 5-bromouracil into cellular DNA may lead to an increase in the total yield of free radicals per unit volume per unit dose, since electron capture by the brominated bases should be principally dissociative and hence irreversible (ADAMS and WILLSON, 1972); furthermore, since the uracilyl radical lies close to the sugar backbone a mechanism exists for the transfer of a radical centre from a pyrimidine base to the sugar i.e., Equation (1.24), a process which may produce a strand break.

BHATIA and SCHULER (1973) have studied the changes in conductivity in pulsed 5-bromouracil solutions and concluded that the anionic precursor of the uracilyl radical must have a short lifetime ($< 10^{-7}$ s). In the case of 5-chlorouracil, however, the electron adduct is sufficiently stable ($t_{1/2} \sim 5\ \mu s$) for protonation to compete with chloride ion elimination below pH $\sim$ 7; protonation competes with dissociation on equal terms at pH = 5.2. Although 5-fluorouracil also reacts rapidly with e^-_{aq}, protonation occurs quickly and no uracilyl radical is produced. [N.B. Dehydrohalogenation follows addition of OH to the 5 position of the halouracils, yielding the 6-yl isobarbituric acid free radical [Formula (13)] (NETA, 1972b) (see I.7.1 and III.3.2)].

The dehydrohalogenation process is complete within a few microseconds (PATTERSON and BANSAL, 1972).

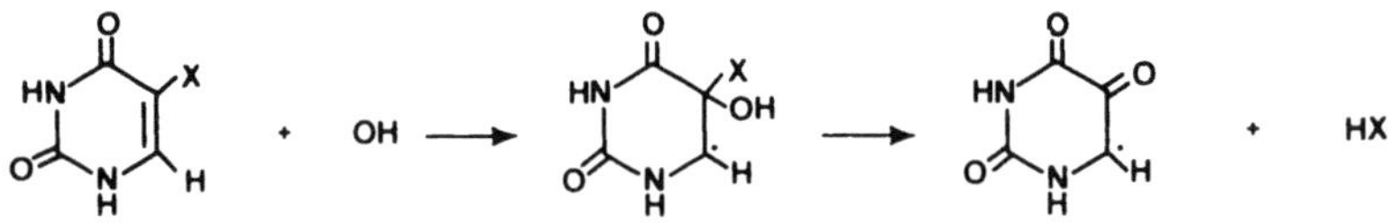

(X = Halogen)

Formula (13)

(1.25)

1.3.4 Reactions of the Intermediates with Added Solutes

1.3.4.1 Oxygen

Molecular oxygen can react rapidly with the organic radicals and radical anions produced in the manner described above. Reactions with the OH adducts of the heterocylic bases were first noted qualitatively in pulsed oxygenated solutions (SCHOLES et al., 1965) and attributed to the process,

$$\dot{P}(OH) + O_2 \longrightarrow P(OH)O_2\cdot \quad (1.26)$$

(where P represents a pyrimidine or purine moiety). WILLSON (1970) examined the absorptions in solutions saturated with N_2O/O_2 mixtures, the relative proportion of the two gases being such that all of the hydrated electrons were trapped by the N_2O (i.e., no O_2^- was present). Where the transient absorptions decayed they were first order in oxygen concentration and therefore consistent with Equation (1.26); the final spectra were assigned to the corresponding organic peroxy radicals. Rate constants for the reaction of oxygen with the adducts of thymine, cytosine and uracil were similar ($2 \times 10^9\ M^{-1}\ s^{-1}$); those of the corresponding nucleosides and nucleotides, and also of adenylic acid, were 50% lower. Oxygen has no effect on the decay rates of the transients in pulsed solutions of deoxyguanylic acid, 5-bromouracil or of DNA, and it was considered possible that the reactions of the OH adducts of these compounds are slower relative to the natural decay, provided, of course, that the adducts and the reaction products have different absorption spectra.

From measurements of the loss of thymine in the γ-irradiation of thymine/O_2 solutions it was concluded (SCHOLES and WILLSON, 1967) that electron attachment to the pyrimidine does not lead to significant base destruction as long as oxygen is in the solution. It was therefore proposed that restitution could occur by the process:

$$(Thy\cdot)^- + O_2 \longrightarrow Thy + O_2^- \quad (1.27)$$

The rate of reaction of the thymine electron adduct with oxygen has been measured [$k_{27} = 4.3 \times 10^9\ M^{-1}\ s^{-1}$ (THEARD et al., 1971)] but the mechanism may involve the intermediate formation of an oxygen adduct, $(ThyO_2^{\cdot})^-$, as suggested by LOMAN and EBERT (1970); indeed, the pyrimidine electron adduct [cf. Formulas (9) and (10) for uracil] may well form peroxy radicals by addition of O_2 at C6.

1.3.4.2 Electron-Transfer Processes

Interest in the mechanism of cellular radiosensitisation has led to many model pulse-radiolysis studies designed to investigate the relevance of electron-transfer processes. Reactions involving various simple DNA derivatives have naturally been considered quite significant in this field. The rates and efficiencies of transfer from radicals and radical anions to a considerable number, and variety, of electron-affinic compounds have been obtained by monitoring the electron-adduct spectra of the latter. Example of these electron-accepting substances (sensitizers) include N-ethyl maleimide and menadione (GREENSTOCK et al., 1970), p-nitro-acetophenone (ADAMS et al., 1972), menaquinone (HAYON and SIMIC, 1973) and nitrofurans (GREENSTOCK and DUNLOP, 1973, 1975). The electron transfer rates from the electron adducts of DNA and its components are high, approaching a diffusion-controlled limit. Electron transfer rates from the corresponding radicals (OH and H adducts) are about an order of magnitude lower and it is concluded that there is a competing reaction between the radicals and the sensitizer, which is presumed to be one of adduct formation.

Attempts have been made to rationalize these electron-transfer processes in terms of redox potentials (RAO and HAYON, 1975). Oxidation of the radicals must lead, in the first instance, to carbocations, and these should then undergo solvolytic substitution and deprotonation (elimination) reactions as discussed above.

Chapter 2

Radiation-Induced Degradation of the Base Component in DNA and Related Substances – Final Products

2.1 Introduction

The rearrangement and the combinations of the transient species described in the preceding chapter give rise to more stable modifications in the chain of DNA. The determination of the chemical nature of these final modifications produced in vivo is a very complex problem because a change as small as that of one nucleotide per million in a large chain of DNA may be sufficient to induce drastic perturbation in the life of the cell. Genetic studies have shown that many enzymatic systems are involved in the repair of these radiation-induced modifications. The altered nucleotides in the chain may be compared to metabolites in classical biochemistry. It would be of great interest to determine the chemical nature of all these defects which are due to ionizing radiation.

Different types of difficulties are encountered in the identification and measurement of the final radiation products of DNA.

1. The best way to measure traces of a substance is to use radioactively labelled compounds. The degradation products due to autoradiolysis may interfere with those which are produced by the external irradiation, at least for the lower doses.

2. Prior to being analyzed, the defects borne by the polynucleotide chain need to be detached from the polymer chain. The radiation products are not always stable in the usual conditions of acid hydrolysis.

3. DNA is a complex molecule with many different chemical groups. Many hypotheses dealing with its radiation-induced modifications may be postulated. Prior to studying the biopolymer itself it is of interest to know the radiation chemistry of the DNA constituents. Furthermore, studies of this are of value as they relate to those of mechanism (see I and II.1).

Most of the research has been carried out in aqueous solutions in which the amounts of radiation product obtained are higher than in the dry state or in frozen solutions.

The work of identification began with free heterocyclic bases and free sugars. At this level the methods used in modern organic chemistry (NMR, mass spectroscopy) are very efficient and account for the recent progress in this field. Most of the radiolysis products of nucleic acid components produced in aqueous solution were characterized during the last decade. It is now possible to draw a clearer picture of the subject and to give general schemes of degradation.

The subject has been reviewed in the past years (NOFRE and CIER, 1964; WEISS, 1964; ALEXANDER and LETT, 1967; FAHR, 1969; SCHOLES, 1968, 1970, 1972; TEOULE et al., 1971; LATARJET, 1972; INFANTE et al., 1973b; BLOK and LOMAN, 1973; MYERS, 1973; WARD, 1975; WANG, 1976).

2.2 Nucleic Acid Constituents in Aqueous Aerated Solutions

The radiosensitivities of the bases decrease as follows: Thy > Cyt > Ade > Gua. The destruction yield G (-base) is dependent on factors such as concentration of solute (MYERS et al., 1965a, b), pH of the solution, presence of salts and nature of the radiation. Densely ionizing radiation (^{210}Po α particles) is less efficient than X or gamma-rays (ANDERSON and HART, 1961). However a similar decomposition pattern is observed in both cases for thymine (DRÁŽIL and RÝZNAR, 1968b).

The extent of base degradation decreases in the following way: free base > nucleoside > nucleotide > nucleotide diphosphate. For example:

Ade (G = 0.51), Ado (G = 0.41), AMP (G = 0.39), ADP (G = 0.37) (SCHACHINGER and HEINDL, 1974).

2.2.1 Free Bases

2.2.1.1 Thymine

The gamma-radiolysis of a 10 mM thymine ^{14}C (*1*)[1] in aqueous aerated solution generates more than 30 products, most of which can be separated by two-dimensional thin-layer chromatography (TEOULE and CADET, 1971). At neutral pH, eight hydroperoxides are detected by spraying the plate with potassium iodide reagent (TEOULE and CADET, 1975). 5(6)-Hydroperoxy-6(5)-hydroxy-5 and 6-dihydrothymine (*2, 3, 4, 5*) (EKERT and MONIER, 1959) which are the major γ-degradation products of thymine (*1*) can be produced in other conditions such as UV photolysis of aqueous solutions (DANIELS and GRIMISON, 1967), γ irradiation in the presence of an N_2O/O_2 atmosphere (LOMAN and EBERT, 1970) and UV or acid homolytic dissociation of hydrogen peroxide and sonolysis (MEAD et al., 1973, 1975). A general mechanism of pyrimidine hydroxyhydroperoxide formation which involves the initial addition of an OH radical to the double bond between C5 and C6 was postulated (SCHOLES et al., 1960; SCHOLES and WEISS, 1959) (see II.1).

The four possible isomeric forms of cis and trans 5(6) hydroperoxy-6(5)-hydroxy-5,6-dihydrothymine (*2, 3, 4, 5*) have been isolated and characterized among thymine radiation-induced products (CADET and TEOULE, 1974).

---

[1] Italic numbers in parentheses refer to encircled numbers in the figures and numbers in tables.

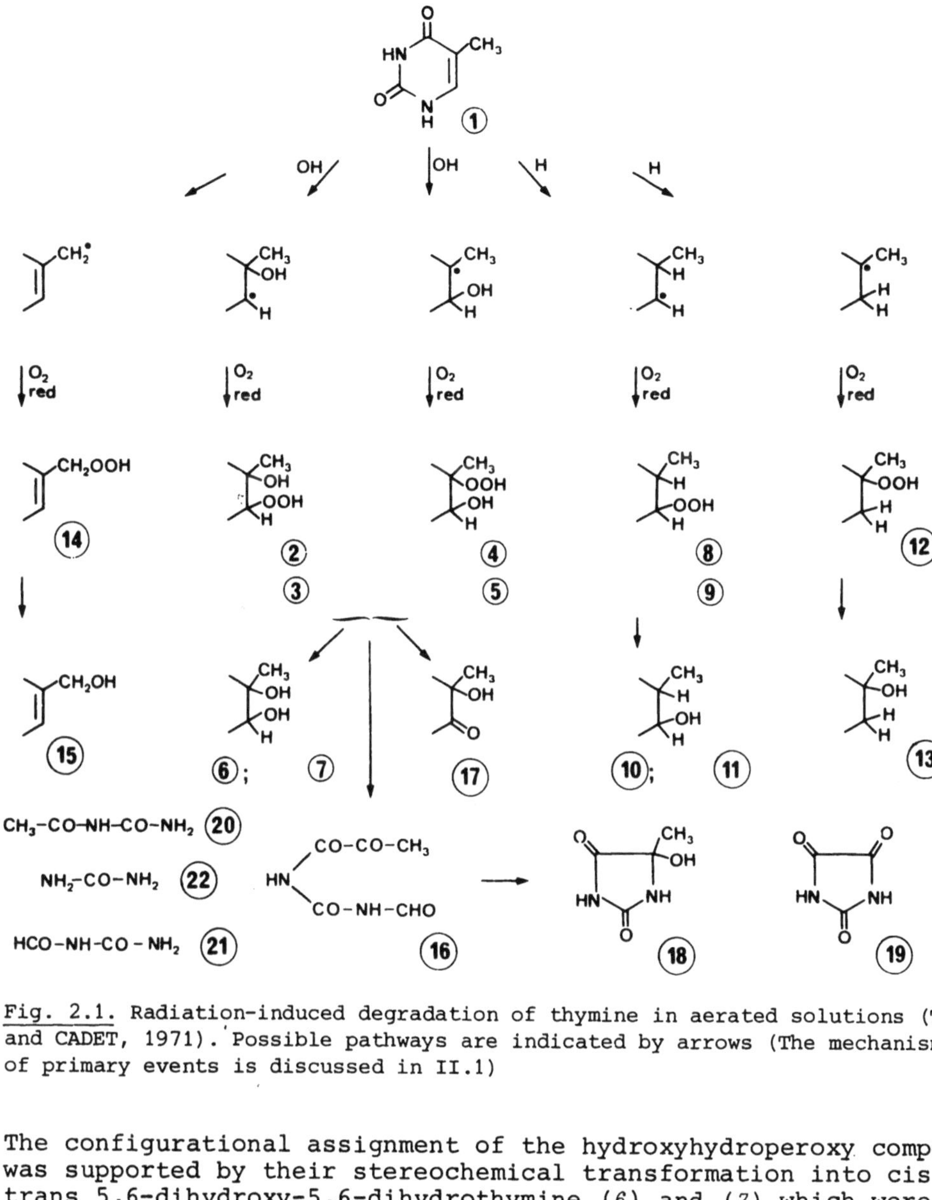

Fig. 2.1. Radiation-induced degradation of thymine in aerated solutions (TEOULE and CADET, 1971). Possible pathways are indicated by arrows (The mechanism of primary events is discussed in II.1)

The configurational assignment of the hydroxyhydroperoxy compounds was supported by their stereochemical transformation into cis and trans 5,6-dihydroxy-5,6-dihydrothymine (*6*) and (*7*) which were unambigously characterized (CADET and TEOULE, 1971a). The heterolysis of the peroxidic bond under the action of reducing agents (KI,H_2S) takes place by nucleophilic substitution at oxygen in a stereospecific way according to a SN_2 rather than an SN_1 mechanism (DAVIES, 1961). This implies a configuration retention of the C-O bond in the peroxide - alcohol transformation. The thermal decay of thymine hydroxyhydroperoxides in aqueous solutions suggests a nonradical reaction. The bimolecular mechanism for the decomposition of peroxides was strongly supported by the low activation energies (41.8-92 KJ/mol) and entropies (-54.4 to -180 J mol^{-1} K^{-1}) (SCHWEIBERT and DANIELS, 1971).

The lack of any splitting for the OH signal in the ^1H NMR spectra (Me_2SO-d_6) or the cis (*2*) and trans (*3*) isomers of 6-hydroperoxy-5-hydroxy-5,6-dihydrothymine is consistent with the presence of a tertiary alcoholic function in the pyrimidine ring. On the other hand, the hydroxyl proton of cis 5-hydroperoxy-6-hydroxy-5,6-dihydrothymine (*4*) displays the expected doublet which collapses into a broad singlet by H6 spin decoupling experiments. Furthermore the specific H1 and H3 deshielding might be correlated with the position of the electronegative peroxide group (WARD and MAIR, 1969). The H1 resonance signal of 6-hydroperoxide is shifted downfield with respect to the corresponding proton of 5-hydroperoxide. An opposite trend was noted for H3 (CADET and TEOULE, 1974).

The first structural determination was based on the halogen substitution reaction of the 5-bromo-6-hydroperoxy-5,6-dihydrothymine (EKERT and MONIER, 1959). In fact, the departure of the bromine ion and the nucleophilic $^-$OOH intramolecular attack take place according to a concerted process involving the transient formation of a bridged cation. As a result, cis 5-hydroperoxy-6-hydroxy-5,6-dihydrothymine (*4*) was produced in a highly stereospecific pathway (CADET and TEOULE, 1973b). A similar anionotropic rearrangement was observed when trans 5-bromo-6-hydroxy-5,6-dihydrothymine was transformed into cis 6-hydroperoxy-5-hydroxy-5,6-dihydrothymine (*2*) in aqueous hydrogen peroxide solutions containing silver oxide. The hydroxyl 1,2 shift from carbon 6 to position 5 was demonstrated through mass spectroscopic analysis of specific ^{18}O labeled derivatives. This peroxy compound can be prepared in a nearly quantitative yield by the treatment of cis 5,6-dihydroxy-5,6-dihydrothymine (*6*) with acidic hydrogen peroxide solutions (CADET and TEOULE, 1973a, 1974). The assignment of the position of the peroxide group is strongly supported by the quantitative loss of heavy oxygen introduced initially in the 6 position of cis glycol (*6*).

The relative yield for the four isomers of thymine hydroxyhydroperoxides is dependent on the pH of the gamma-irradiated aqueous solution (TEOULE and CADET, 1974). The formation of 6-hydroperoxy-5-hydroxy-5,6-dihydrothymine (*2*) and (*3*) is shown to be preponderant with respect to 5-hydroperoxy isomers (*4*) and (*5*) in neutral or slighty acidic aqueous solution. An opposite trend is observed at pH values lower than 1.7. In both cases trans isomers are preferentially generated. On the basis of these results, the OH homolytic attack at the 5,6 pyrimidine double bond would take place mainly on C5 in neutral solution. At pH below 1.7, 6-hydroxy-5,6-dihydrothymine-5-yl radical would be preferentially generated in place of the 5-hydroxy-5,6-dihydrothymine-6-yl one. Similar conclusions can be reached from the result of ESR flow experiments (ORMEROD and SINGH, 1966) or spin trapping studies using tert-nitroso butane (LAGERCRANTZ, 1973).

The cis and trans isomers of 6-hydroperoxy-5,6-dihydrothymine (*8*, *9*) and 5-hydroperoxy-5,6-dihydrothymine (*12*) are produced in low yields (Table 2.1). The low yield of (5 or 6)-hydroperoxy-5,6-dihydrothymine can be correlated with the H atom scavenging properties of oxygen; for the usual experimental conditions (2 mM thymine, 0.2 mM oxygen) a simple competition calculation shows that only 1 or 2% of H atoms can react with thymine. However, these hydroperoxides are the major radiolysis products of 5,6-dihydrothymine aqueous oxygenated solutions. The cis and trans-6-hydroperoxide compounds (*8*, *9*) can be prepared from cis and trans thymine "hydrate" (*10*) and (*11*). The configurational assignment of trans 6-hydroperoxy-5,6-dihydrothymine (*9*) is supported by the weak $J_{5,6}$ coupling value (1.6 Hz) in the ^{1}H NMR spectrum which can be correlated with a preferential 5,6 di-equatorial hydrogen conformation (KATRITZKY et al., 1969).

5-Hydroperoxymethyluracil (*14*) displays a characteristic absorption in the 260 nm region of the UV spectrum (CADET and TEOULE, 1971a). This 5,6 unsaturated hydroperoxide is obtained chemically by $^-$OOH substitution of the highly reactive hydroxyl function of 5-hydroxymethyluracil (*15*). This compound is also produced in strongly alkaline solutions since hydroxyl radical attack then occurs mainly at the methyl group (MYERS et al., 1965b).

Thermal decomposition of hydroperoxides takes place generating more stable compounds. Three main ways of hydrolytic thymine hydroxyhydroperoxide degradation were observed (Fig. 2.1). The formation of the cis and trans isomers of 5,6-dihydroxy-5,6-dihydrothymine (*6*, *7*) which are the main stable gamma radiolysis products of thymine solution (NOFRE and CIER, 1966; TEOULE and CADET, 1970, 1971) results from the loss of the peroxidic oxygen according to a stereospecific mechanism. The cis configurational assignment was first based on the specific chemical synthesis with $KMnO_4$ and confirmed recently by X-ray structure determination (FLIPPEN, 1973). The mechanism of cis-trans isomerization has been studied using ^{18}O specifically labelled precursors in the 5 or 6 positions. The partial loss of heavy oxygen in the hydroxyl group at C6, subsequent to the epimerisation of C6 in water, is consistent with a ring chain tautomerism involving the formation of an aldehydic acyclic compound (ULRICH et al., 1969; SUBBARAMAN et al., 1975; CADET et al., 1975b).

The loss of water by degradation of the cis or trans isomers of 6(5)-hydroperoxy-5(6)-hydroxy-5,6-dihydrothymine is one of the major processes in aqueous solutions (15 to 20%); as a result, 5-hydroxy-5-methylbarbituric acid (*17*) is generated. The OH radical oxidation of 5,6-dihydroxy-5,6-dihydrothymine (*6*) and (*7*) leads in part to the formation of this barbituric derivative which is highly reactive to OH^- nucleophilic attack at the 6 position. The puckering within the 5,6 saturated pyrimidine ring of these compounds seems to be a determining factor in initiating the endonuclease excision repair of base damages in γ irradiated DNA. A half chair conformation of the six-membered ring has been proposed on the basis of IR and ^{1}H NMR experiments in agreement with the theoretical evaluation of the deformation energy (CHABRE et al., 1966; ROUILLIER et al., 1966).

The hydrolytic opening of the 5,6 bond of thymine hydroxyhydroperoxides produces N^1-formyl-N^2-pyruvylurea (*16*) (TEOULE et al., 1970) which can be considered to be in equilibrium with its cyclic hydantoin form in aqueous solution. This diureid is the major compound generated by the γ irradiation of 5,6-dihydroxy-5,6-dihydrothymine (*6*) and (*7*) aqueous aerated solutions.

Table 2.1. G values of the products of radiation-induced degradation of thymine in aerated aqueous solutions (pH 4, 2.0 mM, 300,000 rad; TEOULE and CADET, 1971; CADET and TEOULE, 1973b)

No. in Figure 2.1	Degradation Products	G
1	Thymine	- 2.6
2	Cis-6-hydroperoxy-5-hydroxy-5,6-dihydrothymine	0.32
3	Trans-6-hydroperoxy-5-hydroxy-5,6-dihydrothymine	0.67
4	Cis-5-hydroperoxy-6-hydroxy-5,6-dihydrothymine	0.08
5	Trans-5-hydroperoxy-6-hydroxy-5,6-dihydrothymine	0.17
6	Cis-5,6-dihydroxy-5,6-dihydrothymine	0.125
7	Trans-5,6-dihydroxy-5,6-dihydrothymine	0.123
8	Cis-6-hydroperoxy-5,6-dihydrothymine	0.027
9	Trans-6-hydroperoxy-5,6-dihydrothymine	0.028
10	Cis-6-hydroxy-5,6-dihydrothymine	} 0.008
11	Trans-6-hydroxy-5,6-dihydrothymine	
12	5-Hydroperoxy-5,6-dihydrothymine	0.062
13	5-Hydroxy-5,6-dihydrothymine	0.016
14	5-Hydroperoxymethyluracil	0.047
15	5-Hydroxymethyluracil	0.017
16	N^1-formyl-N^2 pyruvylurea	0.460
17	5-Hydroxy-5-methylbarbituric acid	} 0.149
18	5-Hydroxy-5-methylhydantoin	
19	Parabanic acid	-
20	Acetylurea	0.010
21	N-Formylurea	0.065
22	Urea	0.075
-	5-Hydroperoxy-5 methylhydantoin	0.005
-	5-Hydroperoxy-5-methylbarbituric acid	0.011
-	Trans-5,6-dihydroperoxy-5,6-dihydrothymine	0.009

N-formylurea (*21*) and N-acetylurea (*20*) result mainly from the radiation-induced decomposition of thymine hydroxyhydroperoxides (*2*, *3*, *4*, *5*) and N^1-formyl-N^2-pyruvylurea (*16*) (TEOULE and CADET, 1971), urea being the degradation compound (DRÁŽIL and RÝZNAR, 1968a). The hydrolysis of N^1-formyl-N^2-pyruvylurea (*16*) gives mainly 5-hydroxy-5-methylhydantoin (*18*). The OH radical oxidation of this hydantoin in aqueous aerated solutions produces parabanic acid which is a common degradation product of most pyrimidine and purine bases. The formation in low yield of 5-hydroxy-5,6-dihydrothymine (*13*) and cis and trans thymine "hydrate" (*10*, *11*) results from hydrolytic decomposition of the corresponding hydroperoxides.

5-Hydroperoxymethyluracil (*14*) is more stable than thymine hydroxy-hydroperoxides. Two main decomposition pathways are observed: the loss of peroxidic oxygen gives 5-hydroxymethyluracil (*15*) and the concurrent dehydration mechanism produces 5-formyluracil which is transformed into uracil by gamma irradiation.

2.2.1.2 Uracil

The study of uracil (*23*) is relevant to that of the radiation-induced degradation of cytosine (*33*). Some products are common for uracil and cytosine radiolysis (see Fig. 2.2). The formation of unstable hydroxyhydroperoxides via short-lived hydroxypyrimidine radicals is observed in irradiated aqueous aerated solutions of uracil (SCHOLES et al., 1960; BARSZCZ and SHUGAR, 1961). The recent identification of the rather stable nitroxide resulting from the trapping of 5-hydroxy-5,6-dihydrouracil-6-yl radical with 2-methyl-2-nitrosopropane (TANIGUCHI and HATAMO, 1975) supports a preferential addition of OH radical at the 5 position of the pyrimidine ring. The resulting cis 6-hydroperoxy-5-hydroxy-5,6-dihydrouracil (*24*) was prepared by chemical synthesis (DUCOLOMB et al., 1973) through the action of a pertrifluoro-acetic acid solution which can be considered as a radiomimetic system. The hydrolytic decomposition of uracil hydroxyhydroperoxides takes place according to three processes: reduction of the peroxidic bond, dehydration, and opening of the 5,6 bond, producing, cis and trans isomers of 5,6-dihydroxy-5,6-dihydrouracil (*25*, *26*) (SMITH and HAYS, 1968), isodialuric acid, N^1-formyl-N^2-glyoxyl urea (*27*) (DUCOLOMB et al., 1971, 1973) and its cyclised derivative, i.e., 5-hydroxy hydantoin (*29*) (PEUZIN et al., 1970) respectively. Alloxan (*31*) mainly results from the hydroxyl radical oxidation of isodialuric acid (*30*). Oxaluric acid and unidentified carboxylic ureids were generated by irradiating alloxan and isodialuric acid. The rather low yield of isobarbituric acid (*32*) observed in the recent gamma radiolysis studies of uracil aqueous aerated solutions can be correlated with mild analysis conditions which reduce the dehydration process of 5,6-dihydroxy-5,6-dihydrouracil (*25*, *26*).

2.2.1.3 Cytosine

Radiation chemical yields (G = 2.08-2.5) of chromophore saturation have been determined in the gamma irradiation of neutral aqueous solutions of cytosine (*33*) (SCHOLES et al., 1960; KRYAKOV, 1972). The OH radicals play an important role in this reaction. Theoretical calculations predict equal reactivities at C5 and C6 sites and both hydroxy radicals are likely to be formed. Further experiments have suggested that one of the positions is preferentially attacked, leading to a single radical species (NICOLAU et al., 1969) but there has been no chemical evidence to illustrate this point. Cytosine hydroperoxides are unstable and hard to isolate in large enough quantities for spectral studies (EKERT and MONIER, 1960; LATARJET et al., 1961; LATARJET et al., 1963). However, kinetic studies have been performed on irradiated cytosine solutions. Hydroperoxides were looked for, using their reactions with I^- at pH 4.0. Measurements carried out either in the presence or absence of molybdate catalyst made it possible to distinguish between H_2O_2 and the hydroperoxides. No hydroperoxide could be detected in neutral and alkaline solutions but at pH 2 a metastable hydroperoxide was formed. The second order rate constant has been evaluated at 1.7 M^{-1} s^{-1} at 25^oC (DANIELS and SCHWEIBERT, 1967), which is a very similar value to that reported for thymine hydroperoxides (1.59 M^{-1} s^{-1} and 7.53 M^{-1} s^{-1}) (SCHOLES and WEISS, 1959). Thermal decomposition of cytosine hydroperoxide at pH 6.5

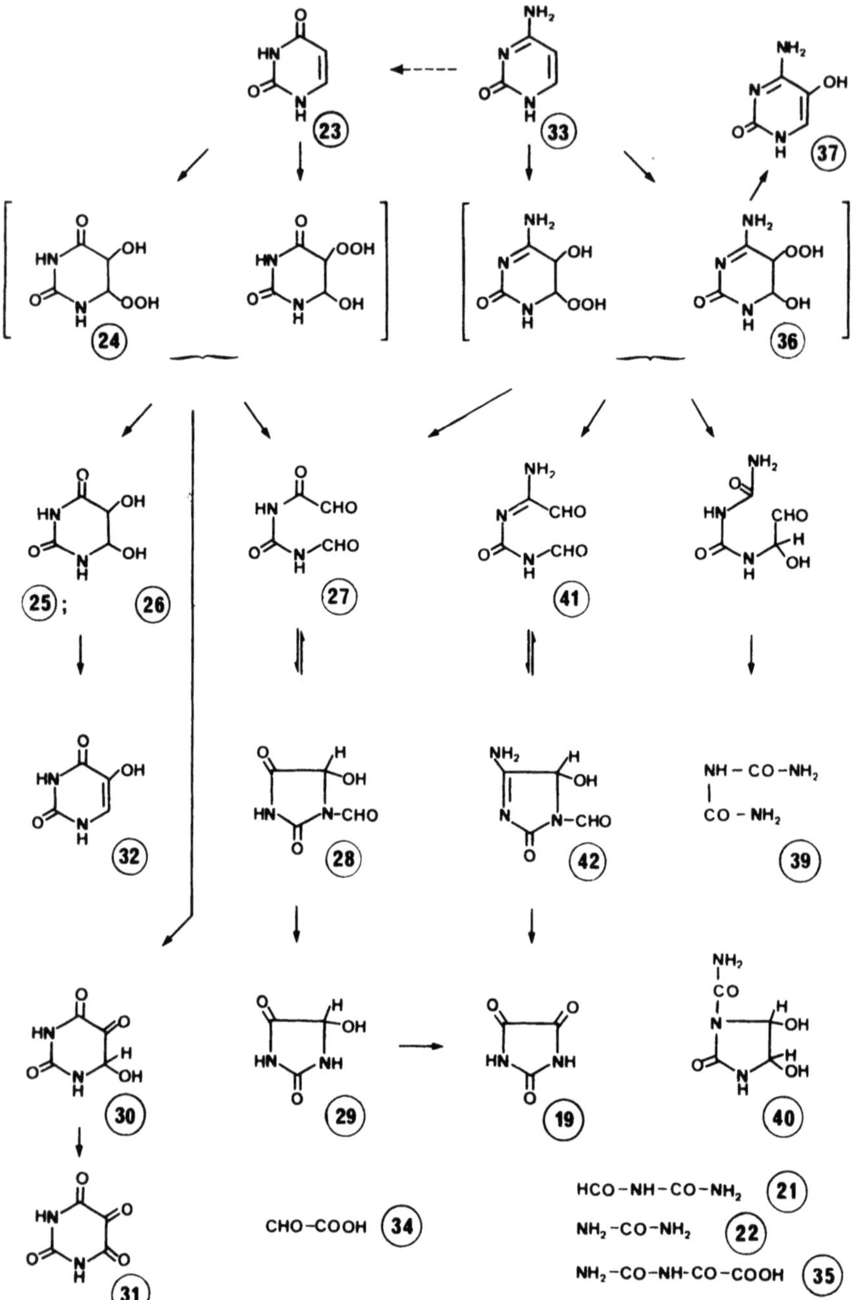

Fig. 2.2. Radiation-induced degradation of cytosine and uracil aerated solutions. (Adapted from DUCOLOMB et al., 1973; POLVERELLI and TEOULE, 1974a, b). Possible pathways are indicated by arrows. (The mechanism of primary events is discussed in II.1)

follows first order kinetics. The temperature dependence (37-52°C) of the effective rate constant for the decomposition of cytosine hydroperoxide in the presence of α-naphtol as inhibitor has the following form: $K = 6.4\times10^5 \exp(-11, 500/RT)\ s^{-1}$ (ZHIZHINA and KRUGLYAKOVA, 1968). The stability depends strongly on the pH and the rapid increase in the decomposition rate occurs close to the pK. As the acid form is the more stable, this suggests that the extra proton on the ring destabilizes the transition state. The formation of cytosine hydroperoxide is relatively independent of cytosine concentration but is much more dependent on oxygen concentration. The G value are respectively 0.11 and 0.58 for an initial concentration in cytosine of 30 mM (DANIELS and SCHWEIBERT, 1967). In spite of these kinetic studies, it has to be noted that the structure of this peroxide has not been established (Fig. 2.2).

The structures (*36*) suggested for cytosine hydroperoxides are hypothetical. These compounds might rapidly yield uracil hydroxyhydroperoxides such as (*24*). These latter products could be key intermediates in the radiolysis of cytosine solution.

Table 2.2. G values of the products of gamma-irradiation-induced degradation of uracil and cytosine in aerated solutions (40 krad, 1 mM, pH 6)

No. in Figure 2.2	Radiation Product	Uracile[a] G = -2.49	Cytosine[b] G = -2.49
25	Cis-5,6-dihydroxy-5,6-dihydrouracil	0.42	0.03
26	Trans-5,6-dihydroxy-5,6-dihydrouracil	0.62	0.10
30	Iso-dialuric acid	0.10	
31	Alloxan	0.06	
27	N^1-formyl-N^2-glyoxylurea	0.72	
29	5-Hydroxyhydantoin	0.18	0.10
32	Isobarbituric acid	0.02	
35	Oxaluric acid + ureids	0.13	0.21
21	N-formylurea	0.27	0.06
22	Urea	0.02	
19	Parabanic acid	traces	0.03
39	Biuret	-	0.06
40	Trans-1-carbamoyl-imidazolidone 4,5 diol	-	0.61
42	4-amino-1 formyl-5 hydroxy-2 oxo-Δ^3 imidazoline	-	0.20

[a] DUCOLOMB et al. (1973)

[b] POLVERELLI and TEOULE (1974b)

About 40% of products are the same as those obtained in the radiolysis of uracil solutions. The structure of uracil glycol was first deduced from its transformation into isobarbituric acid (*32*) in acid solution. Both cis and trans isomers of uracil glycol (*25*, *26*) characterized by comparison with authentic samples, were prepared by chemical synthesis (DUCOLOMB et al., 1973). Isodialuric acid (*30*) which has been found by polarographic studies (PLETICHA-LANSKY and WEISS, 1966) has been recently isolated (POLVERELLI and TEOULE, 1974a, b). However, dialuric acid and other compounds such as alloxan and alloxantin which could result from the equilibrium: dialuric acid + alloxan ⇄ alloxantin have not been actually obtained. 5-Hydroxyhydantoin (*29*) probably results from the cyclisation of N^1-formyl-N^2-glyoxylurea (*27*) (PEUZIN et al., 1970; DUCOLOMB et al., 1971, 1973). Parabanic acid (*19*), formylurea (*21*), oxaluric acid (*35*) and glyoxylic acid (*34*) must be considered as end products of the radiation-induced degradation of cytosine and they are mainly produced at very high doses (POLVERELLI and TEOULE, 1974a, b). It is worthwhile noting that the above compounds are identical to those obtained during gamma radiolysis of uracil. However, uracil (*23*) is found only in very small quantities (KHRYAKOV, 1972; PONNAMPERUMA et al., 1962).

The hydroxyhydroperoxides of cytosine (*36*) could yield cis and trans-5,6-dihydroxy-5,6-dihydroxycytosine by a reductive process. In fact these compounds are not observed in irradiated aerated cytosine solutions. They are probably too unstable to be detected by chromatography but the corresponding product resulting from dehydration, namely 5-hydroxycytosine (*37*), is found. The compounds fall into two groups according to the rupture process of the cytosine peroxide ring. The structure of the major product, trans-1-carbamoyl-imidazolidone 4,5-diol (*40*) (HAHN et al., 1973), may be explained by the rupture of the C4-C5 bond. The ^{1}H NMR coupling constants $J_{5,6}$ of both cis and trans isomers were not observed and the determination of the structures required X-ray analysis (FLIPPEN, 1973). On irradiation of the product (*40*), biuret (*39*) is produced. The other pathway (POLVERELLI and TEOULE, 1973) involving the rupture of the C5-C6 bond is of less importance and leads to the formation of 4-amino-1-formyl-5 hydroxy-2 oxo-Δ^3 imidazoline. This process is very similar to that observed for uracil which leads to N^1-formyl-5-hydroxyhydantoin (*28*).

2.2.1.4 5-Bromouracil

Cells containing DNA in which thymine has been partially replaced by 5-bromouracil (*43*) are more sensitive to killing by ionizing radiation (KAPLAN et al., 1962; SETLOW and SETLOW, 1972; SZYBALSKI, 1974). Isodialuric acid (*30*), isobarbituric acid (*32*), alloxan (*31*), diisobarbituric acid and urea (*22*) were identified as the main products formed on irradiation of aqueous aerated 5-bromouracil solutions (VOLKERT et al., 1967); (see III.3).

The formation of isodialuric acid (*30*) which is the major gamma degradation product, involves OH addition across the 5,6 double bond followed by peroxidation and nucleophilic substitution in the 5 position.

2.2.1.5 Adenine

8-Hydroxyadenine (*46*) is the major compound produced by irradiation of both aerated and deaerated aqueous adenine (*44*) solutions

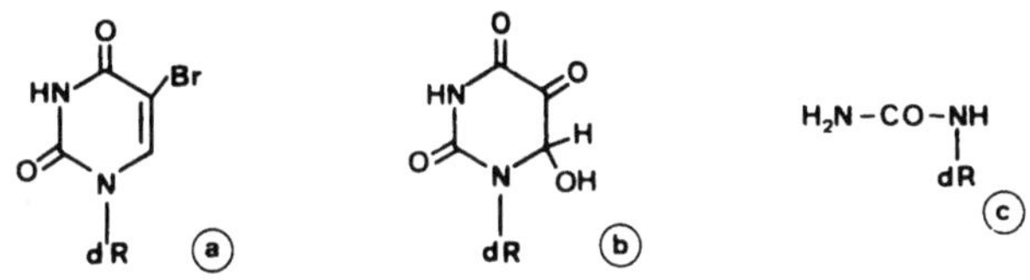

dR = H a: (43) d R = deoxyribosyl a: (71)
b: (30) b: (72)
c: (22) c: (73)

Fig. 2.3. Some radiation products obtained in 5-bromouracil and 5-bromo-2'-deoxyuridine gamma-induced decomposition in aqueous solutions (GILBERT et al., 1967; KOURIM et al., 1971a)

(CONLAY, 1963; VAN HEMMEN and BLEICHRODT, 1971). The same primary reaction takes place in both cases and the results provide evidence for attack by the OH radical at the N7-C8 double bond. This result agrees with quantum chemical calculations of adenine spin densities (PULLMAN and PULLMAN, 1963; MÜLLER and HÜTTERMANN, 1973). 8-Hydroxadenine which is produced at low doses seems to be a key product for other radiolysis compounds of adenine. This product is also formed when adenine reacts with Fenton's reagent (NOFRE et al., 1961).

The peroxidation of the 4,5 double bond has been suggested as another mode of adenine degradation (SCHOLES, 1963), however no peroxides have been detected in irradiated purine solutions under any pH conditions (SCHOLES, 1963). This peroxidation may be subsequent to the hydroxylation of C8 and the formation of degradation compounds recently investigated (Fig. 2.5) is in favour of

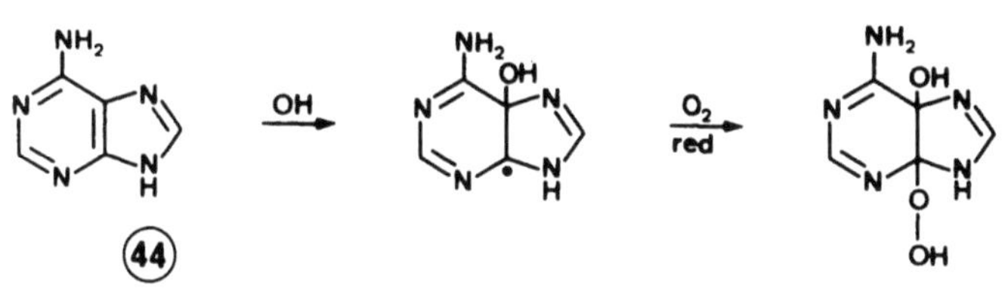

Fig. 2.4. Hypothesis for adenine peroxidation (SCHOLES, 1963)

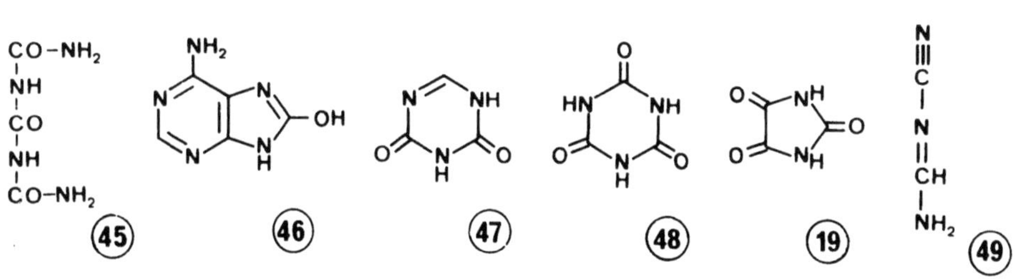

Fig. 2.5. Products of the gamma radiolysis of aerated adenine solutions (MARIAGGI and TEOULE, 1974)

such a mechanism. Some of these substances do not appear at the beginning of the irradiation but may be obtained by further degradation of 8-hydroxyadenine (*46*). Allantoxaidine (*47*), cyanuric acid (*48*), parabanic acid (*19*), biuret (*39*), formylurea (*21*) and triuret (*45*) may be obtained from radiolysis of 8-hydroxyadenine, by the opening of the 4,5 double bond (MARIAGGI and TEOULE, 1974). It seems reasonable to suppose that these substances are produced via the peroxides which would be too unstable to be detected. Furthermore deamination and rearrangement of the 4,5 adduct would account for the yields of ammonia observed ($G \sim 0.5$ at pH 7; SCHOLES and WEISS, 1952). The mechanism of cyanoformamidine (*49*) formation is so far unexplained.

2.2.2 Nucleosides and Nucleotides

A linear decrease of optical density versus the absorbed dose has been observed for pyrimidine and purine nucleosides. The changes in the shape of UV absorption curve for purines and their deoxyribonucleosides were similar. For example, the isobestic point near 273 nm exhibited by adenine was also observed for deoxyadenosine (ULIANA, 1969). This suggests that the UV absorptions depict mainly the changes which occur in the base chromophore component. In fact, such a linear decrease is the result of more complex modifications.

The optical properties exhibited in CD and ORD by nucleosides are due to the chirality of the sugar moiety. Consequently, each 5,6 saturated pyrimidine nucleoside derivative having at least one asymetric carbon within the base ring may be separated into its 5R and 5S (or 6R and 6S) diastereoisomers. The isolation and the characterization of both compounds allow the stereochemistry of addition reactions at the 5,6 double bond to be determined. Furthermore, the C6 absolute configuration of 6-hydroxylated nucleosidic compound induces different conformational properties within the sugar ring between the two diastereoisomers, as demonstrated by 250 MHz ^{1}H NMR analysis (CADET and DUCOLOMB, 1974).

2.2.2.1 Thymidine

The gamma radiolytic decomposition of thymidine (*50*) aqueous aerated solutions essentially results from the action of OH radicals which preferentially react (80%) with the pyrimidine moiety. The attack of the 5,6 ethylenic bond which is the more reactive site leads to the formation of the eight cis and trans diastereoisomers of 5(6)-hydroperoxy- 6(5)-hydroxy-5,6-dihydrothymidine (*51*) to (*58*). The identification of these peroxides was based on their stereospecific reduction into corresponding 5,6-diols and their chemical specific synthesis (CADET, 1973; CADET and TEOULE, 1973b; CADET et al., 1975b). Attack by OH on the exocyclic methyl group is a minor process and leads via the 5-(2'-deoxy-uridilyl) methyl radical to the formation of 5-hydroperoxymethyl-2'-deoxyuridine (*64*) (CADET and TEOULE, 1972b).

The relative yields of the different thymidine hydroxyhydroperoxides depend on the pH of the irradiated solutions. The formation of 6-hydroperoxy-5-hydroxy-5,6-dihydrothymidine (*51, 52, 57, 58*) is still predominant at neutral and slighty acidic pHs. However, this trend is lower than in the case of thymine. Similar conclusions concerning the hydroxyl adducts of thymidine, produced by the Ti Cl_4 - H_2O_2 systems were reached from ESR flow experiments (DERTINGER and NICOLAU, 1970). A possible explanation of this

Table 2.3. G values of the products of thymidine gamma radiation-induced degradation in aerated solutions (1 mM, 150 krad, pH 6) (CADET and TEOULE, 1975c)

		R6	R5	No in the text	G
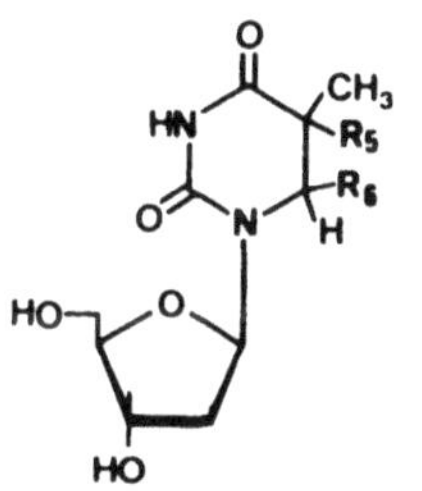	Cis +	OOH	OH	51	0.044
	Cis -	OOH	OH	52	0.048 (52–54)
	Cis +	OH	OOH	53	
	Cis -	OH	OOH	54	
	Trans +	OH	OOH	55	0.070
	Trans -	OH	OOH	56	0.130 (56–58)
	Trans +	OOH	OH	57	
	Trans -	OOH	OH	58	
	Cis +	OH	OH	59	0.092
5,6 peroxides and diols	Cis -	OH	OH	60	0.096
	Trans +	OH	OH	61	0.154
	Trans -	OH	OH	62	0.134
			R1	No in the text	
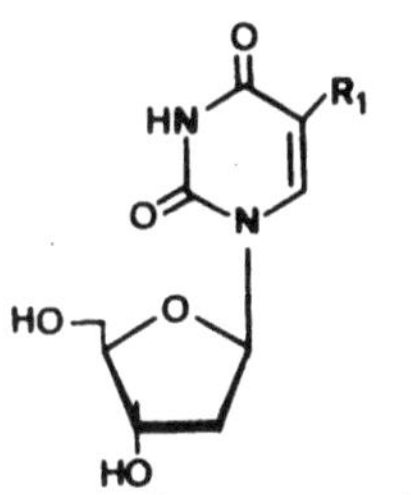			CH_2OOH	63	0.031
			CH_2OH	64	0.044
			CHO	65	0.046
Products from methyl group oxidation					
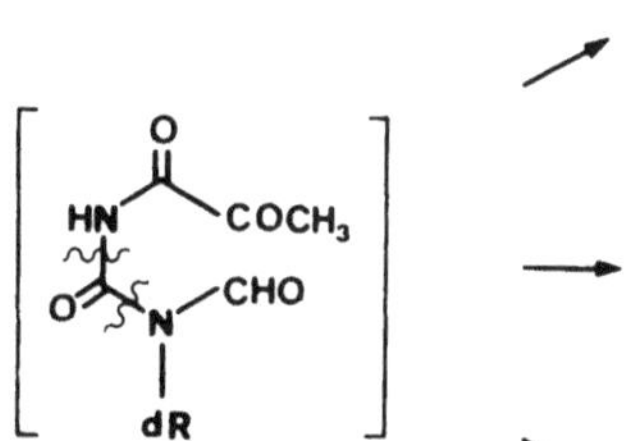			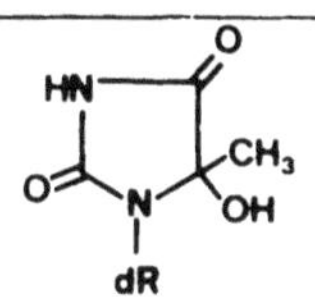	(-) 66	0.025
				(+) 67	0.025
			$CH_3-CO-CO-NH_2$	68	0.490
			CO_2	-	-
			HCO-NH-dR	69	0.380
dR = deoxy-2'-ribosyl			HCO-NH-CO-NH-dR	70	0.100
Pyrimidine ring opening derivatives.					
	Thymine and thymine derivatives			1 to 22	o.450

loss of the selectivity of OH addition in the 5 position lies with the modification of the charge distribution due to the electronegative sugar substituent. In acidic solution (pH < 1.6) 5-hydroperoxy-6-hydroxy-5,6-dihydrothymidines (*53, 54, 55, 56*) are preferentially generated (CADET, 1973; CADET and TEOULE, 1975a, b, c)

The main stable products result from the thermal decomposition of unstable thymidine hydroxyhydroperoxides (*51*) to (*58*) (T 1/2 ≃ 5 days in aqueous solution at 37°C). The two degradation processes which were evidenced showed, in part, some analogies with 6(5)-hydroxy-5(6)-hydroperoxy-5,6-dihydrothymine decomposition. The reduction of thymidine hydroxyhydroperoxides (*51*) to (*58*) into 5,6-diol (*59, 60, 61, 62*) (35%) takes place according to a stereospecific mechanism (CADET and TEOULE, 1975a, b, c). The opening of the 5,6 pyrimidine bond (65%) is followed by three kinds of competitive reactions. The formation of the 5R* and 5S* diastereoisomers (*66, 67*) of the 1-(2-deoxy-β-D-ribofuranosyl)-5-hydroxy-5-methylhydantoin (CADET and TEOULE, 1972a, b) is a minor process (5%). The hydrolysis of the unstable diureid gives N-(2-deoxy-β-D-ribofuranosyl) formylurea (*70*) (20%). The more important decomposition pathway (40%) is not observed in the radiation-induced degradation of thymine. The formation of N-(2-deoxy-β-D-ribofuranosyl) formamide (*69*), which is the major product of the irradiation of an aqueous solution of thymidine, results from the rupture of the 1,2 bond (CADET and TEOULE, 1975a, b, c). Furthermore the decomposition of the pyrimidine fragment which is released gives pyruvamide (*68*) and carbonic acid. The irradiation of stable nucleosidic derivatives which mainly produces the rupture of the N-glycosidic bond involves a preferential OH attack of the sugar moiety. 5-Hydroperoxy-methyl-2'-deoxyuridine (*63*) is relatively stable in aqueous solutions at 37°C (T 1/2 ≃ 5 days); 5-hydroxymethyl-2'-deoxyuridine (*64*) and 5-formyl-2'-deoxyuridine (*65*) are generated by radiation-induced and hydrolytic decompositions (CADET, 1973).

The respective yields of these major degradation processes i.e., decomposition of thymidine, release of free base and thymine derivatives, and production of modified nucleosides versus radiation doses are shown in Figure 2.6. The rupture of the N-glycosidic bond (≃ 20%) is due to the OH radical attack (25%) of the sugar ring.

2.2.2.2 *5-Bromo-2'-Deoxyuridine*

The gamma-radiation induced degradation of 5-bromo-2'-deoxyuridine (*71*), a structural thymidine analogue, shows some similarities with that of 5-bromouracil (*43*) (see Fig. 2.3). The major decomposition product resulting from the transformation of the unstable peroxide intermediate was identified as 1-(2-β-D-deoxyribofuranosyl) isodialuric acid (*72*) by comparison with an authentic sample (KOURIM et al., 1971a, b). The large release of HBr was measured by acid-base titration. N-(2-D-deoxypentosyl) urea (*73*), and products resulting from the rupture of the N-glycosidic bond i.e., 2-D-deoxyribose, urea and 5-bromouracil (*43*), were also characterized. The initial G values of 5-bromo-2'-deoxyuridine

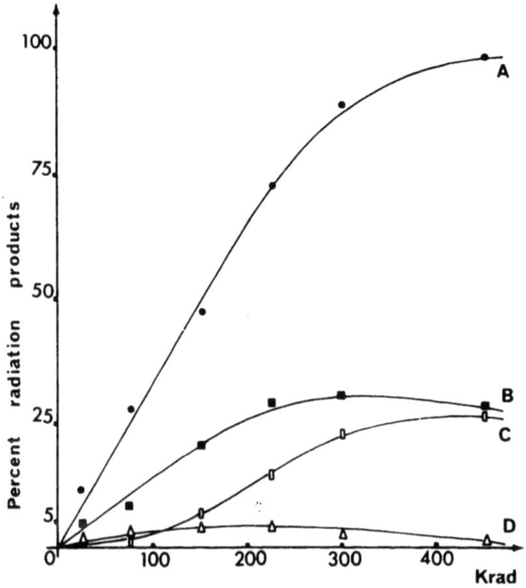

Fig. 2.6. Radiolysis of aerated solutions of thymidine (1 mM, pH 6). *A*: thymidine degradation; *B*: 5,6-saturated nucleoside derivatives of thymidine (*51*) to (*67*); *C*: 5,6-saturated base derivatives of thymine; *D*: thymine. Thymine (*D*) results from the rupture of N glycosidic bond. Compounds *B* result from the saturation of 5,6 pyrimidine double bond. Compounds *C* result from the rupture of N glycosidic bond and saturation of 5,6 double bond (CADET and TEOULE, 1975c)

(*71*) decomposition, HBr and 1-(2-β-D-deoxyribofuranosyl) isodialuric acid (*72*) formations are dependent upon the concentration of 5-bromo-2'-deoxyuridine (*71*). These facts are explained in terms of competition reactions of reductive species between the nucleoside and molecular oxygen. The predominant reaction of e^-aq with the pyrimidine ring at high nucleoside concentration leads to the destruction of 5-bromo-2'-deoxyuridine (*71*) subsequent to the loss of the bromine atom (KOURIM et al., 1971a, b).

2.2.2.3 *Uridine*

The major gamma radiation-induced decomposition product of uridine in aqueous aerated solutions was identified as N(β-D-ribofuranosyl) formamide (*74*). As expected for an N substituted formamide, two conformers resulting from the hindered rotation of the C-N amide bond were observed in the ^{1}H NMR spectra in Me_2SO-d_6 or aqueous solutions. The four cis (-), cis (+), trans (+) and trans (-) diastereoisomers of 5,6-dihydroxy-5,6-dihydrouridine (*75*) (Fig. 2.7) are the main 5,6 saturated nucleoside derivatives. The hydrolytic decomposition of the unstable uridine hydroxyhydroperoxides shows common features with that of 6(5)-hydroperoxy-5(6)-hydroxy-5,6-dihydrothymidine. The dehydration rearrangement seems to be a minor process. The formation of 1-(β-D-ribofuranosyl) isodialuric acid was not detected in the thermal degradation of cis

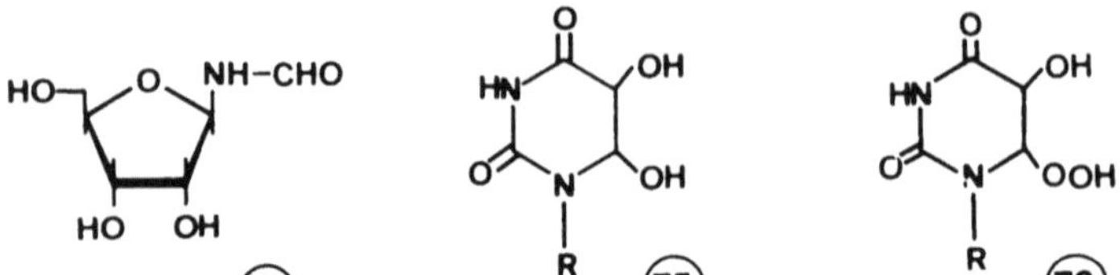

74 75 76

R = ribosyl

Fig. 2.7. Some uridine derivatives produced in aerated solutions (DUCOLOMB et al., 1974b)

6-hydroperoxy-5-hydroxy-5,6-dihydrouridine (*76*) generated by the pertrifluoracetic oxidation of uridine (DUCOLOMB et al., 1974b).

2.2.2.4 Deoxyadenosine

Deamination reactions occur in X-irradiation of purine nucleosides and nucleotides and depend on pH (SCHOLES and WEISS, 1952). In aerated solutions, the major deoxyadenosine radiolytic product besides adenine is 7,8-dihydro-8-oxo-2'-deoxyadenosine (Fig. 2.8) (*77*). This experiment parallels the observation made in the case of adenine (*44*) (VAN HEMMEN and BLEICHRODT, 1971; MARIAGGI and TEOULE, 1976).

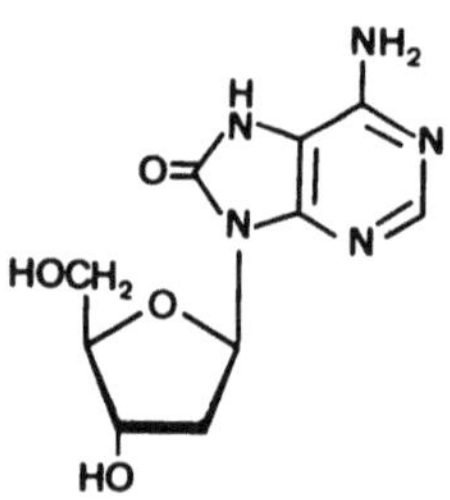

Fig. 2.8. 7,8-Dihydro-8-oxo-2'-deoxy-adenosine (MARIAGGI and TEOULE, 1976)

2.2.2.5 Nucleotides - Thymidylic Acid and Uridylic Acid

The gamma radiolysis of TMP and UMP in aqueous aerated solutions exhibits some important analogies. The pyrimidine ring of both nucleotides was determined as the predominant site of hydroxyl radical attack, in agreement with earlier investigations (SCHOLES et al., 1960). However the sugar moiety of UMP is destroyed to a larger extent (40%) (DUCOLOMB et al., 1974a) than expected. In both cases there was no evidence for the presence of nucleosides. The thermal decomposition of the very unstable thymidylic acid or uridylic acid hydroxyhydroperoxides gave rise to N-formamido nucleotide derivatives (G ≈ 1.2 for TMP) (DUCOLOMB et al., 1974b). This predominant degradation mechanism (CADET et al., 1974) explains the important loss of radioactive label

($^{14}CO_2$) observed in the irradiation of aerated aqueous thymidine and thymidylic acid solutions (Table 2.3; WARD, 1971, 1972).

2.2.2.6 Dinucleotides = Thymidylyl-(3'→5')-Thymidine

Irradiation of oxygenated aqueous solutions of TpT shows its behaviour to resemble that of thymidine more than that of thymidylic acid. Preliminary results show the presence of both diols and formamide defect types in nearly similar yields and the release of free 3' and 5'-phosphate mononucleotide and thymine is observed.

2.3 Nucleic Acid Constituents in Deaerated Aqueous Solutions

2.3.1 Free Bases

2.3.1.1 Thymine

The 5,6 saturated thymine derivatives resulting from the reaction of reducing species (e^-_{aq}, H atoms) have been characterized and separated (TEOULE and CADET, 1969). The common radiation-induced degradation compounds produced in both the presence and absence of oxygen have been identified as cis and trans isomers of 5,6-dihydroxy-5,6-dihydrothymine (*6, 7*) and 5-hydroxymethyluracil (*15*). EKERT (1962) was successful in identifying 5,6-dihydrothymine (*78*) (Table 2.4). The formation of 5-hydroxy-5,6-dihydrothymine (*13*) earlier hypothesized by KHATTAK and GREEN (1966b) was confirmed by NOFRE et al. (1965b), NOFRE and CIER (1966), and later by DRÁŽIL and RÝZNAR (1968a, b). The cis isomer of 6-hydroxy-5,6-dihydrothymine (*10*) which was assumed to be very unstable was prepared by chemical synthesis and characterized among thymine gamma degradation products (NOFRE and OGIER, 1966; CADET and TEOULE, 1971b). The low quantum yield for photohydration rather than thermal instability accounts for the formation in small amounts of this cis "hydrate" (*10*) in UV irradiated aqueous solutions (FISHER and JOHNS, 1973). The trans isomer of 6-hydroxy-5,6-dihydrothymine (*11*) was also produced in gamma irradiated solutions and synthetized from the cis one (*10*) (CADET and TEOULE, 1971b). The trans configurational assignment was supported by the low 5,6 ^{1}H NMR splitting ($J_{5,6} \simeq 2.1$) reflecting a preferential diequatorial conformation for these vicinal protons within the puckered pyrimidine ring.

The formation of the different 5,6 saturated thymine compounds is likely to be accounted for by the general dismutation mechanism proposed recently (HAYSOM et al., 1972). Electron transfer between two hydropyrimidyl radicals produces an ion pair. Protonation of the negative ion can lead to 5,6-dihydrothymine (*78*) and solvolytic substitution of the carbonium ion yields 5-hydroxy-5,6-dihydrothymine (*13*) or cis and trans thymine "hydrate" (*10, 11*). In a similar way the solvolysis of the positive ion derived from 5(6)-hydroxy-5,6-dihydrothymyl-5(or 6) radical produces cis and trans isomers of 5,6-dihydroxy-5,6-dihydrothymine (*6, 7*). The unknown compounds (G ≃ 0.2) would result from a dimerisation reaction which appears as a minor process in the case of thymine. A radical recombination may be involved in the formation of 5,5'-dimethyl-5,5' dibarbituric acid from irradiation of thymine glycol (HAYASHI et al., 1973).

2.3.1.2 Uracil

The 5,6 saturated derivatives produced in irradiated oxygen-free aqueous solutions of uracil were first identified as 5,6-dihydrouracil (*79*) 6-hydroxy-5,6-dihydrouracil (*80*) 5-hydroxy-5,6-dihydrouracil (*81*) and both isomeric forms of uracil glycol (*25*, *26*) (KHATTAK and GREEN, 1966b; SMITH and HAYS, 1968). The main radiation-induced decomposition products were recently isolated by paper chromatography by SHRAGGE et al. (1974b) who showed their dimeric nature. Two possible structures for these dimers were postulated on the basis of mass spectrometric analysis which displays a characteristic fragmentation pattern from the observed molecular ions. However, no definite evidence was obtained for the positions of the covalent bond between the two pyrimidine rings. A mechanism involving competitive electron transfer reaction and radical combination has been suggested to explain the nature and the yield of the radiation products (SHRAGGE et al., 1974a). The pH dependence on dimer formation and uracil decomposition was discussed in terms of dissociative effects due to the low pK_a values for pyrimidine radicals ($pK \approx 7$). In slightly alkaline solutions (pH 8.5) the odd electron of the dissociated OH or e^-aq adduct could be repelled by the negative charge. Consequently, electron transfer would be enhanced over dismutation and would lead to both monomer formation and parent uracil restitution (Table 2.4).

2.3.1.3 Cytosine

At neutral pH both the hydrated electron and the ·OH radical react with cytosine Cyt (*33*) to give the hydropyrimidyl radical ·Cyt H and the hydroxypyrimidyl radical ·Cyt OH. Secondary reactions involving ·Cyt H and ·Cyt OH lead to regeneration of cytosine and a decrease in the observed destruction of cytosine (HOLIAN and GARRISON, 1966). These radicals produce dihydrocytosine derivatives which subsequently give rise by deamination to uracil (*23*) and 5,6-dihydrouracil derivatives (KAMAL and GARISSON, 1965). Ammonia produced in the process was assayed by the Conway diffusion method, so estimating the influence of scavengers. On dehydration of 5,6-dihydroxy-5,6-dihydrocytosine, 5-hydroxycytosine (*37*) is found as the major compound with only traces of 6-hydroxycytosine (*82*) (KHATTAK and GREEN, 1966a); 5,6-dihydroxycytosine (*83*) (isouramil) was identified by comparison with an authentic sample. Dihydrouracil (*79*) has been found when sodium formate was used as scavenger (KAMAL and GARRISON, 1965) but so far 6-hydroxy-5,6-dihydrouracil (*80*) and 5-hydroxy-5,6-dihydrouracil (*81*) have not been detected (Table 2.5).

Table 2.4. G values of irradiation products of deaerated solutions of 2,4-dioxopyrimidines. The amounts of optical diastereoisomers are similar. Ex: (+) dihydrothymidine, G = 0.05; (-) dihydrothymidine, G = 0.05

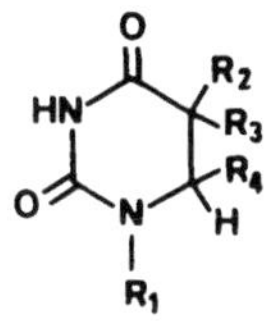

Thymine derivatives R_1 = H; R_2 = CH_3	No. in the text	Vacuum, neutral pH; 1.0 mM G = -0.85 9-135 krad (1)	N_2 bubbling, neutral, pH: 0.1 mM 45 krad (2)
R_3 = R_4 = H	78	0.30	0.04
R_3 = H; R_4 = OH; cis	10	0.17	0.09 to 0.18
trans	11	0.039	
R_3 = OH; R_4 = H	13	0.32	0.04
R_3 = OH; R_4 = OH cis	6	0.041	0.34
trans	7	0.012	0.06
5-Hydroxymethyluracil	15	0.057	-
dimers	-	0.2	-
Uracil derivatives R_1 = R_2 = H		**N_2 bubbling pH 5; 1.0 mM G = 3.3 (3)**	**N_2 bubbling pH 8.3; 1.M (G = -0.8 (4)**
R_3 = R_4 = H	79	0.1	0.1
R_3 = H R_4 = OH	80	0.1	0.2
R_3 = OH R_4 = H	81	0.1	0.1
R_3 = OH R_4 = OH cis	25		
trans	26	0.3	0
dimers		2.4	0.4
Thymidine derivatives R_1 = 2' deoxyribosyl; R_2 = CH_3		**Vacuum; neutral pH; 0.5 mM G = 0.6 (5)**	
R_3 = R_4 = H	88	0.112	
R_3 = H_4 R = OH cis	89	0.058	
trans	90	0.029	
R_3 = OH; R_4 = H	91	0.035	
R_3 = OH; R_4 = OH cis	59,60	0.030	
trans	61,62	0.11	
5-hydroxymethyl-2'-deoxyuridine	64	0.005	

References: (1) CADET and TEOULE (1971b); (2) NOFRE and CIER (1966); (3) SHRAGGE et al. (1974a; (4) SHRAGGE et al. (1974a); (5) CADET (1973); for radical scavengers: INFANTE et al. (1973b)

Table 2.5. G values of irradiation products of deaerated cytosine solutions (10 mM, pH 6, 10^{22} eV/ml) (KHATTAK and GREEN, 1966a, b)

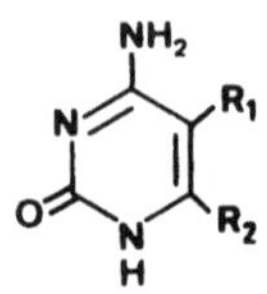

	No. in the text	G
Cytosine	33	-1.8
R_1 = OH R_2 = H	37	0.46
R_1 = H R_2 = OH	82	Traces
R_1 = R_2 = OH	83	Traces
Uracil	23	0.30
Ammonia	-	0.4

2.3.1.4 Adenine

The formation of the radiation products so far identified may be accounted for by attack at the C8-N7 double bond (HEMS, 1960; CONLAY, 1963; PONNAMPERUMA et al., 1962). The H(˙AH) and OH (˙AOH) adducts to adenine probably have the structures shown in Figure 2.9 (VAN HEMMEN and BLEICHRODT, 1971; MÜLLER and HÜTTERMANN, 1973).

From experiments carried out with radical scavengers it could be concluded that 8-hydroxyadenine (*46*) is formed from OH radicals. No solvated electrons are required in the production of 4,6-diamino-5-formamidopyrimidine (*84*) and 6-amino-8-hydroxy-7,8-dihydropurine (*85*) is only obtained under conditions where electron adducts of adenine are produced (VAN HEMMEN and BLEICHRODT, 1971). Reconstitution reactions involving adenine radicals have been suggested to explain the low G values (CONLAY, 1963; VAN HEMMEN and BLEICHRODT, 1971) of adenine degradation in deaerated solutions.

2.3.2 Free Bases in the Presence of a Second Solute

2.3.2.1 Bases and Ethanol

The two main products of the γ irradiation of a 10 mM deaerated aqueous thymine solution containing ethanol (0.5 M) were identified by spectroscopic methods as the cis and trans isomers of 6-α-hydroxyethyl-5,6-dihydrothymine (*86*) (BROWN et al., 1966; Fig. 2.10). In these experimental conditions the bulk of OH radicals and to a lesser extent H atoms are scavenged by the alcohol (SCHOLES and WILLSON, 1967) whereas $e^{-}aq$ react preferentially with thymine. The reactivity of hydroxyethyl radicals is approximately five powers lower than that of $e^{-}aq$ for thymine.

A reasonable mechanism for the formation of ethanol addition products is based on the combination reaction of the anionic thymine radical, or its protonated derivative, with a second alcohol radical. As expected, the disproportionation reaction between 5,6-dihydrothymyl radicals gives rise to 5-6-dihydrothymine (*78*).

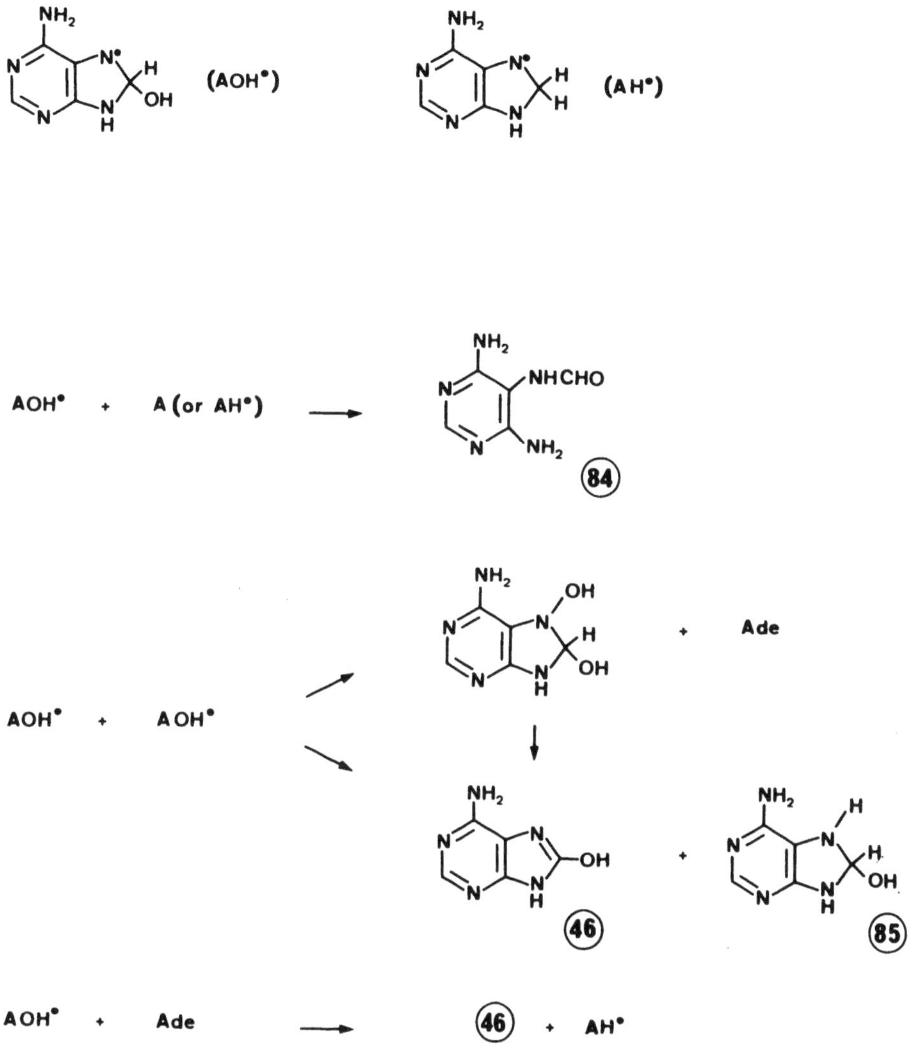

Fig. 2.9. Suggested mechanism for the formation of radiation products of deaerated adenine solutions. (Adapted from VAN HEMMEN and BLEICHRODT, 1971)

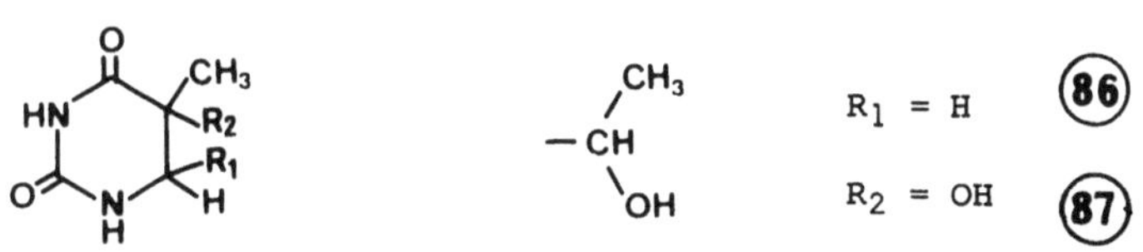

Fig. 2.10. Ethanol adducts of thymine (BROWN et al., 1966; ZAREBSKA and SHUGAR, 1972)

ZAREBSKA and SHUGAR isolated and identified 6-α-hydroxyethyl-5-hydroxy-5,6-dihydrothymine (*87*) which resulted from the reaction between the 5-hydroxy-5,6-dihydrothymine-6-yl and hydroxyethyl radical (ZAREBSKA and SHUGAR, 1972). The radiolysis of aqueous solutions of adenine and isopropanol leads to the addition of an alcohol moiety [$(CH_3)_2$ COH] at C8 (STEINMAUS et al., 1971).

2.3.2.2 Thymine and Radiosensitizers

The use of radiosensitizing agents is a possible method of overcoming hypoxic radioresistance of most solid tumors to X or γ-ray. On the basis of numerous pulse radiolysis and ESR experiments two main mechanisms for radiosensitization of nucleic acid constituents have been suggested: electron transfer from radiation-induced radicals in the DNA component producing oxidized products and the binding of sensitizing drugs to organic free bioradicals. Steady state radiolysis studies of thymine solutions in the presence of the sensitizers metronidazole, nitrofuroic acid, N-ethylmaleimide (NEM), 2,2,6,6-tetramethyl-4-piperidone-N oxide (or TAN) and menadione have been recently reported (CADET et al., 1975a) (see III.3.2).

Chromatography of the different irradiated solutions showed the presence of common products. The low yield in the formation of 5,6-dihydrothymine (*78*) 5-hydroxy-5,6-dihydrothymine (*13*) and thymine "hydrates" (*10*) and (*11*) was consistent with the effective electron scavenging properties of the radiosensitizers. The main radiation-induced oxidation products formed via carbonium ion subsequent to electron transfer reaction, were characterized as cis and trans 5,6-dihydroxy-5,6-dihydrothymine (*6, 7*), 5-hydroxy-5-methyl barbituric acid (*17*), N^1-formyl-N^2-pyruvylurea (*16*) and 5-hydroxy-5-methylhydantoin (*18*). Menadione was ineffective in binding to the thymine radicals. The ratio of binding to oxidation appeared to be zero for menadione, 1.6 for NEM and 4.4 for TAN. Both the latter compounds showed an important binding capacity in agreement with earlier experiments. The major TAN addition product was identified as the trans isomer of 5-hydroxy-6 (2,2,6,6-tetramethyl-4-oxo-piperidinoxy) 5,6-dihydrothymine on the basis of IR, ESR, ^1H NMR and mass spectrometry analysis. The formation in low yield of the unsaturated compound resulting from the binding of triacetoneamine-N-oxyl to 5-methyluracil-5-yl radical was observed (CADET et al., 1975a).

2.3.2.3 Pyrimidine Bases and Surfactants

Miscellar systems provide a simple environment model for radiation-induced radical reactions. Steady-state radiolysis studies of aqueous solutions of thymine and uracil in the presence of three types of surfactants viz., cationic hexadecyltrimethylammonium bromide (CTAB), anionic sodium dodecyl sulfate (NaLS) and neutral polyoxyethylene (15) nonylphenol (Igepal CO-730) were initiated by INFANTE et al. (1973b). These surfactants were selected for their low critical micelle concentration in order to minimize radical scavenging and direct radiation effects in the aggregation studies. The marked decrease of the overall product formation above the critical micelle concentration in both oxygenated and deaerated irradiated thymine or uracil solutions cannot be accounted for in terms of competition for simple radical species between pyrimidine bases and surfactants. The decrease of the thymine or uracil decomposition yield does not correlate with the different scavenging ability of the surfactants (FENDLER et al., 1973).

2.3.2.4 Pyrimidine Bases and Amino Acids

Gamma rays induce binding of nucleic acid bases with aromatic and thio-amino acids. The most efficient are tryptophan, phenylalanine cysteine, cystine, histidine (YAMAMOTO, 1973). This type of reaction may be of interest because DNA is surrounded by proteins in vivo conditions.

2.3.3 Nucleosides and Nucleotides

2.3.3.1 Thymidine

In deaerated aqueous solutions, the four diastereoisomeric forms of 5,6-dihydroxy-5,6-dihydrothymidine (*59*) to (*62*) are produced in lower yields than in the presence of oxygen. The hydrogenated derivatives have been identified as 5R and 5S 5,6-dihydrothymidine, 5R* and 5S* 5-hydroxy-5,6-dihydrothymidine. Warming a cis (-) or cis (+) form of thymidine "hydrate" in pyridine water mixture gives rise to the three other isomers (CADET and DUCOLOMB, 1974).

The initial rupture of the C6-N1 bond, due to a ring chain tautomerism followed by a keto-enolisation reaction, accounts for hydrogen exchange in the 5 position. A similar mechanism is likely to be involved in the epimerization and deuteration of the important photobiological uridine "hydrates" in D_2O solutions (CADET and DUCOLOMB, 1974).

2.3.3.2 Purine Nucleosides and Nucleotides

The radiation-induced degradation of aqueous nucleoside and nucleotide solutions leads to three types of reactions:

1. Breakage of the N-glycosidic bond followed by the liberation of the base (SCHOLES and WEISS, 1952).

2. Change in the structure of the base component without rupture of the N-glycosidic bond. The chemical aspects of radiolytic degradation of the base moiety of deoxyadenosine solutions do not differ much from those observed with adenosine. Comparison of radiation-induced degradations of guanine and guanosine is not possible because no results are available for guanine. Ring opening of the purine nucleosides and nucleotides was recognized as long ago as 1952 (SCHOLES and WEISS, 1952). The irradiation of deaerated solutions of guanosine, guanylic acid, adenosine and adenylic acid gives rise to some opening of the imidazole ring probably via the addition of OH radicals to the N7-C8 double bond

to produce the corresponding 4-amino-5-formamidopyrimidine riboside or ribotide. Acid hydrolysis of these compounds releases the sugar residue forming the corresponding 4-amino-5-formamidopyrimidine (HEMS, 1958, 1960b).

The formamido derivatives are subsequently deformylated into triamino compounds in acid medium (CONLAY, 1963). The experiments were carried out with both a pulsed beam of 15 MeV electrons and a gamma source; 8-hydroxy-7,8-dihydro adenosine is likely to be formed but so far, it has not been isolated.

3. The reactions involving the sugar and the base moieties. 8,5' Cyclisation has been found in gamma irradiation of deaerated solutions of adenosine 5' monophosphate. The structure of 8,5'-cyclodeoxyadenylic acid was unambiguously demonstrated by periodic oxidation, hydrolysis and borohydride reduction (KECK, 1968; KECK et al., 1966). 8,5'-cyclo-2-deoxyadenosine has been characterized in gamma-irradiated solution of 2-deoxyadenosine (MARIAGGI et al., 1976).

2.4 Nucleic Acid Constituents in Solid State or in Frozen Solutions

In order to study the reactivity of the purine ring, the cleavage of tritium from crystalline adenine $^3H2$ and adenine $^3H8$ was determined. The threshold doses for the release of tritium under gamma irradiation appeared to be within the range of 2×10^7 rad for adenine $^3H2$ and 3×10^5 rad for adenine $^3H8$. These values are higher than for thymine-methyl-(^3H) (MERWITZ, 1972).

Thymine dimers are produced under gamma irradiation of thymine in the solid state (NOFRE et al., 1965a). The gamma irradiation of frozen thymine solutions at 77 K gives cis syn thymine dimer as a major product (CADET, 1976).

2.5 DNA in Aerated and Deaerated Solutions

Ionizing radiations may produce a variety of chemical base damage in DNA structure. Some of the modified bases remain attached to the DNA phosphate sugar backbone and the other ones are eliminated.

2.5.1 Modifications of the Base Moieties in the Polynucleotide Chain

2.5.1.1 UV Absorption and IR Spectra

The occurrence of chemical modifications, deamination and ring fission in the heterocyclic bases was established as long ago as 1949 (SCHOLES et al., 1949). The loss of UV absorption has been used as a reasonable approximation for the measurement of the extent of base destruction (SCHOLES, 1963). It is well known that the observed optical densities are the sum of the increase due to the breakage of hydrogen bonding between the bases and the decrease due to the saturation of double bonds (SCHOLES et al., 1960; COX et al., 1955). The optical densities of irradiated DNA solutions at pH 3.2 have been found to decrease linearly with the radiation dose. In these conditions, the polymer is denatured and only the influence of the saturation of chromophore groups is taken into account (SCHOLES et al., 1960). The G values of base degradation are decreased by proteins contained in the DNA solutions (JUST and SCHWARZ, 1969). IR spectroscopy was shown to be applicable for studying chemical modifications in DNA and deoxyribonucleo-protein (YOSHII and TCHOE, 1966; ZHILTSOVA et al., 1973).

2.5.1.2 Formic or Perchloric Acid Hydrolysis - Base Destruction

Direct measurements of the extent of the base destruction in the DNA chain has been performed by formic or perchloric acid hydrolysis at 175°C. The liberated bases were separated by chromatography and the amount of the bases not destroyed was determined by UV absorption of the eluates (SCHOLES et al., 1960; HEMS, 1960a; SCHOLES, 1963). At small doses, the radiosensitivity of the bases in DNA solutions increases in the following order:

Thymine > Cytosine ∿ Adenine > Guanine (Fig. 2.11).

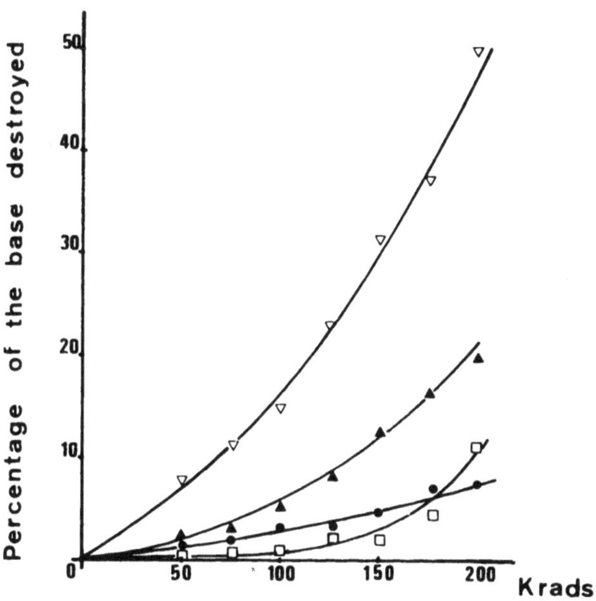

Fig. 2.11. Percentage of bases destroyed in DNA against dose during the gamma irradiation of aqueous DNA. ▽ Thymine; ▲ Cytosine; ● Adenine; □ Guanine. (ULIANA et al., 1971) Solutions (pH 7.1 M NaCl, optical density of DNA, 1.9 max)

The percentage of protection provided by the polymeric structure of nucleic acid may be defined as:

$$P = \frac{N - n}{N} \times 100$$

Where

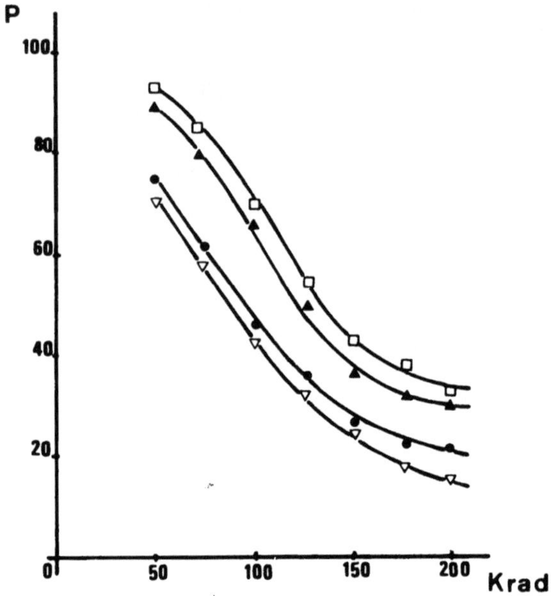

Fig. 2.12. Percentage of protection provided by the polymeric structure of DNA against dose (10^{-4} M, pH 7, 1 M NaCl). □ Guanine; ▲ Cytosine; ● Adenine; ∇ Thymine. (Adapted from ULIANA et al., 1971)

N = number of base molecules destroyed in free nucleotides for a given dose
n = number of base molecules destroyed in DNA for the same dose.

It should be noted that these results were obtained at low DNA concentration (Fig. 2.12).

The degradation of DNA in solution takes place by an indirect process (SHARPATYI et al., 1968). It is for this reason that the attack of purine and pyrimidine bases becomes more efficient when DNA strands are separated. This protection seems to be dependent on hydrogen bonding between bases. DNA is subject to local structural fluctuations which result in frequent opening and closing or "breathing" of base pairs. In double-stranded DNA, the bases are located inside the double helix, and the adenine-thymine bonds which are more labile are responsible for the smaller polymeric protection of thymine and adenine (ULIANA et al., 1971). The use of synthetic homopolyribonucleotides enabled the careful study of the effects of polymerization and organization upon the radiosensitivity of the constitutive bases in the macromolecular structure (EKERT and TISNE, 1966; WARD and URIST, 1967; BARSZCZ and SHUGAR, 1972). Quantitative results of base destruction occurring when the polymers are irradiated in the random state differ only slightly from the values determined in monomer irradiations. In the double or triple helical states the heterocyclic bases are protected to an extent ranging from 25 to 50% (WARD and URIST, 1967).

2.5.1.3 Peroxidation of Pyrimidines

In aerated solution, the initial event which takes place in the destruction of pyrimidine bases in DNA is the hydroxyhydroperoxidation of the 5,6 double bond (SCHOLES et al., 1956; DANIELS et al., 1957; SCHOLES and WEISS, 1959; SCHOLES, 1963; LATARJET et al., 1963; LATARJET, 1972). These peroxides have been shown to be attached to the polynucleotide chain because they are precipitated in trichloroacetic acid (SCHOLES and WEISS, 1959; SCHOLES et al., 1960). The radio-induced peroxides in DNA, decay according to two concurrent first order processes. When the thermal decay rate constants k and activation energies ΔE obtained for the hydroperoxides of the monomer constituents are compared with those measured in aqueous DNA it can be seen that values obtained for deoxycytidylic acid are in good agreement with the first decaying component of DNA. The activation energy ΔE of the slow decaying component is close to that obtained for thymine and thymidylic acid hydroperoxides. It has been suggested that the two hydroperoxides detected in DNA are derived from thymine and cytosine bases (ZHIZHINA et al., 1964; SCHWEIBERT and DANIELS, 1971). Post irradiation processes taking place in air-saturated DNA solutions such as additional increases of final product concentration were explained by hydroperoxide decay and radical appearance (ZAKATOVA and SHARPATYI, 1971). A hyperweak luminescence took place during degradation processes of peroxides (ZHIZHINA et al., 1965).

2.5.1.4 Indirect Determination of the Fairly Stable Products

A reductive assay for the quantitative determination of thymine peroxides, diols and "hydrates" produced by the radiolysis of aqueous DNA solutions has been suggested (HARIHARAN and CERUTTI, 1971). The method is based on the reduction of these compounds with sodium borohydride to 2-methylglycerol (*92*), 1-3 dihydroxy-2-methyl-propane (*93*), and to the urea derivatives (HARIHARAN and CERUTTI, 1972).

The quantity of diols produced by reduction of an irradiated sample is a measurement of the radiolysed thymine. Unirradiated deoxynucleotides are inert towards sodium borohydride. The method is not absolutely specific because various degradation compounds are likely to react but it is sensitive. The procedure has been applied to the study of repair damage in *Micrococus radiodurans* (HARIHARAN and CERUTTI, 1971, 1972). It has been shown that nucleic acid base damage rather than strand break is probably the predominant lesion caused by ionizing radiation. The same procedure has been applied to the study of the repair processes of gamma irradiated chinese hamster ovary cells (MATTERN et al., 1973).

A simpler and more sensitive assay has been recently developed to measure the products of the 5,6-dihydroxy-5,6-dihydrothymine type (ROTI ROTI and CERUTTI, 1974). Acetol (*94*) is formed from such substances by base-acid treatment (Fig. 2.13). The acid precipitable fraction in CCl_3COOH obtained from irradiated nucleoprotein preparations was dissolved in 0.2 N KOH and kept 1.5 h

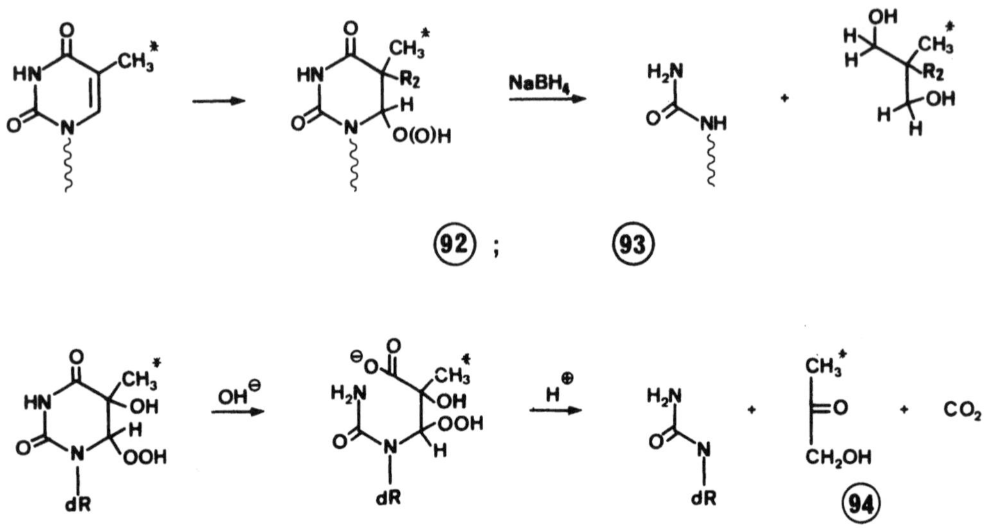

Fig. 2.13. Assays for DNA thymine damage induced by ionizing radiations (HARIHARAN and CERUTTI, 1971, 1972)

at room temperature. The mixture was then acidified and incubated at 70°C for 15 min. The yield of the radioactive fragment produced per ring saturated product was determined to be 20%. The influence of the shielding of DNA in chromatine by chromosomal proteins was measured. The relative reactivities observed were free DNA >> chromatin > reconstituted DNA - histone complexes > reconstituted DNA nonhistone chromosomal proteins. This type of study could shed some light about the binding sites of chromosomal protein in the minor or in the major groove of DNA (ROTI ROTI et al., 1974).

2.5.1.5 *Direct Determination of the Fairly Stable Defects*

Thymine Component. The direct acid hydrolysis of gamma-irradiated solutions of DNA labelled $^{14}CH_3$ on the thymine component has been recently investigated (TEOULE et al., 1974a, b; TEOULE, 1975). The method is based on the fact that the N glycosidic bond of 5,6 saturated thymine derivatives are more easily hydrolysed than that of thymine from the DNA-phosphate-sugar backbone. A good selection for the experimental conditions must be realised. For example, at too low a temperature, thymine glycols (*6*) and (*7*) are not liberated from the polynucleotide chain whereas at too high a temperature they would be destroyed. Warming for 16 h at 90°C in 98% formic acid was found to be convenient. The liberation of 5,6-dihydrothymine (*78*) and 5-hydroxy-5,6-dihydrothymine (*13*) is carried out at a lower temperature (TEOULE et al., 1976). The products resulting from the hydrolysis were separated by thin layer silicagel chromatography and compared with control samples.

In aerated solution, 5,6-dihydroxy-5,6-dihydrothymine-DNA type defect (*95*) and 5-hydroxy-5-methylhydantoin-DNA type defect (*96*)

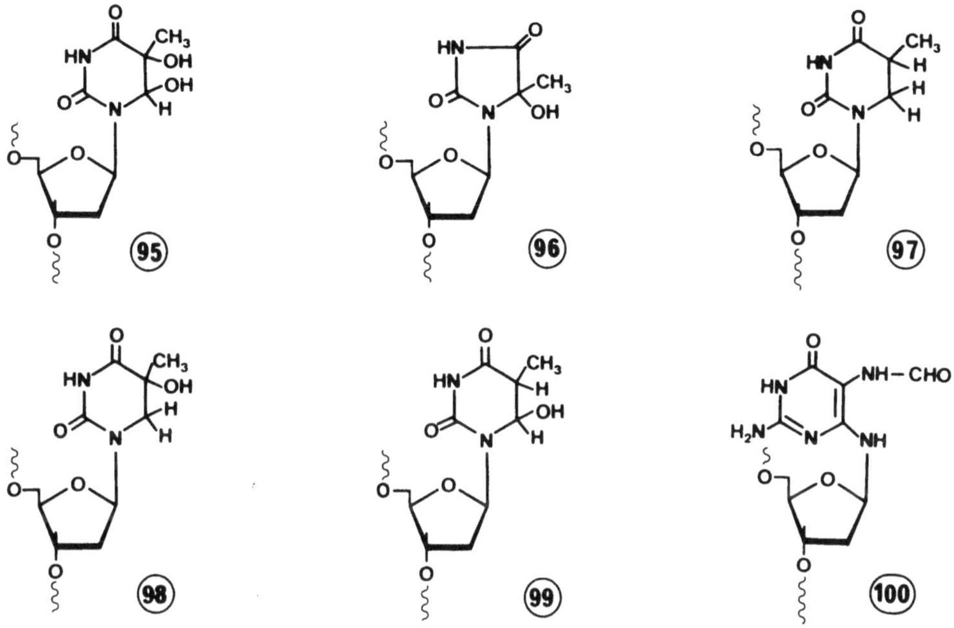

Fig. 2.14. Defects characterized by direct hydrolysis of the irradiated polynucleotide chain (HEMS, 1960a; TEOULE et al., 1974a)

are produced. It is worthwhile noting that under the acid experimental conditions of hydrolysis used, cis-trans isomerisation occurs in 5,6-dihydroxy-5,6-dihydrothymine (*6*) and (*7*) and thus the relative percentage of the two forms is unknown. However, both isomers are likely to be produced as DNA-thymine-defects because cis and trans (*6*) and (*7*) forms were always observed in the radiolysis of solutions of thymine, thymidine, thymidylic acid and TpT (Fig. 2.14).

In gamma-irradiated deaerated solutions, the DNA defects of the types (*97*) to (*100*) have been observed. By acid hydrolysis the following substances have been identified: cis and trans 5,6-dihydroxy-5,6-dihydrothymine (*6, 7*), 5,6-dihydrothymine (*78*), 5-hydroxy-5,6-dihydrothymine (*13*). By treating DNA irradiated in deaerated solution with a crude extract of *E. coli* the same products of thymine damage (*6, 7, 13*) and (*78*) were released from the polymer chain. In addition, 6-hydroxy-5,6-dihydrothymines (*10*) and (*11*) which are unstable in hot acid solutions (Figs. 2.1 and 2.14) were identified.

Cytosine Component. The cytosine component has been recently investigated in gamma irradiated solutions of cytosine-^{14}C-2 labelled DNA. However, under the acidic conditions used for thymine no uracil glycols could be detected since uracil glycol (or cytosine glycol) attached to the DNA chain is likely to be dehydrated into isobarbituric acid, and the C-N glycosidic bond which links the resulting unsaturated compound is too stable to be easily hydrolysed (POLVERELLI et al., 1976).

Guanine Component. 2,4-Diamino-5-formamido-6-hydroxypyrimidine which arose from the attack on the guanine moiety was produced through warming of the irradiated deaerated DNA solution at pH 7, 80°C for 1 h (Fig. 2.14). This product has been thought to be attached to the sugar (*100*) by a labilised glycosidic linkage. The corresponding formamidopyrimidine obtained from the attack on adenine has not been found in the irradiated solution of deoxyribonucleic acid. It is however possible that the amounts of this compound were too small and escaped detection (HEMS, 1960a).

Radiation-Induced Binding of Protein Constituents with Nucleic Acids. In chromatin, bridging is likely to take place between DNA and protein or between bases of the chain (LEHNERT and OKADA, 1965). The radiation induced binding of tryptophan to DNA in solution appears markedly higher than with soluble RNA (YAMAMOTO, 1973). The nucleic acids addition reactions and their biological effects have been reviewed in a recent book (K.C. SMITH, 1976).

2.5.2 Products Released from the Polynucleotide Chain

2.5.2.1 ^3H-H_2O from DNA Thymine Methyl-(^3H)

Tritium abstraction in the form of (^3H)-H_2O from thymine-methyl-(^3H) by OH radicals resulting from water radiolysis leads to the formation of tritiated water in DNA, chromatin and mammalian cells (CERUTTI, 1974; SWINEHART et al., 1974), probably through the transient formation of 5-methylene-uracil radicals.

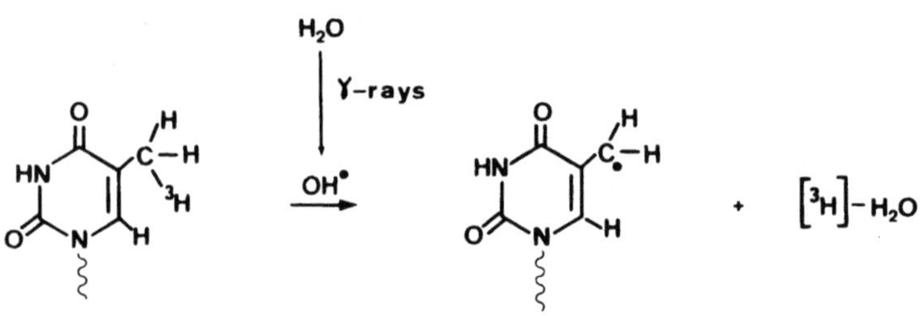

Fig. 2.15. Formation of ^3H_2O by irradiation of DNA-thymine-methyl-^3H (SWINEHART et al., 1974)

The susceptibility of the thymine-methyl groups to indirect radiation action is related to the amount of (^3H)-H_2O produced. Using this simple relationship it is possible to assess the effects of DNA strandedness. In contrast to single stranded ϕX 174 DNA, the overall shapes of the curves for double stranded DNA *E. coli* are sigmoidal (SWINEHART et al., 1974) (Fig. 2.16).

However, at low absorbed doses, the dose-response curve was linear. It has been demonstrated that the reactivity of the thymine methyl group is about 1000 times lower in human lung fibroblasts W_1 38 cells than in free DNA (CERUTTI, 1974).

2.5.2.2 Ammonia

An irradiated solution of DNA was steam distilled at pH 9.2 under vacuum and the ammonia in the distillate measured by the method of NESSLER (SCHOLES and WEISS, 1952; SCHOLES et al., 1960). The

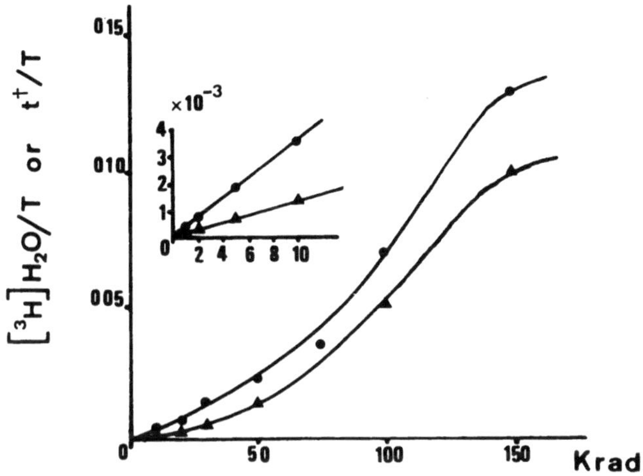

Fig. 2.16. Gamma-ray induced formation of ^3H_2O from DNA thymine-methyl-^3H (30 μM, TMP residue in DNA, 10 μM NaCl). ▲ Radiation products (t) measured by the borohydride reaction. ● Formation of H_2O (T: thymine radioactivity). Inet: dose-effect curves up to 10 krad (SWINEHART et al., 1974)

high deamination yield (G NH_3 ∿ 20) can be explained by the participation of excited water molecules or by the decomposition of degradation products. G (NH_3) measured by the Conway method is reduced by a factor of about 2 in the presence of propylgallate or inositol which are efficient inhibitors of free radical processes (EMANUEL, 1963) and also OH radical scavengers. The contribution of ·OH to the deamination process is about 50% whilst the remaining 50% has been attributed to O_2^- for DNA concentrations below 2 mM (ZAKATOVA and SHARPATYI, 1971). Deamination is one of the main processes of DNP degradation (0.055%) in aqueous solutions [G(NH_3) = 4 ± 0.5], and has been shown to be most important in the histone fragment of DNP. Inositol reduced the G(NH_3) value by a factor of 2 (SHARPATYI et al., 1973).

2.5.2.3 *Free Bases*

The release of free bases results from attack on the sugar moiety (II.3). G values for adenosine and cytosine released from calf thymus DNA oxygen-free solution irradiated with 15 MeV electrons are slightly higher than those for guanine and thymine (HEMS, 1960). The number of bases released is approximately one quarter of the bases destroyed. The amounts of the compounds without charge i.e., free bases and nucleoside-like products, increase linearly under gamma irradiation of N_2-saturated solutions of DNA (500 μg/ml) in a dose range from 0 to 700 krad (ULLRICH and HAGEN, 1971). From regression analysis of the curve slope the G values are evaluated as 0.16 ± 0.01 when the DNA is kept neutral and as 0.18 ± 0.015 when it is treated with alkali after irradiation. There is slight release at zero rad (HEMS, 1960; ULLRICH and HAGEN, 1971; TEOULE et al., 1974a, b) which may be attributed to autoradiolysis products having a labile N-glycosidic bond due to the oxidation of the sugar component. The percentage of the various types of products with respect to the total amount is given in the following table (ULLRICH and HAGEN, 1971).

Nature	Percentage
Guanine	13
Cytosine + Deoxyguanosine	6.7
Adenine + Deoxyadenosine	39.1
Thymine	4.7

No traces of thymidine or deoxycytidine were found in oxygen free or aerated solutions of irradiated DNA (TEOULE et al., 1974a, b; POLVERELLI et al., 1976).

2.5.2.4 Products of Base Component Degradation - Release of Carbon Dioxide from the Thymine Fragment

The 2,4-diamino-5-formamido-6-hydroxypyrimidine arising from the guanine moiety is formed when calf thymus DNA is irradiated with 15 MeV electrons (HEMS, 1960a). A small amount of uracil glycol is released when *E. coli* DNA cytosine ^{14}C is irradiated by gamma rays (POLVERELLI et al., 1976).

The elimination of thymine radiolysis products from the polynucleotide chain due to gamma irradiation of aerated *E. coli* DNA thymine ^{14}C solution has been studied by chromatography. Microreactions performed on the products which were eluted from chromatography spots, when compared with authentic samples, allowed the characterisation of six radiolysis products, namely pyruvamide (*68*) cis-5,6-dihydroxy-5,6-dihydrothymine (*6*), trans 5,6-dihydroxy-5,6-dihydrothymine (*7*), 5-hydroxy-5-methylhydantoin (*18*), 5-hydroxy-5-methylbarbituric acid (*17*) and N^1-formyl-N^2-pyruvylurea (*16*) (TEOULE et al., 1974a, b; TEOULE, 1975). To study the chemical mechanism of their release from the polynucleotide chain, their relative percentages were plotted against absorbed dose. The extrapolation of the curves to zero dose shows that the relative percentage of thymine with respect to the sum of the degradation products released is then close to 80% and that of pyruvamide to 20%. It is worth noting that no pyruvamide was obtained when free base thymine was irradiated in aerated aqueous solutions. On the contrary, pyruvamide was produced when thymidine and thymidylic acid were irradiated and this shows that pyruvamide is not liberated via thymine but directly from the polynucleotide chain. The formamido group remains linked on the sugar phosphate backbone and carbon dioxide provided by the carbon 2 of thymine fragment is immediately released under in vitro irradiation of DNA. In vivo, the measurement of carbon dioxide should be a very simple and sensitive test to follow the degradation of the DNA thymine fragment (Fig. 2.17).

In irradiated deaerated solutions, thymine was the only compound released at doses close to zero, and the other compounds, namely 5,6-dihydrothymine (*78*) and 5-hydroxy-5,6-dihydrothymine (*13*) require two elementary reactions in order to be liberated from the polynucleotide chain.

Chapter 3

Radiation-Induced Degradation of the Sugar in Model Compounds and in DNA

3.1 Introduction

In the present chapter the indirect effect on the sugar in aqueous solution is treated since only a negligible amount of chemical-analytical work has been published with respect to the direct effect. The indirect effect on DNA originates from the attack of the water radicals (OH, e^-_{aq}, H) on the substrate. Solvated electrons (e^-_{aq}) and H atoms are scavenged by the base moieties (see II.1) but an appreciable part (10-20%) of the OH radicals is expected to react with the sugar moiety by hydrogen abstraction as indicated by studies on model compounds (SCHOLES et al., 1960; JUNG et al., 1969; HOARD et al., 1974). Although at first sight this small percentage might seem negligible, important processes such as strand breaks and base release are induced by the sugar radicals. Hence it appeared worthwhile to study the radiation chemistry of the sugar moiety. This has been done by using model compounds such as 2-deoxy-D-ribose, D-ribose, D-ribose-5-phosphate, thymidine and nucleotides. With the knowledge obtained from the study of the model compounds it became possible to investigate part of the sugar damage in DNA itself. The analysis of the carbohydrate products was made possible by recent developments of analytical techniques, especially GC-MS and the borodeuteride reduction method (DIZDAROGLU et al., 1974).

3.2 Free Radical Reactions of Sugars and Sugar Phosphates

Carbohydrate free radicals are formed in hydrogen abstraction reactions caused by reactive radicals, e.g., OH and H. Solvated electrons are not very reactive toward carbohydrates (PHILLIPS et al., 1966) or phosphate esters. OH radicals and H atoms abstract carbon-bound hydrogen to give the primary carbohydrate radicals. The studies on the radiation chemistry of carbohydrates have provided good evidence for a series of free radical reactions typical for this class of compounds.

1. Radicals from vicinal diols rapidly eliminate water, e.g.:

$$\dot{C}HOH\text{-}CH_2OH \longrightarrow H_2O + CHO\text{-}\dot{C}H_2 \qquad (1)$$

This reaction was initially investigated in the case of ethylene glycol by ESR (BULEY et al., 1966; LIVINGSTON and ZELDES, 1966; BAZHIN et al., 1966; GILBERT et al., 1972; STEENKEN et al., 1974), product analysis (SEIDLER and VON SONNTAG, 1969; VON SONNTAG and THOMS, 1970) and pulse radiolysis (BANSAL et al., 1973) and has been noted to be one of the governing processes in the radiation chemistry of carbohydrates.

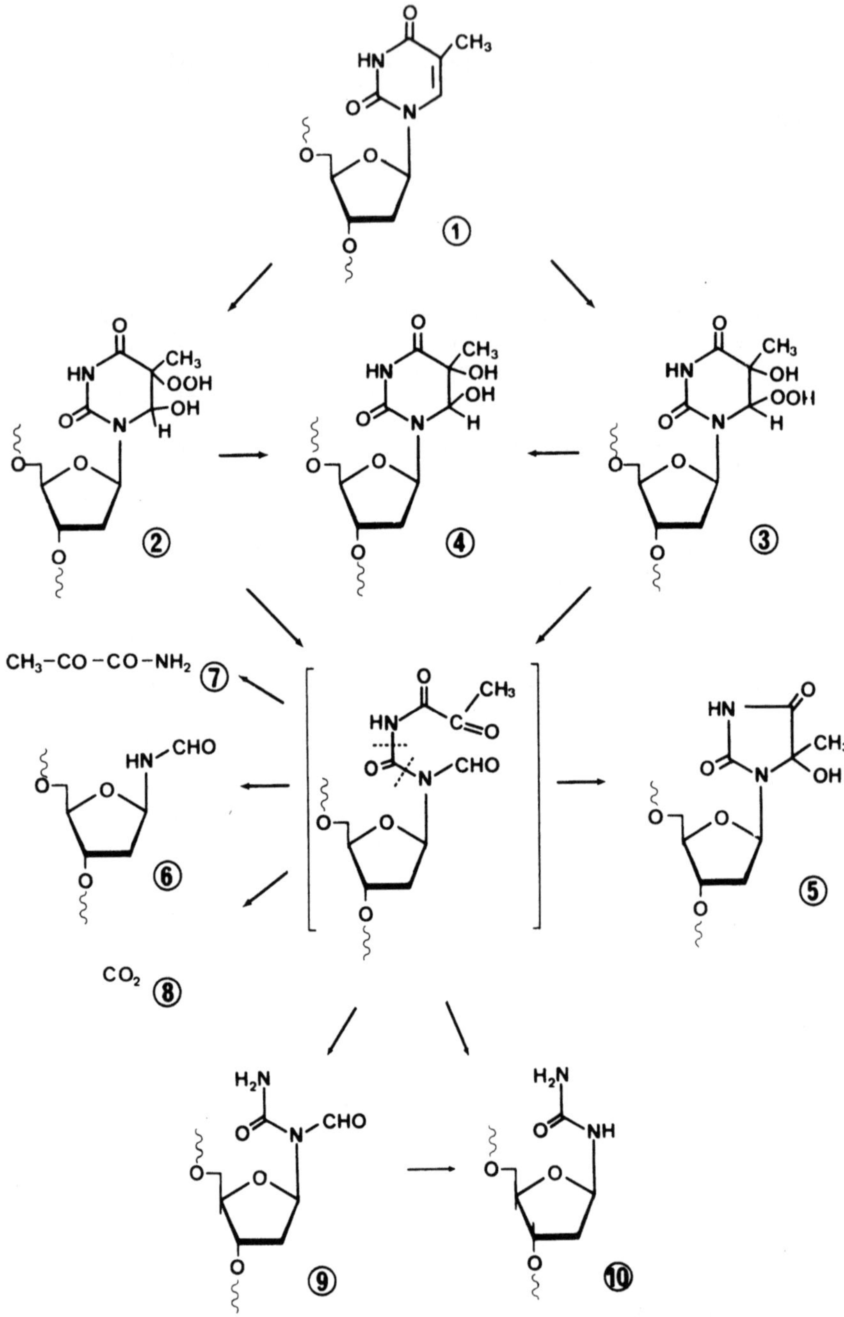

Fig. 2.17. Scheme of the chemical modifications of the thymine moiety in DNA gamma irradiated in aerated solutions: (1) thymine (in DNA chain); (2), (3) thymine peroxides (in DNA chain); (4) thymine glycol (in DNA chain); (5) 5-hydroxy-5-methylhydantoin (in DNA chain); (6) formamide (in DNA chain); (7) pyruvamide (released); (8) carbon dioxide (released); (9) N-formyl urea (in DNA chain); (10) urea (in DNA chain)

2. There are many examples to show that radicals of the type $\dot{C}HOH-CH_2X$ eliminate HX (GILBERT et al., 1972; MATSUSHIGE et al., 1975). For the sugars and sugar phosphates it is of special interest to note that the O-R group (as encountered in the acetal function) and especially the phosphate group are good leaving groups (EDGE et al., 1971; SAMUNI and NETA, 1973; STEENKEN et al., 1974; STELTER et al., 1974; DIZDAROGLU et al., 1976a).

$$R\dot{C}OH-CH_2OPO_3H^- \longrightarrow R-CO-\dot{C}H_2 + H^+ + HPO_4^- \qquad (2)$$

There is evidence that a radical cation is an intermediate in reactions where the OH group is substituted by an ether function (GILBERT et al., 1972). In other systems the same type of mechanism is expected to occur.

3. In the radiolysis of D-ribose-5-phosphate in aqueous solution and of crystalline carbohydrates, products have been observed whose formation is best explained if allyl-type radicals are their precursors, e.g., reaction (3) (STELTER et al., 1975c; VON SONNTAG et al., 1976)

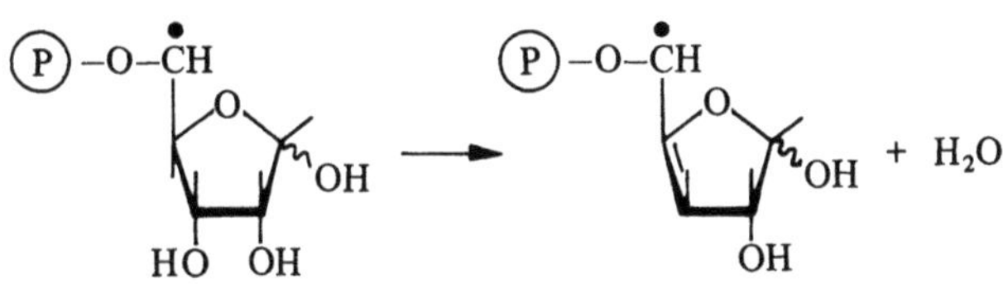

(3)

4. Another type of free radical reaction quite often encountered in the radiation chemistry of carbohydrates is a rearrangement involving the ring oxygen, e.g.:

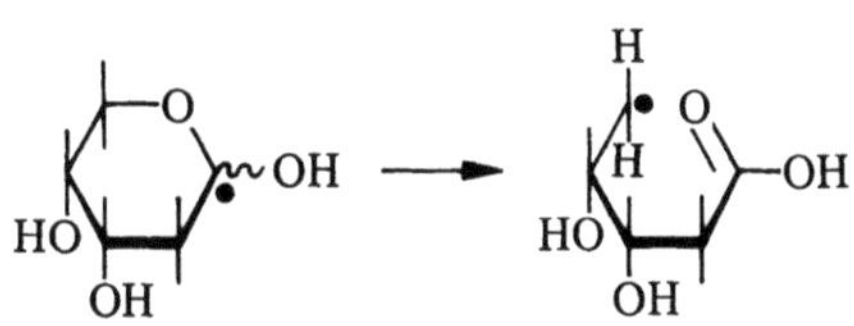

(4)

A rearrangement of this type has been observed in a radical derived from 2-methoxytetrahydropyran (HUYSER, 1960) and in a series of carbohydrates (e.g., HARTMANN et al., 1970; VON SONNTAG and DIZDAROGLU, 1973; DIZDAROGLU et al., 1973; VON SONNTAG et al., 1974; DIZDAROGLU et al., 1975a). In the crystalline state, processes of this type can be propagating steps in radical chain reactions leading to 5-deoxy-aldonic acids.

5. The fast hydrolysis of a glycosidic bond has been postulated for cases where the free spin of the radicals is at the carbon next to this bond, e.g.:

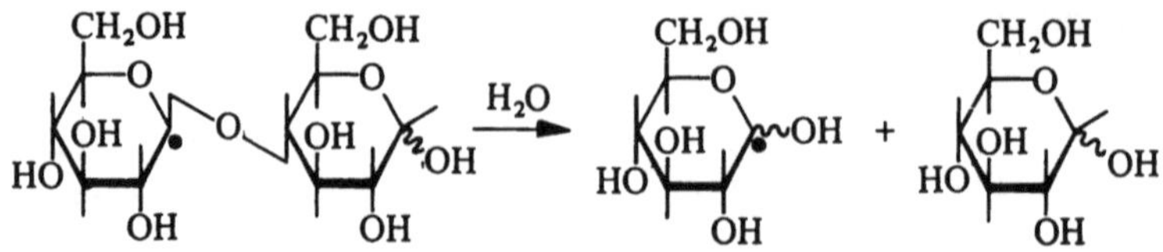
(5)

This mechanism has been postulated to account for a number of products observed in the γ-radiolysis of cellobiose aqueous solutions (VON SONNTAG et al., 1976). A similar reaction may have to be considered in the case of liberation of unaltered bases from irradiated nucleosides, nucleotides and nucleic acids.

6. Radicals of the type RCHOH-$\dot{C}$O readily eliminate carbon monoxide (STEENKEN and SCHULTE-FROHLINDE, 1973):

$$R\text{-}CHOH\text{-}\dot{C}O \longrightarrow CO + R\dot{C}HOH \qquad (6)$$

This reaction can take place in D-ribose after attack at C1, when the ring is opened in a similar way as presented in reaction (5).

7. In the presence of oxygen processes 1-5 are often overtaken by the fast reaction of oxygen with the primary radicals to give peroxyl radicals:

$$R\text{-}\dot{C}OH\text{-}R' + O_2 \longrightarrow R\text{-}\underset{\displaystyle O\text{-}O^{\bullet}}{\underset{|}{C}}OH\text{-}R' \qquad (7)$$

Products then arise from the reactions of peroxyl radicals as given below [reactions (13)-(15)].

The primary carbohydrate radicals and those generated in the processes 1-7 undergo a series of reactions leading to the end products:

8. Disproportionation of highly substituted hydroxyalkyl radicals such as 2-hydroxy-propyl-(2) radicals (e.g., BLANK et al., 1975) or the primary radicals derived from carbohydrates by H abstraction is more favoured than a combination of these radicals.

This disproportionation leads to a compound with an alcohol function (possibly under inversion at an optically active centre (KIRSCH et al., 1975) and to a compound which contains a carbonyl function [reaction (8a)]. Enols can also be formed as intermediates [reaction (8b)] (e.g., BLANK et al., 1975) and hence in carbohydrates stereoisomers are found among the reaction products (DIZDAROGLU et al., 1975a).

$$2\ \text{-}\dot{C}OH\text{-}CHOH\text{-} \longrightarrow \text{-}CHOH\text{-}CHOH\text{-} + \text{-}CO\text{-}CHOH\text{-} \qquad (8a)$$

$$2\ \text{-}\dot{C}OH\text{-}CHOH\text{-} \longrightarrow \text{-}CHOH\text{-}CHOH\text{-} + \text{-}COH{=}COH\text{-} \qquad (8b)$$

9. Radicals derived from the water elimination process dimerize or react with reducing species if present in sufficient amounts.

They are readily reduced by the primary carbohydrate radicals [reaction (9)] as well as by Fe^{2+} [see reaction (12)] giving rise to deoxy sugars.

The majority of the dimeric products are formed by combination of this type of radical (VON SONNTAG and THOMS, 1970; DIZDAROGLU et al., 1972).

$$-\dot{C}H\text{-}CO- + -\dot{C}OH\text{-}CHOH- \longrightarrow -CH_2\text{-}CO- + -CO\text{-}CHOH- \qquad (9)$$

10. The abstractive power of $-\dot{C}H\text{-}CO-$ type radicals is in some cases strong enough to abstract hydrogen atoms from the starting material, thus initiating chain reactions [reaction (10)] (SEIDLER and VON SONNTAG, 1969; VON SONNTAG and THOMS, 1970; BURCHILL and PERRON, 1971; VENTER et al., 1972).

$$-\dot{C}H\text{-}CO- + -CHOH\text{-}CHOH- \longrightarrow -CH_2\text{-}CO- + -\dot{C}OH\text{-}CHOH- \qquad (10)$$

A similar behaviour is shown by the radicals formed in process (4).

11. Oxidation of radicals by various oxidants (OX) can lead to carbonium ions:

$$^{\cdot}RH + OX \longrightarrow {}^{+}RH + {}^{\cdot}OX^{-} \qquad (11)$$

Metal ions and also radiolytically formed hydrogen peroxide or even oxygen may serve as oxidants (KOCHI, 1973; O'NEILL et al., 1975). The role of radiosensitizers in the liberation of phosphate from nucleotides has been discussed on this basis (RALEIGH et al., 1973; RALEIGH et al., 1973a). There is reasonable evidence that radicals themselves may disproportionate not only by hydrogen atom transfer but also by electron transfer (HAYSOM et al., 1972).

12. Reduction of radicals by various metal ions may be rather selective. Fe^{2+} appears to reduce radicals of the type $-\dot{C}H\text{-}CO-$ but not the primary carbohydrate radicals ($-\dot{C}OH\text{-}CHOH-$) [reaction (12)] (STELTER et al., 1976).

$$-\dot{C}H\text{-}CO- + Fe^{2+} + H^{+} \longrightarrow -CH_2\text{-}CO- + Fe^{3+} \qquad (12)$$

The rate constant for the reaction of Fe^{2+} with $\dot{C}H_2\text{-}CHO$ has been determined to be $4 \cdot 10^5$ M^{-1} s^{-1} (GILBERT et al., 1973).

13. The peroxyl radicals formed in the presence of O_2 can eliminate HO_2:

$$-\underset{\substack{|\\ O\text{-}O^{\cdot}}}{C}OH\text{-}CHOH- \longrightarrow -CO\text{-}CHOH- + HO_2^{\cdot} \qquad (13)$$

The decay of α-hydroxy peroxyl radicals has been studied by pulse radiolysis (STOCKHAUSEN et al., 1969; STOCKHAUSEN et al., 1970; STOCKHAUSEN and HENGLEIN, 1971; RABANI et al., 1974).

14. The second order decay of the peroxyl radicals can lead to C-C bond fragmentation via oxyl radicals [reactions (14) and (15)] (LINDSAY et al., 1973; VON SONNTAG et al., 1975a).

$$2\ \text{-}\underset{\substack{|\\ \text{O-}\dot{\text{O}}}}{\text{C}}\text{OH-CHOH-} \longrightarrow O_2 + 2\ \text{-}\underset{\substack{|\\ \text{O}^{\bullet}}}{\text{C}}\text{OH-CHOH-} \qquad (14)$$

$$\text{-}\underset{\substack{|\\ \text{O}^{\bullet}}}{\text{C}}\text{OH-CHOH-} \longrightarrow \text{-}\underset{\substack{\|\\ \text{O}}}{\text{C}}\text{OH} + \dot{\text{C}}\text{HOH-} \qquad (15)$$

3.3 Sugars

3.3.1 2-Deoxy-D-Ribose

2-Deoxy-D-ribose crystallizes in the β-pyranose form (FURBERG, 1960; KRÜGER et al., 1977). On dissolving in water mutarotation occurs. At equilibrium the pyranose form still prevails (70-80%), α- and β-forms contributing to an equal extent (HARTMANN et al., 1970; ANGYAL and PICKLES, 1972). The I.R. spectrum of a concentrated (70%) aqueous solution indicates that the open-chain form might be present in the equilibrium to about 3% (HARTMANN, 1970).

3.3.1.1 Aqueous Solutions

The γ-radiolysis of 2-deoxy-D-ribose in deoxygenated N_2O-saturated aqueous solutions has been studied and a number of products have been identified (Table 3.1 and Scheme 3.1) (HARTMANN et al., 1970; SCHULTE-FROHLINDE and VON SONNTAG, 1972). Formation of the products is explained on the basis of the free radical processes discussed above. The dideoxy compounds are formed by the water elimination process (1) [reactions (14), (19), and (23) Scheme 3.1] and by the rearrangement involving the ring oxygen [reaction (6), Scheme 3.1)].

Another rather reactive radical appears to be the uracilyl-(5) radical generated in the UV photolysis and γ-radiolysis of 5-bromo-uracil (see III.3). From published data (GILBERT et al., 1971; GILBERT and WAGNER, 1972) its reaction rate can be estimated to be only about one order of magnitude less than diffusion controlled. The uracilyl-(5) radical is believed to play an important role in the radiation induced strand breaks of 5-bromo-uracil substituted DNA by attacking a neighbouring 2-deoxy-D-ribose moiety (KÖHNLEIN and HUTCHINSON, 1969; WARD, 1972; HUTCHINSON, 1973; CAMPBELL et al., 1974a, b). From experiments in the presence of oxygen (solutions saturated with N_2O/O_2 80/20 v/v) only qualitative results are available (VON SONNTAG et al., 1975b). No dideoxy compounds are formed. Fragmentation products, not observed in the absence of O_2, are significant (∿ 10% of the total product yield). 2-Deoxy-D-glycero-tetrodialdose (Product 11, Table 3.1) is the most prominent fragment product.

Table 3.1. γ-Radiolysis of 2-Deoxy-D-ribose, thymidine and DNA

No	Product		2-Deoxy-D-ribose Aqueous solutions N_2O[a]	2-Deoxy-D-ribose Aqueous solutions N_2O/O_2[b]	2-Deoxy-D-ribose Crystalline State[c]	Thymidine Aqueous solutions[a] N_2O	Thymidine Aqueous solutions[a] N_2O/O_2	DNA Aqueous solutions N_2O[e]	DNA Aqueous solutions N_2O/O_2[f]
1	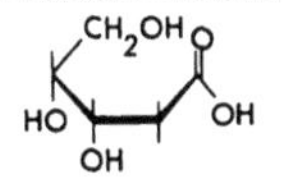	2-Deoxy-D-ribonic acid (γ- and δ-lactone)	2.5	+	1.5	0.02	0.07	-	-
2	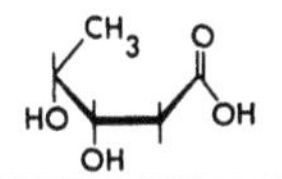	2.5-Dideoxy-D-*erythro*-pentonic acid	0.5	-	> 650	-	-	-	-
3	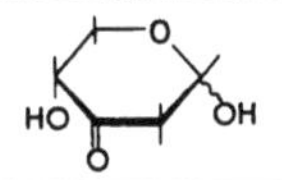	2-Deoxy-D-*glycero*-pentos-3-ulose	}	+		-	-	-	-
4	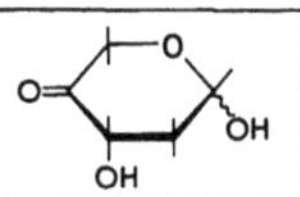	2-Deoxy-D-*glycero*-pentos-4-ulose	} 0.2	+		0.1	0.2	~0.02	~0.004
5	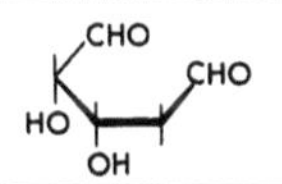	2-Deoxy-D-*erythro*-pentodialdose	}	?				-	-
6	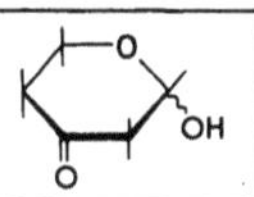	2.4-Dideoxy-pentos-3-ulose	1.15	-		0.03	-	-	-
7	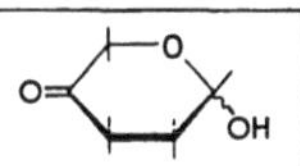	2.3-Dideoxy-pentos-4-ulose	1.0	-		0.01	-	~0.005	-
8	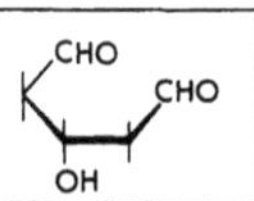	2.4-Dideoxy-pentodialdose	0.55	-		0.02	-	-	-
9	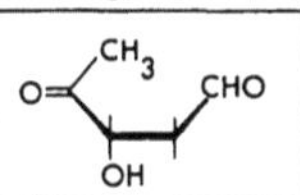	2.5-Dideoxy-D-*glycero*-pentos-4-ulose		-		0.01	-	0.03	-
10	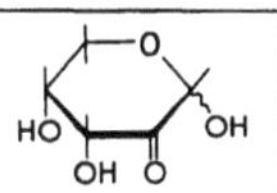	D-*erythro*-Pentosulose	-	?		-	-	-	-
11	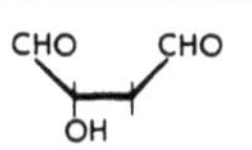	2-Deoxy-D-*glycero*-tetrodialdose	-	+		-	0.03	-	~0.005
12	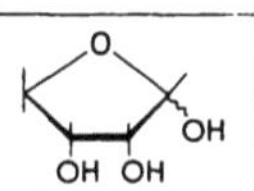	D-Erythrose	-	+		-	-	-	-
13	CHO-CH_2-CHO	Malonaldehyde	≤0.1	?		-	-	0.004[g]	
14	(Thymine)					≤0.2	≤0.4	0.06[g,h]	

+, compound present; -, compound absent; no symbol, compound not determined. Although the starting sugar belongs to the D-series not all products may have retained this configuration.

References: [a] HARTMANN et al. (1970); [b] VON SONNTAG et al. (1975b, unpublished); [c] VON SONNTAG et al. (1974); [d] DIZDAROGLU et al. (1976b); [e] DIZDAROGLU et al. (1975c); [f] DIZDAROGLU et al. (1975b); [g] ULLRICH and HAGEN (1971); [h] free bases.

Scheme 3.1. γ-Radiolysis of deoxygenated N_2O-saturated aqueous solutions of 2-deoxy-D-ribose

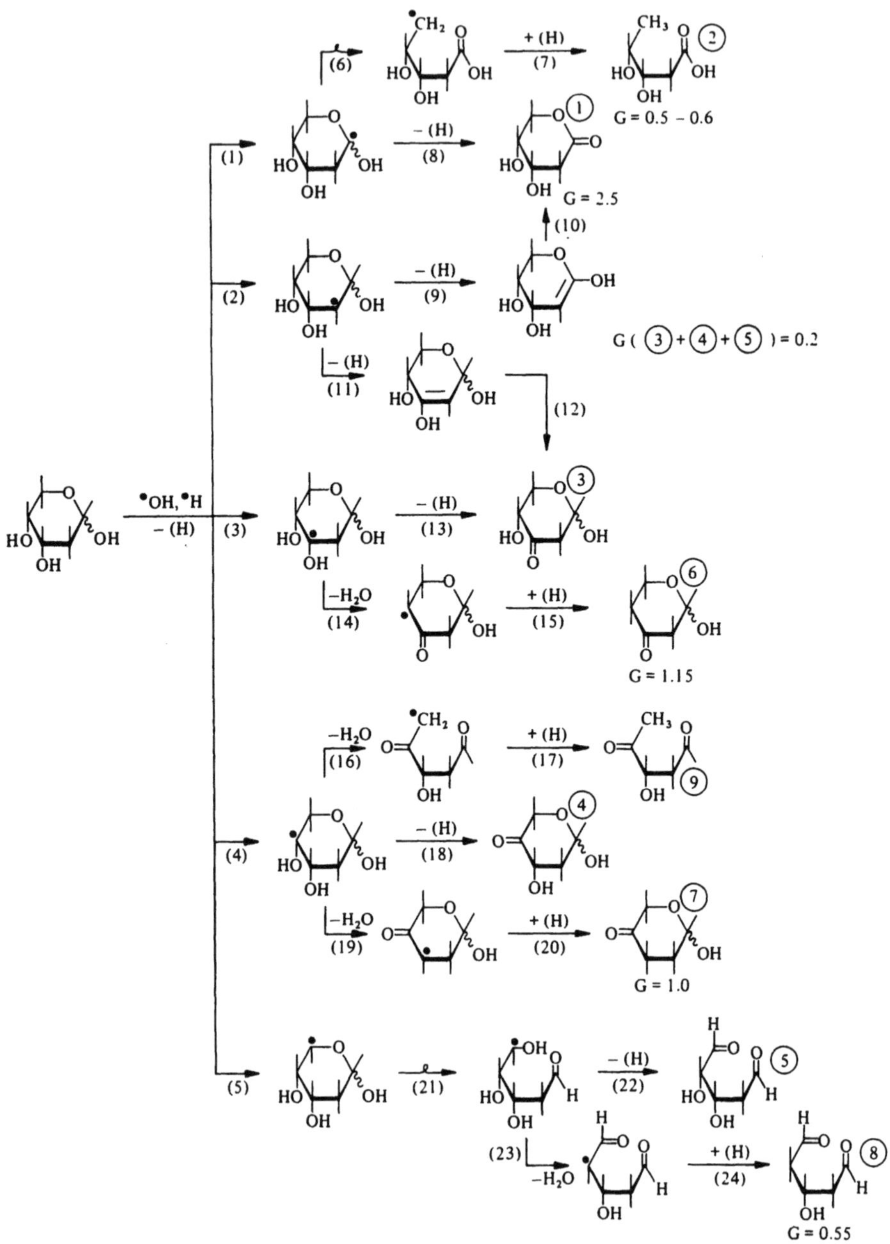

According to HARTMANN et al. (1970) (H) denotes a hydrogen atom transferred in a disproportionation reaction.

3.3.1.2 Crystalline State

2-Deoxy-D-ribose in the solid state is known to be one of the compounds most sensitive to radiation. It decomposes via a chain reaction (VON SONNTAG et al., 1974). The chain starts from a radical at C1 which can be formed either directly through the action of the radiation or through hydrogen abstraction by a H atom [cf. reaction (1), Scheme 3.1]. Hydrogen is a product in the radiolysis of crystalline 2-deoxy-D-ribose (JACOBS, 1962). Rearrangement (4) gives rise to a radical at C5 [radical 15 in reaction (16)]. Hydrogen abstraction by the rearranged radical 15 from a neighbouring 2-deoxy-D-ribose molecule gives rise to 2,5-dideoxy-D-erythro-pentonic acid and a further 2-deoxy-D-ribosyl-[1] radical which can propagate the chain. The initial G-value of 2,5-dideoxy-D-erythro-pentonic acid is > 650 (Product 2, Table 3.1). The chain is terminated if radical 15 abstracts a hydrogen atom from a product molecule 2 instead of from a neighbouring 2-deoxy-D-ribose.

In this case radical 15 is converted into 16:

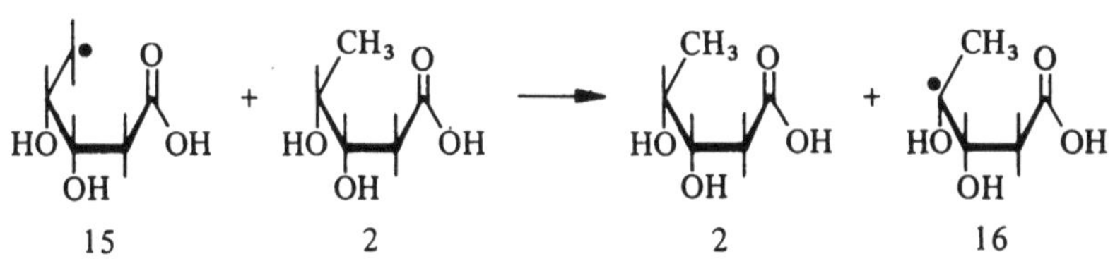

(16)

Radical 16 is a "dead" radical, and it has been shown in ESR work on γ-irradiated single crystals of 2-deoxy-D-ribose that radical 16 is the only detectable species (HÜTTERMANN and MÜLLER, 1969a, b) (see I.3).

3.3.2 D-Ribose

In aqueous solution D-ribose is in an equilibrium mixture consisting of about 6% α-furanoside, 18% β-furanoside, 20% α-pyranoside and 56% β-pyranoside form (LEMIEUX and STEVENS, 1966). No attempt has been made to determine precisely the traces of open-chain form. In the crystalline state D-ribose is present in the β-pyranose form.

3.3.2.1 Aqueous Solution

The radiation chemistry of D-ribose has received some attention (PHILLIPS et al., 1962; PHILLIPS and CRIDDLE, 1962; SCHERZ, 1970; KOCHETKOV et al., 1973).

Qualitative results are presented in Table 3.2. According to the established free radical reactions given above a reaction scheme similar to that for 2-deoxy-D-ribose (Scheme 3.1) can be derived which accounts for the major products in deoxygenated aqueous solution (Table 3.2). Preliminary quantitative measurements in-

Table 3.2. γ-Radiolysis of D-ribose

No.	Product	Aqueous Solution N_2O	N_2O/O_2	Crystalline State
17	D-ribonic acid	+	+	+
18	D-arabinonic acid	+	-	
19	D-erythro-pentos-2-ulose	+	+	
20	D-erythro-pentos-3-ulose	+	+	
21	D-erythro-pentos-4-ulose	+	+	
22	Ribo-pentodialdose	+	+	
1	2-deoxy-D-ribonic acid	+	-	+
23	3-deoxy-D-glycero-pentos-2-ulose	+	-	
3	2-deoxy-D-glycero-pentos-3-ulose	+	-	
24	4-deoxy-D-glycero-pentos-3-ulose	+	-	
25	3-deoxy-D-glycero-pentos-4-ulose	+	-	
26	5-deoxy-D-erythro-pentos-4-ulose	+	-	
5	2-deoxy-D-erythro-pentodialdose	?	-	
27	2-deoxy-D-glycero-tetrose	+	-	
12	D-erythrose	+	-	
28	D-threose	+	-	
29	D-erythro-tetrodialdose	-	+	
30	D-glyceraldehyde	+	+	
31	D-erythronic acid	-	+	
32	D-threonic acid	-	+	
33	5-deoxy-D-erythropentonic acid	-	-	+
34	Carbon dioxide	-		+

Legend: +, compound present; no symbol, compound not determined. Although the starting sugar belongs to the D-series not all products may have retained this configuration.
For some quantitative data see: C. VON SONNTAG and M. DIZDAROGLU: The reactions of OH radicals with D-ribose in deoxygenated and oxygenated aqueous solution. Carbohyd. Res. 58, 21-30 (1977).

dicate that no complete material balance can be obtained on the basis of the products containing five or less carbon atoms. This is thought to be due to the fact that under the experimental conditions (Table 3.2) reducing radicals are not present in sufficient concentration to prevent the dimerization of the radicals of the -ĊH-CO-type. In the presence of N_2O/O_2 formation of the deoxy compounds (Products 1, 3, 5, 23-27, Table 3.2) is suppressed and that of the dicarbonyl compounds (Products 17, 19-22, Table 3.2) and fragment products (Products 29, 31, 32, Table 3.2) is enhanced.

3.3.2.2 Solid State

The self-decomposition of freeze-dried ^{14}C-labelled D-ribose has been studied by PHILLIPS et al. (1962) who found that D-ribonic acid, D-erythrose, D-erythronic acid (Products 17, 12, 31, Table 3.2) and some three-carbon and two-carbon fragments are formed. In γ-irradiated crystalline D-ribose, carbon dioxide (Product 34), 2-deoxy-D-ribonic acid (Product 1) and 5-deoxy-D-ribonic acid (Product 33, Table 3.2) have been identified (VON SONNTAG et al., 1975b). The sum of G-values of the latter two is < 10 indicating that if chain reactions take part in the formation of deoxyribonic acids the chain lengths are apparently short and not comparable to those encountered in 2-deoxy-D-ribose (VON SONNTAG et al., 1974) and α-lactose monohydrate (VON SONNTAG and DIZDAROGLU, 1973)

3.4 Sugar-Phosphates

3.4.1 D-Ribose-5-Phosphate

D-Ribose-5-phosphate is present in the furanose form. The radiation induced dephosphorylation has received considerable interest because mechanisms leading to dephosphorylation in D-ribose-5-phosphate may give rise to strand breaks in the nucleic acids. Studies have been carried out in deoxygenated N_2O-saturated solutions as well as in the presence of oxygen (STELTER et al., 1974, 1975a, b, 1976; KOCHETKOV et al., 1974).

In deoxygenated solutions the material balance between liberated phosphate and phosphate free products containing five or less carbon atoms is poor owing to the fact that the radicals of the type -ĊH-CO- disappear by combination, and disproportionation products are only formed in small quantities [cf. reaction (9)]. However, if these radicals are reduced by Fe^{2+} as described in reaction (12), a better material balance is obtained as the case is in the presence of oxygen. Table 3.3 lists the products and their G-values.

3.4.1.1 Deoxygenated Solutions

In deoxygenated N_2O-saturated solutions the major carbohydrate products free of phosphate containing up to five carbon atoms are ribo-pentodialdose (Product 22) and 5-deoxy-D-erythro-pentos-4-ulose (Product 26, Table 3.3); Product 22 is thought to be formed by OH radical attack at C5 followed by a further oxidation at this position. Other carbohydrate radicals or radiolytically formed hydrogen peroxide may serve as oxidants. It has been noted that $G(H_2O_2)$ is much smaller under these conditions than the expected value of 0.75. When adding an oxidant (e.g., Fe^{3+}) G(ribo-pentodialdose) is increased to about 0.6 (cf. Table 3.3).

The precursor radical of 5-deoxy-D-erythro-pentos-4-ulose (Product 26, Table 3.3) is the radical at C4. Phosphate is eliminated in a manner similar to reaction (2). The enhancement of G(5-deoxy-D-erythro-pentos-4-ulose) in the presence of Fe^{2+} ions (cf. Table 3.3) is thought to be due to a reduction of the phosphate free radical [cf. reaction (12)] thus preventing the dimerization.

Products 36 and 37 (Table 3.3) are formed only in the absence of iron ions. Deuterium labelling experiments (STELTER et al., 1976) indicate that 36 has an enol as its precursor. An allyl-type radical as given in reaction (3) is probably the intermediate formed. The structure of Product 37 also indicates the possibility of an allylic radical being an intermediate. Iron ions could

Table 3.3. γ-Radiolysis of D-ribose-5-phosphate (According to STELTER et al., 1975b, 1976). Products and G-values

No	Product		N_2O	N_2O/Fe^{2+}	N_2O/Fe^{3+}	N_2O/O_2
35	H_3PO_4	Inorganic phosphate	1.3			0.6
22	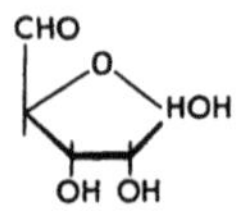	*ribo*-Pentodialdose	0.22	0.45	0.62	0.10
36	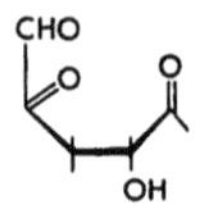	2-Hydroxy-4-oxo-glutaraldehyde	0.06	absent	absent	absent
26	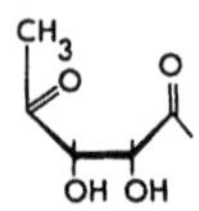	5-Deoxy-*D-erythro*-pentos-4-ulose	0.10	0.80	0.42	absent
37	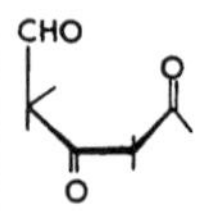	3-Oxo-glutar-aldehyde	0.06	absent	absent	absent
27	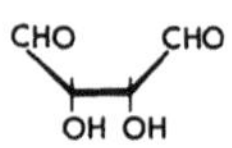	*erythro*-Tetro-dialdose	absent	absent	absent	0.40
38	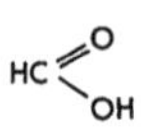	Formic acid	absent	absent	absent	0.41
39	Minor phosphate-free carbohydrate products		0.09	absent	0.12	0.08

interfere with the primary radical thus preventing the formation of allylic radicals and hence of Products 36 and 37.

3.4.1.2 *Oxygenated Solutions*

In the presence of oxygen the elimination of phosphate occurs by a completely different route. There, the radical at C5 appears to be the most important precursor. The addition of oxygen [cf. reaction (7)] gives rise to the corresponding peroxyl radical. The mechanism of formation of erythro-tetrodialdose (Product 27, Table 3.3) will be discussed in II.3.6 which deals with strand breal processes [reactions (28)-(31)] where similar reactions lead to the corresponding product 2-deoxy-tetrodialdose (Product 11, Table 3.1). However, it may be noted that with the model compound a good material balance with respect to Product 27 and formic acid, Product 38, is obtained (cf. Table 3.3), thus confirming the mechanism proposed in the reaction sequence (28)-(31). Ribopentodialdose is the other major product under these conditions even though the value of G is only 0.1 (Product 22, Table 3.3).

The results with Fe^{3+} and with O_2 agree well, both suggesting that the primary attack at C5 is about G = 0.6. The attack at C4 is somewhat higher, G being at least 0.8 as is the case in the presence of Fe^{2+} (Product 26, Table 3.3).

3.5 Nucleosides

3.5.1 Scission of the N-Glycosidic Linkage in Thymidine

The radiation chemistry of nucleosides in aqueous solution is largely governed by the reactions of the water radicals with the base moiety. Solvated electrons and H atoms react exclusively with this part of the molecule. Although the majority of OH radicals add to the double bonds of the base moieties, some OH radicals abstract hydrogen atoms from the sugar part (SCHOLES et al., 1960; HOARD et al., 1974). Products with an altered sugar part and an intact base have not been isolated as yet, but in cases where the base is eliminated it is possible to identify the carbohydrate products and to propose mechanisms to account for the scission of the N-glycosidic linkage. Investigations have been restricted to thymidine (CADET and TEOULE, 1972b; DIZDAROGLU et al., 1976b).

3.5.1.1 Deoxygenated Solutions

In the absence of oxygen, products 1, 4, 6-9 and thymine 14 (Table 3.1) have been identified. Apparently, several routes lead to the scission of the N-glycosidic linkage. One route might start with the radical at C4' generated by hydrogen abstraction from the C4' position of the thymidine molecule, and would lead to products 7 and 9. The mechanism is expected to be similar to that discussed in II.3.6 on strand break processes [reactions (17)-(24)]. The elimination of the OH group is possibly proton assisted [cf. reaction (1)] whereas the phosphate elimination [cf. reaction (2)] in the sugar phosphates and in DNA is not. Hemiaminals as indicated in reactions (23) and (24) are thought to be intermediates in the release of the unaltered base.

Similar mechanisms proceed from the radical at C5' to thymine 14 and the sugars 6 and 8 (Table 3.1).

The mechanism of formation of 2-deoxy-D-glycero-pentos-4-ulose (Product 4, Table 3.1) and 2-deoxy-D-ribonic acid (Product 1, Table 3.1) is less well understood. Their precursors are believed to be the radicals at C4' and C1', respectively. If, in the disproportionation reactions, carbonium ions [cf. reaction (11)] are intermediates, the elimination of the base may be similar to that described above.

3.5.1.2 Oxygenated Solutions

In the presence of oxygen the dideoxy compounds 6-9 (Table 3.1) are absent. Since the primary radicals are scavenged by oxygen, [cf. reaction (7)] elimination of the unaltered bases must then proceed via peroxyl radicals.

The mechanism of formation of 2-deoxy-D-glycero-tetrodialdose (Product 11, Table 3.1) is thought to be similar to that of its formation in the case of DNA [reactions (28)-(31)], and to that of erythro-tetrodialdose (Product 27, Table 3.3) from D-ribose-5-phosphate (see II.3.3.1). Thymine (Product 14, Table 3.1) is expected to be eliminated from the unstable hemiaminal of 2-deoxy-D-glycero-tetrodialdose.

The mechanism of formation of Products 1 and 4 (Table 3.1) is not yet fully understood. Precursor radicals of these compounds are thought to be those at C1' and C4'.

According to our present knowledge, the substitution of an OH group by a phosphate ester group alters some of the reaction routes in the free radical chemistry of this class of compounds. Hence, the phosphate ester analogs of the products 6 and 8 (Table 3.1) are not expected to be formed in the γ-radiolysis of aqueous DNA.

3.6 Nucleotides

3.6.1 Phosphate Release

The radiation chemistry of nucleotides received special interest because this class of compounds may serve as a model regarding radiation-induced strand breaks. The mechanisms of the elimination of inorganic phosphate from nucleotides are expected to be the same as those which operate in the chain breaking process. One of the advantages of the nucleotide model compounds is the fact that inorganic phosphate is easier to determine than the phosphate ester end groups released in the DNA strand breaking process. Inorganic phosphate is released (1) immediately during irradiation, (2) on standing at pH 7 and (3) by alkaline hydrolysis (DANIELS et al., 1955, 1956; SCHOLES, 1963; WARD, 1972; WARD and KUO, 1973).

3.6.1.1 Mononucleotides

G-values of phosphate release from various mononucleotides in the presence and the absence of oxygen at pH 7 and after alkaline treatment are compiled in Table 3.4. In the presence of N_2O the yields are generally doubled indicating that the OH radicals are responsible for the lesions. There are some discrepancies between the values obtained in different laboratories. The reason for this is thought to be largely due to impurities in the form of metal ions, e.g., iron ions which have been shown to alter substantially the amount of phosphate release (WARD, 1971). As can

Table 3.4. G(inorganic phosphate) from nucleotides at pH 7 and after alkaline treatment

	3'-Nucleotides		5'-Nucleotides				
	Neutral		Neutral		Alkaline treatment	Neutral	Alkaline treatment
	N_2	N_2O	N_2	N_2O	N_2	O_2	O_2
AMP	0.53[a]	0.97[d]	0.13[a] 0.17[d]	0.28[d]		0.39[b]	0.59[b]
GMP	0.31[a] 0.25[d]	0.41[d]	0.15[a]	0.22[c] 0.21[d]		0.28[b] 0.19[c]	0.54[b]
UMP	0.48[a]	0.74[h]	0.23[a]	0.44[d]			
CMP	0.38[a] 0.34[d]	0.65[d]	0.15[a] 0.17[d]	0.37[d]		0.60[b]	0.95[b]
d-AMP				0.33[d]		0.34[b] 0.34[e,f]	0.51[b] 0.37[e,f]
d-GMP			0.17[c] 0.17[f]	0.22[d]	0.18[f]	0.29[c] 0.28[b] 0.20[e,f]	0.52[b] 0.21[e,f]
TMP	0.47[a]		0.14[a] 0.15[e,f]		0.31[e,f]	0.25[e,f,g]	0.32[e,f,g]
d-CMP			0.22[f,h]	0.36[d]	0.27[f,h]	0.43[e,f] 0.54[b]	0.52[e,f] 0.89[b]
d-UMP			0.14[c]	0.35[d]		0.40[c]	

[a] RALEIGH et al. (1973b); [b] SCHOLES (1963); [c] RALEIGH et al. (1973a); [d] GREENSTOCK and SHIERMAN (1975); [e] WARD (1972); [f] WARD (1971); [g] 0.48 N_2O/O_2 (4/1) saturated, 0.81 after alkaline hydrolysis; [h] 0.51 N_2O saturated 0.64 after alkaline treatment.

be seen in Table 3.4, 3'-nucleotides eliminate larger amounts of inorganic phosphate in deoxygenated solutions [G(i.p.) = 0.38-0.53] than the corresponding 5'-nucleotides [G(i.p.) = 0.11-0.23]. If one accepts the mechanisms of phosphate release as proposed for the case of D-ribose-5-phosphate by STELTER et al. (1976), this behaviour is to be expected. In the 5'-nucleotides, phosphate can be released from the primary radicals at C5' and C4'. In the case of the 3'-nucleotides the primary radicals at C4', C3' and (at least for AMP, GMP, UMP and CMP) at C2' can give rise to phosphate release. Furthermore, hydrogen abstraction at C5' is expected to be less likely than at the other positions as is indicated by the results from the radiolysis of 2-deoxy-D-ribose (HARTMANN et al., 1970; SCHULTE-FROHLINDE and VON SONNTAG, 1972) and, in part, D-ribose-5-phosphate (STELTER et al., 1976).

WARD (1971) has studied the effect of cysteamine on the phosphate release from dCMP. Cysteamine is capable of competing with phosphate release. The efficiency of cysteamine was independent of dose rate, and it was concluded that its competition inhibited a first order phosphate eliminating process. Mechanisms which can account for this observation have been proposed by STELTER et al. (1974, 1975a, 1976) in the case of D-ribose-5-phosphate (see above) and have been shown to operate in DNA itself (DIZDAROGLU et al., 1975c).

3.6.1.2 Thymidine-3',5'-Diphosphate

The mode of phosphate elimination has been carefully studied by WARD and KUO (1973) using thymidine 3',5'-diphosphate which is a better model compound to study the formation of strand breaks than the monophosphates. Results are summarized in Table 3.5. Damage to the sugar moiety gives immediate phosphate and phosphate which is released by leaving it to stand for 50 h at pH 7.

Table 3.5. Reactions which lead to release of inorganic phosphate from thymidine 2',5'-diphosphate. Sites of damage, resulting reactions and yields in the presence and absence of oxygen. (According to WARD and KUO, 1973)

Site of damage	Resulting reaction	Yield (G value)		
			O_2	N_2
(a) Sugar	B—S*(P)(P) ⟶ BS* + 2Pi	Immediately	0.05	0.06
		Hydrolysis at pH 7	0.07	0.05
(b) Base	B*—S(P)(P) ⟶ B*—S + 2Pi	Immediately	0.04	0.04
		Hydrolysis at pH 7	0.09	0.03
(c) Base	B**—S(P)(P) —Alkali⟶ B-S + 2Pi		0.25	0.35
(d) Base damage leading to loss of 2C atoms	S(P)(P) —Alkali⟶ S* + 2Pi		0.6	0

No mononucleotides were observed and it has been concluded that both phosphate groups are eliminated in these processes. The results from γ-irradiated DNA (DIZDAROGLU et al., 1975b, c) indicate the nature of the reactions which may occur (see II.3.6). However, damage of the base moiety also seems to contribute to this type of phosphate release.

On treating the irradiated solutions with base, further phosphate is eliminated. This damage is entirely due to alterations in the base. The formation of alkali-labile phosphate is more pronounced in the presence of oxygen. There, it has been observed (WARD and KUO, 1973) that from ^{14}C 2-labelled thymidine 3',5'-diphosphate a ^{14}C-labelled volatile product is formed which is absent in deoxygenated solutions. It has been suggested that the breakdown of the altered base liberated the sugar diphosphate which is not stable under these alkaline conditions.

3.6.2 Scission of the N-Glycosidic Linkage

D-Ribonic acid 5-phosphate from uridine-5'-phosphate (DUCOLOMB et al., 1974a) and 2-deoxy-D-ribonic acid 5-phosphate from thymidine 5'-phosphate (CADET et al., 1974) are products which result from the process of scission of the N-glycosidic linkage (cf. nucleosides).

3.6.3 Formation of Cyclo-Nucleotides

In oxygen-free solution, radicals at C5' of purine nucleotides can add to the base moiety as has been shown by KECK (1968). The resulting products are cyclo-nucleotides, e.g. product 40 from AMP:

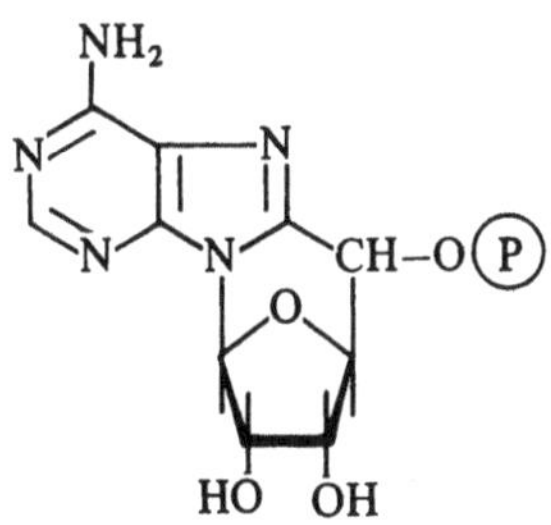

40

Corresponding products have also been obtained with IMP and GMP.

3.7 DNA

3.7.1 Product Release from DNA and the Mechanism of Strand Breaks

Two chemical methods have been applied so far in order to elucidate the mechanism involved in the formation of strand breaks of DNA in aqueous solution. The first method is isolation and identification of small molecules released from DNA during or after irradiation, and the second the determination of end groups. The results from the first method allow conclusions to be drawn on the mechanisms involved. In principle, two routes leading to strand breaks are possible. One is the splitting of the phosphate ester bond, and the other is a C-C-bond rupture within the sugar skeleton. Whereas the first route is observed in the presence and in the absence of oxygen, the C-C-bond rupture appears to occur only in the presence of oxygen. In aqueous solution OH radicals and H atoms can lead to chain breaks in DNA whereas solvated electrons and O_2^- radicals are ineffective (BLOK and LOMAN, 1973; WARD, 1975; ACHEY and DURYEA, 1974).

3.7.1.1 Release of Malonaldehyde

The formation of malonaldehyde (Product 13, Table 3.1) by irradiation of DNA in aqueous solution has been reported (KRUSHINSKAYA and SHAL'NOV, 1967; KAPP and SMITH, 1970; ZAKATOVA and SHARPATYI, 1971; ULLRICH and HAGEN, 1971; WARD, 1972). Recently the results were summarized and thoroughly discussed by WARD (1975). It appears that in the absence of oxygen G(malonaldehyde), obtained without heating the sample, is as low as 0.004, but after heating G(malonaldehyde) rises to 0.21 (ULLRICH and HAGEN, 1971). ULLRICH and HAGEN therefore suggested that the formation of malonaldehyde is not associated with the production of immediate strand breaks in the absence of oxygen, and so its formation is not further discussed in detail.

3.7.1.2 Release of Sugars and Unaltered Bases

Deoxygenated Solutions. After γ-irradiation of DNA in aqueous solution (500 mg/l) saturated with N_2O, three sugars have been isolated and identified: 2,5-dideoxy-pentos-4-ulose, 2,3-dideoxy-pentos-4-ulose and 2-deoxy-pentos-4-ulose (Products 9, 7 and 4, Table 3.1) (DIZDAROGLU et al., 1975c). They were obtained from herring sperm DNA as well as from calf thymus DNA.

The formation of these products is explained as follows: OH radicals and H atoms produced in the aqueous N_2O-saturated solutions in the ratio of 9:1 attack the DNA. Nearly all H atoms and the majority of the OH radicals add to the double bonds of the bases but 10-20% of the OH radicals are expected to react with the sugar moiety by hydrogen abstraction (SCHOLES et al., 1960). About one-fifth of these will abstract the hydrogen atom bound to C4' of the deoxyribose moiety to form radical 41 (HARTMANN et al., 1970; SCHULTE-FROHLINDE and VON SONNTAG, 1972).

For the cleavage of the phosphate ester bond the following route has been postulated on the basis of studies on model compounds as described above. Radical 41 eliminates the phosphate ester anion at position 3' leading to the radical cation 42 [reaction (18)] followed by fast addition of water with loss of a proton giving 43 or 44 [reaction (19)]. In reaction (18) the DNA chain is broken. 43 then eliminates the second phosphate ester group leaving the radical cation 45 which also adds water with loss of a proton yielding radical 46 [reaction (20) and (21)]. In a disproportionation reaction with other radicals ($\dot{R}H$), radical 46 forms compound 47 [reaction (22)]. 47 ring opens [reaction (23)] and yields 48. Compound 48 as a hemiaminal is unstable and eliminates an unaltered base leaving 9 (Table 3.1) which is the product isolated. The release of base is observed in the radiation chemistry of DNA (HEMS, 1960a; ULLRICH and HAGEN, 1971). The formation of product 7 (Table 3.1) can be explained in a similar

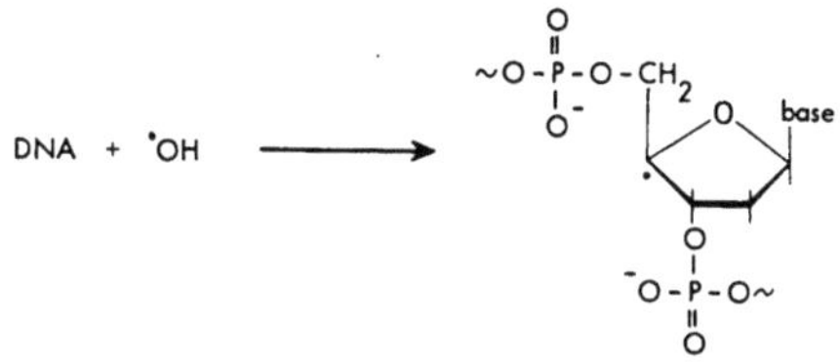

41

(17)

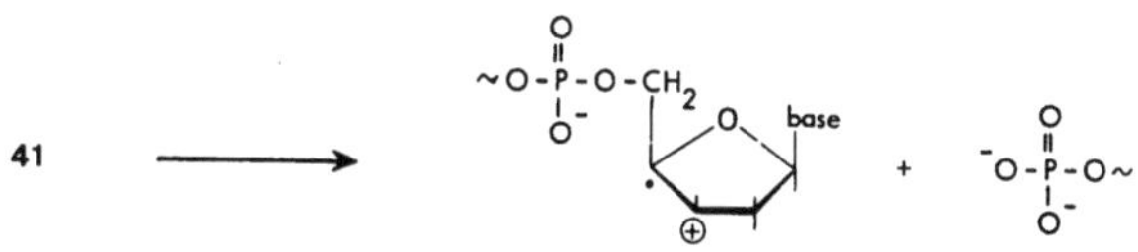

42

(18)

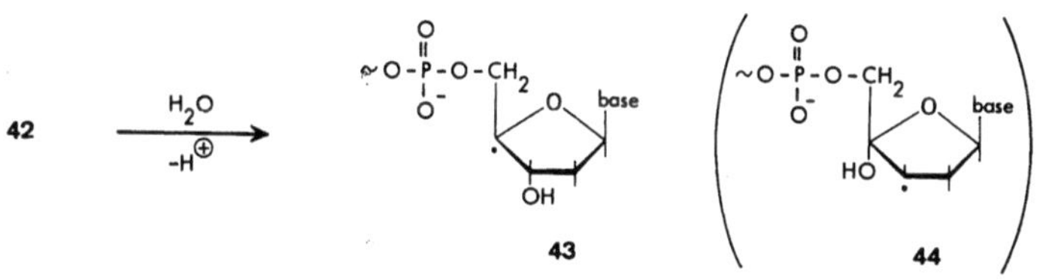

43 44

(19)

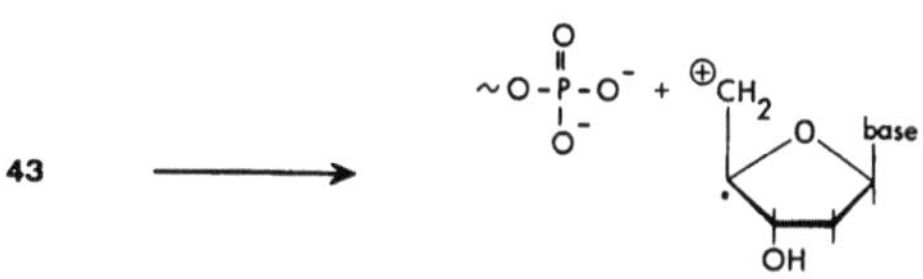

45

(20)

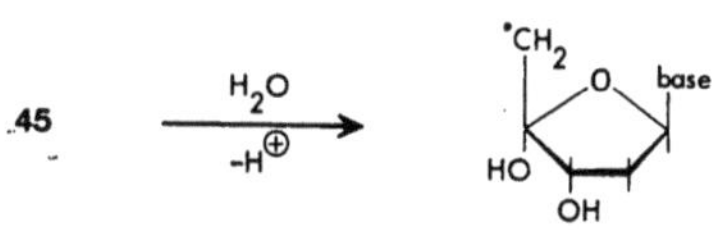

46

(21)

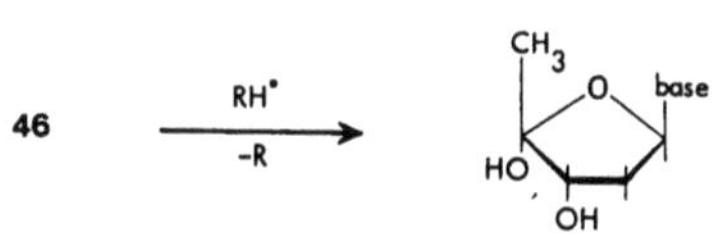

47

(22)

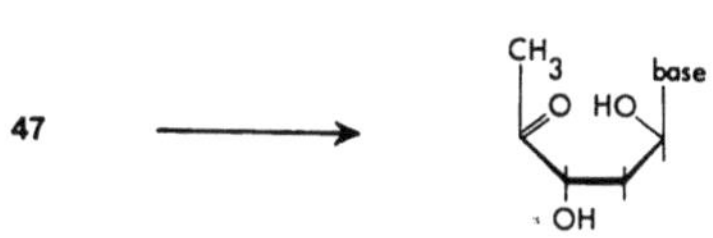

48

(23)

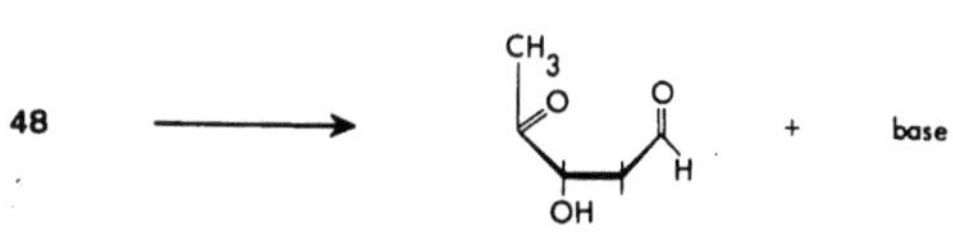

9

(24)

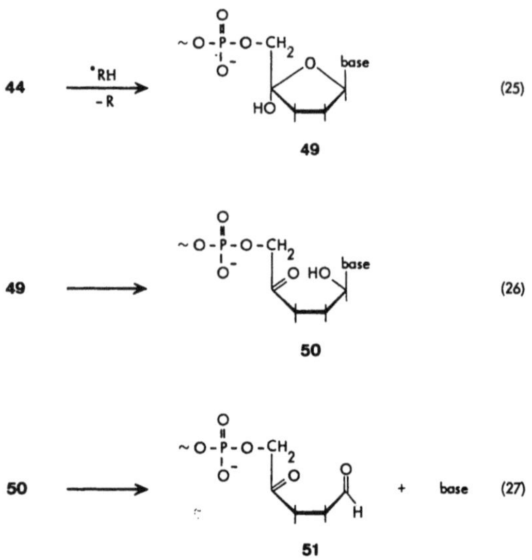

way. In this case radical 41 first eliminates the phosphate ester at C5' followed by corresponding reactions.

Radical 44 in a disproportionation reaction with other radicals (˙RH) can also take up an H atom to form 49 [reaction (25)]. Compound 49 is unstable and the ring opens to yield 50 [reaction (26)]. 50 loses an unaltered base leaving 51. In this case the altered DNA chain is broken and a base released. However, the altered sugar is still bound to the DNA chain.

The formation of 4 (Table 3.1) can be explained by an addition of water at the C5' position of the radical cation 45 followed by an oxidizing step similar to the effect of Fe^{3+} on the product formation of D-ribose-5-phosphate (STELTER et al., 1976).

From the model studies it is known that OH radicals abstract H atoms at every position in the sugar moiety and not only at C4' (HARTMANN et al., 1970). Since the isolated products originate from radicals at C4' the question arises as to what are the reactions and the products related to the other sugar radicals produced in the DNA. It is very probable that H atom abstraction at the other positions can also lead to a chain break or to a change in the sugar moiety by a set of similar reactions as shown above, but these lesions do not necessarily lead to the release of small molecules. In order to elucidate such lesions other techniques will have to be developed.

Oxygenated Solutions. In the presence of oxygen, the irradiation of DNA in aqueous solution (saturated with N_2O/O_2 4 : 1) leads to two sugars: 2-deoxy-pentos-4-ulose and 2-deoxy-tetrodialdose (Products 4 and 11, Table 3.1) (DIZDAROGLU et al., 1975b). The dideoxy sugars 7 and 9 (Table 3.1) which have been identified in deoxygenated N_2O-saturated aqueous solutions are not found in the presence of oxygen.

The mechanism of the formation of product 11 (Table 3.1) has been formulated in a way similar to that proposed for the formation of erythro-tetrodialdose in the γ-radiolysis of oxygenated aqueous solutions of D-ribose-5-phosphate (STELTER et al., 1975b). In the N_2O/O_2-containing solution the OH radicals produced will

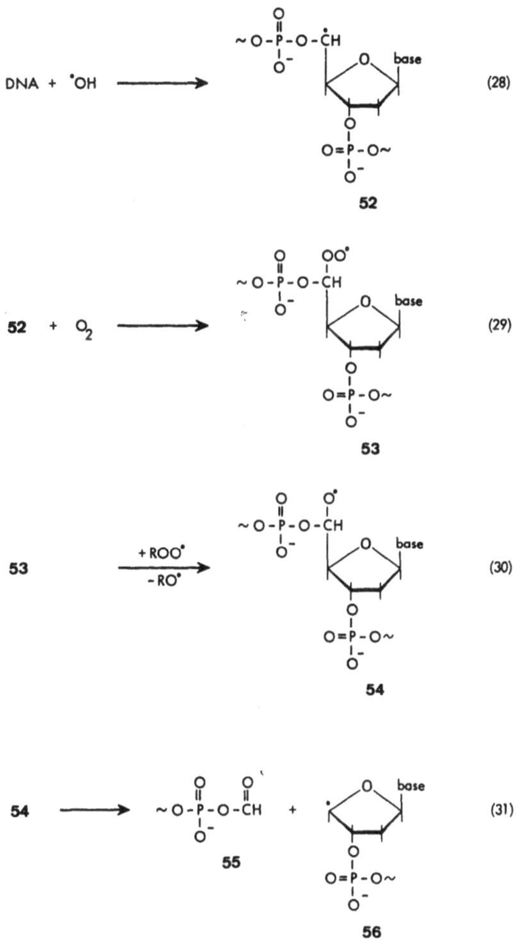

attack the sugar moieties by abstracting hydrogen atoms. Some of the OH radicals abstract a hydrogen atom from C5' of the 2-deoxyribose in the DNA chain to give radical 52 [reaction (28)]. The hydrogen atom abstraction reaction [reaction (28)] is followed by addition of oxygen to form peroxyl radical 53 [reaction (29)]. Radical 53 can react with another peroxyl radical leading to the oxyl radical 54, molecular oxygen and another oxyl radical [reaction (30)]. Such a bimolecular decay has been proposed for peroxyl radicals derived from diethyl ether (LINDSAY et al., 1973; VON SONNTAG et al., 1975a). The oxyl radical 54 undergoes β-fragmentation to give a DNA fragment with a formyl phosphate end group 55 and radical 56 [reaction (31)]. In this step the DNA chain is broken by C-C bond splitting between the carbon atoms 4' and 5' of the sugar moiety; 55 is labile and is hydrolysed into formic acid and a DNA strand with a 3'-phosphate end group. The phosphate ester of radical 56 can be broken in an analogous reaction as given by reaction (18). The remaining radical in the presence of oxygen leads to product 11 (Table 3.1) and unaltered base. Free bases have been observed in the γ-radiolysis of DNA in oxygenated aqueous solution.

The formation of 2-deoxy-pentos-4-ulose (Product 4, Table 3.1) shows that the DNA strand is broken but the carbon skeleton of the sugar is preserved. In this case the strand break is caused only by phosphate ester cleavage. Strand breaks by phosphate ester cleavage with the carbon skeleton of the sugar remaining intact have already been observed in the absence of oxygen.

The formation of strand breaks of DNA caused by C-C-bond rupture as shown in this section appears to be a specific effect of oxygen. The results suggest that the strand breaks occurring in the presence of oxygen could have different origins. However, this might be true only for strand breaks which are accompanied by sugar release.

At present, there is no direct information available on the mechanism of strand breaks which do not lead to the release of small molecules.

3.7.2 End Group Determinations on Broken DNA Strands

A chemical analysis of the nature of the end groups has not been carried out as yet. Attempts have been made to characterize the end groups by means of several enzymes reacting specifically with hydroxyl or phosphate groups at the 3' or 5' end of the broken strands. In Table 3.6 the various possibilities for analysing end groups are summarized but not all of the reactions have been studied in detail so far.

3.7.2.1 5' End Groups

The 5' terminals of broken DNA strands can be determined by means of the combined action of phosphatase and polynucleotide kinase. The latter enzyme transfers the γ-phosphate of ATP to a 5' hydroxyl group (BOPP and HAGEN, 1970). With DNA irradiated in solution only a few end groups are phosphorylated with kinase, either before or after alkaline treatment. After dephosphorylation most of the end groups become reactive. This means that about 90% of the 5' ends in the actual strand breaks and in the alkali-induced strand breaks carry phosphate groups. These observations lead to the conclusion that in aqueous solutions radical attack at the sugar moiety splits the phosphoester bond predominantly at C3'.

When strand breaks in DNA of irradiated cells are analyzed as above, again only a few per cent (10%) of 5' hydroxyl end groups are found. However, 40% of 5' phosphate end groups are observed instead of the 90% found in DNA in aqueous solution (LENNARTZ

Table 3.6. Reactions of enzymes on γ-irradiated DNA

Enzyme	Reaction	End group detected
Polynucleotide kinase	transfers phosphate from ATP to 5' OH end groups	5' hydroxyl group
Phosphatase followed by polynucleotide kinase	dephosphorylates 5' phosphate groups to enable phosphorylation by kinase	5' phosphate group
Exonuclease III	dephosphorylates 3' phosphate end groups	3' phosphate group
Exonuclease III	acts exonucleolytically on 3' end groups	free 3' terminals
Terminal nucleotidyl transferase	polymerizes, starting from 3' hydroxyl end groups	3' hydroxyl end
Polynucleotide ligase	joins 3' hydroxyl ends with 5' phosphate	3' hydroxyl and 5', phosphate ends in vicinity
DNA polymerase I	polymerizes starting from 3' hydroxyl end groups	3' hydroxyl end groups in intact double-stranded DNA

et al., 1975). The rest of the 5' ends cannot be characterized in this enzymatic test.

3.7.2.2 3' End Groups

For characterization of the 3' end group, several enzymes can be used: Exonuclease III, terminal nucleotidyl transferase, DNA polymerase I and polynucleotide ligase (cf. Table 3.6). With exonuclease III the number of 3' phosphate end groups was determined to be 20% in DNA irradiated in solution (BOPP et al., 1973). Through the exonucleolytic activity of exonuclease III, nucleotides are split off from the 3' terminals of the actual strand breaks, i.e., damaged nucleotides will be liberated as well as intact 3' OH terminal end group. In connection with the results of ULLRICH and HAGEN (1971) where damaged nucleosides were found to be liberated in irradiated DNA solutions, it can be assumed that for some strand breaks phosphate end groups on both sites (5' and 3') are present. This result is in agreement with the observation that phosphate-free sugars are formed during irradiation of DNA in aqueous solution (see II.3.6.1).

Working with the polymerizing enzymes on the 3' terminal, it should be kept in mind that the DNA polymerase requires an intact double-stranded structure whereas the terminal transferase adds nucleotides only on the single-strand end. After treatment of the end groups in the DNA of irradiated cells with terminal transferase, about 70% of the actual strand breaks were found to carry 3' hydroxyl end groups (LENNARTZ et al., 1975). These

3' hydroxyl end groups, however, are not able to serve as a primer for the DNA polymerase: no nucleotide incorporation could be measured with the enzyme (LANDBECK and HAGEN, 1973). Moreover, irradiated DNA serves as a strong inhibitor for the action of DNA polymerase on enzymatically degraded DNA. This means that the DNA polymerase I is bound to the broken end but is unable to start the DNA synthesis. It can be assumed that the secondary structure of the double helix is destroyed at the 3' end of the radiation-induced break, possibly accompanied by a loss of base, whereas the hydroxyl group acting as a primer for the terminal transferase is still present. Furthermore it follows from the chemical studies (see II.3.6.1) that the phosphate sugar bond is broken only when the sugar is altered. It is conceivable that the polymerase is not able to work on the altered sugar.

The presence of oxygen during irradiation enhances the formation of strand breaks by a factor of 2-4, compared to deoxygenated conditions. Such an oxygen effect was observed in solution (FREY and HAGEN, 1974), in dry DNA (LÜCKE-HUHLE et al., 1970) or in irradiated thymocytes (LENNARTZ et al., 1975). For DNA in irradiated cells it was found that the nature of the detectable specific end groups is not changed by the presence of oxygen. It should be pointed out that in order to detect an oxygen effect on strand breaks, the composition of the gas applied during irradiation has to be controlled carefully. Under deoxygenated conditions the oxygen concentration of the solution should be lower than 10^{-6} M O_2.

To summarize, several types of radiation-induced end groups in DNA can be detected by means of enzymatic assay. In addition, however, other end groups with unknown sugar moieties are formed which have to be analyzed by sensitive chemical methods.

Chapter 4

Changes in the Secondary and Tertiary Structures of DNA after Irradiation

When DNA is dissolved in aqueous solutions containing small amounts of salt (e.g., 10^{-3} M NaCl and over) it possesses a double helical structure (see I.2). This helical structure is held together by two types of forces, base-pairing due to hydrogen bond formation between A-T pairs and between G-C pairs, and also by vertical interaction between the stacked bases. Any breakdown of this structure is usually referred to as denaturation.

Changes in the secondary structure are not a direct result of radiation, but follow damage to the primary structure of the DNA molecule i.e., attack on the sugar moiety and on the bases (discussed in the earlier part of this section). This article will consider how such changes occur and the manner in which they may be observed.

4.1 Hyperchromicity and Melting Curves

Denaturation of DNA can be followed by changes in optical density measured at wavelengths around 260 nm. On heating a solution of DNA in a moderately strong salt solution, (10^{-1} M NaCl), the optical density stays constant up to about 70°C and then increases rapidly to a maximum at about 95°C (the hyperchromic effect). The change in optical density depends somewhat on the A-T to G-C ratio of the DNA but is of the order of 35-40% of the initial value. The temperature at which half the total change occurs is referred to as the "melting temperature" of the DNA (usually abbreviated to T_m), and is markedly dependent on the salt concentration and the pH. For neutral salt solutions of calf-thymus DNA, T_m is 83°C in 10^{-1} M NaCl, and 53°C in 10^{-3} M NaCl.

It has been observed by several authors, BELMAN et al. (1962); HAGEN and WILD (1964); TSEITLIN et al. (1964); BOHACEK and BLAZICEK (1965); RAFI and WHEELER (1973), that on irradiation the fairly sharp melting curve of DNA changes, becoming progressively flatter (Fig. 4.1), and that a marked change in T_m occurs, as shown by Tables 4.1 and 4.2.

There is an increase in optical density at room temperature and a progressive shift in the wavelength of maximum absorption is seen toward that of non-hydrogen bonded bases. The obvious inference from such effects is that radiation has caused partial denaturation to take place, and that in addition the remaining hydrogen bonded pairs are more easily disrupted by heating. The extent of the radiation-induced increase depends upon dose, and eventually after a high enough dose all hyperchromicity disappears. At the same time as this change is occurring other reac-

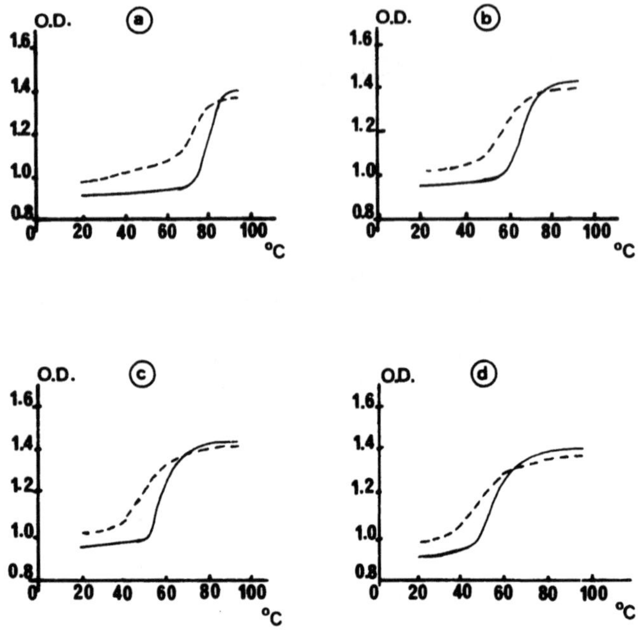

Fig. 4.1a-d. Effect of [NaCl] on the melting curves of irradiated DNA solution (a) 10^{-1} M NaCl; (b) 10^{-2} M NaCl; (c) 3×10^{-3} M NaCl; (d) 10^{-3} M NaCl. ——: Unirradiated DNA. ----: Irradiated DNA (2 krad). (RAFI and WHEELER, 1973)

Table 4.1. Irradiation of aqueous solutions of calf-thymus DNA (0.1% w/v) variation of melting temperature with dose (HAGEN and WILD, 1964)

Dose krad	T_m °C
0	75
1	74
2	72
4	69
8	67
16	64
32	64
64	61
128	54

Table 4.2. Irradiation of aqueous solutions of calf-thymus DNA (0.0075% w/v) effect of salt concentration on the melting temperature (dose = 10 krad) (RAFI, 1968)

NaCl concentration during heating	T_m Unirradiated	T_m Irradiated
10^{-1} M	83	63
10^{-2} M	68	52
3×10^{-3} M	60	49
10^{-3} M	53	46

tions are taking place including chain scission and base destruction, and due to this latter reaction if the DNA is now heated to above its melting temperature a decrease in optical density is now seen.

The loss of hyperchromicity is not a linear function of dose. This is not surprising since if any part of the DNA which has lost its hyperchromicity is subject to further radical attack then no further change will be observed. This situation is analogous to the killing of bacteriophage, and under such conditions it is usual to plot the logarithm of surviving fraction against dose. Thus if one plots the logarithm of the remaining fraction of hyperchromicity against dose, a linear graph is obtained from the slope of which the G values can be found. Examples are shown in Figure 4.2

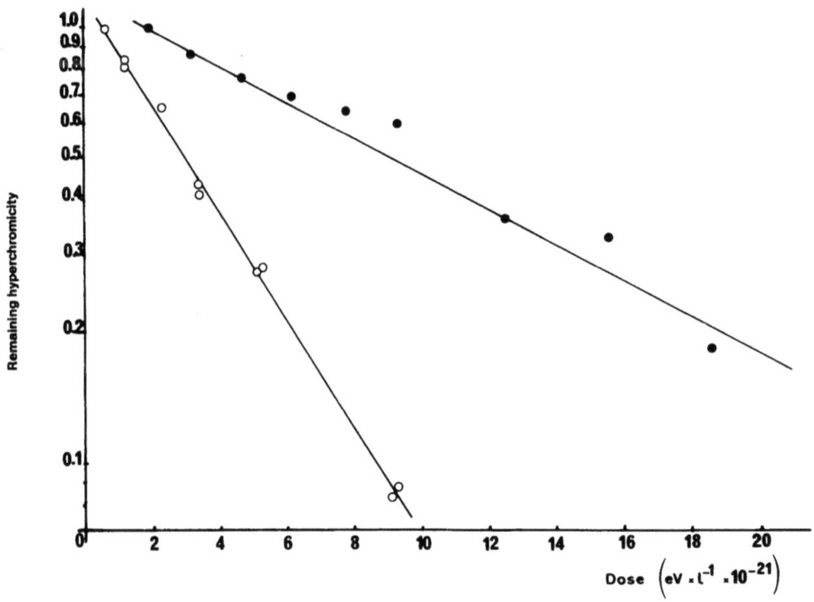

Fig. 4.2. Irradiation of oxygenated aqueous solutions of calf thymus DNA. Plot of the logarithm of remaining hyperchromicity versus dose. -o-: RAFI (1968); -●-: HAGEN and WILD (1964)

An alternative approach to the measurement of hyperchromicity is to make use of the fact that below pH ∿ 3 DNA is denatured in a manner similar to that caused by heat (COX et al., 1955; COLLYNS et al., 1965). Again G-values can be calculated from the loss of hyperchromicity which occurs.

There are large variations obtained by different workers for G-values of radiation-induced loss of hyperchromicity (OKADA, 1970). There are several possible reasons for such discrepancies, and these include:

1. Irradiation in the presence of high concentrations of salts or buffers, (JELLUM, 1966; WARD and URIST, 1967; WARD and KUO, 1970; BARSZCZ and SHUGAR, 1972) leading to scavenging of primary radicals from the water and thus to low G-values.

2. The definition of the base-line from which changes in hyperchromicity can be measured, since DNA can be considered native under a variety of conditions e.g., between pH 4 and 10 in moderately strong salt conditions, and up to 70°C at neutral pH in 10^{-1} M NaCl. From Figure 4.1d it can be seen that the remaining hyperchromicity is much less at 70°C than at 20°C for irradiated DNA. Thus the G-value of loss of hyperchromicity will be much higher at 70°C than at 20°C, even though in both cases the base line for measurement was native DNA. Similar considerations apply to measurements taken at different pH (cf. PEACOCKE and PRESTON, 1960; COLLYNS et al., 1965).

3. The interpretation of the optical density changes. The exact relationship between optical density and changes in molecular configuration (i.e., changes in stacking of bases or base-pair separation) is not known. It is now accepted that hyperchromicity is rather more associated with base stacking.

Taking into account the above mentioned factors and treating the new optical density data according to the method described in Figure 4.2, G-values for the loss of hyperchromicity are as collected in Table 4.3.

Table 4.3. Denaturation of calf-thymus DNA in irradiated aqueous solutions

Reference	G-value for loss of hyperchromicity	[NaCl]	[DNA]
COLLYNS et al. (1965)	10.9	none	0.1%
PEACOCKE and PRESTON (1960)	14.0	none	0.43%
RAFI et al. (1968)	9.0	10^{-3} M	0.075%
HAGEN and WILD (1964)	7.5	10^{-2} M	0.1%
RAFI and WHEELER (1973)	9.3-9.8	10^{-3} M	0.0075%

These G-values are concerned with the number of separated base pairs. The calculations have been made on the basis of linearity between changes in optical density and H-bond breakage and also on the basis of exponential kinetics (see RAFI and WHEELER, 1973). However it is now recognized that hyperchromicity is mainly associated with base stacking and G-values of this table are not known with accuracy.

It can be seen that the values are much higher than [G(-base) + G(chain breakage)] ≈ 2, showing that small changes in the primary structure of DNA lead to much larger changes in the secondary structure. From the table it appears that the highest G-values are obtained at low salt concentration and high DNA concentration.

These values of course, refer to results obtained at room temperature; if a G-value is calculated from data at 70°C in 10^{-1} M NaCl then a value of 30 is obtained.

A study has also been made of the effect of original base composition on the radiation-induced hyperchromicity of DNA (RAFI et al., 1968). It was found that there was a considerable variation in the G-value for loss of hyperchromicity varying from 11.4 for Cl. perfringens (69 mol % A-T), to 5.0 for M. luteus (29 mol % A-T). This variation shows the stability which G-C pairs confer on the secondary structure of DNA, this being in accordance with the well-known variation in T_m with G-C content. From an examination of the spectral characteristics of the hyperchromicity remaining after irradiation it was concluded that hyperchromicity due to A-T interaction was more sensitive to radiation damage.

Similar conclusions were reached by other authors (SIMONOV et al., 1969), who found that the decrease in stability shown by lowering of T_m was essentially the same for calf thymus, *E. coli* and *M. luteus* DNA, although the initial denaturation depended on nucleotide composition.

Another approach to the subject of loss of hyperchromicity is to use synthetic polyribonucleotides such as poly A-poly U or poly A-poly A (EKERT and TISNE, 1966; WARD and URIST, 1967; BARSZCZ and SHUGAR, 1972). However, a complication arises because of the behaviour of single strands containing purine bases. A possible scheme for the breakdown of polynucleotide chains involves the following steps:
Double helix ⟶ stacked single strand ⟶ unstacked single strand ⟶ oligonucleotides ⟶ mononucleotides,
all of which changes show hyperchromicity. The stacking of polynucleotide single strands means that due to vertical forces between bases, single strands do tend to have some structure and indeed much work has been done on this subject. It appears that single chains containing long runs of purine bases are apt to stack and under these conditions they show a melting curve very much like DNA, although the melting is spread over a large temperature range, so that even at room temperature the chains are partially destacked. This effect is more obvious for polyribonucleotides than for polydeoxyribonucleotides such as DNA, in which both the substitution of deoxyribose for ribose, and the more or less random nature of base sequence lessen the degree of stacking, so that there would tend to be a marked decrease in the extent of stacking of any single strands formed as a consequence of irradiation. Such, however, is not the case for a material such as poly A-poly U which has one strand composed entirely of poly A and one of poly U. Upon irradiation and subsequent separation of the two strands, as evidenced by the increase in optical density at 25°C (EKERT and TISNE, 1966) the molecule would be in a state where any separated poly U strands would be unstacked, but where poly A strands would be more or less stacked depending on how much radiation damage they had received. Thus one is faced with a complex situation in which two different processes are leading to hyperchromicity and this makes interpretation of the results very difficult. However the same authors did note that

irradiation of poly A-poly U, or irradiation of either or both of the single strands followed by mixing with the complementary strand, led in every case to a marked lowering of T_m.

In another study (WARD and URIST, 1967) the effects of radiation on single strand and double strand poly A was investigated. (The change from single strand to double strand can be easily brought about by changing the pH from 8.0 to 5.0 or below). The results show that most of the loss of hyperchromicity can be attributed to loss of stacking rather than loss of hydrogen bonding. G-values can be calculated from the data, being about 1 base pair separated for double strand poly A, or in the case of single strand poly A about 2 bases destacked. At the other end of the scale the poly U-poly U double helix which forms at $0^{o}C$ shows a G(loss of hyperchromicity) of about 25, poly U of course showing no stacking. The intermediate case of poly A-poly U showed a G-value of about 4, but because of the stacking of poly A it is difficult to place any significance on this figure in comparison to the values shown by DNA.

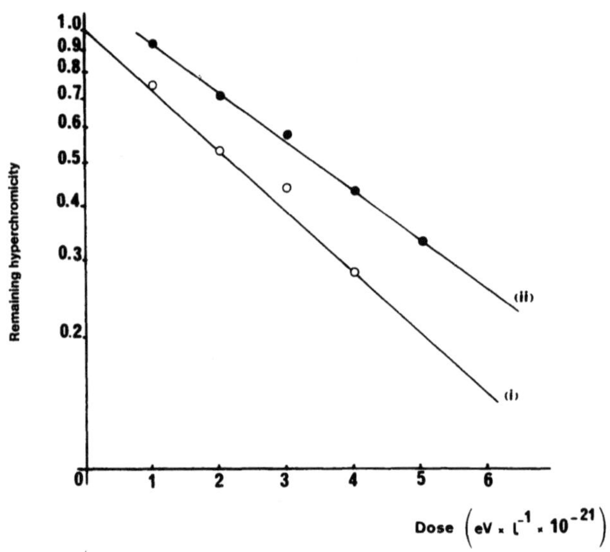

Fig. 4.3. Irradiation of oxygenated aqueous solutions of polynucleotides. Plot of the logarithm of remaining hyperchromicity versus dose. -o-: Poly A; -●-: Poly A-poly A (WARD and URIST, 1967)

BARSZCZ and SHUGAR (1972) have also looked at the effects of radiation on poly A and they observed radiation-induced denaturation of poly A whether irradiated at acid pH, or at neutral pH, and then put into acid conditions. This was shown by the marked changes occurring in the melting curves of poly A-poly A, these changes being somewhat larger when the poly A was irradiated in neutral conditions and then made acidic to form the double strand form.

4.2 Titration

Breakdown of the DNA double helix exposes the amino groups of the bases, which can then be titrated with acid. Since acid itself denatures DNA, native DNA shows a titration curve between pH 6 and 3. On irradiation of DNA this titration curve is displaced (COX et al., 1955; PEACOCKE and PRESTON, 1960) and from the amount of displacement the number of base pairs separated can be calculated. Using this method a G-value of 38 base pairs separated was obtained, which differs from that obtained from hyperchromicity measurements at room temperature, but is of the same order as that measured at 70°C. This is not surprising since in the titration experiments the native DNA is taken to the edge of denaturation, and the effects of radiation are measured from this point, a situation similar to that at 70°C in 10^{-1} M NaCl for measurement of hyperchromicity. The same authors also suggested from the shape of the titration curve that A-T base pairs were denatured faster than G-C base pairs, and this greater sensitivity of the A-T configuration has been substantiated, as noted above (RAFI et al. 1968).

4.3 Other Methods

In the denatured form of DNA, cytosine and adenine are reducible by polarography whereas in native DNA this is not so. Use of this fact has been made by LUKASOVA and PALACEK (1971) by applying the technique of differential pulse polarography. Native DNA, however, does show a peak (called peak II by the authors) which is smaller and appears at a different potential to that, (called peak III), shown by thermally denatured DNA. The presence of peak II in native DNA is attributed to labilized regions within the DNA and can be increased in size by various treatments such as by DNase or sonic degradation. Upon irradiation with X-rays, the first changes seen are only in peak II which increases with dose up to 0.5 eV/P atom, and it is only above this dose that peak III, i.e., denatured DNA, appears, although at smaller doses of about 0.2 eV/P atom peak III can be made to appear by heating to submelting temperatures and re-cooling. The fact that peak II appears in native DNA presumably means that the secondary structure of DNA is not static but that a certain amount of "breathing" takes place allowing penetration of the helix. From the increase in peak II it appears that the initial effects of radiation is simply to increase the amount of "breathing" which takes place, (the authors call the material associated with peak II "intermediate DNA"), and not until later does the helical structure begin to break down. In addition, at low doses no attack by an enzyme specific for denatured DNA could be detected.

The denaturation of DNA can be studied by optical rotatory dispersion (ORD) and circular dichroism spectra (CD). It is well known that heat denaturation causes marked changes in these spectra (BRAHMS and MOMMAERTS, 1964; SAMEJIMA and YANG, 1964) and use has also been made of them in γ-irradiation of DNA (ULIANA et al., 1971). The ORD spectrum of DNA shows peaks at 228 and 290 nm and a trough at 257 nm and on irradiation show linear variations with dose. Given that changes at 228 and 257 nm reflect modifications in the helical structure of the molecule, it was concluded that, under the conditions used, about 30% of the helical structure is lost with a dose of about 150 krad. The CD spectrum also showed a linear decrease in the positive band with radiation.

The work of the previous authors on the radiation-induced effects on the circular dichroism spectrum of DNA has been amplified to include changes induced by very low doses of radiation (VORLICKOVA and PALECEK, 1974). They found that the positive band showed an increase at very small doses (up to 2×10^4 rad) and that a decrease was only evident after 10^5 rad.

An alternative approach to the measurement of radiation-induced denaturation (FREY and HAGEN, 1975) makes use of the fact that only denatured DNA reacts with a water-soluble carbodiimide derivative (CMEC). They showed that two effects occurred simultaneously; the formation of single strand breaks allowing only a relatively small number of nucleotides to react with CMEC, and local denaturation in which 20-50 nucleotide pairs were separated. This latter effect was not linear with dose, but increased at a faster rate than the dose. At the highest dose used (38 krad) it can be calculated that the G-value for reaction with CMEC is about 12, in good agreement with values quoted earlier.

4.4 Conclusions

Denaturation of DNA is of course a consequence of irradiation, but the answers to two important questions are still a matter for conjecture;

1. how does denaturation occur
2. how much of it occurs in relation to the other types of damage to DNA such as base damage, base release, single and double strand breaks?

From the work on hyperchromicity, differential pulse polarography, circular dichroism and the reaction with CMEC, there appears to be a lag in the onset of denaturation. The bases may be somewhat screened from the effects of radiation through lying on the inside of the helix. Since bases are rather more reactive than sugars towards radicals (SCHOLES, 1968), then after any preliminary opening out of the helix, base damage would increase rapidly leading to destacking and loss of hyperchromicity.

At higher doses it appears that a rather different mechanism occurs whereby large segments of oligonucleotides become detached from the DNA thus leading to a much faster apparent rate of denaturation. Thus the lag in hyperchromicity evident from the figures (apart from single-strand poly A), and the lag in base-destruction shown by poly A-poly A and poly A-poly U could be accounted for (WARD and URIST, 1967), as would the results of pulse radiolysis which show an increase in the rate of attack of radicals on the bases, and a larger base signal, as radiation proceeds (SCHOLES et al., 1969).

As to the amount of denaturation caused by irradiation, if one takes only the limited number of studies made under neutral conditions at room temperature then a reasonably consistent picture emerges, that after an initial lag, about 10-14 base pairs are separated per 100 eV.

One approach which so far does not seem to have been used for looking at irradiation-induced denaturation is that of electron microscopy. It has been shown that single and double strands of DNA can be distinguished under the electron microscope (INMAN, 1966) and it may prove a fruitful exercise to examine this method, and using the same sample of DNA under the same conditions, compare it with as many of the other methods as practicable. In this way it may be possible to obtain some self consistent results as to the mechanism and extent of denaturation.

References

ACHEY, P., DURYEA, H.: Production of DNA strand breaks by the hydroxyl radical. Intern. J. Radiat. Biol. 25, 595-601 (1974)

ADAMS, G.E.: The general application of pulse radiolysis to current problems in radiobiology. In: Current Topics in Radiation Research. Ebert, M., Howard, A. (eds.). Vol. III, pp. 35-93. Amsterdam: North-Holland, 1967

ADAMS, G.E., GREENSTOCK, C.L., VAN HEMMEN, J.J., WILLSON, R.L.: Radical oxidation mechanisms in cellular radiosensitization: Electron transfer in the pulse radiolysis of aqueous nucleotide solutions. Radiat. Res. 49, 85-95 (1972)

ADAMS, G.E., WILLSON, R.L.: On the mechanism of BUdR sensitization: a pulse radiolysis study of one electron transfer in nucleic acid derivatives. Intern. J. Radiat. Biol. 22, 589-597 (1972)

ALEXANDER, P., LETT, J.T.: Effects of ionizing radiations on biological macromolecules. In: Comprehensive Biochemistry. Florkin, M., Stotz, E.H. (eds.). Vol. XXVII, pp. 267-356. New York-Amsterdam: Elsevier, 1967

AL-YAMOOR, K.Y., GARNIER, A., SCHOLES, G.: (In preparation) (1977)

ANDERSON, C., HART, E.J.: Molecular product and free radical yields in the decomposition of water by protons, deuterons and helium ions. Radiat. Res. 14, 689-704 (1961)

ANGYAL, S.J., PICKLES, V.A.: Equilibria between pyranoses and furanoses. Australian J. Chem. 25, 1711-1718 (1972)

BANSAL, K.M., GRÄTZEL, M., HENGLEIN, A., JANATA, E.: Polarographic and optical absorption studies of radicals produced in pulse radiolysis of aqueous solutions of ethylene glycol. J. Phys. Chem. 77, 16-19 (1973)

BANSAL, K.M., SELLERS, R.M.: Polarographic studies of short-lived radicals in aqueous solutions. In: Fast Processes in Radiation Chemistry. Adams, G.E., Fielden, E.M., Michael, B.D. (eds.) pp. 259-263. London Inst. Physics: Wiley, 1975

BARSZCZ, D., FIELDEN, E.M.: Transient and stable products in the pulse radiolysis of some dihydropyrimidine solutions. Intern. J. Radiat. Biol. 25, 539-553 (1974)

BARSZCZ, D., SHUGAR, D.: Radiation chemistry of nucleic acids and some derivatives. Acta Biochim. Pol. 8, 455-464 (1961)

BARSZCZ, D., SHUGAR, D.: Effects of ionizing radiation of poly A and its constituents and dependence of radiosensitivity on secondary structure. Acta Biochim. Pol. 19, 25-42 (1972)

BAZIN, N.M., KUZNETSOV, E.V., BUBNOV, N.N., VOEVODSKII, V.V.: Reaction of a hydrogen atom in the system H_2O + H_2SO_4 + $FeSO_4$. III. Reaction with saturated organic compounds. Kinet. Katal. 7, 732-734 (1966)

BELMAN, S., LEVENE, E., TROLL, W.: The effects of some chemical and physical agents on the melting (T_m) of DNA. Federation Proc. Am. Soc. Exp. Biol. 21, 374 (1962)

BHATIA, K., SCHULER, R.H.: Uracilyl radical production in the radiolysis of aqueous solutions of the 5-halouracils. J. Phys. Chem. 77, 1888-1896 (1973)

BLANK, B., HENNE, A., LAROFF, G.P., FISCHER, H.: Enol intermediates in photoreduction and type I cleavage reactions of aliphatic aldehydes and ketones. Pure Appl. Chem. 41, 475-494 (1975)

BLOK, J., LOMAN, H.: The effects of γ-irradiation in DNA. Current Top. Radiat. Res. Q. 9, 165 (1973)
BOHACEK, J., BLAZICEK, G.: The course of thermal denaturation of deoxyribonucleic acids after γ-irradiation. Biophysik 2, 223-238 (1965)
BOPP, A., CARPY, S., BURKART, B., HAGEN, U.: Action of exonuclease III on γ-irradiated DNA. Biochim. Biophys. Acta 294, 47-56 (1973)
BOPP, A., HAGEN, U.: End group determination in γ-irradiated DNA. Biochim. Biophys. Acta 209, 320-326 (1970)
BRAAMS, R., EBERT, M.: The influence of changes in conformation of a macromolecule on reaction rates. In: Radiation Chemistry (Advances in Chemistry Series). Gould, R.F. (ed.). Vol. I, pp. 464-471. Washington, D.C.: Am. Chem. Soc., 1968
BRAHMS, J., MOMMAERTS, W.F.H.M.: A study of conformation of nucleic acids in solution by means of circular dichroism. J. Mol. Biol. 10, 73-88 (1964)
BROWN, P.E., CALVIN, M., NEWMARK, J.F.: Thymine addition to ethanol. Induction by gamma irradiation. Science 151, 68-70 (1966)
BULEY, A.L., NORMAN, R.O.C., PRITCHETT, R.J.: Electron spin resonance studies of oxydation. Part VIII. Elimination reactions of some hydroxyalkyl radicals. J. Chem. Soc. B 849-852 (1966)
BURCHILL, C.E., PERRON, K.M.: Radical-induced rearrangement of ethylene glycol in aqueous solution. Can. J. Chem. 49, 2382-2389 (1971)
CADET, J.: Thesis Univ. Grenoble, France (1973)
CADET, J.: Radiolyse γ de solutions aqueuses de thymine à 77 K et 195 K. Identification du dimère de type cyclobutane cis syn. Tetrahedron Lett. 4275-4277 (1976)
CADET, J., DUCOLOMB, R.: Préparation, isomérisation et étude conformationelle des "hydrates" d'uridine et de thymidine. Table Ronde Intern. Montpellier (1974)
CADET, J., DUCOLOMB, R., TEOULE, R.: Radiolyse gamma de la thymidine-5'-monophosphate en solution aqueuse aérée. Identification du N (phospho-5-desoxy-2-β-D-erythro-pentofuranosyl) formamide. C.R. Acad. Sci. (Paris) Ser. C 279, 549-551 (1974)
CADET, J., GUTTIN-LOMBARD, M., TEOULE, R.: Gamma radiolysis of thymine and thymidine in deaerated aqueous solutions containing electron-affinic radiosensitizers. Formation of stable adducts. Intern. Symp. (ASP) Williamsburg (1975a)
CADET, J., TEOULE, R.: Peroxydes formés par action du rayonnement gamma sur la thymine en solution aqueuse aérée. Biochim. Biophys. Acta 238, 8-26 (1971a)
CADET, J., TEOULE, R.: Irradiation gamma de la thymine en milieu désaéré: isomère cis et trans de l'hydroxy-6-dihydro-5,6-thymine. Intern. J. Appl. Radiat. Isot. 22, 273-280 (1971b)
CADET, J., TEOULE, R.: Synthèse de produits de radiolyse de la thymidine. Tetrahedron Lett. 3229-3232 (1972a)
CADET, J., TEOULE, R.: Irradiation gamma de la thymidine en solution aqueuse aérée. Tetrahedron Lett. 3225-3228 (1972b)
CADET, J., TEOULE, R.: Site d'attaque du radical hydroxyle sur le cycle pyrimidique-Peroxydes de thymine et de thymidine. C.R. Acad. Sci. (Paris) Ser. C 276, 1743-1746 (1973a)
CADET, J., TEOULE, R.: Sur une nouvelle méthode de synthèse des hydroxydroperoxydes dérivés de la thymine et de la thymidine. Tetrahedron Lett. 4245-4248 (1973b)
CADET, J., TEOULE, R.: Radiation chemistry of nucleic acids. Characterization of thymine hydroxyhydroperoxides. Biochem. Biophys. Res. Commun. 59, 1047-1052 (1974)
CADET, J., TEOULE, R.: Radiolyse gamma de la thymidine en solution aqueuse aérée. I. Identification des hydroxyhydroperoxydes. Bull. Soc. Chim. France 879-884 (1975a)

CADET, J., TEOULE, R.: II-Caractérisation des produits stables. Bull. Soc. Chim. France 885-891 (1975b)

CADET, J., TEOULE, R.: III-Aspects quantitatifs et mécanismes. Bull. Soc. Chim. France 891-896 (1975c)

CADET, J., ULRICH, J., TEOULE, R.: Izomerization and new specific synthesis of thymine glycol. Tetrahedron 2057-2061 (1975b)

CAMPBELL, J.M., SCHULTE-FROHLINDE, D., VON SONNTAG, C.: Quantum yields in the UV photolysis of 5-bromouracil in the presence of hydrogen donors. Photochem. Photobiol. 20, 465-467 (1974a)

CAMPBELL, J.M., VON SONNTAG, C., SCHULTE-FROHLINDE, D.: Photolysis of 5-bromouracil and some related compounds in solution. Z. Naturforsch. 29b, 750-757 (1974b)

CERUTTI, P.A.: Effects of ionizing radiation on mammalian cells. Naturwissenschaften 61, 41-59 (1974)

CHABRE, M., GAGNAIRE, D., NOFRE, C.: Conformation de dihydropyrimidinediones II-Etude par résonance magnétique nucléaire. Bull. Soc. Chim. France 108-113 (1966)

COLLYNS, B., OKADA, S., SCHOLES, G., WEISS, J.J., WHEELER, C.M.: Chain scission and hydrogen bond breakage on irradiation of DNA. Radiat. Res. 25, 526-536 (1965)

CONLAY, J.J.: Effect of ionizing radiation on adenine in aerated and deaerated aqueous solutions. Nature (London) 197, 555-557 (1963)

COX, R.A., OVEREND, W.G., PEACOCKE, A.R., WILSON, S.: Effects of gamma rays on solution of sodium deoxyribonucleate. Nature (London) 176, 919-921 (1955)

DANIELS, M., GRIMISON, A.: Photolysis of the aqueous thymine system. I/Excitation at 1849 Å. Biochim. Biophys. Acta 142, 292-303 (1967)

DANIELS, M., SCHOLES, G., WEISS, J.: Studies on the action of X-rays on aqueous solutions of nucleic acids and some nucleotides. Experientia 11, 219-221 (1955)

DANIELS, M., SCHOLES, G., WEISS, J.: Chemical action of ionising radiations in solutions. Part XVI. Formation of labile phosphate esters from purine and pyrimidine ribonucleotides by irradiation with X-rays in aqueous solution. J. Chem. Soc. 3771-3779 (1956)

DANIELS, M., SCHOLES, G., WEISS, J.J., WHEELER, C.M.: Chemical action of ionizing radiations in solutions. Degradation of deoxyribonucleic acid in aqueous solution by irradiation with X-rays (200 KV). J. Chem. Soc. 226-234 (1957)

DANIELS, M., SCHWEIBERT, M.C.: Cytosine hydroperoxide formation in γ-radiolysis of cytosine in aqueous solution. Biochim. Biophys. Acta 134, 481-483 (1967)

DANZIGER, R.M., HAYON, E., LANGMUIR, M.E.: Pulse radiolysis and flash photolysis study of aqueous solutions of simple pyrimidines: Uracil and bromouracil. J. Phys. Chem. 72, 3842-3849 (1968)

DAVIES, A.G. (ed.): O-O Heterolysis: intermolecular nucleophilic substitution at oxygen. Organic Peroxides, pp. 128-142. London: Butterworths, 1961

DERTINGER, H., NICOLAU, C.: An electron paramagnetic resonance study of short-lived free radicals from nucleic acids constituents and related compounds in aqueous solutions. Biochim. Biophys. Acta 199, 316-321 (1970)

DIZDAROGLU, M., HENNEBERG, D., SCHOMBURG, G., VON SONNTAG, C.: Radiation chemistry of carbohydrates-VI. γ-Radiolysis of glucose in deoxygenated N_2O saturated aqueous solution. Z. Naturforsch. 30b, 416-425 (1975a)

DIZDAROGLU, M., HENNEBERG, D., VON SONNTAG, C.: The mass spectra of TMS-ethers of deuterated polyalcohols. A contribution to the structural investigation of sugars. Org. Mass Spectr. 8, 335-345 (1974)

DIZDAROGLU, M., LEITICH, J., VON SONNTAG, C.: Conversion of D-fructose into 6-deoxy-D-threo-2,5-hexodiulose by γ-irradiation. A chain reaction in the crystalline state. Carbohyd. Res. 47, 15-23 (1976a)

DIZDAROGLU, M., NEUWALD, K., VON SONNTAG, C.: Radiation chemistry of DNA model compounds - IX. Carbohydrate products in the γ-radiolysis of thymidine in aqueous solution. The radical induced scission of the N-glycosidic bond. Z. Naturforsch. 31b, 227-233 (1976b)
DIZDAROGLU, M., SCHERZ, H., VON SONNTAG, C.: Strahlenchemie von Alkoholen - XVI. γ-Radiolyse von *meso*-Erythrit in wässriger Lösung. Z. Naturforsch. 27b, 29-41 (1972)
DIZDAROGLU, M., SCHULTE-FROHLINDE, D., VON SONNTAG, C.: Strand breaks and sugar release by γ-irradiation of DNA in aqueous solution. The effect of oxygen. Z. Naturforsch. 30c, 826-828 (1975b)
DIZDAROGLU, M., VON SONNTAG, C.: Strahlenchemie von Kohlenhydraten-IV. γ-Radiolyse von Cellobiose in N_2O gesättigter wässriger Lösung. Teil I. Identifizierung der Produkte. Z. Naturforsch. 28b, 635-646 (1973)
DIZDAROGLU, M., VON SONNTAG, C., SCHULTE-FROHLINDE, D.: Strand breaks and sugar release by γ-irradiation of DNA in aqueous solution. J. Am. Chem. Soc. 97, 2277-2278 (1975c)
DIZDAROGLU, M., VON SONNTAG, C., SCHULTE-FROHLINDE, D., DAHLHOFF, W.V.: Notiz über die Darstellung von 5-Desoxylactobionsäure auf strahlenchemischem Wege. Liebigs Ann. Chem. 1592-1594 (1973)
DORFMAN, L.M., ADAMS, G.E.: Reactivity of the hydroxyl radical in aqueous solution. Monograph on chemical kinetics for the National Standard Reference Data System. Natl. Bureau of Standards. U.S. (1973)
DRAGANIC, I.G., DRAGANIC, Z.D.: The Radiation Chemistry of Water. New York: Academic Press, 1971
DRÁŽIL, V., RÝZNAR, L.: A contribution to the problem of formation of urea in the radiolysis of thymine. Collect. Czech. Commun. 33, 3068-3072 (1968a)
DRÁŽIL, V., RÝZNAR, L.: Comparison of alpha and gamma rays. Radiolysis of thymine. IAEA Report form Biophysical Aspects of Radiation Quality Second Panel Report 17-25 (1968b)
DUCOLOMB, R., CADET, J., TEOULE, R.: Effet des rayons gamma sur l'uracile et l'acide uridylique en solution aqueuse aérée. C.R. Acad. Sci. (Paris) Ser. D 273, 2647-2649 (1971)
DUCOLOMB, R., CADET, J., TEOULE, R.: Effet du rayonnement gamma sur l'uracile en solution aqueuse aérée. Bull. Soc. Chim. France 3, 1167-1174 (1973)
DUCOLOMB, R., CADET, J., TEOULE, R.: Irradiation gamma de l'acide uridylique. Rupture de la liaison N-glycosidique. Intern. J. Radiat. Biol. 25, 139-149 (1974a)
DUCOLOMB, R., CADET, J., TEOULE, R.: Gamma irradiation of uridine and uridylic acid in aerated aqueous solutions. Identification of the major product. Z. Naturforsch. 29c, 643-644 (1974b)
EDGE, D.J., GILBERT, B.C., NORMAN, R.O.C., WEST, P.R.: Electron spin resonance studies. Part XXVIII. Oxidation of enols, enol ethers, and related compounds with the hydroxyl and the amino radical. J. Chem. Soc. B, 189-196 (1971)
EIGEN, M.: Proton transfer, acid-base catalysis and enzymatic hydrolysis. Angew. Chem. Intern. Ed. England 3, 1-19 (1964)
EKERT, B.: Effect of γ-rays on thymine in deaerated aqueous solutions. Nature (London) 194, 278-279 (1962)
EKERT, B., MONIER, R.: Structure of thymine hydroperoxide produced by X-irradiation. Nature (London) 184, BA 58-59 (1959)
EKERT, B., MONIER, R.: Effects of X-rays on cytosine in aerated aqueous solutions. Nature (London) 188, 309-310 (1960)
EKERT, B., TISNE, M.R.: Effets des rayons gamma sur les acides polyriboadényliques, polyribouridylique et leurs associations bi- et trifiliaires. Biochim. Biophys. Acta 114, 481-490 (1966)
EMANUEL, N.M.: The role of free radicals in radiobiological processes. Tr. Mosk. Ova. Ispyt. Prir. 73 Otd. Biol. 7, 73-83 (1963)

FAHR, E.: Chemical investigations of the molecular origin of biological radiation damage. Angew. Chem. Intern. Ed. England 8, 578-593 (1969)
FEL, N.S., ZAOZERSKAYA, L.A., DOLIN, P.I.: Pulse radiolysis studies of some nucleic acid fragments in aqueous solutions. Radiat. Eff. 9, 145-152 (1971)
FENDLER, J.H., INFANTE, G.A., JIRATHANA, P., FENDLER, E.J.: Radiolysis of pyrimidines in aqueous solutions. Micellar effects on product yields formed in the interaction of e^-_{aq}, H, OH and Cl_2^- with thymine and uracil. J. Chem. Soc. Faraday Trans. 69, 1171-1179 (1973)
FIELDEN, E.M., STEVENS, G.C., PHILLIPS, J.M., SCHOLES, G., WILLSON, R.L.: Effects due to nonequilibrium pH conditions in the radiolysis of aqueous systems. Nature (London) 225, 632-634 (1970)
FISHER, G.J., JOHNS, H.E.: Thymine hydrate formed by ultra-violet and gamma irradiations of aqueous solutions. Photochem. Photobiol. 18, 23-27 (1973)
FLIPPEN, J.L.: The crystal and molecular structure of reaction products from gamma irradiation of thymine and cytosine: *cis* thymine glycol $C_5H_8N_2O_4$ and *trans*-1-carbamoylimidazolidone-4,5-diol. Acta Cryst. Sect B 29, 1756-1762 (1973)
FREY, R., HAGEN, U.: Oxygen effect on γ-irradiated DNA. Radiat. Environ. Biophys. 11, 125-133 (1974)
FREY, R., HAGEN, U.: Changes in DNA secondary structure after γ-irradiation. Radiat. Environ. Biophys. 12, 111-118 (1975)
FURBERG, S.: The structure of 2-deoxy-ribose. Acta Chem. Scand. 14, 1357-1363 (1960)
GILBERT, B.C., LARKIN, J.P., NORMAN, R.O.C.: Electron spin resonance studies. Part XXXIII. Evidence for heterolytic and homolytic transformations of radicals from 1.2-diols and related compounds. J. Chem. Soc. Perkin Trans. II, 794-802 (1972)
GILBERT, B.C., NORMAN, R.O.C., SEALY, R.C.: Electron spin resonance studies. Part XXXIX. A kinetic investigation of the role of radical reduction in metal-ion-hydrogen peroxide systems. J. Chem. Soc. Perkin Trans. II, 2174-2180 (1973)
GILBERT, E., VOLKERT, O., SCHULTE-FROHLINDE, D.: Strahlenchemie wäßriger, sauerstoffhaltiger Lösungen von 5-Bromuracil. Identifizierung der Radiolyseprodukte. Z. Naturforsch. Teil B 22, 477-480 (1967)
GILBERT, E., WAGNER, G.: Reaktionen des Uracilradikals mit primären und sekundären Alkoholen in wässriger, sauerstoffgesättigter Lösung. Z. Naturforsch. 27b, 644-648 (1972)
GILBERT, E., WAGNER, G., SCHULTE-FROHLINDE, D.: Photolyse von 5-Joduracil in wässriger sauerstoffgesättigter Lösung in Gegenwart von Methanol. Z. Naturforsch. 26b, 209-213 (1971)
GREENSTOCK, C.L., ADAMS, G.E., WILLSON, R.L.: Electron transfer studies of nucleic acid derivatives in solutions containing radiosensitizers. In: Radiation Protection and Sensitization. Moroson, H.L., Quintiliani, M. (eds.), pp. 65-71. London: Taylor and Francis, 1970
GREENSTOCK, C.L., DUNLOP, I.: Pulse radiolysis studies of nitrofurans: Chemical radiosensitization. Radiat. Res. 56, 428-440 (1973)
GREENSTOCK, C.L., DUNLOP, I.: Reactions of nucleic acid radicals with radiation modifiers. In: Fast Processes in Radiation Chemsitry. Adams, G.E., Fielden, E.M., Michael, B.D. (eds.), pp. 247-258. The Institut of Physics. London: Wiley, 1975
GREENSTOCK, C.L., HUNT, J.W., NG, M.: Pulse radiolysis studies of uracil and its derivatives. Trans. Faraday Soc. 65, 3279-3287 (1969)
GREENSTOCK, C.L., NG, M., HUNT, J.W.: Pulse radiolysis studies of reactions of primary species in water with nucleic acid derivatives. In: Radiation Chemistry (Advances in Chemistry Series). Gould, R.F. (ed.), Vol. I, 397-417. Washington, D.C.: Amer. Chem. Soc., 1968)
GREENSTOCK, C.L., SHIERMAN, E.: Radiosensitivity and hydroxyl radical reactivity of phosphate esters as measured by radiation-induced dephosphorylation. Intern. J. Radiat. Biol. 28, 1-12 (1975)

HAGEN, U., WILD, R.: The radiation sensitivity of DNA - Sensitivity of the hydrogen bridges. Strahlentherapie 124, 275-283 (1964)
HAHN, B.S., WANG, S.Y., FLIPPEN, J.L., KARLE, I.L.: Isolation and characterization of glycols of 1-carbamoylimidazolidone. J. Am. Chem. Soc. 95, 2711-2712 (1973)
HARIHARAN, P.V., CERUTTI, P.A.: Repair of gamma-ray induced thymine damage in *Micrococcus radiodurans*. Nature (New Biol.) 229, 247-249 (1971)
HARIHARAN, P.V., CERUTTI, P.A.: Formation and repair of gamma ray induced thymine damage in *Micrococcus radiodurans*. J. Mol. Biol. 66, 65-81 (1972)
HART, E.J., GORDON, S., THOMAS, J.K.: Rate constants of hydrated electron reactions with organic compounds. J. Phys. Chem. 68, 1271-1274 (1964)
HARTMANN, V.: Strahlenchemie von 2-Desoxy-D-ribose in wässriger Lösung. Dissertation, Univ. Karlsruhe (1970)
HARTMANN, V., VON SONNTAG, C., SCHULTE-FROHLINDE, D.: γ-Radiolyse von 2-Desoxyribose in wässriger Lösung. Z. Naturforsch. 25b, 1394-1404 (1970)
HAYASHI, T., OGISO, O., NAMIKI, M.: The dimerization of thymine glycol and the splitting of the dimer by gamma-irradiation in an aqueous system. Bull. Chem. Soc. Japan 46, 3896-3898 (1973)
HAYON, E.: Optical absorption spectra of ketyl radicals and radical anions of some pyrimidines. J. Chem. Phys. 51, 4881-4892 (1969)
HAYON, E., SIMIC, M.: Addition of hydroxyl radicals to pyrimidine bases and electron transfer of intermediates to quinones. J. Am. Chem. Soc. 95, 1029-1035 (1973)
HAYSOM, H.R., PHILLIPS, J.M., RICHARDS, J.T., SCHOLES, G., WILLSON, R.L.: Pulse radiolysis of aqueous solutions of dihydropyrimidines: the role of carbonium ions. In: Fast Processes in Radiation Chemistry. Adams, G.E., Fielden, E.M., Michael, B.D. (eds.), pp. 241-246. The Institute of Physics. London: Wiley, 1975
HAYSOM, H.R., PHILLIPS, J.M., SCHOLES, G.: Formation of carbonium ions from dihydropyrimidyl radicals in the γ-radiolysis of aqueous solutions of dihydropyrimidines. J. Chem. Soc. Chem. Commun. 1082-1083 (1972)
HEMMEN, J.J. VAN: Some reactions of the hydroxyl adduct of adenine. Intern. J. Radiat. Biol. 27, 403-407 (1975)
HEMMEN, J.J. VAN, BLEICHRODT, J.F.: The decomposition of adenine by ionizing radiation. Radiat. Res. 46, 444-456 (1971)
HEMS, G.: Effects of ionizing radiation on aqueous solutions of guanylic acid and guanosine. Nature (London) 181, 1721-1722 (1958)
HEMS, G.: Chemical effects of ionizing radiation of deoxyribonucleic acid in dilute aqueous solution. Nature (London) 186, 710-712 (1960a)
HEMS, G.: Effects of ionizing radiation on aqueous solutions of inosine and adenosine. Radiat. Res. 13, 777-787 (1960b)
HERAK, J.N., GORDY, W.: An ESR study of nucleosides and nucleotides bombarded with hydrogen atomes. Proc. Natl. Acad. Sci. 56, 7-11 (1966)
HERFORD, N., SUEOKA, N.: Chromosomal location of antibiotic resistance markers in Bacillus subtilis. J. Mol. Biol. 51, 267-286 (1970)
HOARD, D.E., HAYES, F.N., GOAD, W.B.: Degradation of nucleic acid in aqueous solution by ionizing radiation. I. Loss of ultra-violet absorption of solutions of thymine or thymine derivatives on X-irradiation. Intern. J. Radiat. Biol. 25, 603-609 (1974)
HOLIAN, J., GARRISON, W.M.: Radiation-induced oxidation of cytosine and uracil in aqueous solution of copper (II). Nature (London) 212, 394-395 (1966)
HOLMES, D.E., INGALLS, R.B., MYERS, L.S., Jr.: ESR spectra of free radicals induced in nucleic acid, purine and pyrimidine bases and selected analogues by exposure to hydrogen atoms. Intern. J. Radiat. Biol. 13, 225-234 (1967)
HÜTTERMANN, J., MÜLLER, A.: Elektronen-Spin-Resonanz an γ-bestrahlten Einkristallen von 2-Desoxy-D-ribose. Z. Naturforsch. 24b, 463-464 (1969a)

HÜTTERMANN, J., MÜLLER, A.: Free radicals in gamma-irradiated 2-deoxy-D-ribose. Electron spin resonance of single crystals and polycrystalline material. Radiat. Res. 38, 248-259 (1969b)
HUTCHINSON, F.: The lesions produced by ultraviolet light in DNA containing 5-bromouracil. Quart. Rev. Biophys. 6, 201-246 (1973)
HUYSER, E.S.: The free radical induced rearrangement of 2-methoxytetrahydropyrane to methyl valerate. J. Org. Chem. 25, 1820-1822 (1960)
INFANTE, G.A., FENDLER, E.J., FENDLER, J.H.: Recent developments in the radiation chemistry of bases in aqueous solutions. Radiat. Res. Rev. 4, 301-339 (1973a)
INFANTE, G.A., JIRATHANA, P., FENDLER, E.J., FENDLER, J.H.: Radiolysis of pyrimidines in aqueous solutions. J. Chem. Soc. Faraday Trans. 69, 1162-1170 (1973b)
INMAN, R.B.: A denaturation map of the γ phage DNA molecule determined by electron microscopy. J. Mol. Biol. 18, 464-476 (1966)
JACOBS, S.S.: Solid state γ-irradiation of crystalline amino acids and sugars. Ph.D. Thesis, Univ. of Colorado (1962)
JELLUM, E.: The role of cysteamine-nucleic acid interactions in protection against X-ray induced damage of DNA. Intern. J. Radiat. Biol. 10, 577-594 (1966)
JUNG, H., HAGEN, U., ULLRICH, M., PETERSEN, E.E.: Structural and functional changes in DNA after exposure to hydrogen atoms and γ-radiation. Z. Naturforsch. 24b, 1565-1573 (1969)
JUST, S., SCHWARTZ, K.: Base destruction and denaturation of X-irradiated *Bacillus subtilis* DNA. Studia Biophys. 18, 219-226 (1969)
KAMAL, M.N., GARRISON, W.M.: Radiolytic degradation of aqueous cytosine. Enhancement by a second organic solute. Nature (London) 206, 1315-1317 (1965)
KAPLAN, H.S., SMITH, K.C., TOMLIN, P.A.: Effect of halogenated pyrimidines on radiosensitivity of *E.coli*. Radiat. Res. 16, 98-113 (1962)
KAPP, D.S., SMITH, K.C.: Chemical nature of chain breaks produced in DNA by X-irradiation in vitro. Radiat. Res. 42, 34-49 (1970)
KATRITZKY, A.R., NESBIT, M.R., KURTEV, B.J., LYAPOVA, M., POJARLIEFF, I.G.: β ureids acids and dihydrouracils. Tetrahedron 25, 3807-3824 (1969)
KECK, K.: Bildung von Cyclonucleotiden bei Bestrahlung wässriger Lösungen von Purinnucleotiden. Z. Naturforsch. 23b, 1034-1043 (1968)
KECK, K., HAGEN, U., FRIEBOLIN, H.: Bildung eines Cyclonucleotids bei Röntgenbestrahlung von Adenosin-monophosphat. Naturwissenschaften 53, 304 (1966)
KHATTAK, M.N., GREEN, J.H.: Gamma irradiation of nucleic acid constituents in deaerated aqueous solutions. I-Cytosine. Intern. J. Radiat. Biol. 11, 131-136 (1966a)
KHATTAK, M.N., GREEN, J.H.: Gamma irradiation of nucleic acid constituents in deaerated aqueous solutions. III-Thymine and uracil. Intern. J. Radiat. Biol. 11, 577-582 (1966b)
KHRYAKOV, A.I.: Radiolysis of cytosine in aqueous solutions. Radiobiologiya 12, 792-795 (1972)
KIRSCH, A., VON SONNTAG, C., SCHULTE-FROHLINDE, D.: Radiation chemistry of carbohydrates. Part V. γ-Radiolysis of *scyllo*-inositol in deoxygenated and oxygenated aqueous solution. J. Chem. Soc. Perkin Trans II, 1334-1338 (1975)
KOCHETKOV, N.K., KUDRYASHOV, L.I., CHLENOV, M.A., GRINEVA, L.P.: Radiolysis of aqueous solutions of sugar phosphates. Carbohyd. Res. 35, 235-241 (1974)
KOCHETKOV, N.K., KUDRYASHOV, L.I., CHLENOV, M.A., LIVERTOVSKAYA, T.Y.: The epimerisation of monosaccharides by γ-irradiation in frozen aqueous solutions. Carbohyd. Res. 28, 86-88 (1973)
KOCHI, J.K.: Oxidation-reduction reactions of free radicals and metal complexes. In: Free Radicals. Kochi, J.K. (ed.), Vol. I, pp. 591-681. London: Wiley, 1973

KÖHNLEIN, W., HUTCHINSON, F.: E.S.R. studies of normal and 5-bromouracil substituted DNA of Bacillus subtilis after irradiation with ultraviolet light. Radiat. Res. 39, 745-757 (1969)

KOURIM, P., BORS, W., SCHULTE-FROHLINDE, D.: Gamma Radiolyse wässriger, sauerstoffhaltiger Lösungen von 5-Bromo-2'-desoxyuridin. I-Identifizierung von Radiolyseprodukten Z. Naturforsch. 26b, 308-311 (1971a); II-G Werte der Produkte und Vergleich mit der Radiolyse des Aglycons. Z. Naturforsch. 26b, 312-314 (1971b)

KRÜGER, C., TSAI, Y.-H., VON SONNTAG, C.: The structure of 2-deoxy-D-ribose (1977) (To be published)

KRUSHINSKAYA, N.P., SHAL'NOV, M.I.: Nature of breaks in the DNA chain upon irradiation of aqueous solutions. Radiobiology 7, 36-45 (1967)

LAGERCRANTZ, C.: Trapping of radicals formed in the photochemical reactions between peroxide and some pyrimidine bases, nucleosides and yeast nucleic acid. J. Am. Chem. Soc. 95, 220-225 (1973)

LANDBECK, I., HAGEN, U.: Action of DNA polymerase I on γ-irradiated DNA. Biochim. Biophys. Acta 331, 318-327 (1973)

LATARJET, R.: Interaction of radiation energy with nucleic acids. Current Top. Radiat. Res. 8, 1-38 (1972)

LATARJET, R., EKERT, B., APELGOT, S., REBEYROTTE, N.: Etudes radiochimiques sur l'ADN. J. Chim. Phys. Chim. Biol. 58, 1046-1049 (1961)

LATARJET, R., EKERT, B., DEMERSEMAN, P.: Peroxidation of nucleic acids by radiation. Radiat. Res. Suppl. 4, 192-212 (1963)

LEHNERT, S., OKADA, S.: Cesium chloride density gradient studies of nuclear proteins. Studies with control and irradiated nucleic. Biochim. Biophys. Acta 109, 557-567 (1965)

LEMIEUX, R.U., STEVENS, J.D.: The proton magnetic resonance spectra and tautomeric equilibria of aldoses in deuterium oxide. Can. J. Chem. 44, 249-262 (1966)

LENNARTZ, M., COQUERELLE, T., BOPP, A., HAGEN, U.: Oxygen effect on strand breaks and specific end groups in DNA of irradiated thymocytes. Intern. J. Radiat. Biol. 27, 577-587 (1975)

LINDSAY, D., HOWARD, J.A., HORSWILL, E.C., ITON, L., INGOLD, K.U., COBBLEY, T., LL, A.: The bimolecular self-reactions of secondary peroxy radicals. Product studies. Can. J. Chem. 51, 870-880 (1973)

LIVINGSTON, R., ZELDES, H.: Paramagnetic resonance study of liquids during photolysis. III. Aqueous solutions of alcohols with hydrogen peroxide. J. Am. Chem. Soc. 88, 4333-4336 (1966)

LOMAN, H., BLOK, J.: On the radiation chemistry of thymine in aqueous solution. Radiat. Res. 36, 1-13 (1968)

LOMAN, H., EBERT, M.: The radiation chemistry of thymine in aqueous solutions. Some reactions of the thymine electron adduct. Intern. J. Radiat. Biol. 18, 369-379 (1970)

LÜCKE-HUHLE, C., BRAUN, A., HAGEN, U.: Oxygen effect in γ-irradiated DNA. Z. Naturforsch. 25b, 1264-1268 (1970)

LUKÁŠOVÁ, E., PALEČEK, E.: Changes in DNA secondary structure after X-irradiation. Radiat. Res. 47, 51-65 (1971)

MARIAGGI, N., CADET, J., TEOULE, R.: Cyclisation radicalaire de la désoxy-2' adénosine en solution aqueuse, sous l'effet du rayonnement gamma. Tetrahedron 1385-1387 (1976)

MARIAGGI, N., TEOULE, R.: Irradiation gamma de l'adénine. Identification des substances de radiolyse. C.R. Acad. Sci. (Paris) Ser. C 279, 1005-1007 (1974)

MARIAGGI, N., TEOULE, R.: Caracterisation de la dihydro-7,8-oxo-8-désoxy-2'-adénosine produite par irradiation gamma de la désoxy-2' adenosine en solution. Bull. Soc. Chim. France 9-10, 1595-1598 (1976)

MATSUSHIGE, T., KOLTZENBURG, G., SCHULTE-FROHLINDE, D.: Pulse radiolysis of aqueous solutions of acetic acid 2-hydroxyethyl ester. Fast elimination of acetic acid from a primary radical. Ber. Bunsenges. Phys. Chem. 79, 657-661 (1975)

MATTERN, M.R., HARIHARAN, P.V., DUNLAP, B.E., CERUTTI, P.A.: DNA degradation and excision repair in gamma irradiated chinese hamster ovary cells. Nature (New Biol.) 245, 230-232 (1973)

MEAD, E.L., SUTHERLAND, R.G., VERRALL, R.E.: The ultrasonic degradation of thymine. J. Chem. Soc. Chem. Commun. 414 (1973)

MEAD, E.L., SUTHERLAND, R.G., VERRALL, R.E.: The ultrasonic degradation of thymine. Can. J. Chem. 53, 2394-2399 (1975)

MERWITZ, O.: The cleavage of tritium from solide adenine-2T and adenine-8T induced by ^{60}Co-gamma irradiation. Biophysik 8, 245-253 (1972)

MICHAELS, H.B., HUNT, J.W.: Reactions of the hydroxyl radical with polynucleotides. Radiat. Res. 56, 57-70 (1973)

MOORTHY, P.N., HAYON, E.: Free-radical intermediates produced from the one-electron reduction of purine, adenine, and guanine derivatives in water. J. Am. Chem. Soc. 97, 3345-3350 (1975)

MÜLLER, A., HÜTTERMANN, J.: Structure and formation of free radicals in irradiated single crystals of nucleic acid constituents. Ann. N.Y. Acad. Sci. 222, 411-431 (1973)

MYERS, L.S., Jr.: Free radical damage of nucleic acids and their components by ionizing radiation. Federation Proc. Federation Am. Soc. Exp. Biol. 32, 1882-1894 (1973)

MYERS, L.S., Jr., HOLLIS, M.L., THEARD, L.M.: Pulse radiolysis of DNA and related pyrimidine compounds: Reactions of the OH free radical. In: Radiation Chemsitry (Advances in Chemistry Series). Gould, R.F. (ed.), Vol. I, pp. 345-367. Washington, D.C.: Am. Chem. Soc., 1968

MYERS, L.S., Jr., HOLLIS, M.L., THEARD, L.M., PETERSON, F.C., WARNICK, A.: Pulse radiolysis of nucleic acid constituents and related compounds. III. Optical spectra and reactivity of organic free radicals formed by reaction of hydroxyl free radicals with pyrimidine bases. J. Am. Chem. Soc. 92, 2875-2882 (1970b)

MYERS, L.S., THEARD, L.M.: Pulse radiolysis of nucleic acid constituents and related compounds. I. Optical spectrum and reactivity of the 5,6-dihydrothyminyl free radical. J. Am. Chem. Soc. 92, 2868-2870 (1970)

MYERS, L.S., WARD, J.F., TSUKAMOTO, W.T., HOLMES, D.E.: Site of attack in the radiolysis of pyrimidine compounds. Nature (London) 208, 1086-1087 (1965a)

MYERS, L.S., Jr., WARD, J.F., TSUKAMOTO, W.T., HOLMES, D.E., JULIA, J.P.: Radiolysis of thymine in aqueous solutions; change in site of attack with change in pH. Science 148, 1234-1235 (1965b)

MYERS, L.S., Jr., WARNICK, A., HOLLIS, M.L., ZIMBRICK, J.D., THEARD, L.M., PETERSON, F.C.: Pulse radiolysis of nucleic acid constituents and related compounds. II. Resolution of a rapidly decaying transient absorption in cytosine solution. J. Am. Chem. Soc. 92, 2871-2874 (1970a)

NETA, P.: Electron spin resonance study of radicals produced in irradiated aqueous solutions of uracil and related pyrimidines. Radiat. Res. 49, 1-25 (1972a)

NETA, P.: Electron spin resonance study of radicals produced in irradiated aqueous solutions of 5-halouracils. J. Phys. Chem. 76, 2399-2402 (1972b)

NICOLAU, C., McMILLAN, M., NORMAN, R.O.C.: An ESR study of short-lived radicals derived from pyrimidines in aqueous solution. Biochim. Biophys. Acta 174, 413-422 (1969)

NOFRE, C., BOBILLIER, P., CHABRE, M., CIER, A.: Effet des rayons gamma sur les bases pyrimidiques à l'état solide. Bull. Soc. Chim. 2820-2827 (1965a)

NOFRE, C., CIER, A.: Effects of hydroxyl free radicals on pyrimidine bases. In: Electornic Aspects of Biochemistry. Pullman, B. (ed.), pp. 397-414. New-York: Academic Press, 1964

NOFRE, C., CIER, A., CHAPURLAT, R., MARESCHI, J.P.: Hydroxy-5-dihydrouracile et hydroxy-5-dihydrothymine. Bull. Soc. Chim. France 332-335 (1965b)
NOFRE, C., CIER, A.: Effet des rayons gamma sur les bases pyrimidiques en solution aqueuse. II-Thymine. Bull. Soc. Chim. France 1326-1333 (1966)
NOFRE, C., LEFIER, A., CIER, A.: Action du radical libre hydroxyle sur les bases puriques. C.R. Acad. Sci. (Paris) 253, 687-689 (1961)
NOFRE, C., OGIER, M.H.: Préparation et propriétés de l'hydroxy-6-dihydrothymine. C.R. Acad. Sci. (Paris) Série E 263, 1401-1403 (1966)
NUCIFORA, G., SMALLER, B., REMKO, R., AVERY, E.C.: Transient radicals of DNA bases by pulse radiolysis. Effects of cysteine and cysteamine as radioprotectors. Radiat. Res. 49, 96-111 (1972)
OKADA, S.: In: Radiation Biochemistry. Altman, K.I., Gerber, G.B., Okada, S. (eds.). Vol. I, p. 55. London-New York: Academic Press, 1970
O'NEILL, P., STEENKEN, S., SCHULTE-FROHLINDE, D.: Formation of radical cations of methoxylated benzenes by reaction with OH radicals, Tl^{++}, Ag^{++} and $SO_4^{\overline{\cdot}}$ in aqueous solution. An optical and conductometric pulse radiolysis and *in situ* radiolysis electron spin resonance study. J. Phys. Chem. 79, 2773-2779 (1975)
ORMEROD, M.G., SINGH, B.B.: Reaction of OH radicals with nucleic acid bases, an ESR study. Intern. J. Radiat. Biol. 10, 533-538 (1966)
PATTERSON, L.K., BANSAL, K.M.: Pulse radiolysis studies of 5-halouracils in aqueous solutions. J. Phys. Chem. 76, 2392-2399 (1972)
PEACOCKE, A.R., PRESTON, B.N.: The action of γ-rays on sodium deoxyribonucleate in solution III denaturation. Proc. Roy. Soc. London, B. 153, 103-110 (1960)
PEUZIN, M.C., CADET, J., POLVERELLI, M., TEOULE, R.: Parabanic acid and 5-hydroxyhydantoin upon gamma irradiation of uracil. Biochim. Biophys. Acta 209, 573-574 (1970)
PHILLIPS, G.O., CRIDDLE, W.J.: Radiation chemistry of carbohydrates. Part X. Action of γ-radiation of aqueous D-ribose solutions in oxygen. J. Chem. Soc. 2740-2744 (1962)
PHILLIPS, G.O., CRIDDLE, W.J., MOODY, G.J.: Radiation chemistry of carbohydrates. Part XI. Self-decomposition of D-[^{14}C]-mannose, D-[^{14}C]-ribose and D-[^{14}C]-fructose. J. Chem. Soc. 4216-4224 (1962)
PHILLIPS, G.O., GRIFFITHS, W., DAVIES, J.V.: Radiation chemistry of carbohydrates. Part XVI. The contribution of OH radicals to the radiolysis of aqueous solutions. J. Chem. Soc. B 194-200 (1966)
PHILLIPS, J.M.: Ph.D. Thesis, Univ. Newcastle Upon Tyne, England (1971)
PLETICHA-LANSKY, R., WEISS, J.J.: Polarography study of the effects of gamma irradiation on cytosine. Anal. Biochem. 16, 510-522 (1966)
POLVERELLI, M., BONICEL, A., TEOULE, R.: Gamma irradiation of cytosine ^{14}C-2 labelled *Escherichia coli* DNA. Identification of the major radiolysis products resulting from the rupture of the *N*-glycosidic bond. J. Radiat. Res. 17, 127-134 (1976)
POLVERELLI, M., TEOULE, R.: Obtention d'un nouveau produit de radiolyse de la cytosine: l'amino-4-formyl-1-hydroxy-5-oxo-2-didehydro-3,4-imidazolidine. C.R. Acad. Sci. (Paris) Ser. C 277, 747-749 (1973)
POLVERELLI, M., TEOULE, R.: Identification of radiolysis products of cytosine resulting from the deamination pathway. Z. Naturforsch. 29c, 12-15 (1974a)
POLVERELLI, M., TEOULE, R.: Kinetic study of the formation of the radiolysis products of cytosine resulting form the deamination pathway. Z. Naturforsch. 29c, 16-18 (1974b)
PONNAMPERUMA, C.A., LEMMON, R.M., CALVIN, M.: Chemical effects of ionizing radiation on cytosine. Science 137, 605-607 (1962)
PONNAMPERUMA, C.A., LEMMON, R.M., CALVIN, M.: The radiation decomposition of adenine. Radiat. Res. 18, 540-551 (1963)
PULLMAN, B., PULLMAN, A.: Quantum Biochemistry. Interscience (1963)

RABANI, J., KLUG-ROTH, D., HENGLEIN, A.: Pulse radiolytic investigations of $OHCH_2O_2$ radicals. J. Phys. Chem. 78, 2089-2093 (1974)
RAFI, A.A.: Ph.D. Thesis, L1100, Univ. Newcastle Upon Tyne, England (1968)
RAFI, A.A., WEISS, J.J., WHEELER, C.M.: Effect of γ-radiation on DNAs of different base composition. Biochim. Biophys. Acta 169, 230-240 (1968)
RAFI, A.A., WHEELER, C.M.: On the radiation-induced hyperchromicity of DNA. Intern. J. Radiat. Biol. 24, 321-323 (1973)
RAO, P.S., HAYON, E.: Redox potentials of free radicals. III. Re-evaluation of the method. J. Am. Chem. Soc. 97, 2986-2988 (1975)
RALEIGH, J.A., GREENSTOCK, C.L., KREMERS, W.: Chemical radiosensitization of phosphate ester cleavage. Intern. J. Radiat. Biol. 23, 457-467 (1973a)
RALEIGH, J.A., GREENSTOCK, C.L., WHITEHOUSE, R., KREMERS, W.: Radiosensitization of phosphate release from 3'- and 5'-nucleotides: correlations between chemical change and biological inactivation. Intern. J. Radiat. Biol. 24, 595-603 (1973b)
ROTI ROTI, J.L., CERUTTI, P.A.: Gamma ray-induced thymine damage in mammalian cells. Intern. J. Radiat. Biol. 25, 413-417 (1974)
ROTI ROTI, J.L., STEIN, G.S., CERUTTI, P.A.: Reactivity of thymine to γ-rays in HELA chromatin and nucleoprotein preparations. Biochemistry 13, 1900-1905 (1974)
ROUILLIER, P., DELMAU, J., NOFRE, C.: Conformation des dihydropyrimidinediones. Structure demi-chaise du dihydrouracile, de la dihydrothymine, de leurs dérivés hydroxyles, halogénés ou méthyles et de la méthyl-1 dihydrocytosine. Bull. Soc. Chim. France 3515-3524 (1966)
SAMEJIMA, T., YANG, J.T.: Optical rotatory dispersion of DNA and RNA. Biochemistry 3, 613-616 (1964)
SAMUNI, A., NETA, P.: Hydroxyl radical reaction with phosphate esters and the mechanism of phosphate cleavage. J. Phys. Chem. 77, 2425-2429 (1973)
SCHACHINGER, L. VON, EINDL, B.: Strahlenwirkung auf NAD I/Adenin und Adenin Derivative. Atomkernenergie 24, 43 (1974)
SCHERZ, H.: Formation of deoxy compounds and malondialdehyde in irradiated aqueous solutions of carbohydrates and related compounds. Radiat. Res. 43, 12-24 (1970)
SCHMIDT, J., BORG, D.C.: Identification of radiation-induced hydrogen-addition radicals in adenine derivatives. Radiat. Res. 46, 36-51 (1971)
SCHOLES, G.: The radiation chemistry of aqueous solutions of nucleic acids and nucleoproteins. In: Progress in Biophysics. Butler, J.A.V., Huxley, H.E., Zirkle, R.E. (eds.), 13, 59-104. London: Pergamon Press, 1963
SCHOLES, G.: Radiolysis of nucleic acids and their components in aqueous solutions. In: Radiation Chemistry of Aqueous Systems. Stein, G. (ed.), pp. 259-285. London: Interscience, 1968
SCHOLES, G.: The radiation chemistry of aqueous solutions. Ann. Rep. Prog. Chem. Sect. A 67, 179-190 (1970)
SCHOLES, G.: Primary events in the radiolysis of nucleic acids and their components. Israel J. Chem. 10, 1107-1122 (1972)
SCHOLES, G.: The radiation chemistry of pyrimidines, purines and related substances. In: Photochemistry and Photobiology of Nucleic Acids. Wang, S.Y. (ed.). New York: Academic Press, 1976
SCHOLES, G., SHAW, P., WILLSON, R.L., EBERT, M.: Pulse radiolysis studies of aqueous solutions of nucleic acids and related substances. In: Pulse Radiolysis. Ebert, M., Keene, J.P., Swallow, A.J., Baxendale, J.H. (eds.) pp. 151-164. New York: Academic Press, 1965
SCHOLES, G., SIMIC, M.: Reactivity of the hydrogen atoms produced in the radiolysis of aqueous systems. J. Phys. Chem. 68, 1738-1743 (1964)
SCHOLES, G., SIMIC, M.: Radiolysis of aqueous solutions of DNA and related substances: reaction of hydrogen atoms. Biochim. Biophys. Acta 166, 255-258 (1968)

SCHOLES, G., STEIN, G., WEISS, J.J.: Action of X-rays on nucleic acids. Nature (London) 164, 709-710 (1949)
SCHOLES, G., WARD, J.F., WEISS, J.: Mechanism of the radiation-induced degradation of nucleic acids. J. Mol. Biol. 2, 379-391 (1960)
SCHOLES, G., WEISS, J.: Chemical action of X-rays on nucleic acids and related substances in aqueous solution. Exp. Cell Res. Suppl. 2, 219 (1952)
SCHOLES, G., WEISS, J.J.: Oxygen effects and formation of peroxides in aqueous solutions. Radiat. Res. Suppl 1, 177-189 (1959)
SCHOLES, G., WEISS, J.J., WHEELER, C.M.: Formation of hydroperoxides from nucleic acids by irradiation with X-rays in aqueous systems. Nature (London) 178, 157 (1956)
SCHOLES, G., WILLSON, R.L.: γ-Radiolysis of aqueous thymine solutions. Determination of relative reaction rates of OH radicals. J. Chem. Soc. Faraday Trans. 63, 2983-2993 (1967)
SCHOLES, G., WILLSON, R.L., EBERT, M.: Pulse radiolysis of aqueous solutions of deoxyribonucleotides and of DNA: reaction with hydroxy-radicals. J. Chem. Soc. Chem. Commun. 17 (1969)
SCHULTE-FROHLINDE, D., VON SONNTAG, C.: Radiation chemistry of ethylene glycol, *meso*-erythritol, 2-deoxy-D-ribose and alkyl phosphates as DNA model compounds. Israel J. Chem. 10, 1139-1150 (1972)
SCHWEIBERT, M.C., McDANIELS, M.: ^{60}Co gamma ray-induced peroxidation of DNA in aqueous solution. Intern. J. Radiat. Phys. Chem. 3, 353-366 (1971)
SEIDLER, F., VON SONNTAG, C.: Strahlenchemie von Alkoholen - VIII. Die Acetaldehydbildung bei der γ-Radiolyse wässriger, N_2O-gesättigter Lösungen von Äthylenglycol. Z. Naturforsch. 24b, 780-781 (1969)
SETLOW, R.B., SETLOW, J.K.: Effect of radiation on polynucleotides. Ann. Rev. Biophys. Bioeng. 1, 293-436 (1972)
SHARPATYI, V.A., KRUGLYAKOVA, K.E., NIKOLAEVA, N.V.: On the radiation chemical properties of DNA. Radiobiologiya 8, 3-6 (1968)
SHARPATYI, V.A., ZAKATOVA, N.V., MINKHADZHIDINOVA, D.R., YAROVAYA, S.M., KUDRYASHOV, L.I.: Role of hydroxyl radicals in the radiolytic transformation of a deoxyribonucleoprotein complex in aqueous solutions. Izv. Akad. Nauk S.S.S.R. Ser. Khim. 5, 1165-1166 (1973)
SHRAGGE, P.C., HUNT, J.W.: A pulse radiolysis study of the free radical intermediates in the radiolysis of uracil. Radiat. Res. 60, 233-249 (1974)
SHRAGGE, P.C., MICHAELS, H.B., HUNT, J.W.: Factors affecting the rate of hydrated electron attack on polynucleotides. Radiat. Res. 47, 598-611 (1971)
SHRAGGE, P., VARGHESE, A.J., HUNT, J.W.: New products in the gamma radiolysis of deoxygenated solutions of uracil and the effect of pH. J. Chem. Soc. Chem. Commun. 736-738 (1974a)
SHRAGGE, P., VARGHESE, A.J., HUNT, J.W., GREENSTOCK, C.L.: Radiolysis of uracil in the absence of oxygen. Radiat. Res. 60, 250-267 (1974b)
SIMONOV, V.V., RYABCHENKO, N.I., POVERENNYL, A.M.: Factors determining a change in the secondary structure of DNA under the influence of ionizing radiation. Radiobiologiya 9, 663-667 (1969)
SMITH, K.C.: Aging, Carcinogenesis and Radiation Biology. The Role of Nucleic Acid Addition Reactions. Smith, K.C. (ed.). New York-London: Plenum Press, 1976
SMITH, K.C., HAYS, J.E.: The response of uracil-2-^{14}C to X-irradiations under nitrogen and oxygen to treatment with ascorbic acid. Radiat. Res. 33, 129-141 (1968)
SONNTAG, C. VON, ANSORGE, G., SUGIMORI, A., OMORI, T., KOLTZENBURG, G., SCHULTE-FROHLINDE, D.: Radiation chemistry of DNA model compounds. Part II. Alkyl phosphate cleavage of aliphatic phosphates induced by hydrated electrons and by OH radicals. Z. Naturforsch. 27b, 471-472 (1972)

SONNTAG, C. VON, DIZDAROGLU, M.: Radiation chemistry of carbohydrates. Part I. Radical chain reactions in crystalline γ-lactose monohydrate. Z. Naturforsch. 28b, 367-368 (1973)
SONNTAG, C. VON, DIZDAROGLU, M., HENNEBERG, D., SCHOMBURG, G.: (1975b) Unpublished results
SONNTAG, C. VON, DIZDAROGLU, M., SCHULTE-FROHLINDE, D.: Radiation chemistry of carbohydrates - VIII. γ-Radiolysis of cellobiose in N_2O-saturated aqueous solution. Part II. Quantitative measurements. Mechanisms of the radical-induced scission of the glycosidic linkage. Z. Naturforsch. 31b, 857-864 (1976)
SONNTAG, C. VON, NEUWALD, K., DIZDAROGLU, M.: Radiation chemistry of DNA model compounds. Part III. γ-Radiolysis of 2-deoxy-D-ribose in the crystalline state. Conversion of 2-deoxy-D-ribose into 2.5-dideoxy-D-*erythro*-pentonic acid *via* a chain reaction. Radiat. Res. 58, 1-8 (1974)
SONNTAG, C. VON, NEUWALD, K., SCHUCHMANN, H.-P., WEEKE, F., JANSSEN, E.: Radiation chemistry of ethers. Part VI. Photolysis at 254 nm of the diethyl ether-oxygen charge transfer complex. J. Chem. Soc. Perkin Trans. II, 171-175 (1975a)
SONNTAG, C. VON, THOMS, E.: Strahlenchemie von Alkoholen - XV. γ-Radiolyse von Äthylenglycol in wässriger Lösung. Z. Naturforsch. 25b, 1405-1407 (1970)
STEENKEN, S., BEHRENS, G., SCHULTE-FROHLINDE, D.: Radiation chemistry of DNA model compounds. Part IV. Phosphate ester cleavage in radicals derived from glycerol phosphates. Intern. J. Radiat. Biol. 25, 205-210 (1974)
STEENKEN, S., SCHULTE-FROHLINDE, D.: Fragmentation of radicals derived from glycolaldehyde and glyceraldehyde in aqueous solution. An EPR study. Tetrahedron Lett. 653-654 (1973)
STEINMAUS, H., ROSENTHAL, I., ELAD, D.: Light and gamma-ray induced reactions of purines and purine nucleosides with alcohols. J. Org. Chem. 36, 3594-3598 (1971)
STELTER, L., VON SONNTAG, C., SCHULTE-FROHLINDE, D.: Radiation chemistry of DNA model compounds - V. Phosphate elimination from ribose-5-phosphate after OH radical attack at C-4. Intern. J. Radiat. Biol. 25, 515-519 (1974)
STELTER, L., VON SONNTAG, C., SCHULTE-FROHLINDE, D.: Radiation chemistry of DNA model compounds - VII. On the formation of 5-deoxy-D-*erythro*-pentos-4-ulose and the identification of twelve further products from γ-irradiated aqueous solutions of ribose-5-phosphate. Z. Naturforsch. 30b, 656-657 (1975a)
STELTER, L., VON SONNTAG, C., SCHULTE-FROHLINDE, D.: Radiation chemistry of DNA model compounds - VIII. Dephosphorylation products from reactions of OH radicals with ribose-5-phosphate in aqueous solution. The effect of oxygen. Z. Naturforsch. 30b, 609-615 (1975b)
STELTER, L., VON SONNTAG, C., SCHULTE-FROHLINDE, D.: Phosphate ester cleavage in ribose-5-phosphate induced by OH radicals in deoxygenated aqueous solution. The effect of Fe(II) and Fe(III) ions. Intern. J. Radiat. Biol. 29, 255-269 (1976)
STOCKHAUSEN, K., FOJTIK, A., HENGLEIN, A.: Pulsradiolytische Untersuchungen einiger Elementarprozesse der Autoxydation: Alkohole, Cyclohexen und Diäthyläther in wässriger Lösung. Ber. Bunsenges. Phys. Chem. 74, 34-40 (1970)
STOCKHAUSEN, K., HENGLEIN, A.: Pulsradiolytische Untersuchungen des Mechanismus der Oxydation und Autoxydation von Formaldehyd in wässriger Lösung. Ber. Bunsenges. Phys. Chem. 75, 833-840 (1971)
STOCKHAUSEN, K., HENGLEIN, A., BECK, G.: Nachweis geladener Zwischenprodukte der Autoxydation des i-Propanols in wässriger Lösung durch Pulsradiolyse. Ber. Bunsenges. Phys. Chem. 73, 567-571 (1969)
SUBBARAMAN, L.R., SUBBARAMAN, J., BEHRMAN, E.J.: The reaction of oxo-osmium (VI)-pyridine complexes with thymine glycols. J. Org. Chem. 38, 1499-1504 (1973)

SWALLOW, A.G.: Radiation Chemistry. An Introduction. New-York: Longmans, Green, 1973
SWINEHART, J.L., LIN, W.S., CERUTTI, P.A.: Gamma-ray-induced damage in thymine, in monoculeotide mixtures and in single and double-stranded DNA. Radiat. Res. 58, 166-175 (1974)
SZYBALSKI, W.: X-ray sensitization by halogenopyrimidines. Cancer Chemother. Rep. 58, 539-557 (1974)
TANIGUCHI, H., HATAMO, H.: Spin trapping of the short-lived radicals formed in gamma irradiated aqueous solutions of dipeptides and nucleic bases. Chem. Lett. 9-12 (1975)
TEOULE, R.: Etude des dommages subis par l'ADN *in vitro* sous l'action des rayonnements ionisants. Bull Inf. Sci. Tech. Commis. Energ. Atom France 203, 5-10 (1975)
TEOULE, R., BERT, C., BONICEL, A.: Thymine fragment damages retained in the DNA polynucleotides chain after gamma irradiation in aerated solutions. Radiat. Res. 72 (in press)
TEOULE, R., BONICEL, A., BERT, C., CADET, J., POLVERELLI, M.: Identification of radioproducts resulting from the breakage of thymine moiety by gamma irradiation of *E. coli* DNA in an aerated aqueous solution. Radiat. Res. 57, 46-58 (1974a)
TEOULE, R., BONICEL, A., BERT, C., CADET, J., POLVERELLI, M., FOUQUE, B.: Hydrolyse acide et enzymatique du fragment thymine modifié et lié à la chaîne polynucléotidique après irradiation gamma *in vitro* de l'ADN de *E. coli*. C.R. Acad. Sci. (Paris) Ser. D 279, 1797-1799 (1974b)
TEOULE, R., CADET, J.: Effet des rayons gamma sur les pyrimidines. Isolement et identification d'une nouvelle substance obtenue par radiolyse de la thymine: la méthyl-5-hydroxy-5-hydantoïne. C.R. Acad. Sci. (Paris) Ser. D 268, 2501-2503 (1969)
TEOULE, R., CADET, J.: Action du rayonnement gamma sur la thymine en solution aqueuse et aérée. Mesures effectuées à 200 rads. Bull. Soc. Chim. France 927-932 (1970)
TEOULE, R., CADET, J.: Radiolysis of thymine in aerated aqueous solutions. J. Chem. Soc. Chem. Commun. 1269-1270 (1971)
TEOULE, R., CADET, J.: Influence of the pH in the OH radical attack of pyrimidine ring. Z. Naturforsch. Teil C 29, 645-646 (1974)
TEOULE, R., CADET, J.: Comparison of radiolysis products of thymine and thymidine with ESR results. Intern. J. Radiat. Biol. 27, 211-222 (1975)
TEOULE, R., CADET, J., POLVERELLI, M., CIER, A.: Effet des radiations sur les bases constitutives des acides nucléiques. Peaceful uses of Atomic Energy (Proc. 4th Intern. Conf. Geneva 6-16 Sept. 1971), Vol. XIII, 447. U.N. and Atomic Intern. Energy Agency
TEOULE, R., CADET, J., ULRICH, J.: Radiolyse de la thymine en solution aqueuse aérée. Caractérisation de la N^1-formyl-N^2-pyruvylurée. C.R. Acad. Sci. (Paris) Ser. C 270, 262-364 (1970)
THEARD, L.M., PETERSON, F.C., MYERS, L.S., Jr.: Nanosecond pulse radiolysis studies of aqueous thymine solutions. J. Phys. Chem. 75, 3815-3821 (1971)
TSEITLIN, P.I., YASHIVITCH, G.P., RYABCHENKO, N.I.: Action of ionizing radiation on hydrogen bond system of macromolecular deoxyribonucleic acid. Radiobiologiya 4, 3-9 (1964)
ULIANA, R.: Radiodénaturation des désoxynucléosides. Bull. Soc. Chim. France 2908-2912 (1969)
ULIANA, R., CREAC'H, P.V., DUCASTAING, A.: Quelques aspects de la radiodénaturation de l'ADN. Biochimie 53, 461-468 (1971)
ULLRICH, M., HAGEN, U.: Base liberation and concomitant reactions in irradiated DNA solutions. Intern. J. Radiat. Biol. 19, 509-517 (1971)
ULRICH, J., TEOULE, R., MASSOT, R., CORNU, A.: Etude de la fragmentation des dérivés de l'uracile et de la thymine par spectrométrie de masse. Org. Mass Spectrom. 2, 1183-1199 (1969)

VENTER, P.J., VAN DER LINDE, H.J., BASSON, K.A.: Chain formation of acetaldehyde in the γ-radiolysis of deaerated ethylene glycol. J. Chem. Soc. Chem. Commun. 187-188 (1972)
VOLKERT, O., BORS, W., SCHULTE-FROHLINDE, D.: Reaktion des OH-Radikals mit Bromuracil. Z. Naturforsch. 22b, 480-485 (1967)
VORLICKOVA, M., PALECEK, E.: A study of changes in DNA conformation caused by ionizing and ultra-violet radiation by means of pulse polarography and circular dichroism. Intern. J. Radiat. Biol. 26, 363-372 (1974)
WANG, S.Y.: Photochemistry and Photobiology of Nucleic Acids. Wang, S.Y. (ed.). New York: Academic Press, 1976
WARD, J.F.: Deoxynucleotides - Models for studying mechanisms of strand breakage in DNA - I. Protection by sulfhydryl compounds. Intern. J. Radiat. Phys. Chem. 3, 239-249 (1971)
WARD, J.F.: Mechanisms of radiation-induced strand break formation in DNA. Israel J. Chem. 10, 1123-1138 (1972)
WARD, J.F.: Molecular mechanism of radiation-induced damage to nucleic acids. Adv. Radiat. Biol. 5, 181-239 (1975)
WARD, J.F., KUO, I.: Effect of chloride ions on the γ-radiation induced. Destruction of DNA, nucleotides and nucleoside in aqueous solution. Intern. J. Radiat. Biol. 18, 381-390 (1970)
WARD, J.F., KUO, I.: Deoxynucleotides - Models for studying mechanisms of strand breakage, in DNA. II. Thymidine 3'5'-diphosphate. Intern. J. Radiat. Biol. 23, 543-557 (1973)
WARD, G.A., MAIR, R.D.: Application of nuclear magnetic resonance to quantitative analysis of mixtures of organic peroxides, hydroperoxides and alcohols. Anal. Chem. 41, 538-540 (1969)
WARD, J.F., URIST, M.M.: γ-Irradiation of aqueous solutions of polynucleotides. Intern. J. Radiat. Biol. 12, 209-218 (1967)
WEISS, J.J.: Chemical effects of ionizing radiations on nucleic acids and related compounds. Prog. in Nucleic Acids Res. Mol. Biol. Davidson, J.N., Cohn, W.E. (eds.). Vol. III, 103-144. New York: Academic Press, 1964
WILLSON, R.L.: Ph.D. Thesis, Univ. Newcastle Upon Tyne, England (1966)
WILLSON, R.L.: The reactions of oxygen with radiation-induced free radicals in DNA and related compounds. Intern. J. Radiat. Biol. 17, 349-358 (1970)
WILLSON, R.L., WARDMAN, P., ASMUS, K.D.: Interaction of dGMP radical with cysteamine and promethazine as possible model of DNA repair. Nature (London) 252, 323-324 (1974)
YAMAMOTO, O.: Radiation-induced binding of nucleic acid constituents with protein constituents and with each other. Intern. J. Radiat. Phys. Chem. 5, 213-229 (1973)
YOSHII, G., TCHOE, Y.T.: The effect of gamma irradiation on deoxyribonucleoprotein. Intern. J. Radiat. Biol. 10, 485-493 (1966)
ZAKATOVA, N.V., SHARPATYI, V.A.: Mechanism of the radiolysis of aqueous solutions of DNA. Dokl. Akad. Nauk SSSR Ser. Phys. Chem. (Engl. Transl.) 200, 926-929 (1971)
ZAREBSKA, Z., SHUGAR, D.: Radioproducts of thymine in the presence of ethanol. Intern. J. Radiat. Biol. 21, 101-114 (1972)
ZHIL'TSOVA, V.M., ZHIZHINA, G.P., KRUGLYAKOVA, K.E.: Investigations of radiation DNA injuries by the method of I.R. spectroscopy. Izv. Akad. Nauk SSSR, Ser. Biol. 429-432 (1973)
ZHIZHINA, G.P., KRUGLYAKOVA, K.E.: On the decomposition mechanism of the thymine and cytosine hydroperoxides. Dokl. Akad. Nauk SSSR 180, 611-614 (1968)
ZHIZHINA, G.P., KRUGLYAKOVA, K.E., EMANUEL, N.M.: On the nature of the hyperweak luminescence in irradiated DNA solutions. Dokl. Akad. Nauk SSSR 163, 931-934 (1965)

ZHIZHINA, G.P., ZIBINA, D.L., KRUGLYAKOVA, K.E., EMANUEL, N.M.: Kinetic characteristics of the decomposition of peroxides in irradiated DNA solutions. Dokl. Acad. Nauk SSSR 158, 935-938 (1964)
ZIMBRICK, J.D., WARD, J.F., MYERS, L.S., Jr.: Studies on the chemical basis of radiosensitization by 5-bromouracil substitution in DNA. 1. Pulse- and steady-state radiolysis of 5-bromouracil and thymine. Intern. J. Radiat. Biol. 16, 505-523 (1969)

Section III
Biological Aspects

Chapter 1

Biological Functions of DNA and Methods of Testing

1.1 Introduction

The investigation of the effects of ionizing radiation on living organisms has a history as long as the existence of apparatus to produce such radiation. Since the end of the 19th century observations of the destructive effects of radiation on living matter have been reported. However, it is only in the last few decades that investigations at the subcellular level have produced some insight into the radiation induced response of the molecular organization of cells. The main reasons for this timing are:

1. The upsurge of molecular biology centered around the functions of DNA;

2. the refinements of mammalian cell culture techniques;

3. the increased availability of radioactively labelled precursors of important biological macromolecules;

4. the development of new and reliable methods for studying biological macromolecules.

Research is now at a stage where attempts can be made to explain malfunctions at the cellular, tissue, and whole body level by reference to radiation-induced changes in the quality and quantity of vital macromolecules. Of these vital macromolecules the DNA is certainly the most important one, because of its vast informational content. These essential DNA molecules are very sensitive to attack by agents like radiation. This sensitivity results from their structure consisting of long thread-like chains with a large surface. Thus radicals formed during radiolysis of water and other reactive species can easily attack nucleic acid molecules (see II.2). The sensitivity is also a consequence of the high specificity of the nucleic acid structure. Thus minor variations like base alterations, which do not influence the overall structure, may result in important modifications of biological functions and genetic continuity. It seems therefore appropriate to begin this third section of the book with a short survey of some of the functions of DNA and with a description of several testing methods and systems.

1.2 Processes in which DNA is Involved

The nucleotide sequence of the DNA molecule (inside cells) contains the genetic information of the cell. Encoded in this sequence is a set of instructions for the cell to carry out all functions. As store of genetic information the molecule must be able to (1) replicate its information, so that after cell division each daughter cell contains a copy of the DNA molecule identical to that originally present in the parental cell; (2) convert the genetic information into a useable form. This requires a great stability of the DNA molecule during cell metabolism and indeed DNA has no turnover as other constituents of the cell. The high stability of genetic information over many generations, as seen in the phenotype appearance of the species, also reflects the stability of the carrier of that information, the DNA molecule.

1.2.1 Semiconservative Replication

One fundamental purpose of DNA is the conservation of the genetic information. This is done by replication of the duplex DNA molecules inside cells. The mode of replication is semiconservative, that is, each daughter duplex contains one strand of the parental duplex, the other being synthesized de novo from single nucleotides. The mechanism of replication is complex and many details remain to be elucidated. It is known that several genes are coding for proteins which are involved in replication. Functions such as initiation of a new round of replication, supply of DNA constituents, recognition of the DNA template, incorporation of the nucleotides into the growing chain and several others are necessary for replication. Of the proteins carrying out those functions the DNA polymerases have been well described. Others are still unknown.

1.2.2 Transcription

The conversion of genetic information into a useable form is effected by the process of transcription and translation. The sequence of individual sections of the DNA molecule (genes) contain the information of the eventual synthesis of proteins, the latter being the molecules which actually carry out the functions of the cell.

The sequence of DNA in a particular gene is copied or transcribed into an RNA molecule with an analogous sequence. At a simplistic level this process has some similarities to replication with a sequence of ribonucleotides being built into a polymeric chain. The enzyme system responsible for these reactions includes RNA polymerase.

The single-stranded RNA chain is released from the template DNA single strand and subsequently translated into the polypeptide chain of a protein using the ribosomal machinery. Each group of three ribonucleotides is coding for an amino acid in the polypeptide chain.

1.2.3 Repair

Although DNA molecules are exceedingly stable, and replication is extremely accurate, the enormous amount of information stored in the DNA molecules (up to 10^{10} nucleotides in a mammalian cell) makes it inevitable that a small number of defects are continually produced in the DNA, either as a result of damage induced by external agents or by rare errors in replication. In order to maintain the integrity of the genetic information, cells have evolved various repair mechanisms for coping with this damage. These include (1) reversal of damage, whereby the alteration is reversed in situ; (2) excision of damage, whereby the damaged section of DNA is removed and replaced by the correct nucleotide sequence; (3) postreplication repair, whereby damage remaining in the DNA is circumvented during or after replication.

1.2.4 Transformation

The above sections describe complex processes occurring totally inside cells. Free DNA molecules are capable of transferring genetic information into certain competent organisms by the process of transformation. Thus DNA extracted from a donor organism with a certain genetic characteristic (x^+) (the transforming DNA) is taken up from solution by a recipient organism which lacks this characteristic (x^-). Some of the recipient organisms may then permanently acquire the characteristic (x^+). These are termed "transformants" (HOTCHKISS, 1953/54).

1.2.5 Transfection

Virulent bacteriophages (bacterial viruses) infect a host bacterium and their genetic information is expressed by using the host's synthetic machinery. This results in the production of multiple copies of the virus and, usually, destruction of the host cell. With some bacteriophages the DNA can be extracted, and under special conditions infection of the host by the free DNA can be carried out with subsequent production of whole virus. This is known as "transfection" (FÖLDES and TRAUTNER, 1964).

1.3 Degrees of Difficulty of Testing DNA Functions

Although replication and transcription are the most important processes in which DNA is involved, studies on the effects of external agents on these processes are very difficult to interpret. Certainly they are inhibited, but the kinetics are complex, there are many interactions, and there are several possible parameters which could be chosen as an end point. Furthermore a study of these processes involves exposure of the whole organism to the agent so that a correlation of changes in these processes with particular kinds of damage to the DNA is very complex, since other molecules apart from the DNA might also be affected.

On the other hand, with the systems of transformation and transfection, although the actual processes are complex, it is possible to obtain a simple quantitative assessment of the effect of the agent on the biological activity of the free DNA molecules. In the case of transformation one measures the effect that exposure of the transforming DNA to the agent has on the number of transformants produced in the recipient cell population. In the case of transfection the phage DNA is exposed to the agent and the effect on the number of host cells producing phage by transfection can be measured.

1.4 Description of Several Simple Systems

1.4.1 Transformation with Bacterial DNA

Bacterial transformation offers a unique approach to study the biological integrity of a DNA molecule after it has been extracted from a donor cell and then exposed to some inactivating agents, like radiation, mutagenic compounds, antibiotics or degrading enzymes. It is easy to compare the effect of the inactivating agent on biological activity and physicochemical integrity of the DNA. This is done, for example, by measuring the number of strand breaks, base alterations or crosslinks and the loss of transforming activity. So far transformation experiments have been successfully carried out in three bacterial systems. With *Bacillus subtilis*, *Diplococcus pneumoniae* and *Haemophilus influenzae* it is possible to find growing conditions under which these bacteria become "competent" for the uptake of transforming DNA (STUY, 1972).

It is well established, that the DNA is taken up as double-stranded molecule by the competent cell (STUY, 1965; ARWERT and VENEMA, 1975). This DNA is irreversibly fixed and insensitive to extracellular DNase (STRAUSS, 1966; MORRISON and GUILD, 1973).

Once the DNA has entered the competent cell its fate depends very much on the system used. It is believed that in *H. influenzae* the donor DNA is integrated into the recipient genom as a double stranded unit (STUY, 1965), whereas in *B. subtilis* and *D. pneumoniae* all the experimental evidence so far accumulated indicates that the DNA is split into single strands immediately after entry (LACKS, 1962; STRAUSS, 1970; MORRISON and GUILD, 1973). One of the strands is degraded and the other strand is incorporated into the bacterial genom. There is no preference for either strand (CHILTON, 1967). With a transformation system any genetic trait can be transferred for which a detection assay exists. Thus independence of amino acids, drug resistance and changes in UV sensitivity have been successfully transformed (SPIZIZEN, 1958; HERFORD and SUEOKA, 1970; HUSTER-ORTNER and KÖHNLEIN, 1971). The number of transformed bacteria depends linearly on the amount of donor DNA at low concentration (ANAGNOSTOPOULOS and SPIZIZEN, 1961). At higher DNA concentrations (0.5 μg/ml) saturation is observed (SAITO and MIURA, 1963). The transforming activity is also a function of the molecular weight (THORSETT and HUTCHINSON, 1971). The minimum molecular weight for transforming DNA of *D. pneumoniae* and *H. influenzae* is $0.3\text{-}0.5 \cdot 10^6$ daltons for the double strand (CATO and GUILD, 1968). With *B. subtilis* DNA the transforming activity is down four orders of magnitude, when the average molecular weight is around 10^6 daltons as compared to the activity of DNA with an average molecular weight of $20 \cdot 10^6$ daltons (MORRISON and GUILD, 1972). For interpreting transformation data it is important to know the average molecular weight of the transforming DNA and the polydispersity of the preparation (NICOLAIEFF and CHEVALLIER, 1970). The long-pending question whether a tenfold decline in observed transformants does mean that 90% of the DNA molecules have been inactivated or that each is 10% as active as before, has been solved in favour for the latter (CATO and GUILD, 1968), thus precluding the existence of a target size which was suggested from experimental survival curves.

1.4.2 Transfection with Phage DNA

Transfection as defined above is a process in which specially treated bacterial cells take up viral nucleic acid leading to the production of complete virus particles. As compared to transformation, transfection has several advantages. 1. In transformation experiments donor DNA has to be integrated into the recipient genom after uptake; in transfection uptake of phage DNA immediately leads to the production of infectious phage particles. 2. The bacterial DNA used in transformation is a preparation with varying degree of polydispersity; this complicates the interpretation. Transfecting phage DNA represents an extremely homogeneous population of molecules which makes it a most ideal tool for studying the correlation between physicochemical and biological parameters, for example number of strand breaks per phage genom and loss of transfection ability. For the production of phage it is necessary that the bacterium receives and incorporates at least one intact and active phage genom. This can be achieved by various means in the known transfection systems.

1.4.2.1 Transfection of Spheroplasts

Here the bacterial cells are first converted to spheroplasts before phage DNA is added. Transfections of spheroplasts have been successfully conducted with *E. coli* and the DNA of rather small *E. coli* phages, like ϕX 174, M 13, and T_1 (GUTHRIE and SINSHEIMER, 1963; BENZINGER et al., 1967; HOTZ and MAUSER, 1969). Transfection with *B. subtilis* spheroplasts and SP 50 DNA has also been reported (TICHY et al., 1968).

1.4.2.2 Transfection of Competent Cells

Bacteria grown under special conditions under which they will become "competent" for the uptake of bacterial transforming DNA are also transfectable by DNA of bacteriophages, which have such cells as hosts in normal infections.

Transfection of competent cells of *B. subtilis* with DNA of various *B. subtilis* phages was observed (TRAUTNER and SPATZ, 1973). Transfection is also possible in the *Haemophilus influenzae* system with *Haemophilus* phage DNA (HARM and RUPERT, 1963). For single marker transformation it was found that the biological effect varies linearly with DNA concentration. This dose response is different in transfection systems. Here one observed with some phage DNA linear dependence with others multipowered relationships between the number of phage yielding cells and DNA concentration. The multipowered dose response can be converted to a linear response if the transfection experiments are carried out in the presence of "helper" phages (ROTTLÄNDER and TRAUTNER, 1970). This technique is also called "Marker rescue" and increases the biological activity of the transfecting DNA considerably (KLOTZ and SPATZ, 1971).

After this rather short and by no means complete description of some of the major processes where DNA is involved, we will proceed to find out how these DNA functions are affected by the action of ionizing radiation.

Chapter 2

Radiation Effects on the Biological Function of DNA

2.1 Introduction

In this chapter an attempt will be made to establish a relation between the chemical alterations induced by radiation in DNA of living organisms and the loss of its biological function at transcription and replication. The inactivation of the viability of bacteriophage and the destruction of the genetic integrity of transforming DNA and transfecting DNA is a result of this functional inactivation. The chemical changes and the structural alterations in the DNA molecule will briefly be summarized first, together with the methods used to detect structural damages.

2.1.1 Main Types of Lesions Induced in DNA by Ionizing Radiation

As already discussed in II.2 and II.3 ionizing radiation will lead to many chemical changes in the various constituents of DNA. In the majority of studies the DNA has been irradiated in solution. In the organized double helix the nucleotide bases will be destroyed but to a lesser extent than by irradiation in free form. Purines are less sensitive than pyrimidines by a factor of about two. In the presence of oxygen the dominant phenomenon is a hydroxy-peroxidation occurring on the 5-6 double bond. The peroxides of the various bases show different stabilities; that of thymine is the most stable. The yield of base destruction in oxygen is higher by a factor of 3 as compared to anoxic irradiation where peroxidation does not occur. Nucleotide bases will also be liberated from irradiated DNA (ULLRICH and HAGEN, 1971). Subsequent reactions lead to a break of the nucleotide strand at this site with concomitant liberation of nucleosides with a damaged sugar. Consequently, various types of end groups are formed at the strand breaks and they can be analysed by means of enzymes recognizing hydroxyl- or phosphate end groups on the 3' and 5' terminals. Such analysis has been performed on DNA irradiated in solution (BOPP and HAGEN, 1970; BOPP et al., 1973) and partly on DNA irradiated in cells (LENNARTZ et al., 1975). Ionizing radiation induces not only direct breaks in the nucleotide chain but also lesions in the sugar moiety which lead to strand breakage under alkaline conditions; these are referred to below as alkali labile bonds or alkali induced strand breaks (alisb). For details about chemical changes in irradiated DNA the reader is referred to II.2 and II.3 and to review articles such as those written by SCHOLES (1963), WEISS (1964) and LATARJET (1972).

The structural damage produced by ionizing radiation can be described as denaturation and degradation of the DNA molecule. Denaturation occurs by destroying the hydrogen bonds between the two nucleotide strands. It results in an increase of the optical absorbance. This increase, however, does not correspond entirely to the number of ruptured bonds as bases are destroyed simultaneously losing their absorbance at 260 nm (see II.4). The radiation-caused denaturation is induced partly in localised regions where the strains remain intact and partly in connection with strand breakage, which facilitates the unzipping of the double helix. As rupture of hydrogen bonds in DNA occurs mostly in aqueous solutions, but not in irradiated organisms, this phenomenon will not be discussed further.

Ionizing radiation produces breaks in the polynucleotide strand of DNA i.e., a degradation of the macromolecule by direct effect and indirect effect (see I.2). After irradiation of dry DNA where the direct radiation effect is predominant, the number of single- and double-strand breaks increases linearly with dose up to 2 Mrad. The linearity of the dose effect curve for double-strand breaks, found also after irradiation in vivo, implies that this lesion is produced by a single absorption event and not by two opposite single-strand breaks. This conclusion is confirmed by the increase of the ratio double-strand breaks to single-strand breaks with the LET of the radiation. The G-value for formation of single-strand breaks in the dry state, in cells, in nucleoprotein gels, or in bacteriophage, has values between 0.7 and 1.1. The G-value for double-strand breaks lies between 0.06 and 0.16 (cf. table in BOHNE et al., 1970).

After irradiation of DNA in dilute aqueous solution the frequency of single-strand breaks increases linearly with dose while that of double-strand breaks rises with the square of the dose. A break in the double helix is produced by two single-strand breaks either exactly opposite or at a certain distance equivalent to h nucleotide pairs apart (HAGEN, 1967). The value of h depends on the stability of the double helix and varies with the ionic strength of the medium (FREIFELDER and TRUMBO, 1969). The G-value of the induction of single-strand breaks in solution amounts to about 0.3 to 0.5. The alkali-induced single-strand breaks contribute about one third to the total number of single-strand breaks measured in DNA after alkaline denaturation procedures.

By irradiation of dry DNA, DNA gels, or of concentrated DNA solutions, crosslinks between the DNA molecules are formed. After higher doses an insoluble DNA gel is produced and the relative gel fraction is dependent on the equilibrium between crosslinks and double-strand breaks formed. For details of this phenomenon see CHARLESBY (1960) and HENGLEIN et al. (1969). An example of such crosslinks is also in the sedimentation diagram of T 1 phage DNA (Fig. 2.1c). After irradiating T 1 phage with low doses (up to 150 krad) a certain fraction of DNA molecules sediments faster than the intact molecules indicating DNA-DNA crosslinks or crosslinks between DNA and the coat protein.

2.1.2 Calculation of the Number of Strand Breaks in DNA

Changes in the structure of DNA such as strand breaks and crosslinks, are detectable by changes in the molecular weight. Differ-

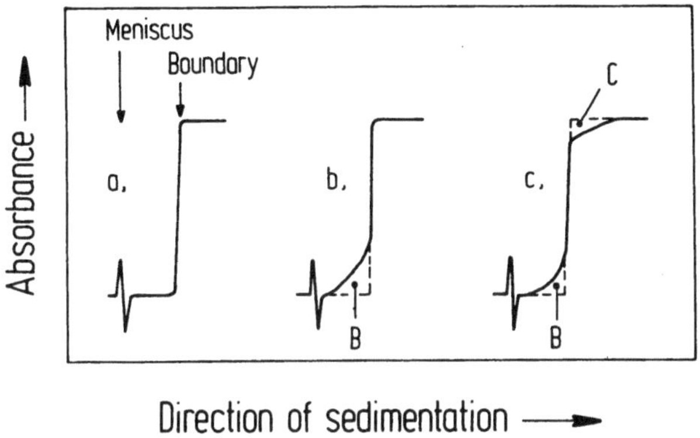

Fig. 2.1. Principle of detection of strand breaks and crosslinks in homogeneous DNA by analytical ultracentrifugation. The figures show a schematic diagramm of the concentration distribution of the DNA in the centrifuge cell, measured by its absorbance at 260 nm. (a) Homogeneous DNA with sharp boundary; (b) partly degraded DNA with a fraction of broken molecules (*B*) moving behind the front of unbroken molecules. The relative absorbance of the unbroken front to the total absorbance gives the fraction N/N_o of unkroken molecules (see text); (c) DNA from irradiated bacteriophage with broken molecules (*B*) and crosslinked molecules (*C*), sedimenting in front of the unbroken molecules

ent methods can be used for the determination of the molecular weight of a macromolecule: light scattering, osmosis, viscosity, diffusion, or sedimentation measurements. One of the most useful technique is the determination of the sedimentation coefficient by ultracentrifugation where the movement of a DNA molecule in a gravitational field is analysed either by optical methods (analytical centrifugation) or is determined by its position in the centrifuge tube at the end of the run. This latter method (preparative ultracentrifugation) needs the labelling of the DNA molecule with a radioisotope or other sensitive methods to measure DNA concentrations.

2.1.2.1 Analytical Ultracentrifugation

The molecular weight, and subsequently, the number of strand breaks after degradation can be evaluated in the analytical ultracentrifuge by boundary sedimentation, band centrifugation or by isopycnic centrifugation. Details of the procedures are described in various review articles or books (SCHACHMAN, 1959; ELIAS, 1961; CHERVENKA, 1969; HAGEN, 1973).

Analysing DNA by boundary sedimentation, the sedimentation coefficient $S_{20,w}$ obtained at the midpoint of the boundary between solution and solvent depends strongly on the concentration c used. This phenomenon obeys the relation:

$$S^{o}_{20,w} = S^{c}_{20,w} \; (1 + F'[\eta]\, c) \qquad (2.1)$$

where S^o is the sedimentation coefficient at zero concentration, $[\eta]$ the intrinsic viscosity of the DNA sample and F' a constant equalling 0.8 for native DNA and 1.6 for denatured DNA. For large DNA molecules several runs at various concentrations (5-50 μg/ml) are advisable and the reciprocal $S^c_{20,w}$ values should be extrapolated to zero concentration.

To calculate the molecular weight of a DNA sample a second hydrodynamic constant must be determined in addition to S^o: the diffusion coefficient D or the intrinsic viscosity [η].

By determining S^o and [η] for DNA molecules of various size in a given ionic medium, a relation between S^o and the calculated $M_{s,\eta}$-value can be obtained showing an upward bent curve in a logarithmic plot of S^o against M. Such calibration curves have been determined for native and denatured DNA in various ionic media. To obtain a straight line in the logarithmic plot of S^o against M, CROTHERS and ZIMM (1965) introduced a constant q_s, which is substracted from S^o. In this way the calibration curves can be described by the equation:

$$S^o - q_s = K_s \cdot M^{a_s} \qquad (2.2a)$$

where q_s, K_s and a_s are constants depending on the ionic conditions used. For example, for native DNA in 0.2 M NaCl, q_s = 2.70, K_s = 0.0152, and a_s = 0.445 (CROTHERS and ZIMM, 1965), when S^o is expressed in SVEDBERG units (10^{-13} sec). For denatured DNA in 0.2 M NaCl + 2% CHOH, q_s = 2.74, K_s = 0.0055, and a_s = 0.52 (HAGEN and COQUERELLE, 1974). In a narrow range of M.W. the S^o/M-relation can be fitted to a straight line by plotting log S^o against log M, i.e., q_s becomes zero and

$$S^o = K_s \cdot M^{a_s} \qquad (2.2b)$$

The value M_s thus obtained is, however, a complex average molecular weight which, in general, cannot be related to the number average M.W. $\bar{M}_n$ nor to the weight average M.W. $\bar{M}_w$. The further calculation of the strand breaks depends on the polydispersity of the DNA sample before and after irradiation. If the DNA molecule is highly degraded, i.e., by more than five breaks per molecule, a random distribution of the molecular size will be obtained, where 2 M_n = M_w, and $M_{s\eta}$ approaches M_w/1.10 (HAGEN and COQUERELLE, 1974). In this way, M_n can easily be calculated from $M_{s\eta}$ and, subsequently, the number of chain breaks:

$$B = m/M_n \qquad (2.3)$$

where B is the frequency of strand breaks per nucleotide and m the average M.W. of a nucleotide in a given DNA species (for example for thymus DNA m = 310). Some authors prefer to express the frequency of strand breaks per 10^6 dalton. In this case B* = $10^6/M_n$. In denatured DNA B_1 represents the frequency of single-strand breaks per nucleotide and in native DNA B_2 the frequency of double-strand breaks per nucleotide.

Usually such a random distribution can be assumed when studying DNA molecules from irradiated cells. However, in the case of studying DNA from bacteriophage, the whole genome is a single DNA molecule, which can be isolated without degradation caused by shearing. Here, all the molecules have the same M.W. $M_{s\eta} = M_w = M_n$ and a sharp boundary is found during sedimentation (Fig. 2.1a). When part of the DNA is broken, the degraded pieces will move behind the uniformly migrating front. Determining the relative fraction of unbroken molecules N/N_0 the number of strand breaks per molecule n is given by

$$N/N_0 = \exp(-n) \tag{2.4}$$

With increasing n a random distribution is approached, and the determination of the unbroken DNA fraction becomes uncertain. In such cases the whole boundary in the centrifuge cell should be analysed step by step resulting in an integral sedimentation distribution curve of s_i against c_i and, subsequently, by use of Equation (2.2) in a M.W. distribution curve allowing one to calculate M_w and M_n (COQUERELLE et al., 1969). According to CHARLESBY (1960) the number of strand breaks per nucleotide can be calculated from M_w and M_n:

$$B_{1,2} = \frac{2\,m}{3}\left(\frac{2}{M_n} - \frac{1}{M_0} - \frac{1}{M_w}\right) \tag{2.5}$$

where B_1 is obtained from the M.W. of denatured DNA and B_2 from those of native DNA; M_0 is the M.W. before irradiation. This formula can also be used when crosslinks are formed by irradiation in addition to the strand breaks. Here the frequency of a crosslink C per nucleotide equals:

$$C = \frac{2\,m}{3}\left(\frac{1}{M_n} + \frac{1}{M_0} - \frac{2}{M_w}\right) \tag{2.6}$$

where for M_n, M_w and M_0 the respective values for native DNA should be taken.

When analysing DNA samples by band centrifugation in the analytical ultracentrifuge the theoretical approach is similar to that for band centrifugation in the preparative ultracentrifuge. In the analytical centrifuge, very small quantities of DNA (0.2 µg in 10 µl solution) can be layered on the top of a sucrose or high ionic salt solution and migrate as a band through the ultracentrifuge cell. The sedimentation coefficient is calculated from the position of the maximum concentration in the band; it is difficult, however, to correct the apparent S-values to $S_{20,w}$ due to the salt gradient in the cell during centrifugation.

Band centrifugation is also useful in observing the separation of molecules with different sedimentation coefficient, for example the supertwisted and the relaxed form of the circular double stranded RF DNA of phage ϕX 174. As the latter generate from the former by one single-strand break per molecule the determination of the relative fraction of both forms allows to calculate the number of single-strand breaks after a given dose according to Equation (2.4).

For isopycnic centrifugation in a density gradient of CsCl, only small quantities of DNA are necessary. The DNA is mixed with CsCl-solution and moves during centrifugation to the position of its own density. The band width, obtained for a given DNA sample in the CsCl-gradient is inversely proportional to the square root of its M.W. (MESELSON et al., 1957). The values obtained by this method are near to M_w for polydisperse DNA samples and have to be corrected for the hydration of the molecule (cf. HEARST and VINOGRAD, 1961). They are especially useful for relative comparison in order to calculate the frequency of strand breaks. The determination of the molecular weight by isopycnic centrifugation is also very suitable for strand break analysis in the light and heavy strands of bromouracil labelled DNA (MÖNKEHAUS and KÖHNLEIN, 1973).

2.1.2.2 Preparative Ultracentrifugation

An analysis of the size of DNA molecules may also be achieved in the swinging bucket rotor of the preparative ultracentrifuge by layering the DNA sample labelled before in vivo with a radioisotope preferably with (^3H) or (^{14}C) thymidine on top of a sucrose gradient. However, the determination of DNA concentration by fluorometry has also been described by KISSAN and ROBINS (1958). This method is useful for the analysis of unlabelled DNA of nondividing cells in the sucrose gradient. The sedimentation coefficients are then determined and finally the M.W. using Equation (2.2).

Furthermore, it is possible to calculate the M.W. M_i of a sample by sedimenting it along with another sample of known M.W. M_o and determining the migration distances d_i and d_o. The treatment of the data implies that S^o is proportional to d, i.e., the DNA molecules move with a constant sedimentation rate through the centrifuge tube. Theoretical and experimental approaches have demonstrated the validity of this assumption if certain precautions are taken, for instance the sucrose gradient in the tube should cover concentrations between 5 and 23% (VAN DER SCHANS et al., 1969).

To determine the extent of degradation after irradiation, various procedures are suitable. DNA molecules of uniform size will give rise to a sharp peak in the sucrose gradient. After induction of strand breaks the degraded molecules move behind the unbroken ones. The frequency of strand breaks is evaluated as described above by determining the relative amount of unbroken molecules (cf. Eq. 2.4).

In the case of a broad band of DNA molecules, heterogeneous in size, the molecular weight distribution should be random. LEHMANN and ORMEROD (1970a) showed that for a random distribution the following relationship holds,

$$\frac{c_i}{M_i \cdot \Delta M_i} = \frac{T}{M_n^2} \exp\ (-\ M_i/M_n) \qquad (2.7)$$

where ΔM_i is the M.W. range spanned by fraction i, c_i the DNA content (e.g., the radioactivity) in this fraction, and T the total DNA content.

Thus a plot of log $(c_i/M_i \Delta M_i)$ against M_i should yield a straight line the slope of which equals $-1/M_n$. This gives a test of the randomness of a M.W. distribution and provides an accurate method of determining the number-average M.W. M_n of a random molecular weight distribution from the data obtained in a sedimentation experiment. Alternative tests for distributions produced by random breakage involve computer simulation of the expected sedimentation profiles with the experimentally derived profiles (e.g., EHMANN and LETT, 1973).

During the run in the tube of a swinging bucket rotor, part of the DNA tends to sediment on to the wall of the cylindrical centrifuge tube and is rapidly removed to the bottom. This fact may lead to serious errors unless the results are carefully corrected for the apparent deficiency in the balance of material. An unimpeded particle migration, comparable to that obtained in the analytical centrifuge, is possible in the zonal centrifuge, developed by ANDERSON and BURGER (1962). Here the solution is centrifuged in a rotating tubular rotor and the particles are quite free to move radially in an outward direction. The application of the zonal centrifugation for the analysis of mammalian DNA was shown by KLUCIS and LETT (1970).

The sedimentation analysis of DNA has been and still is widely used. It has, however, several disadvantages, especially when the number of strand breaks is very high; here the accuracy is reduced. When the number of strand breaks produced by radiation is comparable or smaller than that resulting from extracting and handling of the DNA, sedimentation analysis can lead to wrong interpretations. For these reasons alternative methods like hydroxylapatite chromatography have been developed in recent years (AHNSTRÖM and EDVARDSSON, 1974; RYDBERG, 1975).

With this technique one can separate double-stranded and single-stranded DNA pieces. Since the rate of strand separation in alkali of DNA is also a function of the number of strand breaks in the DNA-molecule, one can correlate the fraction of DNA that remains double-stranded after alkali treatment and sonication with the number average M.W. and finally arrive at a figure for the number of radiation induced strand breaks. A rather new and absolute technique for measuring M_w of large DNA has been developed by WEISSMAN et al. (1976). Here the M.W. are determined by fluctuation spectroscopy. In this method, one measures the spontaneous, inherent, thermodynamic concentration fluctuations of solute macromolecules in a given volume. These fluctuations are related in a simple way to the number of molecules, n, per unit volume. By measuring independently the concentration (weight per unit volume) of solute molecules, C, the weight average M.W. is then obtained from the relation $M_w = C/_n$. The advantage of the method is that it gives rapid and precise results, it requires no shear forces and does not depend on DNA structure, whether linear or circular. The fluctuation spectroscopy can be applied for DNA molecules of M.W. ranging from 10^7 up to 10^{11} dalton. It is of course applicable to any other macromolecule.

After this description of the most commonly used techniques to detect structural alterations in the DNA molecule, an attempt will be made to correlate structural alterations with biological consequences.

2.2 Inactivation of Phage DNA

2.2.1 The Phage and Its Advantages

For investigating the biological consequences of radiation-induced alterations in DNA, the bacteriophage is a very useful biological organism. It is a simple system, allowing a detailed analysis of the lesions in the structure of its constituents and of their consequences for each step in the reproductive cycle.

All known phage contain nucleic acid and protein. The nucleic acid consists of one molecule of single- or double-stranded DNA or single-stranded RNA. Nearly all of the protein is in the capsid, the coat of the virus particle which surrounds the nucleic acid molecule, protects it, and facilitates its entry into the host bacteria. It is built up by self-assembly of a large number of identical protein subunits. Some shells are more complex and contain different proteins as well as lipids and carbohydrates. Another protein component is the internal protein complexed with DNA inside the head. In addition, phage heads contain polyamines and water, bound to DNA. The degree of hydration in the phage varies (30-50% water). It is a determinant factor for the mechanism of DNA inactivation by radiation, as the indirect effect depends on the concentration of diffusible radicals in close vicinity of the DNA. Some phages possess a tail, capable of recognizing attachment sites on the bacterial membrane and guiding the injection of viral nucleic acid through the wall into the cell.

The repoductive cycle begins with the adsorption of the phage on the bacterial membrane and the subsequent injection of the nucleic acid molecule into the bacterium. The phage DNA makes use of the metabolic systems of the host cell to perform its growth cycle. Some phage carry in their DNA genetic information which enables them to synthesize phage specific enzymes. The infected bacteria stop their own metabolism and synthesize the enzymes necessary for the phage progeny, i.e., DNA replication and phage proteins. Double-stranded phage DNA is replicated semiconservatively. In phage with circular single-stranded DNA, such as ϕX 174, after infection the DNA is converted to an intracellular circular double-stranded DNA, the so-called replicative form (RF). This parental RF (pRF) is attached to the cell membrane. Several copies of this double-stranded DNA are produced by semiconservative replication (daughter RF), which then act as templates in the cytoplasm for the synthesis of single-stranded DNA. The final step of the phage development is the synthesis of the capsid protein. Each DNA molecule is packed into a capsid. After maturation of the phage, the host bacteria are usually lysed and the new phage are released.

The advantage of phage for radiation experiments is that lesions can be induced in the nucleic acid and the effects on the progeny can be measured without affecting the host cell metabolism. It is possible to irradiate the phage under various conditions and to study progeny as well as the physicochemical alterations to the isolated DNA (see III.2.1.2). The isolated DNA of some phage remains infectious to spheroplasts, that is, bacteria without cell walls prepared by treatment with lysozyme. In this way it is possible to irradiate free phage DNA without the protein coat and use spheroplast infectivity as a measure of radiation damage. Infectious DNA has been described for a number of single- and double-stranded phage.

The criterion of inactivation of whole bacteriophage by radiation is the loss of the plaque forming ability (PFA). About 20 min after infection with one bacteriophage, the bacterium lyses and liberates 100-200 newly formed phages able to infect the surrounding bacteria. When this process happens on nutrient agar overgrown with bacteria, each phage produces a hole or a "plaque" in the bacterial layer. These plaques can be counted easily. The inactivation curve is a plot of the logarithm of the remaining infective fraction of phage (N/N_0) versus dose. When dry phage or phage suspended in nutrient broth are irradiated, this relationship is linear, i.e., the inactivation process obeys the relation:

$$N/N_0 = \exp(-kD) \qquad (2.8)$$

where the dose D is expressed in rads and k is the inactivation rate constant or target volume. This relation implies that one hit in the target inactivates the system. The critical target in the bacteriophage is the nucleic acid molecule and the lesions leading to inactivation are the lesions in the nucleic acid structure. For comparison, the dose at which 37% survival occurs (D_{37}) is commonly used. At this point $N/N_0 = \exp(-1)$. The mean hit number kD at the D_{37} is 1.

2.2.2 Irradiation of Nucleic Acids Isolated from Phage

The test for infectivity of isolated DNA allows correlation between the integrity of the DNA sample and the damage to the DNA structure after irradiation in vitro (see II.3 and II.4). Irradiation can be performed in the dry state or in aqueous solution so that the direct or indirect effect of radiation can be studied. The addition of radical scavengers to the solution will influence the effect of radiation on both the infectivity and structural damage to the DNA molecule.

2.2.2.1 Biological Inactivation of Single-Stranded DNA

Most of the experiments on the inactivation of single-stranded phage DNA have been with DNA isolated from the phage ϕX 174. It is a circular DNA molecule with a M.W. of $1.7 \cdot 10^6$ dalton. The irradiation of freeze-dried ϕX 174 DNA in vacuo with ^{60}Co-γ-rays gives an exponential dose-response curve (DERTINGER and JUNG, 1970). The M.W. calculated from the D_{37} of 320 krad amounts to $1.8 \cdot 10^6$ dalton, a value in good agreement with the molecular weight determined in other ways. This means that one single primary ionization with an average energy transfer of 60 eV would be sufficient to inhibit the formation of progeny. The direct radiation effect can also be studied by irradiating infectious DNA in 1% tryptone broth at $-60^{o}C$. Under these conditions isolated ϕX 174 DNA as well as the intact phage ϕX 174 show the same D_{37} dose for inactivation (380 krad) (LYTLE and GINOZA, 1968). This is the best evidence that the sensitive target in the virus is the DNA molecule.

When DNA is irradiated in aqueous solution, the system is influenced by the presence of impurities. The inactivation depends on the concentration of the DNA as well as on the presence of other solutes. In solution, the primary radiolytic products are $H^{\cdot}$radicals, $OH^{\cdot}$ radicals and e^{-}_{aq}, as well as the mol. products H_2 and H_2O_2. BLOK (1967) has performed careful studies to determine the contribution of the various radiolytic products of water to DNA inactivation. H_2 does not react with DNA. H_2O_2 reacts very slowly with DNA, nucleotides and bases and may be responsible for a small after effect of DNA inactivation. Although $H^{\cdot}$ radicals react rather rapidly with DNA components, the major inactivation of DNA, however, comes from $OH^{\cdot}$ and e^{-}_{aq}, reacting preferably with the bases (see II.1 and II.2). $OH^{\cdot}$ is more effective than e^{-}_{aq}, since the DNA inactivation is increased by N_2O which converts hydrated electrons to $OH^{\cdot}$ radicals. In addition, KI, an $OH^{\cdot}$ scavenger, is a good protector.

Parallel with the biological inactivation of phage DNA, the frequency of the radiation-induced strand breaks has been measured under various conditions. For measurement of the strand breaks in circular single-stranded DNA the difficulty arises that under neutral conditions one break in the molecule does not much influence its rate of sedimentation. A decrease of the s-value is observed only after two or more breaks per molecule leading to short pieces of linear DNA. In this case, the average number of breaks per molecule p can be evaluated by band sedimentation (see III.2.1.2), determining the fraction F with unchanged sedimentation coefficient in relation to the total DNA. F represents the fraction of unbroken molecules (exp. - p) and that with just one break [p(exp - p)], i.e., F = (p + 1)(exp - p). From F, p can be derived. Centrifuging circular single-stranded DNA, however, in alkaline sucrose gradients, circular and linear molecules can be separated (LÜCKE-HUHLE et al., 1974), allowing a direct evaluation of the fraction of unbroken molecules.

Strand break analysis in irradiated ϕX 174 DNA was compared with the loss of PFA to determine the contribution of strand breaks to inactivation. Per se, each break is lethal (FIERS and SINSHEIMER, 1962); it was of interest, therefore, to determine the contribution of the radiation-induced strand breaks to biological inactivation. In experiments of LÜCKE-HUHLE et al. (1974) the DNA of ϕX 174 was irradiated in the dry state and centrifuged in alkaline sucrose gradients. From the slope of the dose effect curves it could be derived, that 88 ± 16% of the inactivated DNA carry at least one strand break. In this value, however, the alkali-induced strand breaks are included, which can be assumed in this case to amount to 22% of the total strand breaks. Irradiating ϕX 174 DNA in solution, the contribution of the actual strand breaks to inactivation was estimated by BLOK (1967) to be between 30 and 50%.

The importance of base damage to the loss of PFA was studied by BLOK et al. (1967) who compared the loss of PFA with the decrease of absorbance of DNA at 260 nm. The G-value for biological inactivation amounts to 0.13 whereas the yield of base destruction is G = 0.92 for DNA irradiated in solution at 59 μg/ml. Since at least 50% of the inactivation is due to strand breakage, it follows that only few of the base damages contribute to lethal radiation effects. The rest is probably repaired in the infected cell.

2.2.2.2 *Biological Inactivation of Double-Stranded DNA*

For analysis of double-stranded infectious DNA the replicative form (RF DNA) of phage ϕX 174 is very suitable. It is a supertwisted closed loop molecule of $3.4 \cdot 10^6$ dalton with a sedimentation coefficient of 21 S (component I). One single-strand break converts the 21 S - molecule to a 17 S, relaxed circular ring (component II). A double-strand break will lead to the linear molecule (component III, 15 S). Similar conversions have been observed in the DNA of PM2- and λ-phage. The three components can be separated by band centrifugation on a sucrose gradient and the fractions obtained can be tested for their biological activity. The linear molecule is no longer infective, indicating that the double-strand breaks are apparently inactivating (VAN DER SCHANS et al., 1973). Components I and II lose their activity with increasing dose.

In such systems, the contribution of base damage to inactivation can be examined separately from that of single-strand breaks: every break transfers a molecule from the component I fraction to the component II fraction. A decrease of the biological activity in fraction I is due, therefore, exclusively to radiation damage without chain breakage, i.e., base damage. The use of double-stranded circular DNA facilitates also the determination of alkali labile lesions in the nucleotide strand (see III.2.1.1) by sedimenting RF DNA in both neutral and alkaline sucrose gradients. In ϕX 174 RF DNA about 20-30% alkali-induced strand breaks are found in relation to the total amount of strand breaks, independently of the irradiation conditions.

When ϕX RF DNA is irradiated in solution, frozen in liquid nitrogen, about three single-strand breaks and 0.058 double-strand breaks are found per inactivation (TAYLOR and GINOZA, 1967). For PM2 DNA in solution, frozen or at room temperature, again only 0.045 double-strand breaks are found per inactivation. In both conditions about 85% of inactivation is due to base damage. The contribution of single-strand breaks obtained by substracting the two other contributions amount to 8.5%, an unexpectedly low value. Experiments with ϕX 174 RF DNA give similar results. Only 2% of single-strand breaks are lethal and some 30% of the base damage cause loss in infectivity (VAN DER SCHANS et al., 1973; VAN DER SCHANS and BLEICHRODT, 1974).

Apparently, base damage plays a predominant role in the inactivation of infectious DNA, but not every base destruction is lethal. Repair processes in the host bacterium can eliminate part of the base damage. STEPHAN (1973) showed that about half of the radiation-induced lesions not leading to actual strand breaks can be repaired by *E. coli* cells possessing the ability of host cell reactivation. The alkali labile lesions were also found to be lethal, their inactivation efficiency, however, is unknown (VAN DER SCHANS et al., 1973; LÜCKE-HUHLE, 1975).

The contribution of the various types of lesions to biological inactivation may depend very much on the system tested. For instance, HOTZ (1973) succeeded in infecting spheroplasts with linear DNA from phage T1. In DNA from T1, irradiated at $0^{o}C$ in concentrated nutrient broth, the relative contribution of the double-strand breaks to biological inactivation is much higher than for the circular double-stranded DNA of RF ϕX 174 or PM2; it amounts to 18-54%, dependent on the concentration of organic material in the medium (5-30%). As the relation of the various types of radiation-induced lesions in T1 DNA can be assumed to be similar to that in circular double-stranded DNA, it follows that damage to the nucleotides is repaired in the linear T1 DNA with a higher efficiency than in circular DNA. This observation demonstrates again that great attention should be paid to the choice of the biological system to be used in biophysical experiments.

2.2.3 Irradiation of Phage

2.2.3.1 Inactivation of Phages with Single-Stranded DNA

The phage ϕX 174 is by far the most frequently investigated among the phage with single-stranded DNA. The absorption of an average amount of energy of 60 eV is sufficient to cause biological inactivation. As mentioned above, the target M.W. calculated from the D_{37} dose of the inactivation curve is in good agreement with the M.W. of the nucleic acid (GINOZA, 1967) leading to the conclusion that the nucleic acid is the radiation-sensitive target. There is little information about the various lesions in the DNA structure after irradiation of the phage ϕX 174, in contrast to free irradiated phage DNA. For instance, no reliable measurements are available, so far, for the contribution of single-strand breaks in DNA to phage inactivation.

A comparison of thymidine damage in free DNA and in the phage, suspended in buffer, was made by SWINEHART and CERUTTI (1975), measuring the radiation-induced formation of thymidine methyl radicals as well as the formation of 5,6-dihydroxydihydrothymine. The formation of the latter product in the phage is comparable to that in free DNA, whereas the yield of methyl radicals is suppressed in the phage by a factor of 9 compared to free DNA. This observation demonstrates that DNA inside a phage reacts differently to ionizing radiation than free DNA in solution.

A linear relationship was observed between the loss of the biological activity of the phage and the formation of both kinds of thymine damage. It could be calculated, that at the D_{37} dose for PFA, 17 thymidine molecules are damaged at the methyl group and 41 are transformed to 5,6-dihydroxy-dihydrothymine per ϕX 174 genome. It is not known whether these products remain in the DNA or are repaired after infection of the host cells.

The inactivation of a phage results from its inability to complete all steps in its normal replication cycle. An analysis of the various steps to produce phage progeny after irradiation gives, therefore, some insight into the mechanism of inactivation. As shown by LYTLE and GINOZA (1969) and by SCHÖN (1973) the attachment of ϕX 174 phage to the host bacterium is scarcely affected after irradiation in frozen nutrient broth. About 70% of the inactivated phage are still able to inject their DNA into the host cell. Subsequently, between 70 and 80% of the injected DNA molecules are converted to parental RF-DNA. This process is accomplished by bacterial DNA polymerase. The next step, the attachment of the pRF to the bacterial membrane is apparently not inhibited (SCHÖN, 1973). The irradiation of the phage prevents, however, the synthesis of the daughter molecules. This synthetic step needs a phage specific DNA polymerase which has to be synthesized after infection and the synthesis depends on the integrity of the phage DNA template. Furthermore, the synthesis of the protein coat is inhibited. Obviously, one of the primary causes of γ-ray inactivation is the physicochemical damage in DNA affecting the development step involved in the expression of phage specific functions.

2.2.3.2 *Inactivation of Phages with Double-Stranded DNA*

Inactivation of double-stranded DNA viruses is described by a single hit dose effect curve when they are irradiated in nutrient broth containing 10^{-3} M histidine. The inactivation efficiency of a primary ionisation amounts to 0.05-0.2, in contrast to the yield of about 1 for the inactivation of phage with single-stranded DNA. Phage irradiated in buffer are more sensitive. Under these conditions the dose effect curves show a shoulder. An example is given in Figure 2.2 for the survival of phage T7. Here, the inactivation of the phage is due to protein damage leading to a loss of the capacity of the phage to absorb onto the host (WATSON, 1952). Actually, the irradiation leads to a rupture of the phage head with subsequent release of the DNA into the solution and degradation by the radiolytic products of water (FREIFELDER, 1965; LUTHJENS and BLOK, 1969). Consequently, in phage irradiated in buffer a much higher frequency of strand breaks per inactivation is found than in phage irradiated in nutrient broth. It is necessary, therefore, to protect the protein coat against the attack of free radicals when analysing the radiation damage induced in DNA of the intact phage.

The frequency of single- and double-strand breaks in phage irradiated in nutrient broth or histidine solution increases linearly with dose. The G-value for inducing double-strand breaks in T1 phage was found to be 0.058 and for single-strand breaks 1.15 (BOHNE et al., 1970). More experimental data on phage T1, T4 and T7 are summarized in Table 2.1. The yield of double-strand

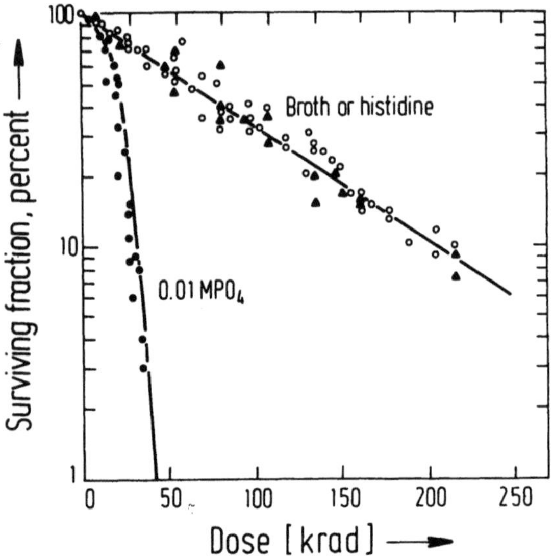

Fig. 2.2. Survival of phage T7 following X-irradiation, suspended in 0.01 M phosphate buffer, pH 7.8 (*solid circles*), nutrient broth (*open circles*), and phosphate containing 10^{-3} M l-histidine (*triangles*) (according to FREIFELDER, 1965)

Table 2.1. DNA strand breakage and biological inactivation of phages with double-stranded DNA after ionizing irradiation

Phage	Gas conditions of irradiation	Y_b	Y_{ds}/Y_b	Y_{ss}/Y_b	Reference
T1	nitrogen	0.34	0.17	3.5	BOHNE et al. (1970)
T7	nitrogen	0.34	0.30	4.0	COQUERELLE et al. (1974)
T7	nitrogen	0.4	0.1	3.0	BREMAN (1972)
T4	nitrogen	0.17	0.25	5.0	VAN DER SCHANS and BLOK (1970)
T7	oxygen	0.33	0.33	5.0	BREMAN (1972)
T4	oxygen	0.12	0.67	12.0	VAN DER SCHANS and BLOK (1970)

Y_b: yield of inactivation, i.e., the inactivation probability per primary ionisation; Y_{ds}/Y_b: relation between double-strand breaks and biological inactivation; Y_{ss}/Y_b: relation between single-strand breaks and biological inactivation.

breaks per biological inactivation amounts to about 0.25 while that of single-strand breaks is found to be between 3 and 5. This does not necessarily mean that single-strand breaks are not lethal. It is possible that some will be repaired in the host or that only part of the DNA genome is sensitive to single-strand breaks. The high radiosensitivity of phage α was explained in this way. It was found (FREIFELDER, 1966) that only one single-strand break in the DNA strand with the higher purine content will inactivate the phage α, the heavy strand serving as template for DNA synthesis.

An unsolved problem exists concerning the effect of oxygen. When the phage suspension is saturated with oxygen, the yield of double-strand breaks increases by a factor of 2 but the phage survival is somewhat protected by oxygen (VAN DER SCHANS and BLOK, 1970; BREMAN, 1972). A corresponding observation was made by FREY and HAGEN (1974) irradiating infectious T1 DNA in solution in the presence or absence of oxygen. Whereas the physicochemical damage was enhanced in presence of oxygen by a factor of 2-3, the loss of PFA was not affected. A detailed explanation of this phenomenon is still under discussion.

Because of the densely packed structure of a phage head, it is possible that radiation-induced crosslinks are formed between the DNA and the protein coat or between the different sections of the folded DNA genome. Such crosslinks are indeed induced leading to a fraction of DNA molecules sedimenting faster than the unbroken ones (cf. Fig. 2.1c). The frequency of crosslinks can be deduced from the coiling properties of the DNA or from the M.W. distribution (see III.2.1.2). Part of the crosslinks in DNA of irradiated phage can be removed by treatment with proteolytic enzymes demonstrating that DNA-protein crosslinks are involved. The contribution of crosslinks to phage inactivation, however, is very low, 3% for DNA-DNA crosslinks and 6% for DNA-protein crosslinks (BOHNE et al., 1970). It should be pointed out that in concentrated gels of isolated DNA with 50% water, many more crosslinks are formed than in the phage head. Apparently the ordered tertiary DNA structure in the phage head prevents radical-radical interactions leading to crosslinks.

The various steps of the cycle of reproduction were studied also with phage with double-stranded DNA. In irradiation conditions affording protection of the protein coat by organic material in the surrounding medium, between 3 and 10% of the inactivation can be ascribed to failure of adsorption (WATSON, 1950; SHARP and FREIFELDER, 1971; COQUERELLE and HAGEN, 1972). In about half of the nonadsorbing phage the head has been ruptured and the DNA released. Also, the injection of the DNA genome is inhibited in irradiated phage. For phage T1, T4, T7 and λ, it was shown, that this event contributes to about one third of the loss of PFA (HARM, 1958; SHARP and FREIFELDER, 1971; COQUERELLE and HAGEN, 1972). The mechanism of the inhibition of injection can be related to the occurrence of the double-strand breaks. For the T1 phage evidence was found that after breakage of the double helix the injection of the whole genome is prevented; only DNA molecules with intact double strand will be injected (COQUERELLE and HAGEN, 1972). For bacteriophage λ, it was claimed that a double-strand break in the genome leads to a partial injection of the first piece of DNA which is of course unable to produce phage progeny (SHARP and FREIFELDER, 1971).

To study the mechanism of the injection after irradiation in more detail and to obtain information on the genetic integrity of the injected phage genome, a correlation between the synthesis of phage specific enzymes and the inactivation of the PFA and the occurrence of strand breaks was attempted (COQUERELLE et al., 1974). In Figure 2.3 the various dose effect curves are summarized. Single-strand breaks (curve A) and 3 times more frequent than

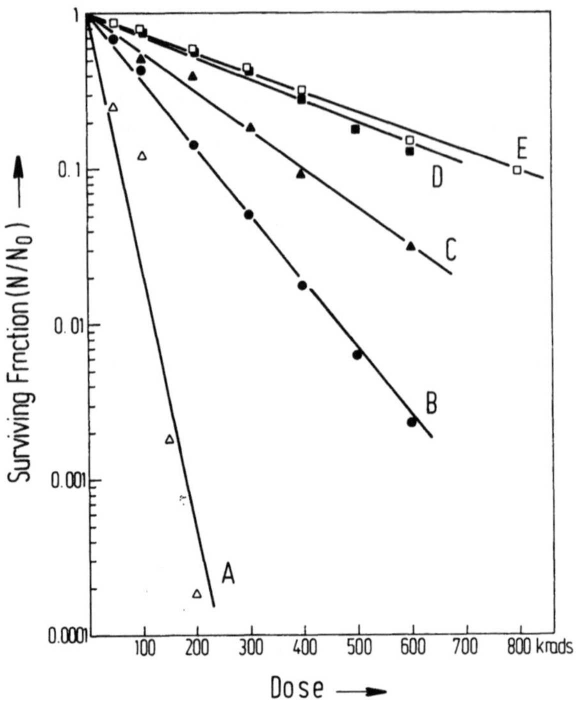

Fig. 2.3. Inactivation of plaque forming activity (PFA), of synthesis of phage-coded enzymes and of molecular DNA alterations in γ-irradiated T7 phages. *A*: Fraction of phages without single-strand breaks in their DNA (Δ); *B*: Fraction of biological active phages (PFA) (●); *C*: Inactivation of the capacity to code for lysozyme synthesis (▲); *D*: Inactivation of the capacity to code for T7 specific RNA polymerase synthesis (■); *E*: Fraction of phages without double-strand breaks in their DNA (□) (according to COQUERELLE et al., 1974)

lethal hits (curve B) and the double-strand breaks (curve E) contribute about 0.3 of the biological inactivation. The loss of ability to inject DNA follows a dose-effect curve similar to curve E, indicating a sensitivity of about 0.3 (curve not drawn in Fig. 2.3). Consequently it may be assumed that only entire DNA molecules, i.e., without double-strand breaks, are injected.

The radiosensitivity of the RNA polymerase gene (curve D in Fig. 2.3) measured by the synthesis of the phage specific RNA polymerase in the host was found to be 0.32 of that of the PFA. This gene is located near the end of the DNA molecule first injected into the host and represents about 8% of the whole genome. Considering the possibility of a partial injection of the broken DNA its sensibility should be much lower (0.06) than the observed value (0.32). The most plausible explanation for this high radiosensitivity is that every double-strand break prevents the injection of the whole phage genome into the host cell.

The relative sensitivity of lysozyme synthesis (curve C in Fig. 2.3) tested as a late phage function is 0.6 of that of PFA. In the host cell the phage gene coding for lysozyme is transcribed by the T7 RNA polymerase. The radiosensitivity of the lysozyme depends, therefore, on the sensitivity of the RNA polymerase gene as well as on the radiation-induced damage in the DNA section of the lysozyme gene including the DNA section transcribed earlier in one piece of messenger RNA. The high sensitivity of this gene demonstrates an amplification of the radiation effect on the various genes involved in lysozyme synthesis.

2.2.4 Effect of Incorporated Radionuclides on the Biological Activity of Phage DNA

The decay of isotopes incoporated in DNA produces local and distant physical and chemical effects. The major difficulty in studying the effect of transmutation is the separation of the reactions, due to emitted particles from the reactions due to local events, such as a change of valence, atomic excitation and nuclear recoil of the resultant atom. These effects lead to a disruption of the DNA molecule.

^{32}Phosphorus (^{32}P): HERSHEY et al. (1951) first demonstrated the lethality of decay of incorporated ^{32}P for phage T4. The number of the surviving phage decreases exponentially with the time between isotope incorporation and biological assay. A single disintegration can inactivate the phage. The equation describing the survival curve is: $N/N_0 = \exp(-\alpha\lambda NT)$, where N/N_0 represents the fraction of phage surviving after time t, λN represents the rate of accumulation of decay per unit of time and α is the killing efficiency of one disintegration.

The bacteriophage ϕX 174 with single-stranded DNA is inactivated by decay of ^{32}P with an efficiency of 1.0 (TESSMAN, 1959). For bacteriophage with double-stranded DNA a killing efficiency of 0.1 was found at + 4°C and of 0.05 at - 196°C (STENT and FUERST, 1955). This lethal effect may be due to those ^{32}P decays in which the recoiling ^{32}S atom in one polynucleotide strand produces a break in the opposite strand resulting in a double-strand break. Measuring the frequency of double-strand breaks after ^{32}P decay and comparing it with the lethal effect on phage, the relative contribution of double-strand breaks depends on the storage conditions of the ^{32}P-labelled phage.

A careful study was made by LEY and KRISCH (1974) in which they compared the lethal effects of ^{32}P with that of ^{33}P on phage T1 and T4. Both isotopes have the same decay scheme except that ^{33}P has a much lower recoil energy (1.7 eV against 20 eV). After storage at -196°C in glycerol-casamino acid, the lethal efficiency per decay is 0.058 for ^{32}P and 0.022 for ^{33}P. Under these conditions, the efficiency of double-strand breaks is 0.057 for ^{32}P and only 0.014 for ^{33}P. This result shows that at the low storage temperature the killing efficiency for ^{32}P is due exclusively to double-strand breaks resultant from nuclear recoil, whereas for ^{33}P other additional damage in DNA contributes to the biological inactivation. Storing the labelled phages at + 4°C, the inactivation efficiency of the isotope decay increases whereas the relative contribution of double-strand breaks decreases, here the phage are inactivated by other mechanisms than the induction of double-strand breaks, e.g., single-strand breaks and base damage.

Tritium (^3H): Hydrogen atoms are less essential to the structural integrity of DNA than phosphorus so that a tritium decay should not cause the same local molecular disruption as ^{32}P. FUNK et al. (1968) have compared the lethal effects of ^3H thymidine in DNA with ^3H-labelled amino acids in the protein of T4 phage. The distribution of the radiation dose was calculated precisely over the phage structure. The results point to the same conclusions as other investigations with *Escherichia coli* that the lethal action of tritium decay in DNA is due only to β-irradiation and not to local events on the tritium position.

^{125}Iodine: The isotope ^{125}I can be incorporated into the phage DNA in form of ^{125}I-labelled 5-iododeoxyuridine (IUDR) as an analogue of thymidine using thymine requiring strains of *E. coli* as hosts. This radioactive compound is used in medical diagnosis and therapy. ^{125}I decays by electron capture followed by vacancy cascades and multiple ionization (Auger-effect, see I.2). This decay causes extensive destruction in organic molecules resulting in a very high killing efficiency for phage and cells. In phage T1, for example, the lethal efficiency is roughly five times greater than observed for ^{32}P decay. The absence of a temperature effect, as well as the observation that radical scavengers do not influence the inactivation, support the view that the ionizing radiation emitted during decay is not responsible for the loss of phage viability. The severe lethal effect of ^{125}I decay in the DNA is due to direct physical disruption of the DNA structure by the Auger-effect with an efficiency of 0.49 per decay (KRISCH, 1972). Every double-strand break in the DNA caused by the electron vacancy cascades (SCHMIDT and HOTZ, 1973) is lethal. In addition one single-strand break is generated per double-strand break whereas after γ-irradiation this relation is 20:1.

The possible consequences of Auger explosions, as the shower of Auger electrons is called following the removal of an electron from the K or L shell of the atom by a proton or by internal conversion, are only now being appreciated. The use of radioisotopes which decay by electron capture for diagnostic purposes has increased steadily. They have been applied for therapeutic uses and in radiobiological research. For quite some time it was believed that the radioisotopes decaying by electron capture like ^{51}Cr, ^{67}Ga, ^{75}Se, ^{85}Sr, ^{125}I, and ^{197}Hg were relatively safe when compared with other β-emitting isotopes. However, experiments with incorporated ^{125}I and the improved microdosimetric calculations indicate that Auger cascades produce excessive damage to cellular and subcellular components due to the highly localized energy deposition and the involved molecular disruption.

It is believed that energy deposition, molecular disruption, and chemical transmutation, which may interfere with repair processes, contribute to the high biological effectiveness of ^{125}I observed in various systems.

As a consequence of the radiation doses delivered by Auger electrons, the attitude of using the mentioned radionuclides in nuclear medicine should certainly be revised.

2.3 Irradiation Effects on the Transforming and Transfecting Activity of DNA

2.3.1 Biological Activity of Transforming DNA Irradiated in vivo

When DNA is irradiated inside the cell and then extracted and tested for biological activity the transforming activity decreases exponentially with dose.

In Table 2.2 a survey of the D_{37}-values obtained for various systems is given.

Table 2.2.

DNA source	Conditions	D_{37} (krad)	Radiation	Reference
B. subtilis spores	dry, N_2	6300	Co γ-rays; 1 MeV Electr.	TANOOKA (1964)
	dry, O_2	2700	dto.	TANOOKA (1964)
B. subtilis vegetative cells	Lyophilized, N_2	1560	1 MeV Electrons	HUTCHINSON and ARENA (1960)
Pneumococcus cells	Lyophilized, N_2	2820-3300	1 MeV Electrons	TANOOKA and HUTCHINSON (1965)
Pneumococcus cells	Wet in air	1500-2000	1 MeV Electrons	HUTCHINSON and ARENA (1960)
Pneumococcus cells	Wet in N_2	1200-1600	1 MeV Electrons	dto.
Pneumococcus cells	Wet in O_2	360- 450	1 MeV Electrons	dto.
Pneumococcus cells	Wet in O_2 + 1% Cysteamin	1000-1300	1 MeV Electrons	dto.
Pneumococcus cells	Lysed and Lyophilized	3600-4200	1 MeV Electrons	dto.

Transforming DNA is very resistant in dry spores (D_{37} = 6.3 Mrad) under anoxic conditions. In oxygen the sensitivity is increased by a factor of 2. DNA in lyophilized cells of *B. subtilis* is about four times as sensitive (D_{37} = 1.6 Mrad) as in spores. In wet cells of *Diplococcus pneumoniae* a higher radiosensitivity of the transforming DNA is observed. This is partly due to the indirect effect of H_2O-radicals. As compared with DNA of lysed and lyophilized cells the sensitivity of DNA of wet cells in O_2 is higher by a factor of 10.

The radiosensitivity of the colony-forming ability of *B.subtilis* spores was less than that of dry vegetative cells by a factor of 4. The D_{37} for tranformation, however, was greater than that for colony forming ability by a factor of 62. This was taken as an indication that the region containing the transformable marker is only a small portion of the total DNA molecule. From D_{37}-values one can calculate M.W. using the target theory (DERTINGER and JUNG, 1970).

For the tryptophan-marker in *B. subtilis* (D_{37} = 6.3 Mrad) one arrives at a target M.W. of about 100,000.

The D_{37} for transformation in dried cells of *Diplococcus pneumoniae* (streptomycin resistance marker) is 3.0 Mrad. With the very simple model that the DNA molecule can be considered as a long thin cylinder of radius r_o = 10 Å and length l and that of this length, a fraction f is concerned with the process of transformation, HUTCHINSON (1960) has given an expression for the inactivation in the dry state. On the assumption that a primary ionisation within the active part of the molecule will inactivate, the following equation holds:

$$S = \exp - D\alpha \cdot \pi \, r_o^2 \, f \, l \, \sigma_n$$

Where S is the fraction surviving a dose, D, α is the number of primary ionisations per unit dose and σ_n is the density of DNA. With a D_{37} = 3 Mrad it can be calculated that the sensitive unit corresponds to a M.W. of about 250,000 dalton.

The additional radiosensitivity in the presence of water is ascribed to the formation of radicals in the water surrounding the DNA. This can be expressed quantitatively by a simple model assuming that radiation-produced radicals can diffuse a mean distance ρ. Thus all radicals produced in the volume of a cylindrical shell of thickness ρ surrounding the DNA molecule contribute with a certain probability to the inactivation of the biological activity. With a radical diffusion length of 10 Å the increase in radiosensitivity from the dry state to the wet state can be understood.

2.3.2 Transforming Activity of DNA Irradiated in vitro

The dose response is quite different from the simple exponential survival curves, when DNA is irradiated in the dry state after extraction and purification. With *Diplococcus pneumoniae* DNA and 10 MeV protons GUILD and DEFILIPPES (1957) found a fast decrease of transformation frequency at low doses; at higher doses the inactivation followed an exponential law. A similar upward concavity of the survival curve for transforming activity of DNA on a semi-logarithmic plot as a function of dose was obtained for other systems and various types of radiation. From the final exponential slope of the curves several authors have calculated the target size for transformation and arrived at M.W. values of around 2 × 10^5 dalton. D_{37}-values derived from the initial part of the dose effect curve vary considerably. A value of 1.6 Mrad has been found by TANOOKA (1964) for *B. subtilis* DNA irradiated in standard saline citrate (SSC). SZYBALSKI and OPARA-KUBINSKA (1964) have investigated the dose dependence for several markers and give D_{37}-values around 0.1 Mrad. Rather large variations are reported by GRUNOW et al. (1969). In one of their papers they state that the inactivation of the transforming ability of DNA obtained from different preparations under identical conditions can vary by almost three orders of magnitude.

There are, however, some common features; dose effect curves for double transformants are steeper than for single marker transformation. This was found for all three major transformation systems.

2.3.3 Contribution of Strand Breakage, Base Damage, and Cross-links to the Inactivation

For an explanation of the unusual dose response curves found for transforming ability it is of great importance to determine which of the radiation-induced alterations in the DNA-molecule are responsible for the loss of biological activity. In dilute solutions the predominant types of radiochemical damage are the formation of single-strand breaks, the chemical alterations of bases and inter- or intrastrand crosslinks (see II.2).

It has been shown by THORSETT and HUTCHINSON (1971) that the loss of transforming activity caused by irradiation in dilute solution is quantitatively accounted for by the production of single-strand breaks. This was demonstrated in the following way. Transforming DNA of *D. pneumoniae* was treated with deoxyribonuclease I for various times, then aliquots were taken for transformation assay and for sedimentation analysis in alkali using the band sedimentation method of VINOGRAD et al. (1965). A dependence of transforming activity and single strand molecular weight was thereby established.

In a different set of experiments DNA was γ-irradiated with doses up to 35 krad. The transforming activity and the single-stranded molecular weight were measured. The results coincided with the curve obtained for DNase-treated DNA.

The commonly accepted G-values for base damage after irradiating DNA in dilute solutions are between 1.3 and 2.3 whereas the yield of single-strand breaks is usually quoted as 0.3 to 0.8. Since changing conditions such as concentration or level of impurities usually alter both yields in the same way, either the low G-values or the high G-values should be compared. There are about four bases chemically altered by the action of gamma rays for each single-strand break.

From the results of THORSETT and HUTCHINSON (1971) it is clear that for bacterial transformation the biological effect of these altered bases is small compared to the effect of single-strand breaks. As a first approximation it can be assumed that a radiochemically altered base has an effect on transforming activity equivalent to a pyrimidine dimer produced by UV light. According to RUPP and HOWARD-FLANDERS (1968) a UV dose of 1 erg/mm^2 produces about 5.6 dimers/genome in *B. subtilis*. Taking a value of 5000 erg/mm^2 for the UV dose resulting in 37% survival, one can calculate that there are 28×10^3 dimers per *B. subtilis* genome of 10^7 nucleotides, which is roughly equivalent to 1 dimer/0.1×10^6 dalton. Since the relative transforming activity of *B. subtilis* DNA of 1×10^6 dalton is 10^{-4} compared to an activity of 1 at 20×10^6 dalton, an average of 10 strand breaks per single strand in a DNA molecule of 10×10^6 M.W. will reduce the transforming ability almost beyond detection. Together with the strand breaks there will be between 40 and 50 base alterations produced. As shown above about 100 dimers are necessary to cause a 37% inactivation in a DNA molecule of 10×10^6 dalton. We can therefore conclude that an effect of the damaged bases will be negligible when compared with the effect of the single-strand breaks. The production of crosslinks in dilute solutions is too small to cause any inactivation of the transforming ability.

One of the most important factors for transforming activity is the length of the DNA molecule. It has been shown for all three transformation systems that transforming activity drops down to very low values upon shearing or sonication of the DNA. Both treatments are known to reduce the double-strand M.W. Since single-strand pieces are integrated into the recipient genome, it is evident that single-strand breaks are the major cause of loss of transforming activity.

The main difficulty in relating length and transforming activity of DNA is that the methods used for fractionation, such as sucrose gradients or methylated albumin columns, do not yield DNA molecules of uniform length. Even a repeated sucrose gradient fractionation does not improve the homogeneity. It seems to be impossible to obtain monodisperse samples of bacterial DNA.

To circumvent these difficulties NICOLAIEFF and CHEVALLIER (1970) have taken advantage of the length distribution measured by electron microscopy to get a direct relation between transforming ability and DNA length. They give the weight average molecular weight for each sample together with transforming activity and polydispersity, and summarize their results by saying that transforming activity is not an all or nothing process but a function of length of DNA. This again indicates the importance of strand breaks for transforming activity.

SUMMERS and SZYBALSKI (1967) have developed a very sensitive method for detection of single-strand breaks in polydisperse DNA samples using a polyethylene glycol-dextran two-phase system. Samples of transforming DNA with natural or chemically induced crosslinks between the complementary strands were irradiated with various doses. The DNA was heat-denatured and the reversibly bihelical fraction was determined using the two-phase system. The term "reversibly bihelical" refers to bihelical DNA which, after melting and rapid cooling returns to the bihelical form - it "zippers up" because of the crosslinks between the complementary strands. The degree of strand breakage was quantitated from the loss of reversibly bihelical DNA. Inactivation of transforming activity was exponential and the mean inactivation dose (D_{37}) was 3.8 krad. Using this value and the average M.W. of the DNA, SUMMERS and SZYBALSKI (1967) calculated that there was one single-strand break per 4.5×10^6 dalton and concluded that their results supported the hypothesis that the main molecular event by which ionizing radiation inactivates a large part of the transforming DNA is single-strand breakage.

Similar experiments have been performed by GRUNOW and GÜNTHER (1971) with special emphasis on the inactivation in the low dose range (up to 2 krad) where the numbers of the various types of lesions are reduced and practically no double-strand breaks are produced. They measured the initial transforming activity and the residual transforming activity after heat denaturation of γ-irradiated *B. subtilis* DNA. The residual transforming activity of denatured DNA is due to a fraction of marker-carrying molecules which have a naturally occurring crosslink between the complementary strands. Thus under denaturation conditions the strands do not completely come apart and return to the bihelical form when denaturing conditions are removed.

The dose effect curves obtained by GRUNOW and GÜNTHER (1971) for relative transforming activity and residual transforming activity of single and double transformants showed the following behaviour: At a dose of 2 krad the relative transforming activity was still 80-100% whereas the relative residual transforming activity was only 30-40% at the same dose. This was found for single and double transformants.

The higher inactivation of residual transforming activity can be explained by loss of single-strand pieces during denaturation due to strand breaks between crosslink and marker, since single strands are not taken up by competent cells.

The same lesion would not necessarily lead to inactivation of native DNA. Here the molecule-carrying strand break is taken up and can transform if the length of the single-strand piece is not below a certain minimal value. There is also a chance that strand breaks are repaired between uptake and integration. The results of GRUNOW and GÜNTHER (1971) are in agreement with the conclusion that single-strand breaks are the inactivation events in transformation. SZYBALSKI and OPARA-KUBINSKA (1964) have found a 1.4- to 2.1-fold sensitization of the transforming activity toward X-rays for unifilarly BrUdr labelled DNA. These figures were roughly doubled for bifilarly labelled DNA. Since it is known that BrUdr-containing DNA is more susceptible to strand breakage, the results have a very simple explanation.

2.3.4 Mathematical Interpretation of the Dose Effect Curves

The upward concavity of the dose effect curves has not been found elsewhere. Usually the surviving fraction decreases exponentially with dose or shows a shoulder at low doses. This is easily explained using the formalism of target theory including repair processes.

The unusual decrease of transforming activity vs. dose has tempted several groups to look more carefully into the process of transformation at the molecular level and to develop biological and mathematical models which might describe the experimentally found inactivation curves.

According to LACKS (1968), marker integration occurs as a result of a recombination process. Prior to recombination, single-stranded donor DNA synapses with homologous recipient DNA. Even numbers of recombination switches then take place between homologous donor and recipient strands, so as to insert portions of the donor strand into the recipient genome. Integration of a donor marker requires an odd number of recombination switches between the marker and each end of the donor DNA fragment. Proximity of the marker to ends of the donor fragment reduces the likelihood of a switch between it and the end thus reducing the probability of its integration.

With this model in mind THORSETT and HUTCHINSON (1971) express the experimentally observed transforming activity T as a product of two factors

$$T = R \cdot Q \qquad (2.9)$$

where R is the fraction of DNA taken up by the cell and Q is the specific transforming activity of the incorporated DNA. According to the model, Q should be a function of the molecular weight. According to CATO and GUILD (1968) one can state

$$Q\ (M) = 0 \text{ for } M \leq Mo \text{ and} \qquad (2.10a)$$

$$Q\ (M) = 1 - \frac{1}{c\ (M-Mo)} + \left(1 + \frac{1}{c\ (M-Mo)}\right) \cdot \exp -[2\ c\ (M-Mo)]$$

$$\text{for } M > Mo \qquad (2.10b)$$

where M is the double-strand M.W., c the recombination switch frequency and Mot the M.W. below which no transformation will be found. Since single-strand breaks are produced at random during irradiation the single-strand DNA pieces will have a random distribution. It is therefore necessary to calculate the quantity

$$Q = \int_{Mo/\overline{M}}^{\infty} Q\ (M)\ (M/\overline{M})\ \exp\ (-M/\overline{M})\ d\ M/\overline{M} \qquad (2.11)$$

Here $\overline{M}$ measures the number average M.W. of the single-strand pieces, and $M/\overline{M}\ \exp\ (-M/\overline{M})\ d\ (M/\overline{M})$ is the weight fraction of DNA with M.W. between M and M + ΔM. Since only one of the strands survives intact after the DNA gets into the cell, M and $\overline{M}$ are the M.W. of the single-strand pieces expressed in terms of the weights of the equivalent double-strand segments. The integral finally yields

$$Q = (\exp -Mo/\overline{M}) \left\{ Mo/\overline{M} \left[1 + \frac{1}{2c\ \overline{M} + 1} - \frac{1}{\overline{M}c} \ln\ (1+2c\overline{M}) \right] + \left(\frac{2c\ \overline{M}}{1 + 2\ c\overline{M}} \right)^2 \right\}. \qquad (2.12)$$

The number of single-strand breaks increases linearly with dose ($b_s \sim D$) and is inversely proportional to the number average M.W. ($b_s \sim \frac{1}{M}$). We can therefore plot the relative transforming activity T versus $\frac{1}{M}$ using Equations (2.9) and (2.12).

THORSETT and HUTCHINSON (1971) could describe their results with *pneumococcal* DNA best by choosing $M_o = 0.3 \times 10^6$ dalton and $c = 0.8/10^6$ dalton, thus obtaining an upward concavity. In a series of papers GRUNOW et al. (1969); GÜNTHER and GRUNOW (1971); GÜNTHER et al. (1975) developed a model to describe quantitatively the loss of the transformation rate with dose for the *B. subtilis* system. They considered the effect of single- and double-strand breaks, base damage and mechanical damage to the DNA on the efficiency of single and double-marker transformation. In their nomenclature transformation rate (TR) can be expressed as

$$TR = const \cdot f1 \cdot f2 \cdot f3$$

where f1, f2 and f3 are rather complex functions of molecular weight, marker distance, switch frequency. With their formalism GÜNTHER et al. (1975) were able to describe the experimental results.

Thus the detailed study of the dose effect curves of transforming DNA has yielded information on the inactivating event and provided a better understanding of the biological process of transformation including recombination and integration.

2.3.5 Biological Activity of Irradiated Transfecting DNA

There are apparently only very few experiments done with irradiated transfecting phage DNA. This might be due to the multipowered dose response found for many *B. subtilis* phages when the number of infective centers is plotted versus DNA concentration. With the DNA of phage ϕ 29, however, a linear response was found (REILLY and SPIZIZEN, 1965).

Consequently SUMMERS and SZYBALSKI (1967) used ϕ 29 DNA to investigate the effect of radiation on transfection. They irradiated the DNA at a concentration of 10 μg/ml with various doses. Portions were then assayed for double-strand breaks by boundary sedimentation. A one-to-one correspondence between the amount of double-strand breakage and loss of infectivity was found. The mean inactivation dose (D_{37}) was 6 krad and the inactivation was "substantially" exponential. SUMMERS and SZYBALSKI (1967) concluded that their results supported the idea that double-strand breaks are the main molecular mechanism responsible for inactivation of the transfectivity of ϕ 29 DNA irradiated in dilute salt solutions. When free phages were irradiated and the DNA extracted and tested for double-strand breaks and biological activity essentially the same one-to-one relation was obtained. Here the mean inactivation dose of the irradiated free phage ϕ 29 or its released infectious DNA was about 7 krad as determined from the exponential part of the dose-effect curve, which shows a shoulder in the low dose region.

Single- and double-strand breaks are produced upon irradiating DNA in dilute solutions. The biological consequences of these types of damage depend on the particular end point that is being investigated.

The biological activity is affected by that type of damage with the highest yield among those reactions which can inhibit the expression of the biological parameters being measured. Transforming activity can be inactivated by double-strand breaks, but it is predominantly inactivated by single-strand breaks which are produced at higher yield. Thus single-strand breaks are the principal chemical event for inactivation of transforming activity. On the other hand, for transfecting ϕ 29 DNA the chain of events resulting in plaque formation is rather insensitive to single-strand breakage. Therefore a double-strand break is required to prevent the expression of this particular end point. Transfecting DNA from large phages which require cooperative infection and some form of recombinational event, may be inactivated by single-strand breaks as speculated by SUMMERS and SZYBALSKI (1967) because of interference with recombinational events, which may be similar to integration of transforming DNA.

2.4 Irradiation Effects on Transcription

The expression of the genetic information of DNA occurs by the biosynthesis of RNA and of protein. When we consider the DNA as important target for the lethal radiation effect in living systems we have to admit that the radiation induced structural alteration of the DNA molecule will lead to a disturbance of the genetic expression, i.e., an inhibition or change of the various steps in transcription and subsequently in translation. Our knowledge about the processes occurring during transciption is not sufficient to predict the effect of specific radiation induced alteration in the DNA on its genetic function. A first approach to understanding the radiation effect on transcription, therefore, is a study of the ability of irradiated DNA to serve as primer in vitro for enzymatic RNA synthesis by RNA polymerase. The results obtained by such studies are clearly useful for interpreting the effect of ionizing radiation on the RNA synthesis in living cells.

2.4.1 Transcription on Irradiated DNA in vitro

The transcription of the genetic code in DNA by RNA polymerase occurs in four steps: (1) binding of the RNA polymerase to DNA on specific binding sites: a reversible process; (2) chain initiation, i.e., the introduction of the first nucleotide of the RNA (in this step, the binding of the enzyme becomes very tight); (3) chain elongation, and (4) release of synthesized RNA and of the enzyme. Each of these steps may be disturbed when we observe a decrease of the RNA synthesis after irradiation of the primer DNA and should be analysed separately.

The total RNA synthesis, i.e., the incorporation of labelled ribonucleotides into the acid-insoluble fraction of an incubation mixture with irradiated DNA and purified RNA polymerase, was measured by several authors (HARRINGTON, 1964; ZIMMERMANN et al., 1964; HAGEN et al., 1970). There is a distinct decrease of the RNA synthesis with increasing radiation doses on the primer. The dose effect curve (in semilogarithmic scale) shows a peculiar upward bent form (Fig. 2.4a). A more detailed analysis of the form of this dose-effect curve is given below as it requires the knowledge of the radiation effect on the various steps of transcription.

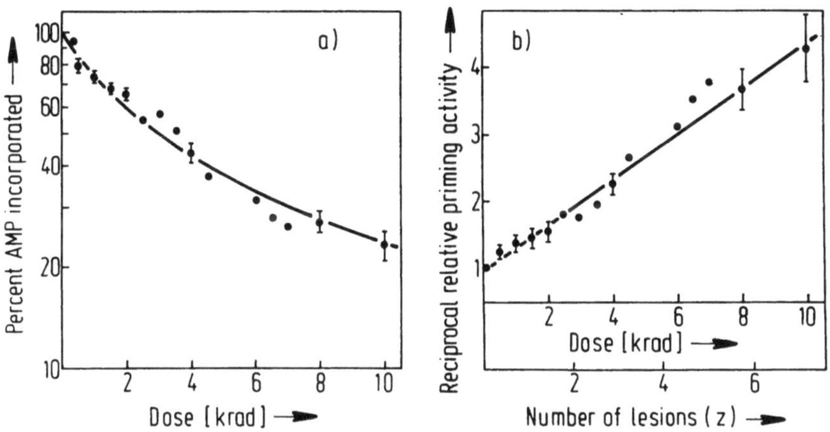

Fig. 2.4a, b. Transcription on irradiated DNA in vitro; (a) Amount of synthesized RNA, determined by the [14C] AMP incorporation into the acid insoluble fraction of an incubation mixture with DNA, RNA-polymerase, and deoxyribonucleotide-triphosphates. The RNA synthesis is expressed in per cent of that on unirradiated DNA. Irradiation of calf thymus DNA at 500 µg/ml in 0.01 M NaCl in air (HAGEN et al., 1970); (b) Relative reciprocal priming activity of irradiated DNA (A_z/A_o), plotted against the irradiation dose as well as against the number of critical lesions (z), formed in a DNA section with the length L_o of 1000 nucleotide pairs. Data of Figure 2.4a (HAGEN et al., 1969)

The binding of the enzyme on DNA can be studied by a special filtration technique as the complex of DNA with enzyme is retained on membrane filters, whereas free DNA is released (HAGEN et al., 1970). It was found that the binding capacity of DNA for RNA polymerase increases with dose, probably due to a formation of tight binding sites as it was observed also by HINKLE et al. (1972) on T7 DNA carrying single-strand breaks induced by DNase I. For this effect, however, much higher doses are necessary than for the inhibition on RNA synthesis. For instance, when the total RNA synthesis is depressed to 20% of the control, the binding capacity is only enhanced by a factor of 1.5. An increase of binding as well as of initiation sites after irradiation has been observed also by CHESSIN and SUMMERS (1970) measuring the transcription on γ-irradiated T7-DNA. Whereas the RNA synthesis on unirradiated DNA is asymmetric, i.e., proceeds only on one strand on irradiated DNA it becomes more and more symmetric, i.e., RNA is copied also on the other nucleotide strand.

The chain initiation on irradiated DNA can be tested using nucleotide triphosphate as precursor labelled with ^{32}P in the γ-position of the phosphate. The first nucleotide in the RNA chain retains its triphosphate property, whereas the subsequent nucleotides are incorporated as monophosphates. Counting the activity of ^{32}P in the acid insoluble fraction a relative value for the number of starting points for RNA chains is obtained. The length of the RNA chains can be determined by measuring in additon the total incorporation of nucleotides by labelling in the base moiety. A study using γ-irradiated DNA showed by this technique (HAGEN et al., 1970) the number of RNA chains synthesized to remain almost unchanged after moderate radiation doses. The length of the RNA chains, however, is distinctly reduced in a degree comparable to the dose effect curve for the overall RNA synthesis. This means that the inhibition of RNA synthesis is almost exclusively due to a shortening of RNA chains whereas the binding of the enyzme and the chain initiation is less effected.

These observations allow the derivation of a model (HAGEN et al., 1969) assuming the RNA synthesis to proceed along the DNA chain and to be stopped by a radiation induced "critical" lesion in the double helix. If we take for the length of DNA section transcribed on unirradiated DNA L_o the RNA chain length will be reduced to the distance between the binding site and the nearest critical lesion produced within the interval L_o. The length L_z is described by

$$L_z = L_o \, [1 - \exp(-z)] \, / \, z \qquad (2.13)$$

where z is the average number of critical lesions on the DNA double helix within the length L_o. The derivation of Equation (2.13) is given in detail by BRESLER et al. (1967) who have used it to calculate the reduction in recombination length of a transforming DNA molecule. Since the amount of RNA synthesized is proportional to the length of the DNA interval transcribed, L_z/L_o represents also the relative priming activity A_z/A_o. A similar formula was obtained also by MICHALKE and BREMER (1969) analysing the RNA synthesis of UV-irradiated bacteria.

A plot of A_o/A_z, i.e., the reciprocal relative priming ability is given in Figure 2.4b where the data of Figure 2.4a has been transformed correspondingly (HAGEN et al., 1969). As predicted by Equation (2.13) there is a linear increase of the dose effect curve allowing also the calculation of the number of lesions produced on a DNA section of length L_o which was found in this case to amount to 1000 nucleotide pairs. A further calculation leads to the result that the G-value for generation of a critical lesion may lie in the range from 0.4 to 0.6, a value useful for further considerations on the mechanism of the radiation effect on transcription. This analysis of the dose effect curve for inhibition of RNA synthesis gives an example that the mechanism of the irradiation effect should be studied in detail before a mathematical analysis of the dose effect relation can lead to meaningful results.

For further description of the radiation effect on transcription, the nature of the critical lesion stopping the chain elongation should be analysed. Mainly, we have to consider in γ- or X-irradiated DNA base damages, strand breaks as well as alkali labile lesions (see II.3). The frequency (G-value) of strand breaks including the alkali labile lesions corresponds to that of the critical lesions. Therefore, the interruption of the strand with a concomitant damage of sugar and base (ULLRICH and HAGEN, 1971) may possibly stop the chain elongation. This may be valid also for the alkali labile sites, which are due partly to a loss of the base moiety (ULLRICH and HAGEN, 1971). This assumption is supported by the observation of MAMET-BRATLEY (1974) that apurinic sites, formed by incubation of DNA in acid solution, stop RNA synthesis.

On the other hand, it cannot be excluded that certain base damages may also stop the transcription; there are twice as many base damages in γ-irradiated DNA than single-strand breaks comparing the G-values for these events. Such an assumption, however, would require more knowledge of the various types of base damages in DNA and their effect on transcription. In this connection so far, only the effect of an alkylation of DNA has been described, a treatment not stopping the chain elongation at the enzymatic RNA synthesis (MAMET-BRATLEY, 1971). It should be noted however, that the pyrimidine dimers, formed in DNA by UV light, are definite chain-terminating lesions for RNA synthesis (SAUERBIER et al., 1970) probably due to the distortion of the double helical structure on this site. A corresponding alteration of the DNA structure by ionizing radiation is not known at the present time.

When favouring the single-strand breaks and alkali labile lesions as the critical lesions for the stop of transcription, we have to consider whether the base damage influences the transcription or not. If a base damage is not a critical lesion for chain elongation it follows that the process of transcription proceeds beyond that. At the point of a damaged base in the DNA the correct corresponding base may be inserted into the RNA or the base configuration is changed in such a way that wrong base pairing occurs.

The first indication that, in fact, such wrong base pairing occurs on irradiated DNA was given by WEISS and WHEELER (1967) who observed a slight change of the base composition of RNA, synthesized on γ-irradiated DNA in comparison to that formed on unirradiated DNA. More evidence for wrong base pairing was obtained measuring enzymatic RNA synthesis on RNA-homopolymers, where the corresponding polymer is synthesized (GODDARD et al., 1970). Using for example polycytidine as primer, mostly guanine-monophosphate is incorporated and negligible amounts of AMP. However, using γ-irradiated polycytidine, much more AMP is incorporated indicating a change of the pairing ability of cytidine to that of thymine. A corresponding observation was made irradiating polyadenine where instead of uracil also guanine is incorporated. Such a change of the pairing ability is quite frequent in irradiated homopolymers; about a third of the base damages may induce this effect.

There is a third possibility: a change of base configuration neither stops the chain elongation nor impairs the base pairing. In this case the atoms in the base moiety forming the hydrogen bonds are unchanged; the radiation induced lesion is not relevant for the biological function of DNA. Considering the structure of hydrogen bonds it is conceivable that for instance an opening of the double bond of the pyrimidine will not change the ketoform of the base; therefore the hydrogen bonds to the purine base are not impaired. Hint of such possibility is given by the observation of JUNG et al. (1969), that most of the base damages induced by atomic hydrogen do not affect the correct phage specific RNA synthesis in ϕX 174 infected host cells. Finally it should be pointed out that γ-irradiation of the RNA polymerase itself also leads to wrong base pairing - in addition to the inactivation of the enzyme - indicating that the intact structure of DNA as well as of the enzyme is necessary for correct base pairing (GODDARD et al., 1969).

2.4.2 RNA Synthesis in Irradiated Cells

Considering the radiation effect on transcription and subsequent protein synthesis in vivo, i.e., in irradiated cells, various processes may be involved that determine the observed radiosensitivity: radiation damages on the DNA target and on the RNA polymerase as well as various metabolic disturbances in the irradiated cells and repair reactions on DNA. The radiation effect can be described measuring the total RNA synthesis, the synthesis of the various RNA species and the induction of specific enzymes. In the latter case the fidelity of transcription and translation will also be described.

Assuming a similar mechanism for the radiation effect on transcription in vivo as observed in vitro, the transcription would be inhibited by certain damages in the DNA terminating the chain elongation of RNA. A corresponding observation was made by MICHALKE and BREMER (1969) in UV-irradiated bacteria, where the pyrimidine dimers are identified as critical terminating lesions but not, so far, in cells, irradiated with ionizing radiation. There are numerous reports on the RNA synthesis in X- or γ-irradiated cells, summarized recently by PAUL and ZIMMERMANN (1975). Most of the authors found a depression of the RNA synthesis; there are few reports describing an increase of RNA synthesis after wholebody irradiation which may be explained by metabolic disturbances in the irradiated animal.

If the radiosensitivity of the RNA synthesis depends on the size of the target, the synthesis of large RNA molecules should be more radiosensitive than that of small RNA molecules. This was shown by KANAZIR (1969) who analysed the synthesis of various RNA fractions in irradiated bacteria. The incorporation of labelled uracil in the 23 S RNA was distinctly inhibited, whereas the synthesis of the smaller RNA with 16 S or 4 S was not impaired by an equal dose. The radiosensitivity of the synthesis of ribosomal RNA in *E. coli* was confirmed by FUJITA and MIYAKOSHI (1973) demonstrating an exponential decrease of the total RNA synthesis as well as of the various RNA species. Analysing the size of the RNA species synthesized there is, however, no shortening of the chain length as expected from the model derived from the in vitro studies. From the exponential dose effect curve a target size of about 1.5 to 2.2 × 10^6 dalton could be derived for the r-RNA synthesis which is of the same order as the size of r-RNA cistron on DNA. This also implies that the r-RNA synthesis is completely inhibited when only one lesion is produced in the cistron. Possibly, incomplete RNA, terminated by a critical lesion, is rapidly degraded in the cell and cannot be found in the irradiated bacteria.

An additional radiation effect is a disturbance of the maturation of the RNA species leading to accumulation of premature RNA (KIM et al., 1970; FUJITA and MIYAKOSHI, 1973) as well as the inhibition of the formation of polyribosomes (PAUL and ZIMMERMANN, 1975). All these observations lead to the conclusion that the radiation effect on the RNA synthesis in the living cell is much more complex than in vitro and cannot be explained alone by radiation-induced lesions in the DNA target. In addition, many lesions on DNA terminating the chain elongation at transcription may be repaired before this DNA section will be transcribed.

Information on the synthesis of specific m-RNA species and the subsequent protein synthesis can be obtained by measuring the induction of enzymes. In convincing experiments of PAULY (1963) the radiosensitivity of *Bacterium cadaveris* was studied. The colony-forming ability is about three times more radiosensitive than the synthesis of RNA and of protein. Moreover, the induction of the lysine decarboxylase was tested, the radiosensitivity being in the order of that of overall RNA- or protein synthesis. Observations of POLLARD and DAVIS (1970) support the conclusion that also in vivo critical, transcription-terminating lesions in DNA are responsible for the radiation effect on enzyme induction. The authors tested the induction of several enzymes in the lac-operon which is transcribed into one single molecule of m-RNA. It was found that the induction of the first enzyme in the operon, the β-galactosidase, is six times more resistant than that of the last enzyme the transacetylase. It can be assumed that each critical lesion in the lac-operon will inhibit the synthesis of the transacetylase, whereas the synthesis of the β-galactosidase is impaired only when terminating lesions are produced within the first section of the operon coding for this enzyme itself.

The induction of enzymes has been measured also in animals after whole-body irradiation where no relationship between the sensitivity for a specific gene and the corresponding target size could be found (STREFFER and SCHAFFERUS, 1971). The authors suggested that in this case specific effects at the translational level of protein biosynthesis contribute to the radiation-induced disturbances.

Summarizing the molecular processes occurring in the course of transcription along irradiated DNA, we may state that some radiation-induced structural changes will stop the RNA synthesis. Single-strand breaks and base-free sites on the sugar phosphate chain may form most of these critical lesions. Other alterations may change the genetic information of the DNA leading to wrong bases in the messenger RNA. In addition there are base damages in DNA, which are not relevant for the correct transcription. They will probably be eliminated in the course of the next replication of DNA. These radiation-induced molecular processes on DNA, however, can explain only part of the events occurring in irradiated cells at RNA- and protein synthesis. Unknown regulation phenomena as well as the various repair reactions may weaken or amplify the primary radiation effect on DNA in living cells.

2.5 Irradiation Effects on DNA Replication

2.5.1 Introduction

It is now well established that radiation energy can be deposited randomly within the structural molecules comprising the cellular sites where the synthesis of the major macromolecules is carried out. Therefore any interpretation of the way in which radiation affects the synthesis of biologically important molecules must take into account radiation damage in neighbouring molecules as well as direct damage to the macromolecules under investigation. This approach is specially relevant when the post-irradiation synthesis of DNA is considered. Much evidence exists that semiconservative DNA replication is dependent upon the close association at the active area of synthesis of cytoplasmic or nuclear membrane and that RNA molecules acting in a synthetic or structural capacity are essential.

2.5.2 Methods of Measuring DNA Replication in Cells

2.5.2.1 Radioactive Labelling

The synthesis of DNA can be followed by observing the uptake of radioactivity into the newly formed DNA molecule supplied by a suitable radioactive labelled compound in the nutrient medium available to the cell. Such compounds range from ^{32}P inorganic sources to the immediate precursor molecules prior to incorporation into the DNA strand, such as the tritiated purine or pyrimidine triphosphates. Depending on the metabolic activity of the cell system under investigation the time of exposure to radioactive labelled compounds can be from seconds to hours. The method of determination of how much label has been taken up into DNA has several variants but consists essentially of cell disruption and precipitation with ice-cold perchloric acid or trichloracetic acid, collection of the DNA containing precipitate and thorough washing procedures to remove all contaminating low M.W. label. When there is a possibility of incorporation of label into macromolecules other than DNA, usually RNA, then a separation process must be used to eliminate such macromolecules, before DNA is determined. To avoid this last difficulty most work is now done with highly pure ^{3}H or ^{14}C labelled DNA bases, nucleosides and phosphorylated nucleosides which are incorporated specifically into newly formed DNA. The estimation of activity in the precipitated DNA is made generally by scintillation counting and occasionally by gas flow counting. By this acid precipitation technique the amount of DNA synthesized after irradiation can be calculated either in relation to the amount synthesized by an unirradiated sample or in terms of the number of nucleotides appearing in the newly synthesized DNA. In order to determine the molecular sizes of thelabelled DNA, gradient centrifugation techniques must be used.

2.5.2.2 Autoradiography

An alternative method for determining the extent of incorporation of labelled precursors into DNA is by the preparation of autoradiographs. Thoroughly washed cells from the radioactive nutrient medium are gently disrupted with detergent, and dialysis performed on the liberated cellular constituents until any low molecular weight radioactive components are removed. The labeled DNA films are filtered off on to plates, dried and covered with sensitive photographic emulsion. After several months exposure the films are developed and scanned photomicroscopically to produce pictures where "grains" produced by this radio activity in the DNA can be counted by eye or by automatic densitometry. This method of determining DNA synthesis gives a good measure of the extension of the DNA chain as well as a measure of total DNA formed, but it is very time-consuming.

2.5.2.3 Replication Synthesis Defined

In addition to describing the methods for the determination of DNA synthesis the nature of DNA replication should be defined in the context of damage by radiation. In the following section it is intended to discuss the effect of radiation on semiconservative replication, that is, the way in which radiation inhibits or stimulates the rate and extent of the normal replicative mechanisms. DNA synthesis required to "fill in" small gaps in damaged parental DNA will not be considered.

The classical experiment to show semiconservative replication consists in prelabelling the DNA in cells with either ^{14}C or ^{3}H activity, washing out the unused radioactive precursor and then allowing further synthesis in the presence of 5-bromouracil. This heavy base will be incorporated into DNA, to an extent depending on the cell line, in place of thymine. The DNA in toto, newly synthesized and labelled parental, is isolated and broken up into small double-stranded pieces by sonication. Finally the distribution of densities is measured on caesium chloride density gradients. Figure 2.5 shows the result of such an experiment where DNA replication has proceeded semiconservatively. The 5-bromouracil DNA fragments appear at a hybrid density arising from a mixture of original DNA acting as template and the new bromine containing strand. If synthesis had proceeded by simple end elongation of parental DNA or by extensive incorporation of 5-bromouracil into both strands of already formed DNA by some replacement or repair process, then DNA fragments of much higher density would show up in the density centrifugation experiments.

When synthesis followed by uptake of precursor is small compared with the parental DNA, then additional evidence that synthesis is semiconservative or not, is required and can be obtained by ex-

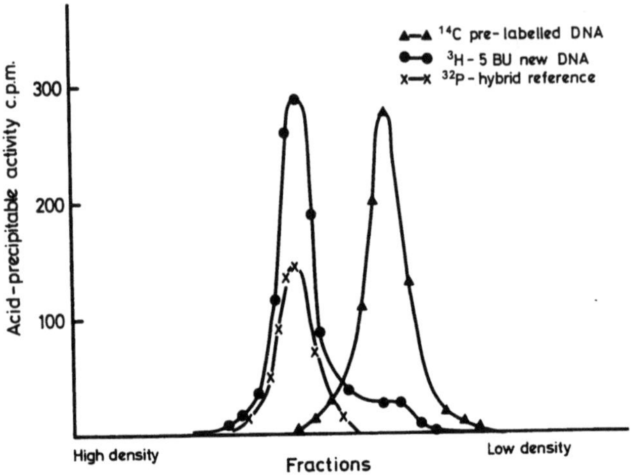

Fig. 2.5. Density analysis of bromouracil-labelled DNA. On denaturation, the bromine-containing DNA separates into two strands banding at heavy and light single strands

amination of the size of the newly made DNA. The techniques for these observations are described in III.2.1.2.2.

That radioactive label is taken up into DNA of normal structure can be easily proved by demonstrating that such newly formed material can be digested by DNA specific enzymes. In some special circumstances where synthesis is followed in vitro, an added proof of the appearance of newly formed complementary DNA in the system is that all four base precursors and specific chemical energy sources are required.

2.5.3 Effect of Radiation on Uptake of Precursors into DNA

2.5.3.1 Bacteria

The effects of ionizing radiation on the synthesis of DNA by bacteria has not been as well investigated as with mammalian cell systems, although much work has been done on the inhibitory effects of UV radiation on bacterial DNA activity. The reader is referred to a general review by POLLARD (1970) on the effects of ionizing radiation on the molecular biology of *E. coli* where DNA synthesis is considered in the light of concomitant DNA degradation, a phenomenon easily observed and measured in bacteria but apparently of negligible proportions in mammalian cells. Also several pertinent references are made in an excellent volume on radiation biochemistry by OKADA (1970).

Various strains of *E. coli* have been used, especially the thymine-requiring strains, in order to facilitate the uptake of labelled thymine or thymidine precursor. However, it is possible by blocking enzyme pathways to induce nonthymine-requiring strains of *E. coli* to take up thymine base precursors.

It was established in early work by BILLEN (1962) that X-irradiation given in vivo or in vitro even at very high doses did not have a direct effect on the enzymes responsible for transport of the four nucleoside triphosphates. The same work also reported that X-irradiation did not adversely affect the biological activity of DNA polymerase I, then considered to be the enzyme responsible for DNA replication. Recently the isolation of at least two more DNA polymerases made it impossible to rule out the direct effect of radiation on these enzymes as the cause of inhibition of DNA synthesis.

Later work by the same author (BILLEN et al., 1965) reinforced earlier conclusions that radiation caused perturbation to primer DNA and although the rate of DNA synthesis was only partially inhibited even at high doses the mode of synthesis was altered post-irradiation. For example, after doses of 2 krad or higher, DNA replication occurred in limited regions of the bacterial chromosome and the cell was induced to replicate at a point or points apparently differing from the original growth point. The interpretation of this work has been made more difficult since the relatively recent discovery that DNA in bacteria, and indeed in mammalian cells, is replicated in opposite directions starting from specific points of initiation.

CRAMP and ELGAT (1972) more recently used experimental conditions whereby the rate of uptake of ^{3}H-thymidine into the bacterial cell and specifically into DNA could be observed very shortly after irradiation had ceased. High dose rates were used so that the duration of irradiation was small compared with the time elapsed before changes in the system were observed. Two strains of *E. coli* were used, one resistant and one sensitive to radiation, and the response of the DNA synthesis was measured after irradiation of the bacterial suspension under either aerated or anoxic conditions. Incorporation of ^{3}H-thymidine into DNA was affected at dose ranges over which cell survival could be determined and consequently correlation between inhibition of incorporation and survival was possible.

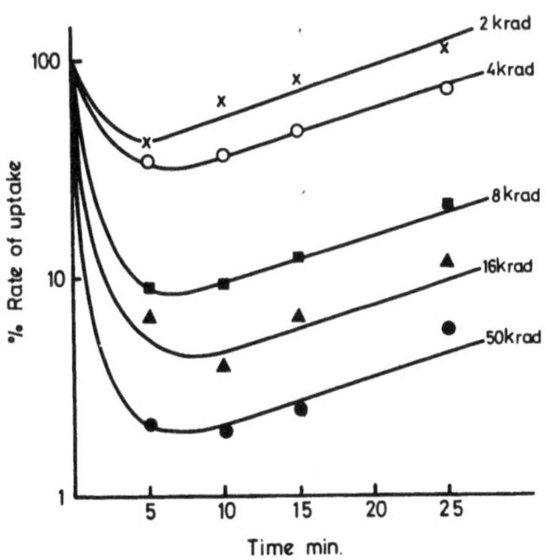

Fig. 2.6. Change with time of the rate of uptake of thymidine into DNA of *E. coli* B/$_r$ irradiated under aerobic conditions

Figure 2.6 illustrates the type of curves obtained in plotting the rate of uptake against time after irradiation.

The authors concluded that it was not possible to correlate survival of the two strains with the general inhibition of the rate of thymidine uptake in the whole population soon after irradiation. At the same level of survival the more sensitive mutant maintains, for a short time, a higher rate of thymidine incorporation into DNA than does the more resistant mutant.

An interesting observation was that at all doses the rate of recovery of thymidine uptake after irradiation under aerated conditions was the same and was exponential with time. This could mean that radiation specifically damages part of the cell which can control its own repair, so that as more undamaged material is regenerated more damage can be repaired. Recovery at the same rate after different doses suggests that radiation does not damage repair systems in the cell. Recovery to, and eventually above the rate of incorporation measured in the unirradiated controls, indicates a recovery mechanism regulated by a faulty, radiation-induced "feed-back" system, such that, for a short time at least, synthesis proceeds at a greater rate than normal. Alternatively, such recovery implies DNA synthesis at several points giving rise to abnormally high utilization of thymidine.

With the present volume of research it is inevitable that advances in the knowledge of DNA replication, particularly of the enzymology and site of activity, require that past work be reinterpreted and new and more illuminating experiments attempted.

2.5.3.2 Mammalian Cells in Culture

It is difficult to write a short account of the effect of ionizing radiation on the uptake of precursors into the DNA of mammalian cells. There has been a great deal of activity in this field in the last 20 years, using many cell lines and a variety of methods of determining DNA replication.

At the outset it must be stated that some controversy exists as to whether radiation affects DNA replication by reducing the available bases for incorporation, as well as having an effect on the replication apparatus at the templating level. There appears to be adequate evidence (WATKINS, 1972) that radiation damage to cell membranes may prevent the uptake of precursors into the cell for ultimate incorporation into the DNA.

Early work by BERRY and his colleagues (1960, 1962) established a biphasic X-ray response since seen by several workers. Using human bone marrow cells in culture and following the inhibition of the incorporation of ^{3}H-thymidine into DNA, these workers showed a steep dose-response curve over a dose range of 500-2000 rad (30% depression for 500 rad) followed by a shallow curve between 2000 and 10,000 rad; over the low dose range they observed that the inhibition of incorporation was reduced by a factor 2.9 if the irradiation was carried out in oxygen-free nitrogen. The same workers (BERRY et al., 1960) and later others (ADELSTEIN and MANASEK, 1967) ruled out the possibility that radiation caused the release of DNA precursors from DNA which would dilute out the added label.

Some elegant work by PAINTER (1967) using HeLa cells and three different DNA precursors over a wide dose range, showed that the inhibitory effect of radiation was very similar despite the fact that the precursors entered into the DNA cycle by different routes. Table 2.3 summarises these results.

Table 2.3.

Dose (rad)	% control ^3H thymidine	^{14}C orotic acid
0	100	100
100	94	90
1000	75	79
10,000	30	32
100,000	2	5

Almost identical results were obtained using ^3H deoxyguanosine or ^{14}C thymidine. An extremely interesting phenomenon reported by WATANABE and OKADA (1968) was that, after an initial post-irradiation depression of incorporation of ^3H thymidine into the DNA of murine leukemia cells, there followed a recovery of the rate of incorporation to a rate that was faster than that of the nonirradiated cells. In addition, the irradiated cells synthesized more DNA than nonirradiated control cells. Doses in the range 1000 to 5000 rad were used, and incorporation was measured by autoradiography. Recent work from the same laboratories using the same cell line showed an interesting all-or-nothing effect. The rate of DNA synthesis at cellular level was depressed immediately after irradiation with a biphasic response curve, (D_{37}s of 1300 and 10,500 rad) whilst the rate of DNA chain elongation measured by autoradiography was only slightly affected (D_{37}:13,000 rad). The authors conclude that the sensitive and resistant responses found at the cellular level are the radiation-induced rehibition of initiation (continuation of synthesis at the same rate as in unirradiated cells or cessation) and the only slightly disturbed chain continuation process, respectively.

A special instance of unequal uptake of DNA precursors has been reported by SANDERS et al. (1969), using starved mouse L cells. These workers reported a radiation-induced cyclic labelling of DNA when either ^3H-deoxy-adenosine or ^3H-adenosine were used. The response consisted of a rapid uptake and loss of label. A similar effect was also observed for ^{32}P labelling DNA. However, with ^3H-thymidine or ^3H-cytidine no cyclic labelling was observed. These authors advance the hypothesis that X-irradiation of mammalian cells produces increased activity of a DNA-ligase enzyme which rejoins single-strand breaks in DNA. The enzyme mechanism requires the formation of a transient DNA-adenylate complex such that labelled adenylate enters and leaves the complex.

It is clear that whilst many workers have identified the inhibition of DNA synthesis with the lack of function and eventual death of the cell, measurements of immediate post-irradiation DNA precursor uptake have not clarified this relationship. It is unfortunate that very few workers have observed radiation effects on cellular systems both in the presence and absence of oxygen. Consequently correlation between these post-irradiation effects and ultimate cell survival, where such gas conditions are routine, is difficult.

2.5.3.3 Cells of Irradiated Animals

The post-irradiation metabolism of DNA, as measured by the incorporation of precursors into DNA of cells of lymphoid tissues after whole-body irradiation, is adequately reviewed by GERBER and ALTMANN (1970).

The immediate and early effects of radiation on precursor uptake are similar to those described for cells in culture. However, for intermediate and late effects, changes in cell population during regeneration of tissues must be taken into account, and make interpretation difficult.

A special situation where DNA replication can be followed by labelled precursors, is where extensive regeneration of the liver is stimulated by partial hepatectomy. After excision of two-thirds of the liver the parenchymal cells begin to divide 24-36 h later, and the organ is restored to nearly its original mass within 2-3 weeks. Regenerating liver represents a fairly uniform, partly synchronous cell population, and is a useful system to study the effects of radiation on rapidly dividing cells.

If irradiation is given 18-24 h after surgery (the time of maximum DNA synthesis before cell division), doses of 2000 rad and below cause significant and immediate depression of uptake of precursor, and mitosis is delayed but not abolished.

LOONEY and CHANG (1965) using regenerating rat liver, examined over the period of the first cycle of synthesis after hepatectomy, showed a biphasic response common to other post-irradiation uptake experiments (see Fig. 2.7).

Rapid depression was observed between 750 and 6000 rad followed by a much lower depression at 40,000 rad. Above doses of 18,000 rad the percentage number of cells labelled decreased rapidly. The latter part of the DNA synthetic period in the regenerating rat liver was the most sensitive.

2.5.4 Influence of Mitotic Cycle on the Irradiation Effects on DNA Synthesis

The influence of the stages of mitosis on DNA synthesis as measured by precursor uptake is well described by work of BRENT et al. (1966).

The incorporation of ^{3}H-thymidine into asynchronous and synchronous HeLa cells was measured after irradiation at different times after mitosis. Typical results are illustrated in Figures 2.8 and 2.9 where irradiation was delivered to the synchronous cells

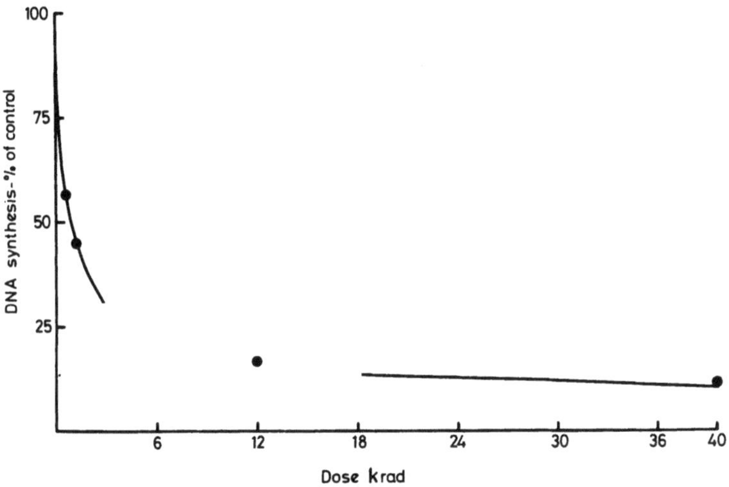

Fig. 2.7. Dose-effect curve for the uptake of ^3H-thymidine into the cells of regenerating rat liver

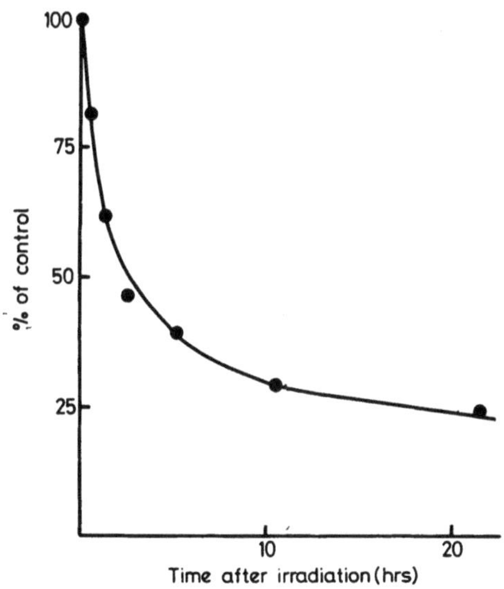

Fig. 2.8. Incorporation of ^3H-thymidine into asynchronous HeLa cells in 15 min at various times after 600 rad

during the DNA synthetic period (S) of the cell cycle (11 h after mitosis). By irradiating the cells at different times after mitosis, the DNA replication mechanism was clearly shown to be more sensitive during the S-period than at other times. The authors concluded that a region of DNA together with the replicating enzyme and other macromolecules required during synthesis are particularly sensitive at this time.

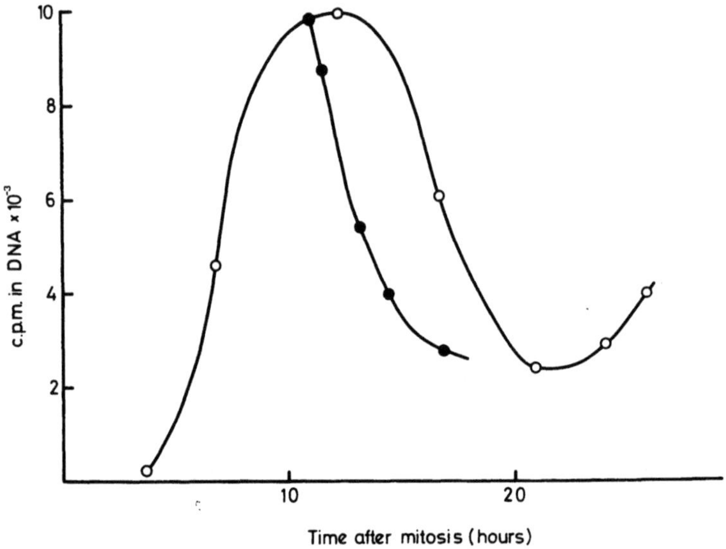

Fig. 2.9. The rate of incorporation of ^{3}H-thymidine into synchronous populations of HeLa cells at various times after mitosis. -o-: control; -•-: culture irradiated with 1000 rad 11 h after mitosis

2.5.5 Form and Interpretation of Dose Effect Curves

The general form of dose vs. effect curves has been illustrated in Figure 2.7. At lower doses a rapid depression of incorporation of precursors is seen, followed by a slower depression of incorporation at higher doses. Very little extra depression occurs at very high, supralethal doses. If incorporation rates are measured (Fig. 2.6) instead of total uptake it is apparent that at least in the bacterial system a recovery of the rate of synthesis occurs. One interpretation of this type of biphasic response is that radiation damages the DNA replicating mechanism in two different ways. Firstly the initiation of DNA synthesis is completely blocked, so that the response is no synthesis at the normal rate. Secondly, at higher doses the DNA replication that has already started continues at a lower than normal rate because it encounters deformations on parental template DNA.

2.5.6 Inhibition of DNA Synthesis and Cell Survival

The most important consideration under this heading is the comparison of the dose ranges in which effects can be measured. If the dose ranges at which inhibition of DNA replication and cell killing are observed (colony forming ability or regrowth of tissue) do not overlap, then correlation between the two responses is difficult.

Figure 2.10 shows dose-effect curves for the rate of incorporation of ^{3}H-thymidine into two bacteria of very widely different sensitivity to the lethal action of radiation.

Quite clearly the wide difference in survival of these two mutants cannot be explained by differences in their immediate post-irradiation DNA synthetic activity.

Similarly with mammalian cells, post-irradiation differences in DNA metabolism do not explain differences in cell survival at any dose range.

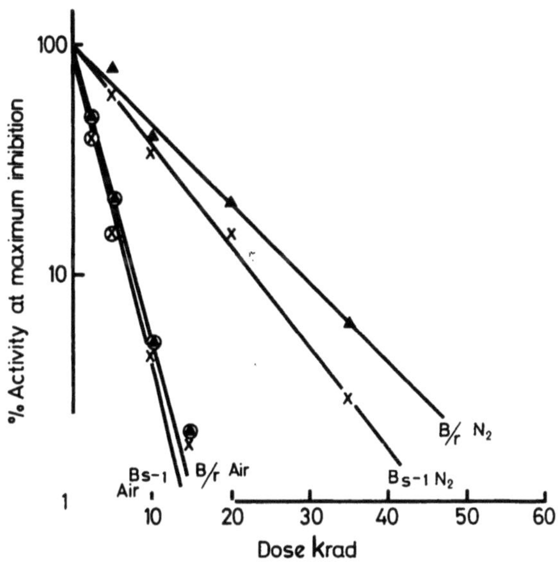

Fig. 2.10. Dose effect curves for the rate of incorporation of ^{3}H-thymidine into *E. coli* B/$_r$ and B$_{s-1}$ at the time of maximum inhibition after electron irradiation

The biphasic response curve of the type illustrated in Figure 2.7 has been found by many authors with a variety of cell lines. There is often some correspondence between the low dose range (0-1000 rad) inhibition of precursor uptake and loss in cell viability. However, at higher doses when most cell lines show an exponential loss in surviving cells, incorporation of label into DNA slows and often remains at a high level at very high doses.

SINCLAIR (1968) compared DNA synthesis with cell survival in some detail for Chinese hamster cells. It is clear from his work that there is no exact coincidence between the rate of synthesis or the number of cells in synthesis, and the maximum level of survival. He demonstrated that DNA synthesis was affected most when the cells were in the S period, but, although the rate of synthesis was reduced and the S period prolonged, the total amount of DNA synthesized remained the same.

In general the experimental results obtained from the examination of post-irradiation synthesis in cultured cell populations, do not indicate that overall changes in the DNA synthesizing ability of the cells are reflected in similar changes in cell survival. The main conclusions are two-fold, either the DNA synthesized after irradiation contains altered sequences of bases and does not provide the correct information for the cell to survive; or, changes in DNA synthesis other than complete blocking are not directly related to cell death and that changes in other structures are more important.

These conclusions must be tempered by the realization that whilst the synthesis of DNA can be observed soon after irradiation, cell survival may depend on changes within the cell not evidenced for much longer times.

2.5.7 Irradiation Effects on DNA Synthesis in vitro

Many cell constituents such as DNA, RNA and proteins have been isolated and irradiated in their purified forms. The post-irradiation changes which have been measured by physicochemical and biological methods indicate that such a simplified approach to the determination of the vital targets in a cell whilst extremely useful does not tell the whole story. For instance, large (supralethal) doses are required before changes can be measured, and the enhancement of the radiation effect by the presence of oxygen, nearly always observed when using cell survival as an end point, is often absent.

In the last seven years, however, subcellular fragments which are responsible for DNA replication have been isolated and characterized. The response of these fragments to radiation suggests that these subcellular systems contain molecular targets, damage to which, brings about cell death.

A complex containing most of the DNA attached to cell membranes can be isolated from *E. coli* (CRAMP et al., 1972a, b). The RNA and soluble enzymes are lost from the cell during lysis and separated from the complex by centrifugation. If a suspension of the DNA-membrane complex is mixed with a suitable solution of the four nucleoside triphosphates, new DNA is synthesized initially at a rate comparable to that in vitro. Such synthesis is followed by measuring the uptake of a labelled triphosphate into the trichloracetic acid precipitable fraction. In addition to DNA and membrane there is reason to believe that the complex contains enzymes and structural RNA necessary for DNA synthesis. This complex contains most, if not all of the macromolecules likely to react to radiation. CRAMP et al. (1972a, b) have shown that when *E. coli* are irradiated either with electrons or neutrons the membrane complexes isolated from the bacteria are much less active in DNA synthesis than from unirradiated control organisms. The dose range over which response is measured coincides with that used to determine lethality over the range 100-0.1% and the presence of oxygen markedly increases the effect of irradiation. There is reason to believe (WATKINS, 1970; ALPER, 1971) that membranes in bacteria and nuclear membranes in eukaryotes are as important targets as nucleic acids, and the results from membrane complexes bear out this contention.

It is interesting to note that if the membrane complexes are isolated and irradiated in a purified state, that is, in the absence of cytoplasmic materials, they prove to be very resistant to radiation; and indeed, DNA synthesis appears to be stimulated by radiation at some doses. Two bacterial strains of different radiation sensitivities (lethality) yield membrane complexes of very similar DNA synthetic activity, after irradiation with electrons, and it is concluded that the measurement of damage to the DNA metabolism does not display the total damage to the cell which causes cell death. However, with radiation of more densely ionizing characteristics (neutrons) the more resistant mutant *E. coli* yields a membrane complex which is much more depressed in DNA synthetic ability than that from the sensitive mutant. The effects of specific antibiotics have been studied in this bacterial system and it has been concluded (CRAMP and BRYANT, 1975) that the membrane complex contains structures which interact with one another after irradiation, and that such interactions can bring about cell death. The different sensitivities of cells to the killing effects of radiation can be explained by postulating that the relative proximities of macromolecules within the membrane complexes are different, as well as advocating differences in repair mechanisms (CRAMP et al., 1972a, b).

Isolated nuclei from mammalien cells suspended in a solution containing the four nucleocide triphosphates, one with an appropriate ^{3}H label, can be shown to synthesize new DNA initially at a rate very close to that in the intact cell. It has also been established with many cell lines that this synthesis is a continuation of the synthesis in the cell before disruption and isolation of the nuclei.

The effect of radiation on the nuclei either directly or in the intact cell has been little investigated. One of the reasons for this neglect is probably that large numbers of cells are required to follow the DNA synthesis reaction. However, MATSUDAIRA and FORUNO (1971) have made some interesting preliminary studies of this mammalian system, and have shown that the effect of X-irradiation on Ehrlich ascites cells in vivo was to increase that incorporation of ^{3}H-thymidine triphosphate into the isolated nuclei to a marked extent. Conversely, LYNCH et al. (1970) using nuclei isolated from irradiated regenerating rat liver cells reported much reduced DNA synthesis.

The availability of a subcellular fragment which apparently continues its specific biological activity outside of the cytoplasmic environment and contains in close juxtaposition the macromolecules central to cell function, provides an ideal system to determine the location and nature of radiation damage responsible for cell death.

Chapter 3

Modification of Radiation Damage

3.1 Introduction

In the preceding chapter the numerous radiation-induced alterations in the DNA molecule and their biological consequences have been discussed in detail at various levels. It is the purpose of this chapter to describe different ways by which radiation damage can be modified, resulting either in sensitization or protection. It has, however, not been attempted to review the wide field of all substances tested and reported in the literature. For further information the readers' attention is directed to a series of detailed papers and reviews (HOWARD-FLANDERS, 1961; NAKKEN, 1965; BRIDGES, 1969; HAGEN, 1972; FRANKE, 1972; LOCHMANN and MICHELER, 1973; HUTCHINSON, 1973; BLOK and LOMAN, 1973; SZYBALSKI, 1974). However, some of the best known compounds may serve as good models for discussion of the basic mechanisms. The discussion will be limited essentially to a consideration of the effects of ionizing radiation on the nucleic acid of bacteriophage, but one has to keep in mind that such studies on phage do not necessarily give a realistic picture of the in vivo situation in a cell or even in a macroorganism.

The aim of experiments carried out in the field of radiosensitization and radioprotection must ultimately be a formulation of a radiobiological basis for radiotherapy, i.e., the clinical applicability in the radiotherapy of tumors. More than ever "the Zeitgeist demands that biomedical reseach be more consumer-oriented" (ALPER, 1974). Consequently, much more basic and quantitative information is urgently needed before the radiotherapist can take over new methods from the radiobiologist.

Chapter III.4 will be mainly devoted to post-irradiation cellular repair of damaged DNA. Therefore the main subject of this section is rather a discussion based on information available on the mechanism of action of some radiosensitizers and radioprotective substances at the molecular level.

3.2 Factors Enhancing Radiosensitivity of DNA

The factors referred to here are phase state, dehydration, temperature and oxygen, the substitution in DNA of halogenated DNA-base analogs, as well as the effects of some other chemical compounds which increase the radiation damage if they are present during irradiation.

3.2.1 Phase State, Dehydration, and Temperature

In addition to primary radicals from the irradiated water, secondary radicals induced directly or indirectly in organic material must be considered as possible agents which might attack DNA irradiated in solution (see II.1). Recently informative results have been derived from experiments on infectious DNA of phage ϕX 174 irradiated in a defined amino acid solution instead of a complex medium (DE JONG et al., 1972). There is little point in giving a review of these primary and secondary radiation-induced reactions in DNA, since this subject has recently been treated by BLOK and LOMAN (1973).

It has often been assumed that the conditions of "frozen" and "dried" biological material were equivalent for radiobiological purposes. This is definitely not true. Table 3.1 gives the results of a series of experiments with coliphage T1 and varying physical parameters. Figure 3.1 shows the influence of temperature on the

Table 3.1. D_{37} of ^{60}Co-gamma-irradiated normal coliphage T1 and 5-bromouracil (BU) substituted T1 under different experimental conditions.

Phage	Conditions		D_{37} (krad)	Phage	Conditions		D_{37} (krad)
	Suspension	300 K	90		Suspension	300 K	55
	Vacuum, dried	300 K	380		Vacuum, dried	300 K	170
	Ice	80 K	625				
	Vacuum, dried	80 K	1080		Vacuum, dried	80 K	580
T1	Suspension CSH	300 K	310	BU-T1	Suspension CSH	300 K	280
	Vacuum, dried CSH	300 K	700		Vacuum, dried CSH	300 K	300
					Vacuum, dried CSSC	300 K	485
					Ice CSH	80 K	510
	Vacuum, dried CSH	80 K	1570		Vacuum, dried CSH	80 K	1010
					Vacuum, dried CSSC	80 K	1300

Material g/ml suspension: 10^{-6} phage, 0.04 Difco nutrient broth, 0.02 cysteamine (CSH), and 0.04 cystamine·2HCl (CSSC), respectively.

radiation effects in experiments with dried infectious DNA of coliphage ϕX 174. Some authors assume that the so-called indirect effects of ionizing radiation, i.e., damage produced by diffusible agents generated by water radiolysis are excluded if irradiation is carried out in media containing about 4% of protein or amino acids. However, the results given in Table 3.1 clearly indicate that careful freeze-drying enhances protection by an additional factor of 4. Partial desiccation always occurs if freezing is performed slowly. Fast freezing results in a reduction of the diffusion of active species in ice. Clearly the protection by desiccation is partly due to a reduced production of toxic agents as a result of water radiolysis inside and outside of phage and free DNA respectively. However, drying the DNA may also result in less

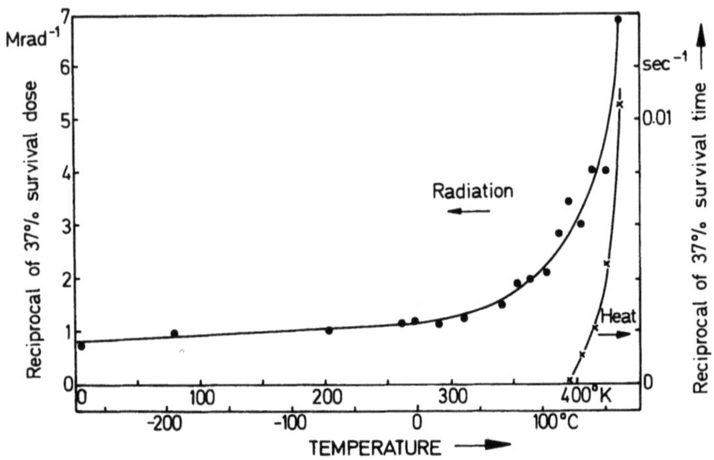

Fig. 3.1. Reciprocal of 37% survival dose and time, respectively, vs. temperature of dry ϕX 174 infectious DNA after ^{60}Co-gamma-irradiation in vacuum (HOTZ and MÜLLER, 1968)

damage to the biomolecule because it alters the course of reaction of DNA-radicals formed by the direct effect of ionizing radiation. This would also explain the additional protection achieved by decreasing the temperature of dried DNA to 80 K during irradiation. The reaction model has been extended (HOTZ and ZIMMER, 1963) by the "Frank-Rabinovich effect". This process results in a cage effect which prevents splitting and favors combination of the radical pair of the biomolecule (e.g., DNA).

A series of experiments with freeze-dried infectious DNA from phage ϕX 174 (HOTZ and MÜLLER, 1968) shows the influence of temperature between 4.2 and 433 K (160°C) and ionizing radiation (^{60}Co-γ) on the biological activity (infectivity) of the DNA. Several conclusions can be drawn: by contrast with the radiosensitivity of DNA enviloped in protein, the inactivation by radiation of dry ϕX 174-DNA is essentially independent of temperature below 50°C. This effect can be explained by an inactivation of DNA-protein complexes in this range of temperature, which is due to protein damage. Substantial energy transfer by small radicals is usually thought to be restricted to aqueous or other liquid systems and is defined as an indirect radiation effect. Taking the infectivity of phage DNA as the criterion of biological activity we are not able to remove the water completely, i.e., including the water shell tightly bound to the DNA-helix, without destroying the activity. On the other hand, there appears to be some evidence to indicate that atomic hydrogen plays a role in the temperature-dependent part of radiation damage in dry biological material (MÜLLER and DERTINGER, 1968; JUNG and KÜRZINGER, 1968).

However, there are too many as yet unknown important parameters which may influence the results. It has been demonstrated, for instance, that the radiosensitivity of dried DNA may be enhanced not only by heat but also by removing the protein or amino acids from the material (suspension medium) surrounding the DNA molecule.

3.2.2 Oxygen

The effect on cells of oxygen when present during ionizing irradiation has been considered to be a "radiobiological law" resulting in a pronounced enhancement of radiation damage in all biological units tested (ALPER, 1955). Therefore, this phenomenon was called "the oxygen effect". The dose-enhancement factor is usually in the range of 2-3.

This is quite different with DNA. It has been observed normally that purified DNA in buffered solution is protected to some extent and DNA in unpurified solutions sensitized by irradiation in the presence of oxygen. This obscure situation became much clearer by the work of BLOK and co-workers (cf. BLOK and LOMAN, 1973) dealing with dilute and concentrated DNA solutions.

The complex reactions of oxygen with the primary radiation induced species $H^{\cdot}$, $HO^{\cdot}$, and e^-_{aq} as well as the transfer of latent damaged sites to peroxy-radicals in DNA are far from being completely understood. This is rather disappointing since as early as 22 years ago ALPER (1955) showed that oxygen protects phage T1 in buffer by elimination of atomic hydrogen. The influence of sulfhydryl compounds on the oxygen effect was demonstrated some years ago when phage had been exposed intra- and extracellular to X-radiation (VAN DYKE, 1960; HOWARD-FLANDERS, 1960). This problem together with the radioprotective agents will be discussed in III.3.2.

When DNA is irradiated in dilute solution, damage to the bases shows an oxygen enhancement ratio of 2 to 3 (LATARJET et al., 1963; WEISS, 1964). A factor of about 2 to 4 is observed for the production of single and double strand breaks in phage DNA as well as for the formation of terminal phosphate end groups (FREY and HAGEN, 1974). Alkali labile breaks were formed only when oxygen was present during irradiation. However, the same authors showed that damage to the biological activity (infectious phage DNA, priming activity) was not enhanced in the presence of oxygen. It was concluded that the occurrence of critical radiation damage leading to biological inactivation is independent of the presence of oxygen during irradiation of purified DNA and phage particles, respectively.

3.2.3 Halogenated Base Analogs

The reasons for referring mainly to the halogenated analogs is that the sensitization by these compounds is well established and some of the analogs have already been introduced in radiotherapy of neoplastic diseases.

The halogenated analogs of thymine include 5-bromo-, 5-iodo-, and 5-chlorouracil. These compounds can be easily incorporated into the DNA of various biological objects; 5-fluorouracil is not incorporated into DNA, it is however used to block the thymidylate synthetase reaction. Incorporation of the analogs results in a sensitization with respect to UV light, ionizing radiation and radioactive decay. As to the photochemical effects the attention of readers is directed to the comprehensive review of HUTCHINSON (1973). The clinical applications of radiosensitization by the halogenated analogs in ionizing radiation and the molecular background of the underlying reactions have been summarized by SZYBALSKI (1974).

The effect of 5-bromouracil (BU) on the radiosensitivity of phage-DNA gamma-irradiated under various conditions was studied with the help of particles of coliphage T1 (HOTZ and ZIMMER, 1963) and ϕX 174 as well as its single-stranded infectious DNA (HOTZ, 1968). It was shown that sensitization by BU occurs to equal extent under three different experimental conditions: irradiation in protein-rich buffer, irradiation in the dry state under vacuum at room temperature, and at the temperature of liquid nitrogen, respectively. When irradiation of dry T1 phage is performed in the presence of a sulfhydryl compound like cysteamine, the protective effect is not influenced by the incorporation of BU. In the wet system, however, a complete protection by cysteamine against sensitization by BU was observed (Fig. 3.2). It was concluded tentatively that this effect is due to an ordinary chemical reaction, which is possible in the wet system but not in the dry state. Early in these studies measurements using the ESR-method showed that the radiation-induced free radicals were formed at a rate about ten times higher in 5-bromouracil than in thymine (MÜLLER et al., 1963). This finding, together with the "abolishing effect" of cysteamine towards sensitization by BU mentioned above, and other studies on photochemical BU-reactions, led to the design of a model for the sensitizing action of BU (HOTZ and REUSCHL, 1967): After absorption of radiation energy by BU-substituted DNA free radicals are induced at the pyrimidine ring. This results in a uracilyl radical and bromide ions (see II.1). The reaction of the bromide ion is not known at the moment. In the absence of a protective compound containing sulfur (H-donor) the uracilyl radical abstracts a hydrogen atom from a neighboring deoxyribose molecule resulting in damage to the DNA sugar as well as to the phosphate ester bonds. As a result, single-strand breaks occur at these sites. The protective action of an H-donating substance like cysteamine consists in furnishing the hydrogen atom to the uracilyl-radical and thus preventing the destruction of deoxyribose. This hypothesis includes the assumption that the occurrence of uracil in DNA is a nonlethal event. Obviously, radiosensitization by halogenated DNA-base analogs is based on (a) in-

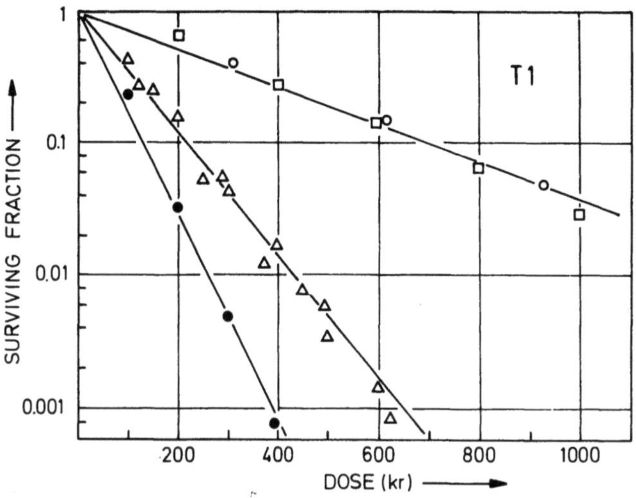

Fig. 3.2. Surviving fraction of the plaque-forming ability of phage T1 after ^{60}Co-gamma irradiation. Phage suspended in anoxic 4% Difco nutrient broth. Control of normal (Δ), and BU-phage (•). Irradiation in the presence of M/10 cysteamine: normal phage (o) and BU-phage (□) (HOTZ and ZIMMER, 1963)

creased radiochemical reaction rate compared with natural bases and (b) the production of irreparable lesions.

3.2.4 Chemical Compounds Acting on Radiation Induced DNA Transients

NEM (N-ethylmaleimide) belongs to a group of chemicals which were shown to be radiosensitizing many years ago. These compounds are known to act on cells by combination with naturally occurring essential cell constituents: the sulfhydryls. Another sulfhydryl binding compound is IAA (iodoacetic acid); it was already described as a sensitizer in 1934. For more general considerations BRIDGES (1969) listed the following hypothetical mechanisms of action for these compounds: (1) Blocking of intracellular protective sulfhydryl groups or (2) cellular repair systems; (3) reaction with transient biomolecular lesions or (4) stabilization of radiation induced radicals, and (5) production of toxic radiolysis products from halogenated sensitizers. However, we shall focus our attention here on pure DNA, which is a biological unit not containing sulfhydryl compounds.

NEM was found to sensitize T-even bacteriophage in the absence of oxygen. Important information on the mechanism of sensitization was contributed by the finding that covalent bonds are formed during the action of ionizing radiation between free DNA in aqueous solution and NEM (JOHANSEN et al., 1968; II.2.2). Evidence has accumulated that at the cellular level also a great number of compounds, including sulfhydryl binding agents, act primarily at an early stage after the energy absorption takes place and lead to a different type of damage in the cell. This may sometimes result in a block to the repair system for radiation damage (cf. BRIDGES, 1969).

Another interesting compound is triacetoneamine-N-oxyl or TAN (EMMERSON and WILLSON, 1968; NAKKEN et al., 1970) which sensitizes infectious DNA of phage T1 (HOTZ, 1973). This compound is a stable free radical combining with DNA in the absence of oxygen via OH-adducts or e^-_{aq} induced radical sites at different DNA subunits. The same reaction should hold for sensitization by other members of this group of nitroxide free radicals.

Some organic compounds like p-nitroacetophenone probably sensitize to the effect of ionizing radiation by electron-trapping from ionized DNA. This reaction may prevent the restitutio ad integrum of biomolecules which otherwise would occur in the absence of the electron trap (sensitizing compound) by recombination of the negatively charged DNA site with the positive charge (ADAMS and COOKE, 1969; cf. ADAMS, 1972). This hypothesis has been a great help in the search for sensitizing compounds which could be of clinical importance. Recently, a number of drugs proved to give rise to a relatively small but significant sensitization of bacterial and mammalian cells in the absence of oxygen. Although some of them, as has been expected from the hypothesis, form adducts with free DNA, we do not intend to review a long list of these drugs. However, because of its exceptionally favorable pharmacological and toxicological properties, metronidazole ("Flagyl") should at least be mentioned as a prototype. All of these compounds belong to the group of nitroimidazoles and nitrofuranes acting according to the mechanism proposed for electron-affinic radiosensitizers and described above for p-nitroacetophenone. It should be stressed again that all these substances are "oxygen mimics" i.e., sensitize under anoxic conditions only. Their lack of interaction with the biomolecules in the presence of oxygen is presumably due to a completion reaction for radiation induced biomolecular transient sites.

3.2.5 Chemical Compounds Combining with DNA Independently of Radiochemical Reactions

Sensitization in many respects is connected to the problem of DNA repair. Several repair systems have been described during the past ten years for different kinds of lesions. We shall try to make an approach to the problem by choosing in the first case some compounds for which mechanisms of action have been suggested. If a chemical blocks cell metabolism completely it will always reversibly inhibit repair processes. Such effects, however, should not be considered in a discussion of "specific repair inhibition".

In III.3.2.3 we have already discussed a group of compounds which can be incorporated into DNA before irradiation replacing a naturally occurring atom or molecule. In III.3.2.4 we mainly discussed chemicals reacting with the DNA molecule during ionizing irradiation, a process resulting in combination with a DNA transient. In this section, however, the compounds mostly have to interact with DNA after irradiation in order to inhibit processes for reactivation of radiation damage.

Atebrin (Quinacrine) has been known for a long time as an effective inhibitor of malaria plasmodia. This compound belongs to the group of acridine dyes and the result of interaction with DNA is probably the inhibition of synthesis of DNA, RNA and protein. The mechanism of action is not precisely known. However, the hypothesis most frequently discussed is "intercalation" (LERMAN, 1961; PRITCHARD et al., 1966). According to this model the planar triple-ring configuration of the organic molecules in question is intercalated between adjacent base pairs of the DNA Watson-Crick double helix.

The action on plasmodia radiosensitization of bacteria has also been observed (FUKS and SMITH, 1971). The underlying mechanism is assumed to be inhibition of single-strand break repair. With mammalian cells the disadvantage of toxicity at doses necessary for radiosensitization probably limits the clinical use of these drugs.

3.3 Sulfhydryl Compounds and Radioprotection by Radical Scavenging

Many radical scavengers have been described in chemical systems. However, not all of these compounds were found to be radioprotectors in biological systems. The reason for this phenomenon is probably the ability of some compounds, e.g., sulfhydryls, to interact not only with products of water radiolysis, i.e., radiation-induced small organic radicals of the solute, but in addition with radical sites on relevant biomolecules of the DNA-type as well as with hydrogen atoms in the dried state (cf. NAKKEN, 1965).

A hypothesis explaining the protective action of compounds of the cysteine-cysteamine group was proposed by HOWARD-FLANDERS for phage (1960). According to this hypothesis, which was extended from thiols to compounds of the disulfide type (e.g., cystamine; HOTZ, 1966) donation of hydrogen from the protector molecule to the radiation-induced bioradical (DNA-radical) results in repair of the damaged site. However, during the last years some information has been published on this subject which makes a complete repair i.e., a restitutio ad integrum rather unlikely. The most probable explanation is that radical damage produced in the presence of a hydrogen donor undergoes restitution reactions. Although hydrogen donated sites obviously preserve their biological activity, the original configuration of the biomolecule will not be restituted in some cases. Radiation-induced phosphate liberation from DNA can also be prevented by hydrogen donation to the corresponding deoxyribose radical.

Summing up this chapter on sensitization and protection of DNA with respect to ionizing radiation we can draw an important conclusion. Obviously, the basic mechanism for both reactions is closely related for many compounds so that protection, a reductive process or repair, means donating electrons to a particular biomolecule and sensitization, an oxidative process, means abstracting electrons from it.

Chapter 4

Repair Processes for Radiation-Induced DNA Damage

4.1 Introduction

DNA inside cells can be damaged by ionizing or nonionizing radiations or by a variety of chemicals. An accumulation of such damage would result in a loss of genetic information and cell death. Cells have evolved several mechanisms by which the damaged DNA can either be restored to its intact state or circumvented during DNA replication, these processes being collectively termed "DNA repair". Although evidence that cells could recover from induced damage had been available for many years, the first direct demonstration of repair at the molecular level came from the discovery by SETLOW and CARRIER (1964) that *Escherichia coli* cells could excise UV-induced pyrimidine dimers from their DNA. Since then many reports have been produced describing several repair processes acting in a variety of cells on damage produced in DNA by many different agents. This chapter will describe the current state of knowledge on the repair of damage induced by ionizing radiation. Repair of other types of damage will be referred to only for comparative purposes. Only direct biochemical and biophysical demonstrations of repair processes will be described, as opposed to evidence based on genetic experiments and survival curves.

As discussed in detail in II, the principal lesions induced by ionizing radiation in the DNA inside cells are chemical damage to the purine and pyrimidine bases and to the deoxyribose sugar, and physicochemical damage resulting in a break in the phosphodiester backbone in one strand of the molecule (single-strand break) or in both strands at the same place (double-strand break). As yet neither intra nor interstrand crosslinks nor DNA-protein crosslinks have been detected as a consequence of X- or gamma-irradiation of whole cells. Both procaryotic and eucaryotic cells are capable of repairing many of these different lesions.

There have been several recent reviews dealing with various aspects of DNA repair after exposure to ionizing radiation: repair in general (SETLOW and SETLOW, 1972; BLOK and LOMAN, 1973), repair of strand breaks in bacteria (TOWN et al., 1973) and in mammalian cells (ORMEROD, 1973, 1975), repair in mammalian cells (CERUTTI, 1974a) and repair of base damage (CERUTTI, 1974b). In addition a multi-author book has recently been published covering all aspects of repair after exposure to chemicals, nonionizing and ionizing radiations (HANAWALT and SETLOW, 1975).

4.2 Repair of Single-Strand Breaks in Bacteria

4.2.1 Measurement of Single-Strand Breaks

By far the most thoroughly studied lesion induced by ionizing radiation in the DNA inside cells is the single-strand break (SSB), resulting from cleavage of the phosphodiester backbone of one of the strands of the DNA double helix. Such breaks result in a decrease in the molecular weight of the single DNA strands, without affecting the size of the molecule in its native form. The number of induced single-strand breaks per dalton is given by $1/M_n - 1/M_o$, where M_o and M_n are the number average M.W. of the single-stranded DNA molecules before and after irradiation (see III.2.2.2). Early measurements of SSBs were beset with problems, because many spurious breaks were introduced into the DNA by mechanical shearing during its extraction from the cell. These difficulties were overcome by the revolutionary discovery by McGRATH and WILLIAMS (1966), that conventional extraction of the DNA was unnecessary. They found that bacterial cells could be lysed in strong alkali layered on top of an alkaline sucrose gradient in a centrifuge tube, and free DNA single strands could be sedimented on subsequent centrifugation. Thus cell lysis, removal of associated molecules and denaturation of the DNA (by the alkali) could be accomplished on top of the sucrose gradient without any handling of the DNA. The M.W. distribution of the DNA strands could be determined from the sedimentation profiles obtained after centrifugation of the DNA, using previously determined relationships between rate of sedimentation and M.W. (BURGI and HERSHEY, 1963; STUDIER, 1965; ABELSON and THOMAS, 1966). The McGrath and Williams' procedure has formed the basis for the majority of subsequent measurements of SSBs in many different cells. It should be pointed out that these SSBs, measured under alkaline conditions, are the sum of true SSBs and other induced lesions which are converted into breaks by the alkaline treatment (see III.4.5.3). Furthermore, double-strand breaks (DSBs) are scored as two SSBs.

With their technique McGRATH and WILLIAMS were able to measure the number of SSBs introduced by X-irradiation in the DNA of *Escherichia coli*. An alternative technique for measuring SSBs in certain plasmid and phage DNAs which exist as double-stranded supercoiled closed circular molecules has been discussed in detail in III.2.1 and 2.2.

This technique has been used to measure the production of SSBs in superinfecting lambda phage DNA molecules (JOHANSEN et al., 1975).

Current estimates for the initial yield of SSBs produced by ionizing radiation in chromosomal DNA and superinfecting lambda phage DNA in *Escherichia coli* are about 3-10 breaks/10^{12} dalton/rad (10-30 eV/break) for oxic irradiation (see Table 1 in JOHANSEN et al., 1975; YOUNGS and SMITH, 1976) and about one-third this amount for anoxic irradiation (JOHANSEN et al., 1971; PATERSON et al., 1973). As discussed in detail below, many earlier estimates (McGRATH and WILLIAMS, 1966; BOYCE and TEPPER, 1968; TOWN et al.,

1973) were in error because of undiscovered rapid repair processes, incorrect interpretation of results, or inaccurate calibration of sedimentation data.

McGRATH and WILLIAMS showed that on incubation of *Escherichia coli* B/r cells after irradiation the molecular weight of the DNA strands increased, reaching the size of that from unirradiated controls after 40 min, indicating that the X-ray-induced breaks could be rejoined in the cell. This rejoining process appeared to be under genetic control, since it took place in the radioresistant strain B/r but no rejoining was observed in a radiosensitive strain B_{s-1}.

Studies aimed at elucidating the mechanisms of rejoining of SSBs in bacteria have been principally involved with investigations of the dependence of the rejoining processes on post-irradiation conditions and with the genetic characterisation of the various rejoining processes by the use of radiosensitive mutants which are partially deficient in their ability to rejoin SSBs.

4.2.2 Different Rejoining Processes

SSBs measured by physical techniques represent in chemical terms a heterogeneous mixture of breaks which can have various end groups. As discussed in II.3, a single-strand break may arise as a result of

1. direct scission of a phosphodiester bond, leaving either 5'-PO_4 and 3'-OH ends, or 5'-OH and 3'-PO_4 ends;

2. destruction of the deoxyribose sugar;

3. depurination or depyrimidination following damage to a base;

4. the production of an alkali-labile lesion;

5. a scission introduced enzymatically subsequent to irradiation.

Unfortunately there have been no reported attempts to measure the different end groups obtained after irradiation of bacteria, although such experiments have been carried out on DNA irradiated in vitro. Because of the differing environment between DNA in solution and inside a bacterium, it is unwise to extrapolate the in vitro data to the situation inside a cell.

Different enzyme systems might be required to carry out rejoining of breaks with different end groups. Thus a break with 5'-PO_4, 3'-OH end groups could be rejoined directly by polynucleotide ligase, whereas, for example, a damaged sugar would need to be removed and replaced before sealing of the break could occur. Although there has as yet been no correlation with the chemistry of the end groups involved, it now seems clear that there are at least two and probably more different mechanisms for rejoining SSBs. The one first discovered by McGRATH and WILLIAMS (1966) is now termed slow repair; as its name suggests, it is relatively slow, affected by post-irradiation conditions and under the control of several genetic loci, indicating that it is a complex process involving several different enzymatic steps. The second process (fast repair) is very fast, much less dependent on post-irradiation conditions and is under the control of only a small number of genetic loci, indicating that it is simpler than the slow repair process. Because of the way in which the experiments of McGRATH and WILLIAMS (1966) were carried out, the fast repair process was not seen on account of its rapidity. Both the slow and fast repair processes may have to be further subdivided, as more data are brought to light.

4.2.3 Slow Repair in *Escherichia coli*

Unless special precautions are taken to inhibit or prevent the rapid repair reaction (see below) it occurs during the irradiation and subsequent handling procedures and only "slow repair" is seen. This section will deal with the slow repair process and it should be borne in mind by the reader that work described relates to the rejoining of that proportion of the induced strand breaks which have not been rejoined by the rapid repair process.

4.2.3.1 Genetic Control

After the original findings of McGRATH and WILLIAMS, it was shown that *Escherichia coli* could also rejoin SSBs in K-12 strains (KAPLAN, 1966), and in superinfecting lambda phage DNA (OGAWA and TOMIZAWA, 1967; BOYCE and TEPPER, 1968).

During the last decade various mutant strains of *E. coli* have been isolated, which, among other properties, have increased sensitivity to the action of UV-light and ionizing irradiation as compared to their wild-type parent strains; some of these are listed in Table 4.1. The existence of mutants sensitive to both ultra-violet and ionizing radiations suggests that some common steps are involved in repair of lesions produced by these different types of radiation. The mutations in many of these strains affect the rejoining of SSBs induced by ionizing radiation.

The strains uvrA and uvrB are more sensitive to UV light than the parent strain, but they have normal X-ray sensitivity and normal ability to rejoind SSBs (KAPP and SMITH, 1970). It is now known that $uvrA^+$ and $uvrB^+$ genes control the production of a UV-specific endonuclease, which specifically nicks DNA in the vicinity of UV-induced pyrimidine dimers (BRAUN and GROSSMAN, 1974). This enzyme does not attack X-irradiated DNA and is not involved in the repair of single-strand breaks.

The strains recA, recB and recC were originally isolated for their reduced ability to undergo genetic recombination, this deficiency being more pronounced in recA than in recB and recC strains. The $recB^+$ and $recC^+$ genes code for the ATP-dependent exonuclease-V. The nature of the $recA^+$ gene product is not yet known. In contrast to the uvrA and uvrB strains, the rec strains are sensitive to both ionizing and UV-irradiation, the recA strains being somewhat more sensitive than recB and recC strains. Correlating with these

Table 4.1. Some important genes and gene products involved in DNA repair in *E. coli* (see TOWN et al., 1973 for further details)

Mutant designation	Gene product	Sensitivity[a] to UV	to ionizing radiation	Other important properties
uvrA, uvrB	UV-specific endonuclease	s	r	
recA	?	s	s	Recombination-deficient
recB, recC	Exonuclease V	s	s	Recombination-deficient
lexA(exrA)	?	s	s	Non-mutable by many agents
polA(resA)	DNA polymerase I	s	s	
polB	DNA polymerase II	r	r	No definite function established
polC(dnaE)	DNA polymerase III	lethal under restrictive conditions - enzyme involved in DNA replication		
lig	DNA ligase	lethal under restrictive conditions - enzyme involved in DNA replication		

[a] s: sensitive; r: resistant (same as wild-type).

increased sensitivities, it was found that recA strains could not repair SSBs after X-irradiation (KAPP and SMITH, 1970), whereas rejoining in recB and recC strains was only partially inhibited (KAPP and SMITH, 1970). The interpretation of the results with recA strains was complicated by the excessive DNA degradation seen in these strains after X-irradiation (see III.4.2.6) which could clearly interfere with any possible rejoining of breaks. However, in a recA recB double mutant, very little post-irradiation degradation took place, yet rejoining of strand breaks was still inhibited (KAPP and SMITH, 1970). In contrast to the findings of KAPP and SMITH (1970) with chromosomal DNA, BOYCE and TEPPER (1968) found that rejoining of SSBs in DNA from superinfecting lambda phage was only partially inhibited in a recA host and not affected in a recB host.

The lexA (or exrA) strains of *E. coli* were isolated for their high sensitivity to X-irradiation. These strains are also nonmutable by UV, X-rays and a variety of chemicals. The strain B_{s-1} used in the original experiments of McGRATH and WILLIAMS (1966) was in fact a triple mutant, one of the mutations being at the exrA locus. It was this mutation which conferred sensitivity to X-rays and inability to rejoin SSBs. SEDGWICK and BRIDGES (1972) showed that in various exrA strains only a small amount of rejoining of SSBs took place, and there was massive degradation of the DNA after irradiation (as in recA strains). On the other hand, if exrA cells were treated with chloramphenicol for 90 min prior to irradiation, so that both protein synthesis and DNA replication had ceased at the time of irradiation, SSB rejoining was normal, no degradation took place and the viability of the cells was greatly increased as compared with cells not exposed to chloramphenicol. It thus appears that the exrA$^+$ gene product is necessary for SSB rejoining only in the presence of an active DNA replication fork and/or protein synthesis (GREEN et al., 1973).

Slow rejoining also seems to require the intact polC$^+$ (SMITH, 1974) and dnaB$^+$ genes (TOWN et al., 1973).

4.2.3.2 Effects of Post-Irradiation Conditions

Slow repair occurs only in growth medium at 37°C after irradiation, and can be inhibited by various drugs. Thus, quinacrine, acriflavine and dinitrophenol irreversibly inhibited rejoining and also sensitized wild-type bacteria to X-rays, while not affecting recA strains (TOWN et al., 1973). This implies that the sensitization is a direct result of the inhibition of the recA$^+$-mediated rejoining process. Chloramphenicol, if present for 90 min before and after irradiation did not inhibit the rejoining process (GREEN et al., 1973), whereas if present only after irradiation some inhibition was seen (GANESAN and SMITH, 1972).

In summary, slow repair seems to require full growth conditions and the presence of several different enzymes.

4.2.4 Fast Repair in *E. coli*

The discovery of a fast repair process in *E. coli* (TOWN et al., 1971) was made with strains of *E. coli* deficient in DNA polymerase I (polA). DE LUCIA and CAIRNS (1969) isolated such a strain (polA1). It was viable, but had increased sensitivity to UV light, X-rays and alkylating agents. TOWN et al. (1971) found that if *E. coli* were X-irradiated at 25°C at pH 6.9 (i.e., under conditions similar to those used for most of the studies described in III.4.2.3), approximately three times as many breaks were observed in the polA1 strain as in the parental strain. On the other hand if the cells were irradiated at pH 8.0 at 0°C, the same number of breaks were found in polA1 and polA$^+$ cells, this number being approximately the same as that in the polA1 strain at the higher temperature. This suggested that in polA$^+$ cells there was a rapid repair process which had sealed about 70% of the breaks at 25°C during the irradiation and handling procedures. This rapid repair was dependent on the presence of DNA polymerase I. In confirmation it was found that if polA$^+$ strains were irradiated at 0°C and pH 8.0, followed by incubation in buffer at room temperature, rapid rejoining of many of the SSBs did indeed take place. This fast repair, which has been observed also with superinfecting lambda phage DNA (JOHANSEN et al., 1971) probably involves only limited base removal and replacement by DNA polymerase I, followed by strand closure mediated by polynucleotide ligase. In the polA mutant or under conditions when the polymerase I-dependent process was inhibited (TOWN et al., 1973) the rapid process did not take place. However a much slower rejoining was observed in buffer at 37°C. This required DNA polymerase III, since it did not occur at the restrictive temperature in a polA polC^{ts} double mutant (YOUNGS and SMITH, 1973a) and was distinct from the slow medium-dependent repair discussed in the previous section.

After irradiation at 0°C, to prevent the occurrence of fast repair, about three times as many breaks were produced by irradiation in oxic conditions as in anoxic conditions in polA1 strains (TOWN et al., 1973). However, either pre-treatment at 52°C for 5-10 min or cold shock before irradiation increased the number of breaks such that almost as many breaks were seen after irradiation in N_2 as in O_2. These results were originally interpreted as evidence for an ultra-fast repair process which sealed 80% of the breaks produced under anoxia, even at 0°C and in the absence of DNA polymerase I (TOWN et al., 1973). It now seems that this interpretation was incorrect. JOHANSEN et al. (1975) irradiated *E. coli* or superinfecting lambda phage with 4 MeV electrons, delivering the radiation dose and lysing the cells within 100 ms, a time period during which a significant amount of repair could not have taken place. The three-fold greater yield of breaks produced in air as opposed to anoxia was still observed, suggesting a genuine difference in the number of breaks produced, and arguing against the existence of ultra-fast repair. The pre-treatments used by TOWN et al. (1973) probably caused changes in the physiological state of the cells, so that the radiation chemistry rather than the repair capacity of the cells was altered (JOHANSEN and BOYE, 1975).

Of the breaks produced without pre-treatments, current estimates suggest that over 90% can be rejoined in buffer at room temperature by the fast polA$^+$-dependent process, and a few additional breaks can be subsequently sealed by the slow repair process discussed in III.4.2.3 (YOUNGS and SMITH, 1976).

There thus seem to be at least two different enzyme systems which seal SSBs in *E. coli* - one requiring DNA polymerase I and presumably DNA ligase, and another requiring the recA$^+$ and various other gene products. It is of interest in this respect that recA$^-$polA$^-$ strains are inviable (GROSS et al., 1971; MONK and KINROSS, 1972) indicating that some ability to rejoin SSBs may be required for the cell to survive, even in the absence of ionizing radiation. On the other hand, a polA exrA double mutant is viable, but more sensitive than either single mutant. It can rejoin a limited number of breaks by a slow buffer-repair system (YOUNGS and SMITH, 1973a).

4.2.5 Rejoining of SSBs in Other Bacteria

The production and rejoining of single-strand breaks has also been shown to occur in *Haemophilus influenzae* (SETLOW et al., 1968) and in *Bacillus subtilis* (TERANO et al., 1969).

At the same time as the work with *E. coli* was being carried out, repair processes in the highly radioresistant bacterium, *Micrococcus radiodurans* were being investigated in detail, principally by DEAN and collaborators. They also concluded that there were several different rejoining processes, based on the differing extents of rejoining under different post-irradiation conditions. Their suggestion of an ultra-fast repair process, inhibitable by EDTA (DEAN et al., 1969) to account for the difference in breakage efficiency after irradiation under O_2 and N_2 is probably incorrect for reasons discussed above (see III.4.2.4). They showed, however, that of the breaks produced under oxygen or anoxia, more than 50% could be rapidly rejoined in 5 min in buffer, the rest being rejoined more slowly under full growth conditons, requiring protein synthesis (DEAN et al., 1970; DRIEDGER and GRAYSTON, 1971) findings which are similar to those subsequently reported for *E. coli* (see III.4.2.4).

4.2.6 Post-Irradiation Degradation

It has long been known that after exposure to ionizing irradiation, exonucleolytic degradation of DNA occurs, presumably from ends of broken strands (POLLARD, 1970). This degradation will only be considered briefly here. It complicates interpretation of strand-rejoining data, since a failure to rejoin strand breaks may be a consequence of this degradation, or alternatively degradation may occur from unrejoined strand breaks (e.g., GREEN et al., 1973). Excessive degradation is seen in *E. coli* in recA, exrA and polA strains (MORIMYO et al., 1968; KATO and KONDO, 1970; PATERSON et al., 1971; SEDGWICK and BRIDGES, 1972; YOUNGS and BERNSTEIN, 1973) and seems to be carried out by the recBC-mediated exonuclease V since its extent is reduced in recB and recC trains (EMMERSON, 1968) and in double mutants containing either recB or recC and either recA, exrA or polA (KAPP and SMITH, 1970; SEDGWICK and BRIDGES, 1972; YOUNGS and BERNSTEIN, 1973). The recA recB double mutant is, however, as sensitive to X-rays as the recA single mutant. These results suggest that some degradation mediated by exonuclease V is essential for recovery from X-ray-induced damage, but, if uncontrolled, may lead to rampant degradation and cell death. Evidence obtained with both *M. radiodurans* and *E. coli* suggests that an inducible inhibitor, synthesized after irradiation, may halt the degradation process. The radioresistance of *M. radiodurans* may lie in the efficiency of the inhibitor in this organism (DEAN et al., 1970).

The above brief description is an oversimplification of the degradative processes, which have been studied in detail by several groups, especially that of POLLARD and co-workers (e.g., see POLLARD and RANDALL, 1973; MARSDEN et al., 1974) but it serves to indicate its importance and interrelationship with other repair processes.

4.3 Rejoining of Single-Strand Breaks in Eucaryotic Cells

4.3.1 Introduction

Soon after the demonstration of rejoining of SSBs in *E. coli* by McGRATH and WILLIAMS (1966), their technique was applied to mammalian cells by LETT et al. (1967) who showed that cultured mouse lymphoma cells (L5178Y) were also capable of rejoining SSBs. Similar findings were subsequently reported for a variety of other mammalian cell lines. However, DNA from unirradiated mammalian cells does not sediment as free DNA molecules, presumably on account of its enormous size (10^{11} dalton of DNA per average chromosome). Interpretation of sedimentation data is therefore rendered difficult. Again, at the outset, I shall summarize the main findings on rejoining of SSBs in mammalian cells, in the hope that the reader will have a better understanding of the problems discussed. There is general agreement that single-stranded DNA molecules smaller than 2×10^8 dalton sediment according to theory (BURGI and HERSHEY, 1963; STUDIER, 1965; ABELSON and THOMAS, 1966). Such molecules are obtained after radiation doses in excess of 5 krad and behave exactly as predicted. Breaks are produced at random by ionizing radiation and are rejoined on subsequent incubation. In the M.W. range $2\text{-}5 \times 10^8$ dalton (obtained by doses of 1-5 krad) DNA molecules sediment anomalously, but under appropriate conditions will behave according to theory. DNA molecules greater than 5×10^8 dalton (obtained after doses of 0-1 krad) do not sediment as free single-stranded DNA molecules. Such DNA will be termed "complexed DNA". The complexed DNA can be disrupted by irradiation, eventually producing free DNA molecules. During post-irradiation incubation complexed DNA is reformed. Most workers agreed on these experimental observations. However, for the last five years a controversy has been raging concerning the nature and significance of this complexed DNA and the mechanism of its disruption. The controversy is concerned with interpretation of experimental data rather than with the actual data obtained. Discussion of various aspects of strand breaks in mammalian cells can be found in ORMEROD (1975).

Post-irradiation exonucleolytic degradation as seen in bacteria (see III.4.2.6) only occurs to a very limited extent in mammalian systems (e.g., HILL, 1969).

4.3.2 End Groups Produced

The comments concerning the nature of SSBs produced in bacteria in III.4.2.2 are equally applicable to those produced in DNA of mammalian cells. There have been a few attempts to determine the nature of the end groups produced in vivo by ionizing radiation. Thus COQUERELLE et al. (1973) and LENNARTZ et al. (1975) irradiated rat thymocytes with very high doses of gamma rays and extracted the DNA. Using a variety of techniques they were able to measure the total number of SSBs, the number of alkali-labile bonds (see III.4.5.3), the 5'OH end groups, the 5'-PO_4 end groups and the 3'-OH end groups. They concluded that at the 5'-end 12% of the true breaks had 5'-OH end groups, 25% had 5'-PO_4 end groups the remaining 63% having neither, and at the 3'-end 65% of the breaks had a 3'-OH end group. These results were completely different from the types of end groups produced by irradiating dry DNA or DNA in solution.

4.3.3 Efficiency of Break Production

SSBs are produced with an efficiency of about 50 eV/break in all cell lines studied (see Table 1 in ORMEROD, 1975). This corresponds to about 2 breaks/10^{12} dalton/rad. There is fairly good agreement between various groups using different cell lines, the differences in efficiencies obtained probably reflecting differences in dosimetry and in calibration of the sedimentation data as much as genuine differences. Some groups have found about one-third as many breaks after anoxic as compared with oxic irradiation (as in bacteria, - TOWN et al., 1973; JOHANSEN et al., 1975), whereas others measured the same number of breaks under oxygen and nitrogen (see ORMEROD, 1975). The discrepancy may be due to differences between cell lines (ORMEROD, 1975).

4.3.4 Rejoining after High Doses in Cultured Cells

4.3.4.1 Rejoining in Different Cells

There is complete agreement by numerous workers using a variety of cultured mammalian cell lines that high doses of ionizing radiation reduce the molecular weight of the DNA strands, and it increases on subsequent incubation, implying that the radiation-induced SSBs are rejoined. About 90% of the breaks are rejoined within one hour of post-irradiation incubation at 37°C (see ORMEROD, 1973, 1975). This is considerably faster than, for example, the excision of UV-induced pyrimidine dimers in cultured human fibroblasts, which takes several hours (REGAN et al., 1968). Recently indications of more than one type of repair process have been reported and, in view of the different end groups produced by radiation, several repair processes might be expected, as in bacteria. However, the lack of repair-deficient mutants in mammalian cells makes unambiguous identification of different pathways rather difficult.

Rejoining of SSBs has now been demonstrated in cultured cell lines derived from human, hamster, mouse, rat and pig tissues (see ORMEROD, 1975); it occurs at the same rate in all phases of the cell cycle (HUMPHREY et al., 1968; LOHMAN, 1968; SAWADA and OKADA, 1970; LETT and SUN, 1970). Breaks could be rejoined in vivo in Ehrlich ascites tumour cells inside live mice but not in dead mice (MATSUDAIRA et al., 1969a).

In other eucaryotic systems rejoining of SSBs has been demonstrated in the slime mould *Physarum polycephalum* (BREWER and NYGAARD, 1972), in the alga *Chlorella* (CABELA et al., 1971) and in carrot protoplasts (HOWLAND et al., 1975).

4.3.4.2 Effects of Inhibitors

Many groups have investigated the requirements of the rejoining process for macromolecular synthesis and energy production by the use of metabolic inhibitors. The following have been found to be without effect: 0.1-2 µg/ml actinomycin D, which inhibits RNA synthesis; hydroxyurea, excess thymidine, mitomycin C and fluorodeoxyridine which all inhibit DNA synthesis; puromycin and cycloheximide, inhibitors of protein synthesis (see SETLOW and SETLOW, 1972; ORMEROD, 1973, 1975). Thus breaks can be rejoined in the absence of macromolecular synthesis, implying that the rejoining enzymes are always present in cultured cells, and are not induced by the radiation treatment.

4.3.4.3 Energy Requirements

Both the rate and extent of rejoining were reduced at low temperatures (MATSUDAIRA et al., 1969b; SAWADA and OKADA, 1970; ORMEROD and STEVENS, 1971), whereas raising the temperature to 42°C or 43°C initially had little effect, but subsequently led to endonucleolytic breakdown of the DNA (ORMEROD and STEVENS, 1971; BEN-HUR and ELKIND, 1974).

Initially rejoining could occur in buffer or growth medium (MATSUDAIRA et al., 1969b; ROOTS and SMITH, 1975), and low doses of dinitrophenol, an uncoupler of oxidative phosphorylation did not inhibit rejoining if added after low radiation doses (SAWADA and OKADA, 1970; ORMEROD and STEVENS, 1971). However pre-treatment with dinitrophenol or KCN to deplete the endogenous pool of ATP led to an inhibition of rejoining (PALCIC and SKARSGARD, 1972b; KLEIJER et al., 1973). Furthermore, MATSUDAIRA and co-workers found that rejoining was inhibited if cells were incubated at 22°C under anoxia in buffer at high cell concentrations, conditions in which the ATP supply in the cell was reduced to 17% of its normal level. Addition of glucose restored the ATP level and overcame the inhibition of rejoining (MATSUDAIRA et al., 1969b; MATSUDAIRA et al., 1970). Thus a persistent energy supply seems to be required for rejoining of SSBs.

In isolated nuclei a limited amount of rejoining of SSBs has been demonstrated, the process requiring ATP, NAD and some cytoplasmic factors (MATSUDAIRA and FURUNO, 1972).

4.3.5 Studies Using Low Doses (< 5 krad) of Irradiation

4.3.5.1 Rotor Speed Anomalies

If cells are exposed to a series of decreasing doses of ionizing radiation and the DNA sedimented in alkaline sucrose gradients, the following phenomena have been observed. As the dose was reduced from, say, 100 krad down to 5 krad, the DNA gradually increased in sedimentation rate and the size and distribution of the molecules were as predicted for the introduction of a decreasing number of random breaks into long molecules. After lower doses (1-5 krad) and centrifugation of the DNA at high rotor speeds (40,000 rpm), deviation from this behaviour was observed. The sedimentation rate did not change appreciably and the distribution became nonrandom, with a sharp edge at the high M.W. side of the profile (also seen with DNA from bacteria). Several groups have suggested that this phenomenon is probably the result of a sedimentation anomaly, which has been mathematically analysed by ZIMM (1974) and is also found with double-stranded DNA in neutral gradients, as discussed in III.4.4.1. The result of this anomaly is that, for any given rotor speed, all molecules greater than a critical size have the same rate of sedimentation, this critical size being lower at higher rotor speeds. Thus it has been found that in the dose-range 1-5 krad (M.W. of DNA 1-5 × 10^8) if the DNA was centrifuged at low rotor speeds (about 10,000 rpm) the DNA molecules did sediment as predicted by theory (ELKIND, 1971; McBURNEY et al., 1971; ORMEROD and LEHMANN, 1971a; PALCIC and SKARSGARD, 1972a, b). It should be pointed out, however, that WHEELER et al. (1974) dispute this interpretation of the data obtained. They suggest that it is the low rotor speed data which are inaccurate, on account of convective and other disturbances occurring during the long centrifugation times employed.

4.3.5.2 Definition of Complexed DNA

If the radiation dose administered is less than 1 krad theoretical behaviour has not been observed at any rotor speed. Instead there is a loss of DNA from the freely sedimenting fractions and the appearance of a peak (or peaks) of complexed DNA whose sedimentation properties are different from those of the free DNA strands. The term "complexed DNA" is used here only as a means of distinguishing it from free DNA strands without implying any particular kind of structure. The sedimentation characteristics of this complexed DNA are extremely dependent on the particular experimental conditions used (lysing agents, lysis time, temperature, pH of gradients, ionic strength, rotor speed etc.). In all cases there is an abrupt transition in sedimentation characteristics between complexed and free DNA. The transition complexed → free DNA occurs with increasing doses of X-rays in the 0-1 krad range, and can be reversed by post-irradiation incubation (LETT et al., 1970; ELKIND and KAMPER, 1970; ORMEROD and STEVENS, 1971; ELKIND and CHANG-LIU, 1972a; McBURNEY et al., 1972; BELLI et al., 1972). This reconversion of free to complexed DNA is presumably a manifestation of some type of repair.

4.3.5.3 Properties of Complexed DNA

The complexed → free DNA transition can be effected either by irradiating cells (LETT et al., 1970; ELKIND and KAMPER, 1970; ORMEROD and STEVENS, 1971; McBURNEY et al., 1972) or by incubating cells in actinomycin D (ELKIND, 1971; ELKIND and CHANG-LIU, 1972b). Alternatively it may be brought about by lysing cells in alkali on top of sucrose gradients for extended periods of time, this process being accelerated by EDTA, visible light and increased temperature (LETT et al., 1970; ELKIND and KAMPER, 1970; ELKIND, 1971; SIMPSON et al., 1973) or by substitution of some of the thymidine residues of the DNA with bromodeoxyuridine (SHIPLEY and ELKIND, 1971).

Despite the alkaline conditions, it has been demonstrated that complexed DNA obtained under several different conditions is in fact double-stranded (AHNSTRÖM and ERIXON, 1973; SIMPSON et al., 1973; JOLLEY and ORMEROD, 1974; CLEAVER, 1974a) as originally suggested by McBURNEY et al. (1972).

4.3.5.4 The Subunit Hypothesis

In 1970 the hypothesis was put forward (LETT et al., 1970) that the discrete monodisperse peak of free DNA obtained after irradiating cells with 1 krad of irradiation or by lysing cells for a few hours in alkaline EDTA (the 165 S subunit) was a subunit of DNA structure (size 5 × 10^8 dalton), and that the complexed DNA represented arrays of these subunits joined together by links or linkers of an unknown nature. Conversion of complexed DNA to free DNA by low doses of radiation represented disruption of the linkers between the subunits to give free DNA molecules, and further degradation by higher doses represented random breakage of DNA within the subunits. Post-irradiation incubation first rejoined breaks to produce the 165 S subunits, and subsequently the subunit linkers were rejoined. An important postulate of the hypothesis is that the latter process is distinct from the former. Arguments in support of the subunit hypothesis are: (a) It provides an explanation for the difference in behaviour of complexed DNA and free DNA; (b) under a variety of post-irradiation conditions the reformation of the complex appears to be inhibited, whereas rejoining of breaks after higher doses seems to be unaffected (SHIPLEY and ELKIND, 1971; ELKIND and CHANG-LIU, 1972b; LEE et al., 1972; WHEELER et al., 1973; BEN-HUR and ELKIND, 1974).

The subunit hypothesis has been subjected to severe criticism. Opponents of the hypothesis have suggested that DNA above a certain size cannot form free DNA molecules because it is incapable of complete denaturation in alkali on account of its enormous size (McBURNEY et al., 1972). Thus disruption of complexed DNA is caused by the introduction of random strand breaks into the DNA, facilitating denaturation and production of free DNA strands. Conversely restitution of the complex is merely a manifestation of rejoining of SSBs and does not represent a distinct repair process. Arguments in favour of this and against the subunit hypothesis are:

1. All treatments which disrupt the complex (ionizing radiation, actinomycin D, prolonged lysis at high temperature in alkali etc.) also eventually produce breaks in DNA (e.g., LETT et al., 1970; ELKIND, 1971).

2. Since the linkers joining DNA subunits are presumably much smaller than the DNA in the subunits, it is difficult to conceive how ionizing radiation, which is thought to produce ionizations at random, can introduce many more breaks specifically into a small linker region than into large DNA molecules.

3. McBURNEY and WHITMORE (1972) overcame the problem of undenaturability of complexed DNA by growing cells for one generation in bromodeoxyuridine. This strand could then be degraded by exposure to fluorescent light which breaks bromuracil-containing DNA but not unsubstituted DNA, thus permitting denaturation. The fate of the thymidine-containing strand alone could then be studied. Under these conditions with increasing dose there was no abrupt transition between complexed and free DNA, but a smooth decrease in sedimentation rate of the DNA with increasing X-ray dose, as would be expected for free single strands of DNA.

Space does not permit a detailed appraisal of the above arguments. Most of the experimental results can be explained equally well with or without invoking the subunit hypothesis. There is evidence for and against the subunit hypothesis. In the view of the author the evidence against is somewhat stronger than evidence in favour, but this is a subjective view, and the question remains to be definitely resolved. Readers should form their own opinions from the original literature.

Irrespective of the interpretation of the data, by using a restoration of the properties of the DNA from unirradiated cells as a criterion of repair, demonstrations of DNA repair have been made in the biological dose range by studying the sedimentation properties of complexed DNA (LETT et al., 1972; ELKIND and CHANG-LIU, 1972a, b; McBURNEY et al., 1972; WHEELER et al., 1973; WHEELER and LINN, 1974), by examining its denaturability on hydroxyapatite (AHNSTRÖM and ERIXON, 1973; AHNSTRÖM and EDVARDSSON, 1974) and by the novel technique of alkaline elution analysis (KOHN and GRIMEK-EWIG, 1973). Within the limits of resolution of these experiments, complete repair has been observed after low doses of irradiation.

4.3.6 Rejoining of SSBs in Nondividing Cells

The empirical relationships between sedimentation rate and M.W. of DNA (BURGI and HERSHEY, 1963; STUDIER, 1965; ABELSON and THOMAS, 1966) are only valid for low concentrations of DNA (< 1 μg/ml). With growing cells such low concentrations can be detected by radioactive labelling of the DNA with ^{3}H-thymidine. With nondividing cells this is not possible and other methods of detecting the DNA must be used. This has usually been done with the fluorometric technique of KISSANE and ROBINS (1958), and in the work of LETT and co-workers (1972), by employing zonal rotors enabling larger quantities of DNA to be sedimented. Using these techniques rejoining of strand breaks has been demonstrated in the DNA in tissues shown in Table 4.2.

4.3.7 Physiological Effects

4.3.7.1 Genetic Diseases

A human genetic disease, the Hutchinson-Gilford progeria syndrome results in premature senescence of the affected individuals. EPSTEIN et al. (1973) reported that fibroblasts from one such patient were unable to rejoin SSBs after a dose of 10 krad. These results were then extended to include a further three patients (EPSTEIN et al., 1974). However, REGAN and SETLOW (1974a) were unable to detect any difference in the rejoining rates of SSBs in cells from the same patients. These contradictory findings remain to be resolved.

Xeroderma pigmentosum is a disease which in most cases is associated with an inability of cells from affected individuals to remove UV-induced pyrimidine dimers from their DNA (CLEAVER, 1974b). Fibroblasts from such patients were able to rejoin SSBs at the

Table 4.2. Rejoining of SSBs in nongrowing cells

Animal	Tissue	In vitro[a]	In vivo[b]	Rejoining[c]	Reference
Rat	Bone-marrow	√		+	
	Spleen	√		+	
	Thymus	√		+	KARRAN
	Thigh muscle (young)	√		+	and
	(old)	√		-	
Chick	Lymphocytes	√		+	ORMEROD
	Erythrocytes (young)	√		+	(1973)
	(old)	√		-	
Rabbit	Retina cells		√	+	WHEELER et al. (1973)
Dog	Cerebellar neurons		√	+	WHEELER and LETT (1972, 1974)
Mouse	Thymus	√	√	+	ONO and OKADA
	Liver	√		-	(1974)
			√	+ (slow)	
Human	Lymphocytes	√		+ (slow)	HASHIMOTO et al. (1975)

Cells removed from animal prior to [a] or after [b] irradiation and incubation. [c]+: rejoining, [c]−: no rejoining.

same rate as cells from normal donors (KLEIJER et al., 1970, 1973). Normal rejoining was also found in lymphocytes from patients with ataxia telangiectasia and Fanconi's anaemia, hereditary diseases giving rise to increased X-ray-induced chromosome aberrations (CERUTTI, 1974a; TAYLOR et al., 1975; VINCENT et al., 1975). On the other hand a claim that faster than normal rejoining occurred in lymphocytes from patients with chronic lymphocytic leukaemia than from normal donors has been made by HUANG et al. (1972).

4.3.7.2 Electrical Activity of Retinas

WHEELER et al. (1973) studied the effects of radiation on the photoreceptor cells of rabbit retinas. The electrical activity a few hours after irradiation disappeared over a narrow dose range, above a critical dose of 4.3 ± 0.3 krad. Below the critical region reconstitution of DNA similar to that in unirradiated cells occurred, but above this dose, initial rejoining in the first few hours was followed by degradation of the DNA. This important finding indicated a correlation between the ability of these cells to reconstitute DNA from broken chains and the maintenance of their physiological function, although the cause/effect relationship of the observations remains to be established.

4.3.7.3 *Aging*

In rat thigh muscle and chick erythrocytes, SSBs were rejoined in young but not in old tissues (KARRAN and ORMEROD, 1973). Furthermore the ability of neurons of the cerebellum of beagles to rejoin SSBs completely, decreased with increasing age of the dogs. This indicates a possible correlation between aging and the ability to rejoin SSBs (WHEELER and LETT, 1974). The age of human fibroblasts in culture, however, did not affect the rate of rejoining of SSBs (CLARKSON and PAINTER, 1974).

4.4 Repair of Double-Strand Breaks

4.4.1 Nature and Measurement of DSBs

Double-strand breaks (DSBs) produce a decrease in the molecular weight of the DNA in its native form. A priori one would expect that these would be much more deleterious to the cell than SSBs since they may result in the molecule falling apart, and/or may lead to a loss of genetic information from both DNA strands at the same place. Double-strand breaks are measured by a technique similar to that of McGRATH and WILLIAMS (1966), but by lysing the cells in detergent below pH 11 and centrifuging the DNA in neutral sucrose gradients. A double-strand break measured in this way may represent two SSBs on opposite strands so close together that the base-stacking forces are insufficient to hold the molecule together under the experimental conditions. Such SSBs may be the result of a single ionizing event or of two separate events. The former will arise linearly with dose, the latter being proportional to the square of the dose. Calculations indicate that the latter case is only significant at very high radiation doses. The DNA molecule inside the cell will not necessarily fall apart at a DSB since ionic conditions etc., may be completely different inside the cell from those used in experiments.

Technically, measurement of DSBs seems to have proved more difficult than SSBs since, under the milder neutral conditions, it is more difficult to remove other cellular constituents from the DNA, a necessary condition for meaningful M.W. measurements (e.g., ORMEROD and LEHMANN, 1971b). Furthermore, as with single-stranded DNA (see III.4.3.5.1) sedimentation of high molecular weight DNA under neutral conditions at high rotor speeds gives rise to a rotor speed anomaly such that all DNA molecules of molecular weight greater than a critical size sediment with the same velocity (slower than predicted), this critical size decreasing with increasing rotor speed. This was first reported for DNA from mammalian cells by LEHMANN and ORMEROD (1970a, b) and for T2 phage DNA by RUBENSTEIN and LEIGHTON (1971), and subsequently by several other groups (KAVENOFF, 1972; PETES and FANGMAN, 1972; LEVIN and HUTCHINSON, 1973; CHIA and SCHUMAKER, 1974). It has been analysed in detail by ZIMM (1974). As a consequence of this effect, if DNA is spun at high rotor speeds, over a wide X-ray dose range sharp nonrandom profiles are obtained, and the sedimentation rate does not change with decreasing dose. The true situation, namely random distributions and a decreasing average M.W. with increasing dose, is revealed at low rotor speeds (< 10,000 rpm) (LEHMANN and ORMEROD, 1970a, b; LEVIN and HUTCHINSON, 1973; ORMEROD, 1975). In order to ensure that the measurements of molecular weights are meaningful, LEHMANN and ORMEROD (1970a, b) have suggested several criteria which must be fulfilled to indicate that the DNA molecules are sedimenting freely.

The current status of production and repair of DSBs in various organisms is summarized in Table 4.3. At this early stage it is not clear whether the variations in measurement of efficiency of DSB production represent genuine differences, or shortcomings in the techniques used by some workers. Using appropriate conditions a linear production of DSBs with dose has been obtained in several systems (e.g., LEHMANN and ORMEROD, 1970a, b; BURRELL et al., 1971; CORRY and COLE, 1973; RESNICK, 1975).

Table 4.3. DSB production and repair in various organisms

Organism	DSB production (eV/break)[a] Oxic	Anoxic	Repair	Reference
E. coli K12	530	1300	-	BONURA et al. (1975)
M. radiodurans			+	KITAYAMA and MATSUYAMA (1968)
	520		-	BURRELL et al. (1971)
B. subtilis	1100		+ (limited)	HARIHARAN and HUTCHINSON (1973)
Saccharomyces servisiae	1700		+	RESNICK (1975)
Physarum polycephalum			+	BREWER and NYGAARD (1972)
Chinese hamster ovary	2700 (γ-irradiation)		+	COLE et al. (1975)
	1700 (α-irradiation)			CORRY and COLE (1973)
Mouse L5178Y	2900		-	LEHMANN and ORMEROD (1970a)
			+	LANGE (1974)
			+	SAWADA and OKADA (1972)
Rat thymocytes	1100-1900	4200	Not examined	LENNARTZ et al. (1973)

[a] Where no values are given, only quanlitative findings reported.

4.4.2 The Production and Repair of DSBs in Bacteria

An early attempt to measure DSBs in *E. coli* K12 was made by KAPLAN (1966) who obtained no evidence for repair of these lesions. Despite the fact that the profiles obtained by KAPLAN (1966) almost certainly did not represent freely sedimenting DNA, his conclusions seem to be correct. More recent work by BONURA et al. (1975) using a properly characterized system measured the production of DSBs by irradiation under N_2 and O_2 (Table 4.3) and again failed to detect repair. In wild-type bacteria, 1-2 DSBs were produced per lethal event suggesting that in wild-type *E. coli* the production of a DSB might be lethal for the cell. On the other hand some DSBs induced by ^{32}P disintegrations in intracellular lambda phage DNA could be rejoined (TOMIZAWA and OGAWA, 1967).

In contrast it is fairly certain that *Micrococcus radiodurans* is able to repair DSBs. This bacterium can survive doses of radiation which produce about 100 DSBs per cell, implying the existence of some functional repair of these lesions. Thus KITAYAMA and MATSUYAMA (1968) showed that the molecular weight of the native DNA was decreased by radiation, and a return to the unirradiated profile was achieved 3 h after irradiation, this process being inhibited at the restrictive temperature in some temperature-sensitive gamma-ray-sensitive mutants (SUHADI et al., 1972). BURRELL et al. (1971) showed that rejoining of DSBs was a complex process involving several membrane components.

HUTCHINSON and co-workers (1973) carried out a careful study of DSB production and repair in *B. subtilis*, taking account of the rotor speed effects discussed above. They obtained a rapidly-sedimenting peak which was converted by radiation into a "main peak", which in turn was broken down in a random fashion as predicted by theory. From the kinetics of the interconversion of these DNA molecules, they found that DSBs were produced with an efficiency of 0.09 breaks/10^{12} dalton/rad. One DSB converted the fast component into the main peak, and 3 DSBs were produced per lethal hit per genome, implying that a limited amount of rejoining was possible. The fast peak could be reconstituted on post-irradiation incubation and the amount of fast-sedimenting DNA that was reformed correlated well with cell survival (LEVIN and HUTCHINSON, 1973; HARIHARAN and HUTCHINSON, 1973).

4.4.3 The Production and Repair of DSBs in Eucaryotic Organisms

Double-strand breaks have been measured in various mammalian systems by CORRY and COLE (1968, 1973), LEHMANN and ORMEROD (1970a), LANGE (1974) and COLE et al. (1975) with results shown in Table 4.3. The ratio of DSB:SSBs produced is about 1/20-1/70. Although rejoining of DSBs was not observed by LEHMANN and ORMEROD (1970) in mouse L5178 Y cells after 20-100 krad, it has been convincingly demonstrated in Chinese hamster cells, at least 80% being rejoined after doses less than 50 krad (CORRY and COLE, 1973; COLE et al., 1975), and has been reported for a different strain of L5178 Y cells by SAWADA and OKADA (1972) and by LANGE (1974).

In yeast, RESNICK (1975) measured DSBs and demonstrated their repair in a 6 h post-irradiation incubation period. This rejoining process did not take place in the gamma-ray-sensitive mutant rad-52 which is also incapable of forming gamma-ray-induced recombinants. Based on these findings (RESNICK (1975) put forward a model for DSB rejoining in yeast involving the formation of recombinant molecules between sister duplexes, and showing how this could lead to the production of gene conversions and reciprocal recombinants.

4.5 Repair of Base and Sugar Damage

4.5.1 Chemical Demonstrations of Repair

The measurement of base damage and its repair after UV irradiation has been relatively easy since the principal lesion, the pyrimidine dimer is stable to the drastic acid hydrolysis conditions used for complete degradation of the DNA. It is therefore possible to measure its frequency directly by chromatographic techniques. Unfortunately, lesions induced by ionizing radiation are rather unstable to acid hydrolysis under normal conditions as well as being much more heterogeneous than UV-induced base damage (see II.2.4). Recently, however, HARIHARAN and CERUTTI have devised techniques for estimating some of the damaged bases after ionizing radiation. Their first experiments (HARIHARAN and CERUTTI, 1971, 1972) demonstrated that bases of the type 5-hydroperoxy-6-hydroxy-5,6-dihydrothymine, or 5-hydroxy-6-hydroperoxy-5,6 dihydrothymine (t^+) were excreted into the medium after gamma-irradiation of *M. radiodurans*. Similar results were obtained with Chinese hamster cells (MATTERN et al., 1973). The yield of t^+ products was similar to that of single-strand breaks, indicating the quantitative importance of base damage.

More recently they have developed an assay for measuring damaged thymine residues in DNA. Using this system they have shown that damage of the 5,6-dihydroxy-dihydrothymine type can be removed from gamm-irradiated DNA or OsO_4-treated poly (dA-dT) by extracts of *E. coli* (HARIHARAN and CERUTTI, 1974a). By using extracts from various mutants, they concluded that this removal needed neither endonuclease I, nor, in contrast to the removal of UV-induced pyrimidine dimers, did it require the UV-endonuclease coded for by the $uvrA^+$ gene. About 8-16 thymine residues were removed for each damaged base excised, and a directly induced single-strand break was not necessary for excision of these damaged bases. Further work suggested that the damaged bases were removed by the 5'-3' exonuclease activity of DNA polymerase I, and the gap left after excision was normally rapidly filled by a DNA polymerase and polynucleotide ligase (HARIHARAN and CERUTTI, 1974b, c). Thus the pattern for excision and repair of damaged bases probably involves a process similar to that for removal of UV-induced pyrimidine dimers. The first incision step, however, seems to involve a different endonuclease. The subsequent steps of excision and repolymerisation by DNA polymerase I and sealing by polynucleotide ligase are probably similar for UV- and gamma-induced base damage in *E. coli*.

Specific excision of this type of damage by nuclei of cultured mammalian cells has also been demonstrated (MATTERN and CERUTTI, 1975). In human fibroblasts this excision ability disappeared in "old" cultures, the disappearance occurring before the decrease in DNA synthesis, and several generations before the cells ceased to divide.

Techniques for identification and estimation of other types of base damage should provide further valuable information on repair of this type of lesion after ionizing radiation.

4.5.2 Endonuclease Sensitive Sites

An alternative, less direct but much simpler and more sensitive method for measuring base damage has taken advantage of lesion-specific nucleases. *Micrococcus luteus* contains at least two such nucleases one of which is highly specific for pyrimidine dimers (GROSSMAN, 1974) and another with a broader range of specificities. Thus crude extracts of *M. luteus* contain both these activities, and will introduce nicks into DNA in the neighbourhood of damaged bases. Such nicks (i.e., SSBs) are much more readily measurable (by the alkaline sucrose gradient techniques discussed above) than are the damaged bases themselves. The disadvantage of the technique is, of course, that the nature of the nuclease-susceptible site (presumptive damaged base) is in general not known.

Using minicells containing plasmid lambda dv DNA, PATERSON and SETLOW (1972) irradiated the cells, incubated for various times, extracted the DNA and measured the direct breaks produced and the nuclease-susceptible sites. Approximately equal numbers were produced after anoxic irradiation and both types of lesion were repaired within 10 min. WILKINS (1973a, b) used a similar technique with chromosomal DNA using whole cell lysates. The nuclease-susceptible sites were detected in osmotically-shocked cell lysates, obviating the need for extraction of the DNA with the associated shearing processes, and in agreement with the minicell system of PATERSON and SETLOW (1972) he found that the sites were removed within 10 min. This excision took place equally rapidly in uvrA$^+$ and uvrA$^-$ strains, again showing that the uvr system was not involved, in accordance with the findings described above of HARIHARAN and CERUTTI (1974a).

Normal human cells are also capable of removing nuclease-sensitive sites after γ-irradiation, but in fibroblasts from three patients with the hereditary disorder, ataxia telangiectasia, this repair process was present only at very low levels (PATERSON et al., 1976). Patients with ataxia telangiectasia are hypersensitive to the effects of ionizing radiation and this sensitivity is reflected at the cellular level (TAYLOR et al., 1975). The defect in excision of base damage may well account for this radiosensitivity.

4.5.3 Repair of Alkali-Labile Lesions

As discussed in III.4.2.1 and 4.2.2, SSBs as measured by alkaline sucrose gradients represent the sum of true radiation-induced breaks and base or sugar damage which is converted into a break by the alkaline treatment. PATERSON et al. (1973) measured the relative amounts of true breaks and alkali-labile bonds by using *E. coli* minicells containing a closed circular plasmid DNA molecule (lambda dv), whose sedimentation rate is different from a nicked circle in both alkaline and neutral gradients. Thus assaying the conversion of closed to open circles by irradiation will give the total SSBs + alkali-labile bonds if the DNA is examined in alkaline gradients, but only true SSBs will be measured in neutral gradients. Using this system, they found that 21% (oxic irradiation) and 34% (anoxic irradiation) of the total breaks were in fact alkali-labile lesions, the remainder being true breaks. In the minicell system the true breaks were rapidly rejoined in 10 min, whereas the alkali-labile lesions were not removed. However, since in wild-type *E. coli* after moderate doses of irradiation, essentially all the measured breaks can be repaired (TOWN et al., 1973), this suggests that alkali-labile lesions, as well as true breaks, can be repaired in whole cells.

In irradiated rat thymocytes (LENNARTZ et al., 1973), the number of alkali-labile lesions was measured by denaturing extracted DNA at 75°C in the presence of formaldehyde and comparing the number of SSBs with those from DNA denatured in alkali. They found 3.7 times as many total breaks produced by irradiation in O_2 as opposed to N_2, of which 28% of the oxic breaks and 49% of the anoxic breaks were in fact alkali-labile lesions rather than true breaks. These results are very similar to those found in *E. coli* by PATERSON et al. (1973).

4.6 Repair Synthesis in Eucaryotic Cells

The filling-in of the gap left after removal of a damaged section of DNA (repair synthesis) has been measured in several ways (CLEAVER, 1974b):

1. Unscheduled DNA synthesis (UDS): the autoradiographic demonstration that agents which damage DNA stimulate cells outside the S phase of the cell cycle, to synthesize small amounts of DNA. This procedure in fact provided the first evidence of a repair process in mammalian cells (RASMUSSEN and PAINTER, 1964).

2. In nondividing cells, residual semiconservative DNA synthesis can be suppressed by hydroxyurea, and then unscheduled DNA synthesis (which is insensitive to hydroxyurea - see CLEAVER, 1974b) measured by the radiation-stimulated incorporation of ^3H-thymidine into the cells. In growing cells, even in the presence of hydroxyurea, the residual semiconservative synthesis is still sufficient to obscure the repair synthesis completely.

3. Repair replication (CsCl). Equilibrium centrifugation in CsCl showed that in damaged cells a small amount of nonconservative DNA synthesis occurred which was termed "repair replication". It was subsequently demonstrated that, after UV irradiation, UDS was in fact measuring this nonconservative repair replication (PAINTER and CLEAVER, 1969), and also that repair replication was in-

deed the result of the filling in of gaps left after excision of damaged bases (CLEAVER, 1974b).

4. Repair replication (BrU photolysis). A further technique for measuring repair replication is the bromuracil photolysis technique devised by REGAN et al. (1971). Repair of damaged DNA is allowed to take place in cells in the presence of bromodeoxyuridine. Subsequent exposure of the cells to light at 313 nm selectively breaks the repaired patches of DNA which contain bromuracil instead of thymine. The kinetics of the breakdown and the resultant M.W. distribution of the broken DNA enable a calculation to be made of the number and size of the repaired patches.

All the above phenomena can easily be detected after sublethal doses of UV irradiation (see CLEAVER, 1974b for detailed discussion), but after gamma irradiation they are much more difficult to observe. Either very high doses must be used, or long exposure times for UDS and several rebands for equilibrium centrifugation must be employed. [The results claiming detection of repair replication after low doses of gamma irradiation were later shown to be erroneous (FOX and FOX, 1973)]. UDS and repair replication have now been reliably measured after doses of ionizing radiation in excess of 1 krad in a variety of cultured mammalian cell lines and other eucaryotic organisms (e.g., CLEAVER, 1969; SPIEGLER and NORMAN, 1969, 1970; BRENT and WHEATLEY, 1971; SCHAEFFER and MERZ, 1971; PAINTER and YOUNG, 1972; REGAN and SETLOW, 1974b). They occurred to an equal extent in fibroblasts from normal donors and xeroderma pigmentosum patients (CLEAVER, 1969; KLEIJER et al., 1970) and in young and aging fibroblast cultures (CLARKSON and PAINTER, 1974). A reduced level of repair replication has been found in cells from some patients with ataxia telangiectasia (PATERSON et al., 1976) in keeping with the findings of a low level of excision of base damage (see III.4.5.2). Neither repair replication nor UDS have been detected in Vicia roots after exposure to chemical mutagens or ionizing radiation (PAINTER and WOLFF, 1973; WOLFF and CLEAVER, 1973).

Using the BrU photolysis technique REGAN and SETLOW (1974b) were able to distinguish two types of repair in human cells. UV-type repair was characterized by the replacement of relatively large patches of DNA by repair replication (about 100 nucleotides) and was defective in xeroderma pigmentosum cells, whereas X-ray-type repair (and that found after exposure to several alkylating agents) gave rise to much shorter patches of about 3-4 nucleotides and took place equally well in normal and xeroderma pigmentosum cells. This estimate of a patch size of 3-4 nucleotides after ionizing radiation agrees well with those determined by repair replication (PAINTER and YOUNG, 1972) and by thymidine incorporation in lymphocytes (SPIEGLER and NORMAN, 1970). These low values also account for the difficulty in measuring UDS and repair replication after ionizing radiation as compared with UV-irradiation and indicate that there are some important differences in the repair of lesions induced by the two types of radiation. None of the work discussed has provided information as to whether these small repair patches are the result of repair of strand breaks, base damage, or both.

4.7 The Enzymology of DNA Repair

As discussed in III.4.2 genetic studies with *E. coli* have implicated several enzymes in the repair of SSBs produced by ionizing radiation: DNA polymerase I, DNA polymerase III, but not DNA polymerase II (YOUNGS and SMITH, 1973b) polynucleotide ligase, the $recA^+$ and $lexA^+$ gene products (as yet unidentified) and exonuclease V. Semi in vitro biochemical studies have implicated DNA polymerase I, polynucleotide ligase and an as yet unidentified endonuclease (which is not the UV-endonuclease coded for by the $uvrA^+$ and $uvrB^+$ gene products), in the excision-repair of base damage. Endonucleases which can specifically nick gamma-irradiated DNA have been isolated from *M. luteus*, *E. coli* and human cells (see CERUTTI, 1974a).

Apurinic sites are produced by gamma-irradiation and enzymes which specifically nick DNA near apurinic sites have been isolated from bacterial, plant and mammalian systems (see CERUTTI, 1974b). Finally extracts of rat liver were able to remove a malonaldehyde-like substance from gamma-irradiated DNA (PAYES, 1974), presumably a product of sugar damage. A detailed review of some of these enzymes is provided by GROSSMAN (1974).

4.8 Conclusions

Both procaryotic and eucaryotic organisms have evolved a variety of systems whereby several different types of lesions induced by ionizing radiation can be repaired, thus ensuring that the integrity of the genetic information can be maintained in the face of such radiations.

Although studies on repair of damaged bases are at an early stage, they indicate that such lesions may be efficiently removed from cells. The considerable work on SSBs suggests that, except in the most sensitive bacterial mutants, the production of such lesions is not very serious for the cells, since they can be rapidly repaired in a wide variety of organisms. What these studies have as yet been unable to reveal is whether (particularly in mammalian cells) a small proportion of unrejoined or incorrectly rejoined breaks remain, and if so whether these may have a deleterious effect on the cell. Certainly bacteria with reduced abilities to rejoin breaks become more sensitive to the effects of ionizing radiation.

With regard to double-strand breaks, it appears that some organisms can rejoin these breaks, but again we would expect an unrejoined double-strand break to have dire consequences for cell survival. As yet, the resolution of available biochemical techniques is a long way from being able to detect a single lesion per mammalian cell, although this may be possible in bacteria in the near future.

Chapter 5

Molecular Aspects of Mutagenesis Due to Ionizing Radiation

5.1 Introduction

At the beginning of this chapter the question may be asked why the induction of mutations by ionizing radiation should be studied. The following reasons are in the author's opinion the most relevant. Firstly, according to the principle that by changing a thing one can hope to get to know the thing itself, mutation studies may be expected to give some insight into the nature of the gene and the structure of the chromosome. It should be admitted, however, that the elucidation of these problems has profited more from other research disciplines than from radiation genetics. Secondly, the induction of mutations can be used merely as a means to investigate the mechanisms of the biological action of ionizing radiation, the fact that one deals with hereditary structures being of secondary interest in this connexion. Thirdly, the enormous practical implications of radiation genetics especially as regards the protection from radiation-induced genetic damage in man, but also concerning other topics like plant breeding, justify this kind of research. Even those studies in radiation mutagenesis that are performed for purely theoretical reasons, can be expected to yield practical results in the long run. The fourth reason may be considered as analogous to "art for art's sake": Ionizing radiation has turned out to be a powerful tool for changing the genetic material. It is therefore tempting to find out as much as possible about the nature of these changes, even if no knowledge beyond this level is gained and irrespective of what one calls this attitude of mind (e.g., "satisfaction of the researcher's curiosity").

This chapter is devoted exclusively to the molecular nature of the genetic alterations produced by ionizing radiation. One reason for this restriction is that the nonmolecular aspects of radiation mutagenesis would not be a suitable topic for a book dealing with the effects of ionizing radiation on nucleic acids. Furthermore, the nonmolecular aspects of radiation mutagenesis (with respect to ionizing radiation) have already been discussed in several recent reviews (AUERBACH and KILBEY, 1971; NEWCOMBE, 1971; UNSCEAR, 1972; REVELL, 1974) whereas a survey specifically concerned with the molecular nature of the genetic alterations produced by ionizing radiation does not seem to have been published so far. However, this article by no means intends to cover the whole literature on this topic and the studies selected represent a subjective choice.

Because of lack of space no explanations will be given of those genetical and radiobiological terms that can be easily found in relevant textbooks and glossaries (e.g., CASARETT, 1968; RIEGER et al., 1968). Furthermore in this survey the representation and interpretation of results are emphasized rather than methodological aspects.

Although we are dealing with mutagenic effects of ionizing radiation, a comment on UV mutagenesis should be made (see WOLFF, 1967a; DRAKE, 1969, 1970; AUERBACH and KILBEY, 1971). After the discovery that pyrimidine dimers are the main premutational lesions produced by UV (as shown especially in the bacterial chromosome), the elucidation of the mechanisms leading to a final mutation seemed to be a simple affair. However, this proved to be an unjustified expectation, because several steps are involved in these processes. The production of a final mutation by UV may even be prevented altogether by the action of various repair mechanisms.

There are reasons not to call the DNA-molecules of prokaryotes "chromosomes" (see WHITE, 1973). Nevertheless, for the sake of simplicity we shall designate the hereditary structure of both prokaryotes and eukaryotes by this term. The following discussions are based on the assumption that the chromosome is "single-stranded", i.e., that it contains only one DNA double helix. One reason for this decision is the plausibility of that model (see e.g., the review by PRESCOTT, 1970, the results of LAIRD, 1971 and the report on the "Cold Spring Harbor Symposia on Quantitative Biology", 1974), although the multistrand model cannot be excluded entirely (see the review by WOLFF, 1969). Moreover the single-strand model represents the simpler assumption, therefore complying better with the requirements of the "economy of thinking" principle than the other model does.

5.2 Molecular Nature of Gene Mutations Produced by Ionizing Radiation

There are two fundamental effects of ionizing radiation on the hereditary material: gene mutations and chromosome breaks. We shall confine ourselves to a discussion of radiation-induced gene mutations because too little is known of the molecular mechanism of radiation-induced chromosome breakage. There are, however, some recent experimental and theoretical studies of the latter which go beyond the usual phenomenistic approach (NEARY, 1965, 1966, 1967; KIHLMAN, 1971; COMINGS, 1974; REVELL, 1974; BENDER et al., 1974).

5.2.1 Terminology

The terms "point mutation" and "gene mutation" are used by many authors synonymously. However, STADLER and ROMAN (1948) and MULLER (1954) attribute to the term "point mutation" a somewhat broader meaning which includes both "true" gene mutations and loss of a gene via an ultraminute deletion. This distinction is a useful one because in experiments with ionizing radiation one often does not know whether a change or the loss of a gene has been induced. These authors therefore discriminate between "point mutations" and "gene mutations", and this distinction will also be used throughout this chapter. The mutations induced at specific loci in the germ cells of wild-type mice (see III.5.2.8) are according to that terminology "point mutations", whereas the genetic alterations induced in *Neurospora* discussed in III.5.2.5 are "gene mutations". Another term used in this survey is "reverse" or "back mutation", meaning a change assumed to restore the original (wild-type) configuration of the gene.

5.2.2 Studies with Phage

In order to answer the fundamental question whether ionizing radiation is able to induce gene mutations (besides gene losses), it is of particular interest to study the induction of mutations in an object with a relatively simple chemical structure.

The recent work of BLEICHRODT and co-workers (1971, 1973, 1974) on the induction of gene mutations by ionizing radiation in the single-stranded bacteriophage ΦX 174 is one of the few studies performed so far on the induction of mutations by ionizing radiation in phage. Note that in phage genetics the term "strand" has another meaning than in chromosome theory. In the former case a "strand" is identical with a polynucleotide chain of the DNA, in the latter case with the DNA double helix itself (see III.5.1). Amber mutants of ΦX 174 were irradiated with γ-rays. These mutants can multiply only in special bacterial strains (*Escherichia coli*) carrying a suppressor for the amber codon UAG. Mutations enabling the phage to grow on a suppressor-less host (*E. coli*), called "reversions to pseudo wild-type", were looked for. The term "pseudo wild-type" was used by the authors because it was not known to what extent suppressors were induced instead of true reversions (changing the amber codon into the corresponding wild-type codon). Strictly speaking this designation should be used only for the former of these two groups.

Based on the single-strand nature of ΦX 174 and on other evidence, the authors concluded that single DNA breaks, repair by excision or recombination and replication of the replicative form of the phage (in the bacterial host) could not be involved in the induction of the "reversions" found by them. Thus we know a number of ways in which "true" and "pseudo-reversions" are apparently not produced, but still remain ignorant of how they are produced.

Interestingly, it was possible to convert by γ-irradiation an ochre mutant into an amber mutant, i.e., to induce a change from ATT (codon UAA) in the parental DNA into ATC (codon UAG) in the progeny DNA. The efficiency of induction of this one base exchange is enhanced when the irradiation is carried out in an atmosphere of oxygen, this being a demonstration of the "oxygen effect" at the molecular level.

Using the double-stranded bacteriophage T4 MUNSON and BRIDGES (1973) have also studied the induction of reversions (more exactly: pseudo-reversions plus true reversions) in an amber mutant, and that by irradiating the mutant strain with radiations (γ-rays, protons, α-particles) of different linear energy transfer (LET).

The authors concluded from the LET-dependence of the effects observed that a mutational lesion was induced by the passage of one ionizing particle through (or close to) one of the two DNA strands. It seems likely that when a true reversion was induced the amber codon itself was hit or - assuming energy conduction over a few nucleotides - that the hit occurred in the neighbourhood of the triplet. At present more far-reaching assumptions would be too speculative.

Conclusion: The phage studies (ΦX 174, T4) show that ionizing radiation is able to induce true gene mutations.

5.2.3 Studies with Bacteria

Compared with our knowledge of the induction of mutations in bacteria by UV, the molecular basis of mutations produced by ionizing radiation is less well understood. We do not intend to discuss all the ways in which a bacterial gene mutation might be produced by ionizing radiation, among them the possibility that incorrectly performed repair processes are involved in mutational events. There is no doubt that ionizing radiation is able to induce gene mutations in *Escherichia coli*, among them probably base transitions (AT → GC) induced by γ-rays at the tryptophan locus (BRIDGES et al., 1968), and various kinds of transversions and transitions induced by γ-rays and β-particles, transforming an arginine dependent strain of *Escherichia coli* into an arginine independent one (PHILLIPS et al., 1972). It is also established (see the review by BRIDGES and MUNSON, 1968) that in *E. coli* both daughter chromatids produced at the first replication after irradiation are usually mutant ("double-stranded segregation"). Are then both DNA strands of the irradiated bacterial chromosome altered simultaneously in a complementary manner, or is at first only one of them affected, the "double-stranded segregation" being the result of secondary steps of the mutational process? These questions cannot be answered definitely but the "repair hypothesis" discussed in III.5.2.9 presents a plausible explanation (BRIDGES, persl. comm.).

Possibly the initial lesion induced by ionizing radiation leading to a bacterial gene mutation is a break of one DNA strand, i.e., of the polynucleotide chain (MUNSON and BRIDGES, 1969). How this single-strand break would then be transferred into a genetic change characterized by "double-stranded segregation" is not known. The "normal" excision repair system, responsible for the removal of UV-induced pyrimidine dimers, does not seem to be involved in that process (see the review by BRIDGES and MUNSON, 1968).

In agreement with postulates formulated in the early days of radiation genetics (TIMOFEEFF-RESSOVSKY and ZIMMER, 1947), the effectiveness of different types of ionizing radiation for the production of gene mutations in *E. coli* (probably single base changes) steadily decreases with increasing LET (BRIDGES and MUNSON, 1968; MUNSON and BRIDGES, 1969). This is to be expected from target hypothetical considerations if the primary lesion is caused by a single energy loss event (production of an ion cluster).

5.2.4 Studies with *Paramecium*

The protozoon *Paramecium aurelia* has various advantages for studies on the induction of mutations by ionizing radiation (AUERBACH, 1962; KIMBALL, 1963): (1) recessive lethal mutations induced in the diploid micronuclei can be detected via induced autogamy (self-fertilization) which makes such lethals homozygous; (2) animals can be obtained at particular stages of the division cycle, i.e., the dependence of genetical radiosensitivity on the cell phase irradiated can be studied; (3) *Paramecium* is very resistant to the killing effects of ionizing radiation, i.e., high radiation doses can be used for mutation studies.

KIMBALL (1963, 1964a, b, 1966) has shown by analysis of the division products of the irradiated *Paramecia* that most of the recessive lethal mutations induced by X-rays in the G_1 stage of interphase affect both DNA strands of the chromosome and, consequently, both of the two new strands produced at the first post-irradiation duplication. Furthermore he found that the mutation frequency can be reduced in several ways, e.g., by various post-irradiation treatments, which, however, have to be performed before the first chromosome duplication, or by increasing the interval between irradiation in G_1 and chromosome duplication.

What is the nature of those premutational reparable lesions induced by X-rays in G_1? (Other types of mutations induced by ionizing radiation in *Paramecium* will not be discussed in this context.) KIMBALL suggests that they are a special type of double-strand break, held in juxtaposition by chromosomal protein until replication. The conditions mentioned above (post-irradiation treatments, interval) might cause these breaks to restitute ("repair of premutational damage"). Otherwise the premutational lesions would be fixed during replication and irreparable mutations would be produced. The nature of this fixation process is not known, but loss of bases seems likely because of the recessive lethal effect of these alterations.

5.2.5 Studies with *Neurospora*

The mould *Neurospora crassa* offers the following advantages to mutation research: it is a eukaryote, i.e., its chromosomes are constructed like those of higher organisms, and it is a microorganism, i.e., microbiological methods can be applied in experiments with *Neurospora*. The molecular nature of mutations induced by ionizing radiation in *Neurospora* has been extensively investigated, especially by DE SERRES and by MALLING. Forward recessive lethal mutations were induced at two specific loci, ad-3A and ad-3B, by irradiating the wild type alleles. Both loci are closely linked and are involved in the control of purine synthesis. The nature of the forward mutations induced was identified by tests of specific revertibility, i.e., by treatment of the mutant loci with specific chemical mutagens and by tests for allelic complementation. (There has been some criticism as regards the specificity of the chemical revertibility tests; see KILBEY et al. (1971); PRAKASH and SHERMAN (1973). Of course the best way to identify the nature of a gene mutation would be to analyse the base sequences of mutationally altered tRNAs or the amino acid sequences of mutationally altered proteins. These procedures, however, would be too time-consuming.)

An analysis of the gene mutations induced by X-rays at the ad-3B locus, based on revertibility tests, e.g., with nitrous acid and methoxylamine, yielded the following results (MALLING and DE SERRES, 1973): base pair transitions, 42%; insertions or deletions of single base pairs, 37%; the rest being mutants reverting only spontaneously and nonrevertible mutants (e.g., transversions?). Most of the reversions induced by chemical treatment seem to be "true" ones because suppressors for ad-3B mutants are rare.

The *Neurospora* results represent considerable progress in our knowledge of the molecular nature of mutations induced by ionizing radiation in eukaryotes.

Besides these intragenic changes two-hit deletions of the ad-3A locus, the ad-3B locus, or of both loci simultaneously have been induced (see e.g., DE SERRES and MALLING, 1969; DE SERRES, 1971). These results, however, will not be considered here.

As regards the "strandedness" of the gene mutations induced by ionizing radiation in *Neurospora*, the "double-stranded segregation" type seems to prevail (see III.5.2.3), although for technical reasons the *Neurospora* system cannot be used to study the problems of "strandedness" in detail (DE SERRES, pers. comm.).

5.2.6 Studies with Maize

In contrast to the compelling evidence for the induction of gene mutations by ionizing radiation in microorganisms, experiments with higher organisms (maize, *Drosophila*, mouse) are less conclusive. STADLER and ROMAN (1948) and MOTTINGER (1970) using a technique which discriminated between gene mutations and small deletions (i.e., gene loss), observed in their X-ray experiments only the latter type of genetical damage. In order to explain this finding and the discrepancy existing between the maize results and the microbiological results, several assumptions can be made (see also MOTTINGER, 1970): firstly, the postmeiotic stages irradiated in maize may be especially resistant to the induction of gene mutations by ionizing radiation, whereas premeiotic stages might have given a positive result. Secondly, a "repair-system" operating in maize cells may prevent the induction of gene mutations by ionizing radiation. (Note, however, that "repair" hypotheses of this kind are so vague and all-embracing that they can be used to explain nearly every phenomenon in radiation genetics.) Thirdly, gene mutations might be induced, but their frequencies would be too low to be detected in the maize experiments so far performed. Whatever the correct explanation is, it seems rather unlikely, as MOTTINGER (1970) states himself, that in maize (and in higher organisms as a whole) gene mutations are not induced by ionizing radiation.

5.2.7 Studies with *Drosophila*

Despite the extensive use of the fruit fly *Drosophila melanogaster* in radiation genetics there are only few results which can be discussed from a molecular point of view. In this category belong the studies devoted to the induction of reverse (back) mutations. Flies carrying a special mutant allele [e.g., forked or white; see LINDSLEY and GRELL (1968) for explanation] are irradiated and their offspring are scanned for individuals apparently carrying the corresponding normal, i.e., "back-mutated", allele. Experiments of this kind have been carried out since the early days of radiation genetics. Although many of the authors were aware of the pitfalls connected with those experiments (e.g., radiation-induced suppression of the function of the normal allele), a too optimistic interpretation of the results obtained was often given, i.e., the effectiveness of ionizing radiation for inducing true reversions in germ cells of *D. melanogaster* was overestimated.

We do not intend to discuss the divergent results and interpretations published so far concerning radiation-induced reversions in *D. melanogaster*, but prefer instead to point out a recent extensive investigation by VOSS and FALK (1973). These authors looked for reverse mutations induced by X-rays in postmeiotic male germ cell stages. Although detailed genetic analyses were performed to check whether true reversions were induced, not a single example was found. There are several ways to explain this lack of evidence for the induction of true reversions by X-rays in *D. melanogaster* (see also the discussion of the maize results in III.5.2.6), e.g., the following ones: either X-rays are completely unable to induce true back mutations in *Drosophila*, or this type of gene mutation is induced so rarely by ionizing radiation (perhaps especially rarely in postmeiotic stages) that it can only be detected by microbiological methods, allowing the analysis of extremely large samples, but not with the *Drosophila* technique. If the frequencies of true reversions induced by ionizing radiation in germ cells of *D. melanogaster* were as low as those observed in microorganisms (and even if, for one or the other reason, they were somewhat higher) the detection of only one true revertant after irradiation with a dose of a few thousand rad would on the average require the analysis of at least one million flies.

Another molecular aspect of *Drosophila* radiation mutagenesis is "strandedness" (see also III.5.2.9). MULLER et al. (1961) have suggested how complementary two-strand changes may arise: Complementary parts of the two opposite strands, small enough to comprise only a few base pairs, would exchange their position. This would result in an intragenic inversion. As another mechanism proposed by these authors the bonds that attach the two bases of a base pair to their sugar groups would break. The base pair would be reattached to the sugar groups after rotating through 180°, the result of this "rotational substitution" being a double-transversion. It would be a pity if nature did not utilize this ingenious idea of "rotational substitution" at least occasionally. However, it seems now that most of those alterations which were explained by this hypothesis, namely X-ray induced "whole body mutations" in *D. melanogaster*, can be more plausibly explained in the following way. In view of the fact that the germ cells of *Drosophila*, especially the postmeiotic stages, are much more prone to the induction by ionizing radiation of small chromosome aberrations than to the induction of gene mutations, most of the "whole body mutants" observed in *Drosophila* are likely to be produced by small chromosomal aberrations affecting both strands of the DNA and simulating the induction of gene mutations.

5.2.8 Studies with Mice

The main reason for carrying out radiation genetical experiments with mice is a practical one, namely to establish guidelines for genetic radiation protection of man. However, these studies have important theoretical implications also, one of them being the problem of the nature of radiation-induced point mutations. These experiments are carried out with the "specific locus method". Wild-type mice are irradiated - mostly with X-rays or γ-rays - in order to induce specific recessive mutations at selected loci. Pioneer results with this technique were obtained by W.L. and L.B. RUSSELL at Oak Ridge (see the reviews by W.L. RUSSELL, 1964, 1965, 1972; VOGEL et al., 1969 and SEARLE, 1974).

The question to what extent gene mutations are induced by ionizing radiation in mouse germ cells and how far they are simulated by gene loss, has not yet been answered definitively. In this context a recent finding is of interest: Thalassaemia, a severe blood disease of man, seems to be caused not by a change but by the loss of a gene, or at least, by loss of a large part of it (OTTOLENGHI et al., 1974; TAYLOR et al., 1974). This finding (besides other evidence) demonstrates that the practical importance of the mouse results does not depend on the answer to the question: gene mutation or gene loss?

WOLFF (1967b) favours the assumption that the mutations induced in mice are mainly minute chromosome aberrations (e.g., deletions) instead of intragenic alterations. One of his arguments is the fact that about three-quarters of the mutations induced in spermatogonia are lethal in individuals homozygous for them, as is to be expected when the mutations are small deletions. An answer to these questions may be obtained from new experimental approaches (e.g., complementation mapping, see L.B. RUSSELL, 1971).

In agreement with results obtained in *Drosophila* (see MULLER et al., 1961) most of the mutations induced in mouse spermatozoa and spermatids produce "whole body mutants", i.e., are of the "double-stranded segregation" type (W.L. RUSSELL, 1964).

5.2.9 Strandedness

"Strandedness", i.e., the involvement of either one strand or both strands of the chromosomal DNA in the mutational event, is an important aspect of radiation mutagenesis. It is a remarkable finding that, irrespective of the organism studied, most gene mutations induced by ionizing radiation are characterized by "double-stranded segregation", i.e., they produce only mutant (viable) daughter cells in unicellular organisms like *E. coli* and *P. aurelia* (see III.5.2.3 and 5.2.4, respectively), and "whole body mutants" in multicellular organisms like *D. melanogaster* and the mouse (see III.5.2.7 and 5.2.8, respectively). Notice, however, that in multicellular organisms true gene mutations may be simulated by small chromosome aberrations. The results obtained from double-stranded phage are still inconclusive.

Following the discussion of strandedness by NASIM and AUERBACH (1967) and AUERBACH and KILBEY (1971) we enumerate four hypotheses accounting for the origin of "double-stranded segregating" gene mutations induced by ionizing radiation (see also DUBININ and SOYFER, 1969).

The dual-mechanism hypothesis assumes that both strands are affected in a complementary fashion (and that another mechanism would account for the production of single-strand alterations leading to mosaicism; hence the term "dual-mechanism"). Two ways in which such complementary two-strand changes may arise have been suggested by MULLER et al. (1961) (see III.5.2.7). Another possibility is the loss of a base pair or of a few base pairs (intragenic deletion).

The following three hypotheses assume that the initial mutational lesion is induced in one strand only. According to the repair hypothesis the "normal" bases of the opposite strand are removed via excision of the corresponding single-strand segment. This would occur because the "repair system" might not differentiate between a normal and a mutant single-stranded DNA segment when taking care that each base in one strand has its complementary, i.e., matching, partner-base in the opposite strand. The missing bases are then replaced by repair replication. When the mutant strand serves as template during this process, an initially one-strand alteration is transferred to a double-strand mutation. (If, however, as may be expected to happen as frequently, the repair system were to remove the wrong bases and use the normal strand segment as template, true repair would be achieved and the initial one-strand lesion would disappear.) Another version of the repair hypothesis has been suggested by FREESE and FREESE (1966).

According to the lethal-hit hypothesis one strand would be damaged mutationally whereas the opposite strand would obtain a lethal hit and could, therefore, not participate in the production of viable daughter cells.

The master-strand hypothesis, finally, postulates that only one of the two strands, the "master-strand", serves as template for replication. If this strand were affected, only mutant daughter cells would be produced, if the other strand were affected no mutation at all would arise.

We would like to confine ourselves to this enumeration of the four hypotheses, without discussing to what extent they may be realized. Such a discussion would be facilitated by further research along these lines.

5.2.10 Mutation Frequency and DNA Content of the Genome

Recently a relationship has been reported to exist between the frequency of point mutations induced in wild-type chromosomes by acute X- or γ-irradiation and the DNA content of the genome (ABRAHAMSON et al., 1973). The mutation frequency is proportional to the DNA content of the genome for the seven species studied so far, namely the bacterium *Escherichia coli*, the yeast *Saccharomyces cerevisiae*, the fungus *Neurospora crassa*, the fruit fly *Drosophila melanogaster* (spermatogonia), the tomato *Lycopersicon esculentum*, the mouse *Mus musculus* (spermatogonia), and the barley *Hordeum vulgare*. Whereas the mutation frequencies per locus per rad obtained for these species vary over three orders of magnitude, they vary only by a factor of three when they are adjusted for the amount of DNA per genome (calculation of the quotient between mutation frequency and DNA content). An "adjusted" mutation frequency close to the values obtained by ABRAHAMSON et al. (1973) has been recently published by LOPRIENO et al. (1975) for the yeast *Schizosaccharomyces pombe* (wild type).

In order to interpret this astonishing relationship one may assume (1) that it is not the gene locus but the nucleus which represents the "target", or (2) that the size of the gene locus is not the same for all the species studied but is proportional to the DNA content of the genome. The former explanation, however, is not in line with the expectation of the target hypothesis, and the latter explanation contradicts the hypothesis that structural genes have the same size in all species.

According to CONGER (1974) the mutation frequency is not only proportional to the DNA content but also to the chromatin mass (DNA plus protein) per cell.

Whatever the answers to these questions will be, this correlation has an interesting implication for man. Extrapolating from the seven organisms studied, one can calculate from the known value of the DNA content of the human genome, that the forward mutation frequency per locus per rad is 2.6×10^{-7} in man.

The great theoretical and practical significance of the relationship reported demands an especially careful search for eventual weak spots of it. Although very unlikely, it is not impossible that the relationship is fortuitous. Therefore an extension of this study to further species would be desirable. Moreover the fact that the forward mutations studied are - depending on the organism - either point mutations or gene mutations (as defined in III.5.2.1) complicates the interpretation of that correlation. Because of these (and other) uncertainties this mutation frequency/ DNA relationship could well have been discussed in the following paragraph.

5.3 Open Questions

Our understanding of the molecular basis of radiation mutagenesis is far from being complete, especially as regards the induction of mutations by ionizing radiation in higher organisms (e.g., insects, mammals).

Besides the elucidation of the molecular nature of the mutations induced by ionizing radiation, it is necessary to find out how these alterations are produced: What molecular processes following the absorption of radiation energy lead to a gene mutation or to a chromosome break? This question touches problems like direct and indirect radiation effects, energy transfer and the validity of the target hypothesis. It is, of course, not a specifically radiation genetical question but concerns the whole field of radiation biology (see ZIMMER, 1961).

Little can be said in molecular terms of the "repair" phenomenon observed in mutation studies carried out on multicellular organisms with ionizing radiation (see SOBELS, 1963). Another future task of radiation genetics will therefore be the elucidation of the molecular nature of repair processes - a proposal which carries the risk that in some cases the repair hypothesis will have to be replaced by a less attractive explanation.

As pointed out in III.5.1, the interpretation of results in this survey is based on the assumption that the chromosome contains only one DNA double helix. The following radiation-genetical observation is still waiting for becoming reconciled with this chromosome model. We have demonstrated (SCHEID and TRAUT, 1970, 1971a, b) in experiments performed with isolated chromosomes of the broad bean *Vicia faba* that X-ray induced "achromatic lesions" are characterized by two parallel "strands" (whatever the nature of these "strands" may be). This is shown in Figure 5.1. "Achromatic lesions" or "gaps", as visualized by the conventional squash and staining techniques, are nonstaining regions traversing the diameter of the chromatid. They are, at least in *V. faba*, the most frequent microscopically visible X-ray induced chromosomal alteration and share several characteristics with "real" mutations (see JAEA, 1966 and the review of COMINGS, 1974). Furthermore our observation is not in line with the suggestion (READ, 1959; BENDER et al., 1974) that an achromatic lesion is the result of a half-chromatid break. (BENDER et al., 1974, equate the "half-chromatid" with a single polynucleotide chain of the DNA). According to that hypothesis an chromatic lesion should reveal one strand only instead of the two strands observed by us.

5.4 Conclusion

As this survey has shown, our knowledge of the molecular basis of mutagenesis due to ionizing radiation is still fragmentary. Molecular radiation genetics, therefore offers future research a lot of interesting problems (e.g., the question of the "strandedness", see III.5.2.9). The prognosis that these problems will be solved in the near future is not too bad, if one extrapolates from the state of affairs as it existed a few years ago, when in a book entitled "Molecular Basis of Mutation" (DRAKE, 1970), only 1 and 1/4 page, and in a review article on the same topic (DRAKE, 1969) only one sentence was devoted to the molecular aspects of mutagenesis due to ionizing radiation.

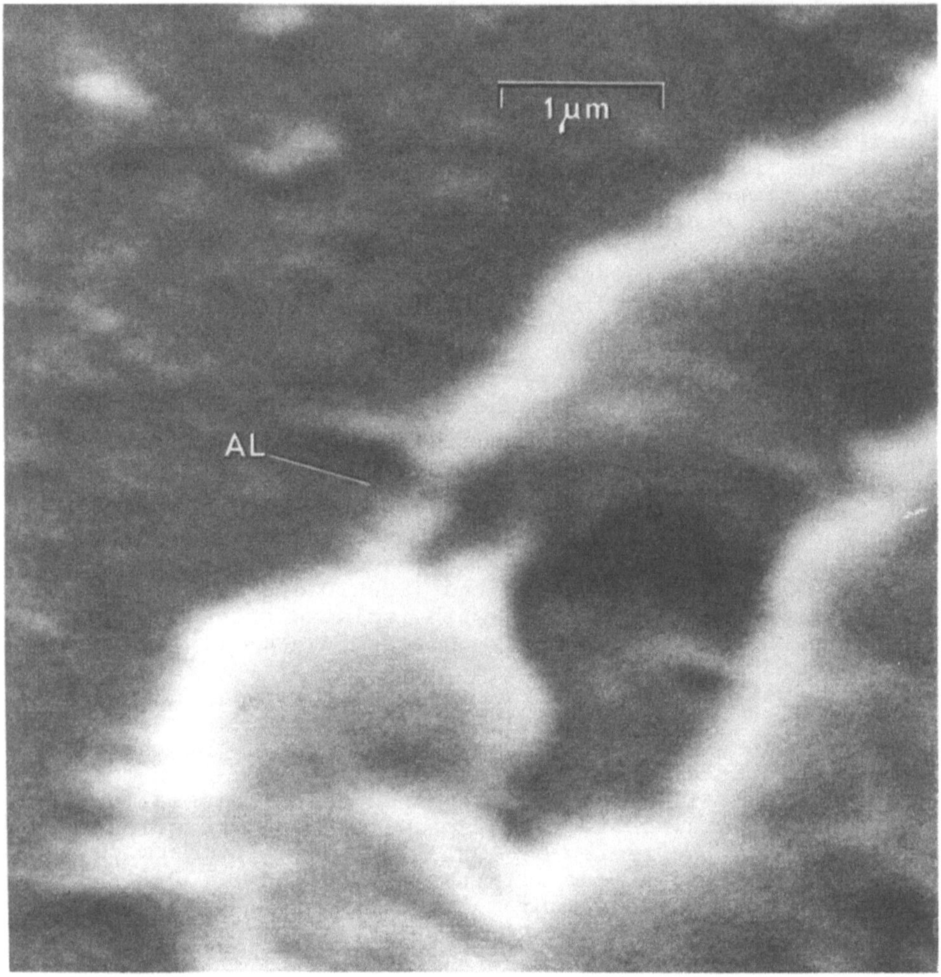

Fig. 5.1. Part of an isolated metaphase chromosome of the broad bean *Vicia faba* with an X-ray induced achromatic lesion (*AL*) exhibiting two parallel strands. Visualization by scanning electron microscopy (SCHEID and TRAUT, 1971a)

Acknowledgements. The author gratefully acknowledges the valuable help of Drs. B.A. Bridges (Falmer, Great Britain), R. Falk (Jerusalem, Israel), R.J. Munson (Harwell, Great Britain), H.B. Newcombe (Chalk River, Canada) and F.J. De Serres (Research Triangle Park, USA).

Note added in Proof. After the article on the molecular aspects of mutagenesis due to ionizing radiation has been written CHARLOTTE AUERBACH'S excellent book "Mutation Research" (London: Chapman and Hall, 1976) was published and the Glossary of RIEGER et al. (see "References") appeared in its fourth edition. The author (H.T.) also wants to mention the following recent papers dealing with the subject:

ABRAHAMSON, S., WOLFF, S.: Re-analysis of radiation-induced specific locus mutations in the mouse. Nature 264, 715-719 (1976)
MAGNI, G.E., PANZERI, L., SORA, S.: Molecular specifity of X-radiation and its repair in *Saccharomyces cerevisiae*. Mutation Res. 42, 223-234 (1977)
SCHALET, A.P., SANKARANARAYANAN, K.: Evaluation and re-evaluation of genetic radiation hazards in man. I. Interspecific comparison of estimates of mutation rates. Mutation Res. 35, 341-370 (1976)

Chapter 6

Conclusions and Perspectives

Most of the existing information about the effects of ionizing radiations on living matter has been obtained from experiments using much larger doses than those which a human population generally receives. More and more interest is therefore nowadays given to the low-dose range problems in terms of cancer induction and other genetic effects. Investigations in dose ranges which only slightly exceed the background level require, however, experiments with large numbers of animals and have to be extended over a long period of time. It is therfore necessary to tackle the problem also by other means. One useful approach is to explore the molecular background of various observed end effects. The hope is that the knowledge of the chain of molecular events taking place after the absorption of radiation would permit predictions regarding extrapolation to lower doses of these end effects.

An example of this molecular approach is given by the case of the hereditary disease Xeroderma Pigmentosum (X.P.). It is well known that UV light induces damage to DNA and that this damage, mainly in the form of the so called pyrimidine dimers, causes mutations in bacteria and animal cells. These pyrimidine dimers can, however, be removed from the DNA by certain enzymes, constituting the repair machinery for genetic and lethal damage in the cell. The X.P. patients were known to be very sensitive towards sun light, especially the UV-component, and in most cases the lesions induced by light developed into tumors. It was now found that in contrast to cells from normal persons the X.P.-cells were unable to repair the UV-induced damage in their DNA (CLEAVER, 1968). The remaining defects obviously caused transformation of normal cells to malignant ones. There exist a number of classes showing the X,P. syndrome, but in a large group the defective repair capability was due to a lack of a specific endonuclease. Without this enzyme the defect could not be "cut out" from the DNA. The X.P. case, however, did not only clarify parts of the important chain in a radiation-induced carcinogenesis, but it was also possible to device simple clinical tests to detect an X.P. child before it was born.

Final evidence for carcinogenicity of the same well-defined change in DNA, the pyrimidine dimer, came from SETLOW and HART (1975). It is based on the principle of product modification by means of a so called photoreactivating enzyme present in the cells of many species, among them fishes. This enzyme can split pyrimidine dimers in situ with the help of visible light, but does nothing else in the cell (as far as current knowledge goes). The experiment was performed by taking thyroid cells from a fish, treat them with UV (254 nm) and/or photoreactivating (PR) light (300-400 nm), reinject the cells into the thyroid gland of the fish (a population from one clone of the fish) and observe whether or not tumors were induced. The results from a statistically satisfying number of experiments may be summarized in a table as follows.

Treatment	*Tumors*
None	No
UV	Yes
UV + PR	No
PR	No
PR + UV	Yes

The conclusion must be that the UV-induced dimers were the cause of thyroid carcinogenesis (albeit with a very small probability of tumor induction per dimer).

A natural question is whether this type of molecular approach, which has been so successful to elucidate the UV-induced carcinogenesis, also would be applicable to ionizing radiation and its effects. In this field, however, considerably less progress has been made so far. One reason for this is that ionizing radiation in contrast to UV light induces a large number of changes in DNA (see II.2 and II.3) as well as in other molecules and structures in the cell and up to now it has not been possible to strictly sort out which changes are lethal, mutagenic or carcinogenic. The present state of knowledge has been discussed in III.2 where some identified radiation-induced changes in DNA are related to survival of phage. It can be shown that for double-strand phage, a single-strand break is tolerated while a double-strand break is lethal. Base damage seems also to contribute to lethality, but in contrast to a double-strand break which always is lethal, the G-value for base destruction is about 15 times higher than the corresponding G-value for biological inactivation (see III. 2.2.2.1). Therefore there may be a specific type of base damage occurring at low frequency which causes the death of the phage, other types of damage being tolerated (or repaired by the host).

The situation becomes even more complex in higher organisms. As already mentioned, repair systems are operating in cells to correct for damage introduced into their DNA and there is evidence that bacteria and animal cells have the capability to handle all types of damage, i.e., no type of damage is irrevocably lethal.

In the case of ionizing radiation, repair capacity is only one factor which influences survival as exemplified in the following. The survival curves after X-irradiation of two strains of *E. coli*, B_{s-1} and B/r, both seem to follow a one hit kinetic, i.e., the plot of the logarithm for the number of surviving bacteria against the dose gives a straight line. The D_{37} dose is, however, found to be several times larger for the B/r strain. UV irradiation of the same two strains has, however, been found to give a strikingly different result. The survival curve for the B_{s-1} strain is still a straight line in the log-lin plot, whereas the B/r strains shows a large shoulder and only at low survival a linear part is observed. This phenomenon is interpreted as follows (HAYNES, 1966): The B_{s-1} strain is incapable of repairing the UV-induced damage and a bacteria is killed as soon as one defect is introduced into its chromosome. The B/r strain on the other hand has the capability to repair UV-induced damage, but at higher doses its enzymatic repair system is saturated and additonal dimers will cause cell death.

It has to be noted that there is a large variation in repair capacity for UV damage among bacteria, i.e., there seem to be quantitative differences in repair capacity depending on the amount of enzyme molecules available.

As mentioned above, the corresponding curves for X-irradiated bacteria (and similarly for mammalian cells) showed no or only a minor shoulder and there was no difference between sensitive and resistant bacteria in this respect. We know that repair of X-ray-induced damage takes place. However, the data discussed here indicate that the repair systems are not saturated and that there is a qualitative difference between bacteria of varying sensitivity.

One could visualize the situation as follows. In some organisms the double-strand break seems to be lethal while the single-strand break has no harmful effect. From an enzymatic point of view the same enzymes would be required to repair single or double-strand breaks. The double-strand break, however, will immediately cause a DNA molecule to break into two parts unless some structural support exists which can hold the two pieces together. In order to permit repair, the DNA molecule must therefore be in a complex with some other structure in the cell to keep its linear integrity intact.

The way in which the DNA is organized in complex with for example RNA, proteins or the whole membrane may therefore be the crucial point determining whether existing repair enzymes can act or not. This hypothesis, of course, also suggests that the double-strand break is a much more important lesion in bacteria and higher cells than all types of single-strand damage which would not require any extra structural support for the repair to take place.

As a general perspective for the future we think that in order to understand the molecular biology of the radiation effects one must study both the isolated molecules on the one hand and functioning biological cells on the other. Phage studies represents a situation between these two extremes, but one must also consider molecular systems with an increasing level of organization and complexity.

The DNA in a bacterial cell is packed according to some regular rules. When the folded chromosome is released under mild lysing conditions it behaves more like a particle than a fiber of DNA. The compact structure of the chromosome seems to be due to the fact that DNA is folded into about 50 loops and each loop is supercoiled. Since the chromosome is unfolded by RNase treatment or by growing the bacteria in rifampicin, RNA-DNA interaction probably at the base of each loop seems to be responsible for this structure (WORCEL et al., 1974; PETTIJOHN and HECHT, 1974).

An experimental approach which points to the importance of structural factors in determining radiosensitivity has been discussed in III.2. It is possible to isolate complexes from *E. coli* containing most of the DNA attached to the cell membrane. These membrane complexes synthesize new DNA when the four nucleotide triphosphates are added. There is therefore reason to believe that the complex contains the enzymes and structural RNA necessary for DNA synthesis. Irradiation of the bacteria before isolation of the complex reduces the DNA synthetic capacity, but there is for example no difference in sensitivity between a sensitive strain *E. coli* B_{s-1} and the resistant one *E. coli* B/r when they are irradiated with low LET radiation, e.g., fast electrons.

In the case of high LET, fast neutrons, B/r is more affected than B_{s-1}. This indicates that the structural organisation of the target is different in the two strains. Evidence that RNA constitutes an important part of the target, at least in the B/r strain, comes from the fact that rifampicin treatment increased the radiosensitivity with respect to the DNA synthetic capacity as well as survival in this strain.

Another example of the possible influence of structural factors on radiosensitivity has also been pointed out in studies on the very radioresistant bacteria *Micrococcus radiodurans*. These bacteria seem to be able to repair all types of damage as indicated by the large shoulder of the survival curve. The repair efficiency has been attributed to existence of multiple sites of attachment between DNA and plasma membranes in the cell (BURRELL et al., 1971), i.e., a structural support which would permit repair of double-strand breaks.

The field of repair enzymes in radiobiology is in many aspects analogous to the development in the field of enzyme research in general: based on the study of the isolated macromolecules one must try to extend the knowledge to the more organized structures in the living cell, such as membrane-protein units, where the interaction between molecules of different structure and functional characteristics makes "the whole apparently more than just the simple sum of the units".

Radiation therapy is one area of practical implication that is likely to profit in the future from a molecular approach. Radiation sensitization and protection have to be understood better, and the possibility to use inhibitors or activators of the repair enzymes have to be explored.

Our discussion also points towards other future possibilities of a molecular approach to such practical problems as examplified by the case of Xeroderma Pigmentosum. In this case of UV-induced skin cancer it has been possible to identify a sensitive group of individuals with a specific molecular defect. Methods have also been developed to detect this defect prenatally. In this case the link between exposure to UV and induction of skin cancer was particularly easy to detect. The molecular basis for UV-induced damage in DNA and its repair was also fairly well understood. Other cases may be expected to be considerably less straightforward. As mentioned earlier the situation concerning ionizing radiation is much more complex both with regard to number of different chemical changes introduced as well as with regard to the problem of repair.

Because of the biological relevance the effects of low doses of ionizing radiation have become a major field of interest. It is possible today to make a rough estimate of the number of genetic defects and cancer cases induced by a certain dose of ionizing radiation; it is also possible to predict on a quantitative basis what will happen if the dose is reduced or increased. We do not know, however, if an increase in dose will lead to an increase in cancer occurring at random in the population or if a large part of the cases originate from a small group of very sensitive individuals. Such a group of individuals could, in analogy to the Xeroderma Pigmentosum patients, have a genetically programmed reduced activity of certain repair enzymes, or their low resistance against lethal and mutagenic changes could be due to other reasons as discussed above. In cases where radioresistance is not just a matter of enzymatic activity, the problem of determining the critical factor has to be tackled on a broad basis. Not only damage to DNA itself has to be investigated, but also for example the influence of structural organisation of its neighbouring molecules.

The last decade of research in radiobiology has been characterized by extensive studies of repair mechanisms. To a large extent the experiments have been performed with microorganisms where it is possible to link repair deficiencies to mutations in specific genes. Only limited efforts have so far been made to include structural considerations into the picture. It is, however, our guess that this more biophysical approach will gain momentum and be of major importance in the near future. This in turn will need extended collaboration between different specialists to bridge the gaps between the various fields of research.

References

ABELSON, J., THOMAS, C.A., Jr.: The anatomy of the T5 bacteriophage DNA molecule. J. Mol. Biol. 18, 262-291 (1966)

ABRAHAMSON, S., BENDER, M.A., CONGER, A.D., WOLFF, S.: Uniformity of radiation-induced mutation rates among different species. Nature (London) 245, 460-462 (1973)

ADAMS, G.E., COOKE, M.S.: Electron affinic sensitization. 1. A structural basis for chemical radiosensitizers in bacteria. Intern. J. Radiat. Biol. 15, 457-471 (1969)

ADAMS, G.E.: Radiation chemical mechanisms in radiation biology. In: Advances in Radiation Chemistry. Burton, M., Magee, J.L. (eds.), Vol. III, pp. 125-208. New York: Wiley-Interscience, 1972

ADELSTEIN, S.J., MANASEK, G.B.: X-irradiation of splemic lymphocytes; thymidine pools and radio thymidine incorporation into DNA. Intern. J. Radiat. Biol. 12, 587-591 (1967)

AHNSTRÖM, G., ERIXON, K.: Radiation-induced strand breakage in DNA from mammalian cells. Strand separation in alkaline solution. Intern. J. Radiat. Biol. 23, 285-289 (1973)

AHNSTRÖM, G., EDVARDSSON, K.-A.: Radiation induced single-strand breaks in DNA determined by rate of alkaline strand separation and hydroxylapatite chromatography: an alternative to velocity sedimentation. Intern. J. Radiat. Biol. 26, 493-497 (1974)

ALPER, T.: Bacteriophage as indicator in radiation chemistry. Radiat. Res. 2, 119-134 (1955)

ALPER, T.: Proc. 2nd Symp. Microdosim. Edited by Ebert, EUR 4452-d-e. Brussels, p. 5 (1970) and Biophysical aspects of radiation quality. (Vienna, I.A.E.A.) SM-14511 (1971)

ALPER, T.: Radiobiological support for radiotherapy. What kind of research? Lancet 1328-1330 (1974)

ANAGNOSTOPOULOS, C., SPIZIZEN, J.: Requirements for transformation in *Bacillus subtilis*. J. Bacteriol. 81, 741-746 (1961)

ANDERSON, N.G., BURGER, C.L.: Separation of cell components in the zonal ultracentrifuge. Science 136, 646-648 (1962)

ARWERT, F., VENEMA, G.: Transformation in *Bacillus subtilis:* Fate of newly introduced transforming DNA. Mol. Gen. Genetics 123, 185-198 (1975)

AUERBACH, C.: Mutation. Part I: Methods. Edingburgh and London: Oliver and Boyd, 1962

AUERBACH, C., KILBEY, B.J.: Mutation in eukaryotes. Ann. Rev. Genetics 5, 163-218 (1971)

BELLI, J.A., COOPER, S., BROWN, J.A.: Sedimentation properties of mammalian cell DNA; evidence that nonspecific molecular aggregation does not occur during cell lysis. Intern. J. Radiat. Biol. 21, 603-606 (1972)

BENDER, A.B., GRIGGS, H.G., BEDFORD, J.S.: Mechanisms of chromosomal aberration production. III. Chemicals and ionizing radiation. Mutation Res. 23, 197-212 (1974)

BEN-HUR, E., ELKIND, M.M.: Thermally enhanced radioresponse of cultured Chinese hamster cells: Damage and repair of single-stranded DNA and a DNA complex. Radiat. Res. 59, 484-495 (1974)

BENZINGER, R., DELIUS, H., JAENISCH, R., HOFSCHNEIDER, P.H.: Preparation and properties of *E. coli* component for infectious DNA from bacteriophage ΦX 174 and M 13 and RNA from bacteriophage M 12. Europ. J. Biochem. 2, 414-428 (1967)

BERRY, R.J., HELL, E., LAJTHA, L.G.: Mechanism of the radiation effect on the synthesis of DNA. Nature (London) 186, 563-564 (1960)

BERRY, R.J., HELL, E., LAJTHA, L.G., EBERT, P.I.: Further studies on the mechanism of inhibition of DNA-synthesis by ionizing radiation. Intern. J. Radiat. Biol. 4, 61-66 (1962)

BILLEN, D.: Alteration in DNA synthesizing capacity in bacteria: an in vivo-in vitro study. Biochim. Biophys. Acta 55, 960-968 (1962)

BILLEN, D., HEWITT, R., JORGENSEN, G.: X-ray induced perturbations in the replication of the bacterial chromosome. Biochim. Biophys. Acta 103, 440-454 (1965)

BLEICHRODT, J.F., BERENDS-VAN ABKOUDE, E.R., VERHEIJ, W.S.D.: Induction of mutations in bacteriophage ΦX 174 by ionizing radiation. Proc. Ned. Akad. Wet., Ser. C 74, 172-182 (1971)

BLEICHRODT, J.F., VERHEIJ, W.S.D.: Reversion of cistron A amber mutants of bacteriophage ΦX 174 by ionizing radiation. Mutation Res. 18, 363-365 (1973)

BLEICHRODT, J.F., VERHEIJ, W.S.D.: Influence of oxygen on the induction of mutations in bacteriophage ΦX 174 by ionizing radiation. Intern. J. Radiat. Biol. 25, 505-512 (1974)

BLOK, J.: Lethal and mutagenic action of gamma rays on bacteriophage DNA. In: Proc. 3rd Intern. Conf. Radiat. Res. pp. 423-437. Silini, G. (ed.). Amsterdam: North Holland, 1967

BLOK, J., LUTHJENS, L.H., ROOS, A.L.M.: The radiosensitivity of bacteriophage DNA in aqueous solution. Radiat. Res. 30, 468-482 (1967)

BLOK, J., LOMAN, H.: The effects of gamma-radiation in DNA. Curr. Top. Radiat. Res. 9, 165-245 (1973)

BOHNE, L., COQUERELL, Th., HAGEN, U.: Radiation sensitivity of bacteriophage DNA: II. Breaks and crosslinks after irradiation in vivo. Intern. J. Radiat. Biol. 17, 205-215 (1970)

BONURA, T., TOWN, C.D., SMITH, K.C., KAPLAN, H.S.: The influence of oxygen on the yield of DNA double-strand breaks in X-irradiated *E. coli* K-12. Radiat. Res. 63, 567-577 (1975)

BOPP, A., HAGEN, U.: End group determination in γ-irradiated DNA. Biochim. Biophys. Acta 209, 320-326 (1970)

BOPP, A., CARPY, S., BURKART, B., HAGEN, U.: Action of exonuclease III on γ-irradiated DNA. Biochim. Biophys. Acta 294, 47-56 (1973)

BOYCE, R.P., TEPPER, M.: X-ray-induced single-strand breaks and joining of broken strands in superinfecting lambda DNA in *Escherichia coli* lysogenic for lambda. Virology 34, 344-351 (1968)

BRAUN, A., GROSSMAN, L.: An endonuclease from *Escherichia coli* that acts preferentially on UV-irradiated DNA and is absent from the *uvrA* and *uvrB* mutants. Proc. Natl. Acad. Sci. 71, 1838-1842 (1974)

BREMAN, H.: Herstel van bakteriofaag DNA na beschadiging door gamma strahling. Thesis, Free Univ. Amsterdam, 1972

BRENT, T.P., BUTLER, J.A.V., CRATHORN, A.R.: Effects of irradiation on synthesis of DNA and mitosis in synchronous cultures of HeLa cells. Nature (London) 210, 393-395 (1966)

BRENT, T.P., WHEATLEY, S.A.: Repair and replication of X-irradiated HeLa cell DNA. Intern. J. Radiat. Biol. 19, 339-348 (1971)

BRESLER, S.E., KALININ, V.L., PERUMOV, D.A.: Inactivation and mutagenesis on isolated DNA. Theory of inactivation of transforming DNA. Mutation Res. 4, 389-398 (1967)

BREWER, E.N., NYGAARD, O.F.: Correlation between unrepaired radiation-induced DNA strand breaks and mitotic delay in *Physarum Polycephalum*. Nature (New Biol.) 239, 108-110 (1972)
BRIDGES, B.A.: Sensitization of organisms to radiation by sulfhydryl-binding agents. In: Advances in Radiation Biology. Augenstein, L.G., Mason, R., Zelle, M. (eds.). Vol. III, pp. 123-176. New York and London: Academic Press, 1969
BRIDGES, B.A., LAW, J., MUNSON, R.J.: Mutagenesis in *Escherichia coli*. II. Evidence for a common pathway for mutagenesis by ultraviolet light, ionizing radiation and thymine deprivation. Mol. Gen. Genetics 103, 266-273 (1968)
BRIDGES, B.A., MUNSON, R.J.: Genetic radiation damage and its repair in *Escherichia coli*. Curr. Top. Radiat. Res. 4, 95-188 (1968)
BURGI, E., HERSHEY, A.D.: Sedimentation rate as a measure of molecular weight of DNA. Biophys. J. 3, 309-321 (1963)
BURRELL, A.D., FELDSCHREIBER, P., DEAN, C.J.: DNA-membrane association and the repair of double breaks in X-irradiated *Micrococcus radiodurans*. Biochem. Biophys. Acta 247, 38-53 (1971)
CABELA, E., WAWRA, E., ALTMAN, H.: Untersuchungen über die Reparatur von DNS-Strangbrüchen in Chlorella-Zellen nach Gamma-Bestrahlung. Biophysik 7, 200-204 (1971)
CASARETT, A.P.: Radiation Biology. New Jersey: Prentice-Hall, Inc., 1968
CATO, A., GUILD, W.R.: Transformation and DNA size. I. Activity of fragments of defined size and a fit to a random double crossover model. J. Mol. Biol. 37, 157-178 (1968)
CERUTTI, P.A.: Effects of ionizing radiation on mammalian cells. Naturwissenschaften 61, 51-59 (1974a)
CERUTTI, P.A.: Excision repair of DNA base damage. Life Sci. 15, 1567-1575 (1974b)
CHARLESBY, A.: Atomic Radiation and Polymers. Oxford-London-New York-Paris: Pergamon Press, 1960
CHERVENKA, C.H.: A manual of methods for the analytical ultracentrifuge. Palo Alto, Calif.: Beckman Inst. Inc. Spinco Div., 1969
CHESSIN, H., SUMMERS, W.C.: Initiation by RNA polymerase on UV or X-ray damaged T7 DNA. Biochem. Biophys. Res. Commun. 38, 40-45 (1970)
CHIA, D., SCHUMAKER, V.N.: A rotor speed dependent crossover in sedimentation velocities of DNAs of different sizes. Biochim. Biophys. Res. Commun. 56, 241-245 (1974)
CHILTON, M.D.: Transforming activity in both complementary strands of *Bacillus subtilis* DNA. Science 157, 817-819 (1967)
CLARKSON, J.M., PAINTER, R.B.: Repair of X-ray damage in aging WI-38 cells. Mutation Res. 23, 107-112 (1974)
CLEAVER, J.E.: Defective repair replication of DNA in Xeroderma Pigmentosum. Nature (London) 218, 652-656 (1968)
CLEAVER, J.E.: Xeroderma pigmentosum: A human disease in which an initial stage in DNA repair is defective. Proc. Natl. Acad. Sci. 63, 428-435 (1969)
CLEAVER, J.E.: Conformation of DNA in alkaline sucrose: The subunit hypothesis in mammalian cells. Biochem. Biophys. Res. Commun. 59, 92-99 (1974a)
CLEAVER, J.E.: Repair processes for photochemical damage in mammalian cells. In: Advances in Radiation Biology, Vol. IV, pp. 1-75. Lett, J.T., Adler, H., Zelle, M. (eds.). New York-London: Academic Press, 1974b
Cold Spring Harbor Symposia on Quantitative Biology. Vol. 38. Chromosome Structure and Function. New York: Cold Spring Harbor Lab., 1974
COLE, A., SHONKA, F., CORRY, P., GRANT COOPER, W.: CHO cell repair of single strand and double strand DNA breaks induced by gamma and alpha radiations. In: Molecular Mechanisms for Repair of DNA, pp. 665-676. Setlow, R.B., Hanawalt, P.C. (eds.). New York: Plenum Press, 1975

COMINGS, D.E.: What is a chromosome break? In: Chromosomes and Cancer. German, J. (ed.). New York, London, Sydney, Toronto: Wiley, 1974, pp. 95-133

CONGER, A.D.: Is DNA the only chromosomal component contributing to target size for mutation induction by ionizing radiation? Mutation Res. 25, 131-134 (1974)

COQUERELLE, T., BOHNE, L., HAGEN, U., MERKWITZ, J.: Strahlenempfindlichkeit von Bakteriophagen-DNS I. Brüche und Vernetzungen nach Bestrahlung in vitro. Z. Naturforsch. 24b, 485-493 (1969)

COQUERELLE, T., BOPP, E., HAGEN, U.: Gene function and strand breaks in DNA of γ-irradiated T7 bacteriophages. Intern. J. Radiat. Biol. 26, 219-226 (1974)

COQUERELLE, T., BOPP, A., KESSLER, B., HAGEN, U.: Strand breaks and 5' end-groups in DNA of irradiated thymocytes. Intern. J. Radiat. Biol. 24, 397-404 (1973)

COQUERELLE, T., HAGEN, U.: Loss of adsorption and injection abilities in irradiated phage T1. Intern. J. Radiat. Biol. 21, 31-41 (1972)

CORRY, P.M., COLE, A.: Radiation-induced double-strand scission of the DNA of mammalian metaphase chromosomes. Radiat. Res. 36, 528-543 (1968)

CORRY, P.M., COLE, A.: Double strand rejoining in mammalian DNA. Nature (New Biol.) 245, 100-101 (1973)

CRAMP, W.A., ELGAT, M.: The effects of ionizing radiation on thymidine uptake into Escherichia coli $B/_r$ and B_{s-1}. Radiat. Res. 51, 121-133 (1972)

CRAMP, W.A., WATKINS, D.K., COLLINS, J.: Effects of ionizing radiation on bacterial DNA-membrane-complexes. Nature (New Biol.) 235, 76 (1972a)

CRAMP, W.A., WATKINS, D.K., COLLINS, J.: The synthesis of DNA by membrane - DNA complexes from E. coli $B/_r$ and B_{s-1}, after exposure to fast electron and neutrons: the measurement of oxygenenhancement ratios and RBE values. Intern. J. Radiat. Biol. 22, 379-387 (1972b)

CRAMP, W.A., BRYANT, P.E.: The effects of rifampicin on electron- and neutron-irradiated Escherichia coli B/r and B_{s-1}: Survival, DNA degradation and DNA synthesis by membrane fragments. Inter. J. Radiat. Biol. 27, 143-156 (1975)

CROTHERS, D.M., ZIMM, B.H.: Viscosity and sedimentation of the DNA from bacteriophages T2 and T7 and the relation to molecular weight. J. Mol. Biol. 12, 525-536 (1965)

DEAN, C.J., LITTLE, J.G., SERIANNI, R.W.: The control of post-irradiation DNA breakdown in *Micrococcus radiodurans*. Biochem. Biophys. Res. Commun. 39, 126-133 (1970)

DEAN, C.J., ORMEROD, M.G., SERIANNI, R.W., ALEXANDER, P.: DNA strand breakage in cells irradiated with X-rays. Nature (London) 222, 1042-1044 (1969)

DERTINGER, H., JUNG, H. (eds.): Molecular Radiation Biology, pp. 155-194. (Heidelberg Science Library, Vol. 12). New York-Heidelberg-Berlin: Springer, 1970

DRAKE, J.W.: Mutagenic mechanisms. Ann. Rev. Genetics 3, 247-268 (1969)

DRAKE, J.W.: The Molecular Basis of Mutation. San Francisco, Düsseldorf, Johannesburg, London, Panama, Singapore, Sydney, Toronto: Holden-Day, 1970

DRIEDGER, A.A., GRAYSTON, M.J.: Demonstration of two types of DNA repair in X-irradiated *Micrococcus radiodurans*. Can. J. Microbiol. 17, 495-499 (1971)

DUBININ, N.P., SOYFER, V.N.: Chromosome breakage and complete genic mutation production in molecular terms. Mutation Res. 8, 353-365 (1969)

DYKE, J.G. VAN: An oxygen effect on X-irradiated intracellular bacteriophage T2Hr. Biochem. Biophys. Res. Commun. 1, 54-58 (1960)

EHMANN, U.K., LETT, J.T.: Review and evaluation of molecular weight calculations from the sedimentation profiles of irradiated DNA. Radiat. Res. 54, 152-162 (1973)

ELIAS, H.G.: Ultrazentrifugenmethoden. München: Beckman Instruments, 1961

ELKIND, M.M.: Sedimentation of DNA released from Chinese hamster cells. Biophys. J. 11, 502-520 (1971)
ELKIND, M.M., CHANG-LIU, C.-M.: Repair of a DNA complex from X-irradiated Chinese hamster cells. Intern. J. Radiat. Biol. 22, 75-90 (1972a)
ELKIND, M.M., CHANG-LIU, C.-M.: Actinomycin D inhibition of repair of a DNA complex from Chinese hamster cells. Intern. J. Radiat. Biol. 22, 313-324 (1972b)
ELKIND, M.M., KAMPER, C.: Two forms of repair of DNA in mammalian cells following irradiation. Biophys. J. 10, 237-245 (1970)
EMMERSON, P.T.: Recombination deficient mutants of *Escherichia coli* K12 that map between *thyA* and *argA*. Genetics 68, 19-30 (1968)
EMMERSON, P.T., WILLSON, R.L.: Pulse radiolysis of aqueous solutions of thymine and triacetoneamine-N-oxyl. J. Phys. Chem. 72, 3669-3671 (1968)
EPSTEIN, J., WILLIAMS, J.R., LITTLE, J.B.: Deficient DNA repair in human progeroid cells. Proc. Natl. Acad. Sci. 70, 977-981 (1973)
EPSTEIN, J., WILLIAMS, J.R., LITTLE, J.B.: Rate of DNA repair in progeric and normal human fibroblasts. Biochem. Biophys. Res. Commun. 59, 850-857 (1974)
FIERS, W., SINSHEIMER, R.L.: The structure of DNA of bacteriophage ΦX 174. III. Ultracentrifuge evidence for a ring structure. J. Mol. Biol. 5, 424-434 (1962)
FÖLDES, J., TRAUTNER, T.A.: Infectious DNA from a newly isolated *B. subtilis* phage. Z. Vererbungsl. 95, 57-65 (1964)
FOX, M., FOX, B.W.: Repair replication in X-irradiated lymphoma cells in vitro. Intern. J. Radiat. Biol. 23, 333-358 (1973)
FRANKE, H.D.: Die Anwendung strahlensensibilisierender Substanzen in der Strahlentherapie. Strahlentherapie 143, 296-317 (1972)
FREESE, E.B., FREESE, E.: Induction of pure mutant clones by repair of inactivating DNA alterations in phage T4. Genetics 54, 1055-1067 (1966)
FREIFELDER, D.: Mechanism of inactivation of coliphage T7 by X-rays. Proc. Natl. Acad. Sci. 54, 128-134 (1965)
FREIFELDER, D.: Inactivation of phage α by single-strand breakage. Virology 30, 328-332 (1966)
FREIFELDER, D., TRUMBO, B.: Matching of single-strand breaks to form double-strand breaks in DNA. Biopolymers 7, 681-693 (1969)
FREY, R., HAGEN, U.: Oxygen effect on γ-irradiated DNA. Radiation Environ. Biophys. 11, 125-133 (1974)
FUJITA, H., MIYAKOSHI, M.: Effect of γ-irradiation on the synthesis of ribosomal RNA and ribosomal subparticles in *E. coli*. Radiat. Res. 56, 186-200 (1973)
FUKS, Z., SMITH, K.C.: Effect of quinacrine on X-ray sensitivity and the repair of damaged DNA in E. coli K12. Radiat. Res. 48, 63-73 (1971)
FUNK, F., PERSON, S., BOCHRATH, R.C., Jr.: The mechanism of inactivation of T4 bacteriophage by tritium decay. Biophysical J. 8, 1037-1050 (1968)
GANESAN, A.K., SMITH, K.C.: Requirement for protein synthesis in *rec*-dependent repair of DNA in *Escherichia coli* after ultraviolet or X-irradiation. J. Bacteriol. 111, 575-585 (1972)
BERGER, G.B., ALTMANN, K.I.: Radiation Biochemistry, Vol. II. London-New York: Academic Press, 1970
GINOZA, W.: The effects of ionizing radiation on nucleic acids of bacteriophages and bacterial cells. Ann. Rev. Microbiol. 21, 325-368 (1967)
GODDARD, J.P., WEISS, J.J., WHEELER, C.M.: Error frequency during in vitro transcription of poly U is increased with γ-irradiated RNA polymerase. Nature (London) 222, 670-671 (1969)
GODDARD, J.P., WEISS, J.J., WHEELER, C.M.: Studies on RNA synthesis primed by damaged templates. II. Homopolyribonucleotide templates damaged by γ-radiation. Biochim. Biophys. Acta 199, 139-146 (1970)

GREEN, M.H.L., GRAY, W.J.H., SEDGWICK, S.G., BRIDGES, B.A.: Repair of DNA damage produced by gamma-radiation in *Escherichia coli* K-12 and a radiation-sensitive *exrA* derivative during inhibition of protein synthesis and normal DNA replication by chloramphenicol. J. Gen. Microbiol. 77, 99-108 (1973)

GROSS, J.D., GRUNSTEIN, J., WITKIN, E.: Inviability of *recA*⁻ derivatives of the DNA polymerase mutant of De Lucia and Cairns. J. Mol. Biol. 58, 631-634 (1971)

GROSSMAN, L.: Enzymes involved in the repair of DNA. In: Advances in Radiation Biology. Vol. IV, pp. 77-129. Lett, J.T., Adler, H., Zelle, M. (eds.). New York-London: Academic Press, 1974

GRUNOW, R., ECKART, W.: Strahleninduzierte Veränderungen an isolierter DNS und deren funktionelle Konsequenzen. VII. Beeinflussung der Transformationsaktivität für die Marker Tryptophan und Histidin von *B. subtilis* nach Röntgenbestrahlung. Stud. Biophys. 18, 235-241 (1969)

GRUNOW, R., GÜNTHER, K.: Strahleninduzierte Veränderungen an isolierter DNS und deren funktionelle Konsequenzen. XII. Zur Inaktivierung der Transformationsaktivität von *Bacillus subtilis* DNS nach Röntgenbestrahlung mit niedrigen Dosen. Stud. Biophys. 27, 21-28 (1971)

GRUNOW, R., GÜNTHER, K., ECKART, W., SCHWARZ, K.: Strahleninduzierte Veränderungen an isolierter DNS und deren funktionelle Konsequenzen. VIII. Modellvorstellungen über den Einfluß strahleninduzierter DNS-Veränderungen auf die Transformationsaktivität. Stud. Biophys. 18, 245-251 (1969)

GÜNTHER, K., GRUNOW, R.: Strahleninduzierte Veränderungen an isolierter DNS und deren funktionelle Konsequenzen. X. Mathematische Auswertung eines Modells über den Einfluß strahleninduzierter DNS-Veränderungen auf die Transformationsaktivität. Stud. Biophys. 27, 1-12 (1971)

GÜNTHER, K., KRUNOW, R., HARTWIG, M., KÖRNER, J.: Strahleninduzierte Veränderungen an isolierter DNS und deren funktionelle Konsequenzen. Bestimmung biologischer Parameter des *Bacillus subtilis* Transformationssystems mittels DNase-geschädigter DNS. Stud. Biophys. 48, 185-201 (1975)

GUILD, W.R., DEFILIPPES, F.M.: Ionizing radiation and ultrasonic evidence for the minimum unit of transforming principle DNA. Biochim. Biophys. Acta 26, 241-251 (1957)

GUTHRIE, G., SINSHEIMER, R.L.: Observations on the infection of bacterial protoplasts with the DNA of bacteriophage ΦX 174. Biochim. Biophys. Acta 72, 290-297 (1963)

HAGEN, U.: Bestimmung von Einzel- und Doppelbrüchen in bestrahlter Desoxyribonukleinsäure durch Molekulargewichtsverteilung. Biochim. Biophys. Acta 134, 45-58 (1967)

HAGEN, U.: Molekulare Grundlagen der Strahlensensibilisierung. Strahlentherapie 143, 237-243 (1972)

HAGEN, U.: Ultracentrifugation. In: Experimental Methods in Biophysical Chemistry. Nicolau, Cl. (ed.). New York-London-Sydney: Wiley, 1973

HAGEN, U., COQUERELLE, T.: Correlation between sedimentation coefficient and molecular weight of denatured DNA. Biochim. Biophys. Acta 374, 271-282 (1974)

HAGEN, U., ULLRICH, M., JUNG, H.: Transcription on irradiated DNA. Intern. J. Radiat. Biol. 16, 597-601 (1969)

HAGEN, U., ULLRICH, M., PETERSEN, E.E., WERNER, E., KRÖGER, H.: Enzymatic RNA synthesis on irradiated DNA. Biochim. Biophys. Acta 199, 115-125 (1970)

HANAWALT, P.C., SETLOW, R.B. (eds.): Molecular Mechanisms for Repair of DNA. New York: Plenum Press, 1975

HARIHARAN, P.V., CERUTTI, P.A.: Repair of gamma-ray-induced thymine damage in *Micrococcus radiodurans*. Nature (New Biol.) 229, 247-249 (1971)

HARIHARAN, P.V., CERUTTI, P.A.: Formation and repair of gamma-ray induced thymine damage in *Micrococcus radiodurans*. J. Mol. Biol. 66, 65-81 (1972)

HARIHARAN, P.V., CERUTTI, P.A.: Excision of damaged thymine residues from gamma-irradiated poly (dA-dT) by crude extracts of *Escherichia coli*. Proc. Natl. Acad. Sci. 71, 3532-3636 (1974a)

HARIHARAN, P.V., CERUTTI, P.A.: The incision and strand-rejoining step in the excision repair of 5,6-dihydroxy-dihydrothymine by crude *E. coli* extracts. Biochem. Biophys. Res. Commun. 61, 375-379 (1974b)

HARIHARAN, P.V., CERUTTI, P.A.: Excision of gamma-ray damaged thymine by *E. coli* extracts is due to the 5'→3' exonuclease associated with DNA polymerase I. Biochem. Biophys. Res. Commun. 61, 971-976 (1974c)

HARIHARAN, P.V., HUTCHINSON, F.: Neutral sucrose gradient sedimentation of very large DNA from *Bacillus subtilis*. II Double-strand breaks formed by gamma ray irradiation of the cells. J. Mol. Biol. 75, 479-494 (1973)

HARM, W.: Multiplicity reactivation, marker rescue and genetic recombination in phage T4 following X-ray inactivation. Virology 5, 337-361 (1958)

HARM, W., RUPERT, C.: Infection of transformable cells of *Haemophilus influenzae* by bacteriophage and bacteriophage DNA. Z. Vererbungsl. 94, 336-348 (1963)

HARRINGTON, H.: Effect of X-irradiation on the priming activity of DNA. Proc. Natl. Acad. Sci. 51, 59-66 (1964)

HARTWIG, M., GRUNOW, R.: Zur Inaktivierung der Transformationsaktivität von DNS nach Röntgenbestrahlung in wässriger Lösung. Stud. Biophys. 38, 41-44 (1973)

HASHIMOTO, Y., ONO, T., OKADA, S.: Radiosensitivities of DNA molecules in lymphocytes from the circulating blood of man. Blood 45, 503-510 (1975)

HAYNES, R.H.: The interpretation of microbial inactivation and recovery phenomena. Radiat. Res. Suppl. 6, 1-24 (1966)

HEARST, J.E., VINOGRAD, J.: The net hydration of T4 bacteriophage deoxyribonucleic acid and the effect of hydration on buoyant behavior in a density gradient at equilibrium in the ultracentrifuge. Proc. Natl. Acad. Sci. 47, 1005-1025 (1961)

HENGLEIN, A., SCHNABEL, W., WENDENBURG, J.: Einführung in die Strahlenchemie. Weinheim: Verlag Chemie, 1969

HERFORD, N., SUEOKA, N.: Chromosomal location of antibiotic resistance markers in *Bacillus subtilis*. J. Mol. Biol. 51, 267-286 (1970)

HERSHEY, A.D., KAMEN, M.D., KENNEDY, J.W., GEST, H.: The mortality of bacteriophage containing assimilated radioactive phosphorus. J. Gen. Physiol. 34, 305-319 (1951)

HILL, M.: The search of X-ray induced DNA degradation in mouse L cells. Intern. J. Radiat. Biol. 15, 483-490 (1969)

HINKLE, D.C., RING, I., CHAMBERLIN, M.I.: Studies of the binding of *Escherichia coli* RNA polymerase to DNA. III. Tight binding of RNA polymerase holoenzyme to single-strand breaks in T7 DNA. J. Mol. Biol. 70, 197-207 (1972)

HOTCHKISS, R.D.: The genetic chemistry of the pneumococcal transformations. Harvey Lect. Ser. 49, 241-251 (1953/54)

HOTZ, G.: Untersuchungen über den Mechanismus des Strahlenschutzes von Thiol- und Disulfidverbindungen am Modell biologischer Elementareinheiten. Z. Naturforsch. 21b, 148-152 (1966)

HOTZ, G.: On the mechanism of radiosensitization by 5-bromouracil. Studies on ^{60}Co-gamma irradiated phage ΦX-174 and its single-stranded infectious DNA. Mol. Gen. Genet. 102, 44-49 (1968)

HOTZ, G.: Infectious DNA from coliphage T1. IV. The action of radicals induced by ^{60}Co-gamma rays in the wet state. Intern. J. Radiat. Biol. 24, 1-13 (1973)

HOTZ, G., MAUSER, R.: Infections DNA from coliphage T1. I. Some properties of the spheroplast assy system. Mol. Gen. Genetics 104, 178-194 (1969)

HOTZ, G., MÜLLER, A.: The action of heat and ionizing radiation on the infectivity of isolated ΦX-174 DNA. Proc. Natl. Acad. Sci. 60, 251-257 (1968)

HOTZ, G., REUSCHL, H.: Damage to deoxyribose molecules and to u-gene reactivation in uv-irradiated 5-bromouracil-DNA of phage T4BO^r as influenced by cysteamine. Mol. Gen. Genetics 99, 5-11 (1967)
HOTZ, G., ZIMMER, K.-G.: Experiments in radiation chemistry of T1 phage. Intern. J. Radiat. Biol. 7, 75-86 (1963)
HOWARD-FLANDERS, P.: Effect of oxygen on the radiosensitivity of bacteriophage in the presence of sulfhydryl compounds. Nature (London) 186, 485-487 (1960)
HOWARD-FLANDERS, P.: Factors effecting radiation injury to DNA in bacteria and bacteriophage systems. In: Brookhaven Symposia in Biology, 1961, Nr. 14, pp. 18-30
HOWLAND, G.P., HART, R.W., YETTE, M.L.: Repair of DNA strand breaks after gamma-irradiation of protoplasts isolated from cultured wild carrot cells. Mutation Res. 27, 81-87 (1975)
HUANG, A.T., KREMER, W.B., LASZLO, T., SETLOW, R.B.: DNA repair in human leukaemic lymphocytes. Nature (New Biol.) 240, 114-115 (1972)
HUMPHREY, R.M., STEWARD, D.L., SEDITA, B.A.: DNA strand breaks and rejoining following exposure of synchronized Chinese hamster cells to ionizing radiation. Mutation Res. 6, 459-465 (1968)
HUSTER-ORTNER, M., KÖHNLEIN, W.: Transformation experiments with UV-sensitivity markers in *Bacillus subtilis*. 8th Ann. Mtg. of Europ. Soc. for Radiation Biology (Basko Polje 1971)
HUTCHINSON, F.: The lesions produced by ultraviolet light in DNA containing 5-bromouracil. Rev. Biophys. 6, 201-246 (1973)
HUTCHINSON, F., ARENA, J.: Destruction of the activity of deoxyribonucleic acid in irradiated cells. Radiat. Res. 13, 137-147 (1960)
IAEA Panel Proceedings Series. Genetical Aspects of Radiosensitivity: Mecnanimsms of Repair. Vienna: I.A.E.A., 1966
JOHANSEN, I., BOYE, E.: Radiation-induced DNA strand breaks in *Escherichia coli* measured within a fraction of a second. Nature (London) 255, 740-742 (1975)
JOHANSEN, I., BRUSTAD, T., RUPP, W.D.: DNA strand breaks measured within 100 milliseconds of irradiation of *Escherichia coli* by 4 M eV electrons. Proc. Natl. Acad. Sci. 72, 167-171 (1975)
JOHANSEN, I., GURVIN, I., RUPP, W.D.: The formation of single-strand breaks in intracellular DNA by X-rays. Radiat. Res. 48, 599-612 (1971)
JOHANSEN, I., WARD, J.F., SIEGEL, K., SLETTEN, A.: X-Ray induced binding of N-ethylmaleimide to DNA in aqueous solution. Biochem. Biophys. Res. Commun. 33, 949-953 (1968)
JOLLEY, G.M., ORMEROD, M.G.: The incomplete separation of complementary strands of high molecular weight DNA in alkali. Biochim. Biophys. Acta 353, 200-214 (1974)
JONG, J. DE, LOMAN, H., BLOK, J.: Inactivation of biologically-active DNA by radiation-induced phenylalanine radicals. Intern. J. Radiat. Biol. 22, 11-21 (1972)
JUNG, H., HAGEN, U., ULLRICH, M., PETERSEN, E.E.: Structural and functional changes in DNA after exposure to hydrogen atoms and gamma radiation. Z. Naturforsch. 24b, 1565-1573 (1969)
JUNG, H., KÜRZINGER, K.: Inactivation of bacteriophage DNA and ribonuclease by thermal hydrogen atoms. Radiat. Res. 36, 369-383 (1968)
KANAZIR, D.T.: Radiation-induced alterations in the structure of deoxyribonucleic acid and their biological consequences. Prog. Nucleic Acid Res. and Mol. Biol. 9, 117-222 (1969)
KAPLAN, H.S.: DNA strand scission and loss of viability after X-irradiation of normal and sensitized bacterial cells. Proc. Natl. Acad. Sci. 55, 1442-1446 (1966)
KAPP, D.S., SMITH, K.C.: Repair of radiation-induced damage in *Escherichia coli*. II Effect of *rec* and *uvr* mutations on radiosensitivity and repair of

X-ray induced single strand breaks in deoxyribonucleic acid. J. Bacteriol. 103, 49-54 (1970)
KARRAN, P., ORMEROD, M.G.: Is the ability to repair damage to DNA related to the proliferative capacity of a cell? The rejoining of X-ray-produced strand breaks. Biochim. Biophys. Acta 299, 54-64 (1973)
KATO, T., KONDO, S.: Genetic and molecular characteristics of X-ray-sensitive mutants of *Escherichia coli* defective in repair synthesis. J. Bacteriol. 104, 871-881 (1970)
KAVENOFF, R.: Characterization of the *Bacillus subtilis* W23 genome by sedimentation. J. Mol. Biol. 72, 801-806 (1972)
KIHLMAN, B.A.: Molecular mechanisms of chromosome breakage and rejoining. Advan. Cell. Mol. Biol. 1, 59-107 (1971)
KILBEY, B.J., DE SERRES, F.J., MALLING, H.V.: Identification of the genetic alteration at the molecular level of ultraviolet light-induced *ad-3B* mutants in *Neurospora crassa*. Mutation Res. 12, 47-56 (1971)
KIM, S.H., KIM, J.H., DJORDJEVIC, B.: Effects of X-irradiation on RNA and protein synthesis in HeLa cells. Radiat. Res. 42, 577-589 (1970)
KIMBALL, R.F.: The relation of repair to differential radiosensitivity in the production of mutãtions in *Paramecium*. In: Repair from Genetic Radiation Damage and Differential Radiosensitivity in Germ Cells, pp. 167-178. Sobels, F.H. (ed.). Oxford, London, New York, Paris: Pergamon Press, 1963
KIMBALL, R.F.: Studies on radiation mutagenesis in micro-organisms. Proc. XI Intern. Congr. Genetics, The Hague, Vol. II, pp. 227-234. Oxford, London, Edinburgh, New York, Paris, Frankfurt: Pergamon Press, 1964a
KIMBALL, R.F.: The distribution of X-ray induced mutations to chromosomal strands in *Paramecium aurelia*. Mutation Res. 1, 129-138 (1964b)
KIMBALL, R.F.: Repair of premutational damage. Advan. Radiat. Biol. 2, 135-166 (1966)
KISSAN, J.M., ROBINS, E.: The fluoremetric measurement of deoxyribonucleic acid in animal tissues with special reference to the central nervours system. J. Biol. Chem. 233, 184-188 (1958)
KITAYAMA, S., MATSUYAMA, A.: Possibility of the repair of double strand scissions in *Micrococcus radiodurans* DNA caused by gamma rays. Biochem. Biophys. Res. Commun. 33, 418-422 (1968)
KLEIJER, W.J., HOEKSEMA, H.L., SLUYTER, M.L., BOOTSMA, D.: Effects of inhibitors on repair of DNA in normal human and xeroderma pigmentosum cells after exposure to X-rays and ultraviolet irradiation. Mutation Res. 17, 385-394 (1973)
KLEIJER, W.J., LOHMAN, P.H.N., MULDER, M.P., BOOTSMA, D.: Repair of X-ray damage in DNA of cultivated cells from patients having xeroderma pigmentosum. Mutation Res. 9, 517-523 (1970)
KLOTZ, G., SPATZ, H.CH.: A biological assay for intracellular phage DNA. J. Mol. Gen. Genetics 110, 367-373 (1971)
KLUCIS, E.S., LETT, J.T.: Zonal centrifugation of mammalian DNA. Analytical Biochemistry 35, 480-488 (1970)
KÖHNLEIN, W.: Transforming activity in both complementary strains of *Bacillus subtilis* DNA. Z. Naturforsch. 29c, 63-65 (1974)
KOHN, K.W., GRIMEK-EWIG, R.A.: Alkaline elution analysis, a new approach to the study of DNA single-strand interruptions in cells. Cancer Res. 33, 1849-1853 (1973)
KRISCH, R.E.: Lethal effects of iodine-125 decay by electron capture in *Escherichia coli* and in bacteriophage T1. Intern. J. Radiat. Biol. 21, 167-189 (1972)
LACKS, S.: Molecular fate of DNA in genetic transformation of pneumococcus. J. Mol. Biol. 5, 66-77 (1962)
LACKS, S.: Theoretical relationship between probability of marker integration and length of donor DNA in Pneumococcal transformation. J. Mol. Biol. 37, 179-180 (1968)

LAIRD, C.D.: Chromatid structure: relationship between DNA content and nucleotide sequence diversity. Chromosoma 32, 378-406 (1971)
LANGE, C.S.: The organization and repair of mammalian DNA. FEBS Lett. 44, 153-156 (1974)
LATARJET, R.: Interaction of radiation energy with nucleic acids. Curr. Top. in Radiat. Res. 8, 1-38 (1972)
LATARJET, R., EKERT, B., DEMERSEMAN, P.: Peroxydation of nucleic acids by radiation: Biological implications. Radiat. Res. Suppl. 3, 247-256 (1963)
LEE, Y.C., BENNETT, L.R., BYFIELD, J.E.: Inhibition of repair of DNA single strand breaks in mouse leukemia cells by actinomycin D. Biochem. Biophys. Res. Commun. 49, 758-765 (1972)
LEHMANN, A.R., ORMEROD, M.G.: The replication of DNA in murine lymphoma cells (L 5178 Y). I. Rate of replication. Biochim. Biophys. Acta 204, 128-143 (1970a)
LEHMAN, A.R., ORMEROD, M.G.: Double strand breaks in the DNA of a mammalian cell after X-irradiation. Biochim. Biophys. Acta 217, 268-277 (1970b)
LENNARTZ, M., COQUERELLE, T., BOPP, A., HAGEN, U.: Oxygen effect on strand breaks and specific end groups in DNA of irradiated thymocytes. Intern. J. Radiat. Biol. 27, 577-587 (1975)
LENNARTZ, M., COQUERELLE, T., HAGEN, U.: Effect of oxygen on DNA strand breaks in irradiated thymocytes. Intern. J. Radiat. Biol. 24, 621-625 (1973)
LERMAN, L.S.: Structural considerations in the interaction of DNA and acridines. J. Mol. Biol. 3, 18-30 (1961)
LETT, J.T., CALDWELL, I., DEAN, C.J., ALEXANDER, P.: Rejoining of X-ray-induced breaks in the DNA of leukaemia cells. Nature (London) 214, 790-792 (1967)
LETT, J.T., GLUCIS, E.S., SUN, C.: On the size of the DNA in the mammalian chromosome structural subunits. Biophys. J. 10, 277-292 (1970)
LETT, J.T., SUN, C.: The production of strand breaks in mammalian DNA by X-rays: At different stages in the cell cycle. Radiat. Res. 44, 771-787 (1970)
LETT, J.T., SUN, C., WHEELER, K.T.: Restoration of the DNA structure in X-irradiated eucaryotic cells: in vitro and in vivo. In: Molecular and Cellular Repair Processes. 5th Intern. Symp. on Mol. Biol., pp. 147-158, Suppl. 1 John Hopkins Med. J. Beers, R.F., Jr., Herriott, R.M., Tilghman, R.C. (eds.). Baltimore: John Hopkins Univ. Press, 1972
LEVIN, D., HUTCHINSON, F.: Neutral sucrose sedimentation of very large DNA from *Bacillus subtilis*. I. Effect of random double-strand breaks and centrifuge speed on sedimentation. J. Mol. Biol. 75, 455-478 (1973)
LEY, R.D., KRISCH, R.E.: Lethality and DNA breakage from ^{32}P and ^{33}P decay in bacteriophage T4. Intern. J. Radiat. Biol. 25, 531-537 (1974)
LINDSLEY, D.L., GRELL, E.H.: Genetic Variations of *Drosophila melanogaster*. Carnegie Inst. Wash. Publ. No. 627, 1968
LOCHMANN, E.-R., MICHELER, A.: Binding of organic dyes to nucleic acids and the photodynamic effect. In: Physico-Chemical Properties of Nucleic Acids. Duchesne, J. (ed.), Vol. I, pp. 223-267. London-New York: Academic Press, 1973
LOHMAN, P.H.M.: Induction and rejoining of breaks in the deoxyribonucleic acid of human cells irradiated at various phases of the cell cycle. Mutation Res. 6, 449-458 (1968)
LOONEY, W.B., CHANG, L.O.: An auto-radiographic and biochemical study of the effects of radiation on DNA synthesis in the intact animal. Radiat. Res. 24, 312-323 (1965)
LOPRIENO, N., BARALE, R., BARONCELLI, S., CAMMELLINI, A., MELANI, M., NIERI, R., NOZZOLINI, M., ROSSI, A.M.: Mutations induced by X-radiation in the yeast *Schizosaccharomyces pombe*. Mutation Res. 28, 163-173 (1975)

LUCIA, P. DE, CAIRNS, J.: Isolation of an *E. coli* strain with a mutation affecting DNA polymerase. Nature (London) 224, 1164-1166 (1969)
LÜCKE-HUHLE, C.: Biological relevance of alkali-labile sites in double-stranded DNA after γ-irradiation. Intern. J. Radiat. Biol. 26, 1-6 (1975)
LÜCKE-HUHLE, C., SEITER, A., JUNG, H.: Contribution of strand breakage to inactivation of γ-irradiated dry ΦX 174 DNA. Z. Naturforsch. 29c, 98-99 (1974)
LUTHJENS, L.H., BLOK, J.: The indirect action of γ-rays on the coat protein of bacteriophage particles. Intern. J. Radiat. Biol. 16, 101-111 (1969)
LYNCH, W.E., BROWN, R.F., UMEDA, T., LANGRETH, S.G., LIEBERMAN, T.: Synthesis of DNA by isolated liver nuclei. J. Biol. Chem. 245, 3911-3916 (1970)
LYTLE, C.D., GINOZA, W.: Frequency of single-strand breaks per lethal γ-ray hit in ΦX 174 DNA. Intern. J. Radiat. Biol. 14, 553-560 (1968)
LYTLE, C.D., GINOZA, W.: Intercellular development of bacteriophage ΦX 174 inactivated by γ-rays, ultraviolet light or nitrous acid. Virology 38, 152-165 (1969)
MALLING, H.V., DE SERRES, F.J.: Genetic alterations at the molecular level in X-ray induced *ad-3B* mutants of *Neurospora crassa*. Radiat. Res. 53, 77-87 (1973)
MAMET-BRATLEY, M.D.: Alkylated DNA as template in the synthesis of RNA in vitro. Biochim. Biophys. Acta 247, 233-242 (1971)
MAMET-BRATLEY, M.D.: Transcription in vitro from a DNA template containing apurinic sites. Biochim. Biophys. Acta 340, 237-243 (1974)
MARSDEN, H.S., POLLARD, E.C., GINOZA, W., RANDALL, E.P.: Involvement of *recA* and *exr* genes in the in vivo inhibition of the *recBc* nuclease. J. Bacteriol. 118, 465-470 (1974)
MATSUDAIRA, H., FURUNO, I.: Increase in the incorporation of tritiated thymidine triphosphate into the DNA of nuclei isolated from Ehrlich ascites-tumor cells irradiated in vivo. Intern. J. Radiat. Biol. 19, 393-397 (1971)
MATSUDAIRA, H., FURUNO, I.: The rejoining of X-ray induced DNA strand-breaks in nuclei isolated from Ehrlich ascites tumour cells. Biochim. Biophys. Acta 272, 202-211 (1972)
MATSUDAIRA, H., FURUNO, I., OTSUKA, H.: Possible requirement of adenosine triphosphate for the rejoining of X-ray-induced breaks in the DNA of Ehrlich ascites-tumour cells. Intern. J. Radiat. Biol. 17, 339-347 (1970)
MATSUDAIRA, H., NAKAGAWA, C., BANNAI, S.: Rejoining of X-ray induced breaks in the DNA of Ehrlich ascites tumour cells in vitro. Intern. J. Radiat. Biol. 15, 575-581 (1969b)
MATSUDAIRA, H., NAKAGAWA, C., HISHIZAWA, T.: Rejoining of X-ray-induced breaks in the DNA of Ehrlich ascites tumour cells in vivo. Intern. J. Radiat. Biol. 15, 95-100 (1969a)
MATTERN, M.R., CERUTTI, P.A.: Age-dependent excision repair of damaged thymine from gamma-irradiated DNA by isolated nuclei from human fibroblasts. Nature (London) 254, 450-452 (1975)
MATTERN, M.R., HARIHARAN, P.V., DUNLAOP, B.E., CERUTTI, P.A.: DNA degradation and excision repair in gamma-irradiated Chinese hamster ovary cells. Nature (New Biol.) 245, 230-232 (1973)
McBURNEY, M.W., GRAHAM, F.L., WHITMORE, G.F.: Anomalous sedimentation of high molecular weight denatured mammalian DNA. Biochem. Biophys. Res. Commun. 44, 171-177 (1971)
McBURNEY, M.W., GRAHAM, F.L., WHITMORE, G.F.: Sedimentation analysis of DNA from irradiated and unirradiated cells. Biophys. J. 12, 369-383 (1972)
McBURNEY, M.W., WHITMORE, G.F.: Molecular weight analysis of mammalian DNA. Biochem. Biophys. Res. Commun. 46, 898-904 (1972)
McGRATH, R.A., WILLIAMS, R.W.: Reconstruction in vivo of irradiated *Escherichia coli* deoxyribonucleic acid; the rejoining of broken pieces. Nature (London) 212, 534-535 (1966)

MESELSON, M., STAHL, F.W., VINOGRAD, J.: Equilibrium sedimentation of macromolecules in density gradients. Proc. Natl. Acad. Sci. 43, 581-588 (1957)
MICHALKE, H., BREMER, H.: RNA synthesis in *Escherichia coli* after irradiation with ultraviolet light. J. Mol. Biol. 41, 1-23 (1969)
MÖNKEHAUS, F., KÖHNLEIN, W.: Single and double-strand breaks in 5-bromouracil-substituted DNA of *B. subtilis* and phage PBSH after irradiation with long-wave length UV and their correlation to intramolecular energy transfer. Biopolymers 12, 329-340 (1973)
MONK, M., KINROSS, J.: Conditonal lethality of *recA* and *recB* derivatives of a strain of *Escherichia coli* K-12 with a temperature-sensitive deoxyribonucleic acid polymerase I. J. Bacteriol. 109, 971-978 (1972)
MORIMYO, M., HORII, Z., SUZUKI, K.: Appearance of low molecular weight DNA in a *Rec*$^-$ mutant of *Escherichia coli* K12 irradiated with X-rays. J. Radiat. Res. (Japan) 9, 19-25 (1968)
MORRISON, D.A., GUILD, W.R.: Structure of DNA on the cell surface during uptake by *Pneumococcus*. J. Bacteriol. 115, 1055-1062 (1973)
MORRISON, D.A., GUILD, W.R.: Activity of DNA fragments of defined size in *Bacillus subtilis* transformation. J. Bacteriol. 112, 220-223 (1972)
MOTTINGER, J.P.: The effects of X-rays on the bronze and shrunken loci in maize. Genetics 64, 259-271 (1970)
MÜLLER, A., DERTINGER, H.: Die Erzeugung atomaren Wasserstoffs in γ-bestrahlten Bakteriophagen und in deren DNS. Z. Naturforsch. 23b, 83-89 (1968)
MÜLLER, A., KÖHNLEIN, W., ZIMMER, K.G.: X-ray-induced unpaired spins in nucleic acid bases and in 5-bromouracil. J. Mol. Biol. 7, 92-94 (1963)
MULLER, H.J.: The nature of the genetic effects produced by radiation. In: Radiation Biology, Vol. I, Part 1, pp. 351-473. Hollaender, A. (ed.). New York, Toronto, London: McGraw-Hill Book Co., 1954
MULLER, H.J., CARLSON, E., SCHALET, A.: Mutation by alteration of the already existing gene. Genetics 46, 213-226 (1961)
MUNSON, R.J., BRIDGES, B.A.: Lethal and mutagenic lesions induced by ionizing radiations in *E. coli* and DNA strand breaks. Biophysik 6, 1-5 (1969)
MUNSON, R.J., BRIDGES, B.A.: The LET factor in mutagenesis by ionizing radiations. I. Reversion to wild type of a bacteriophage T4 *amber* mutant. Intern. J. Radiat. Biol. 24, 257-273 (1973)
NAKKEN, K.F.: Radical scavengers and radioprotection. Corr. Top. Radiat. Res. I, 49-92 (1965)
NAKKEN, K.F., SIKKELAND, T., BRUSTAD, T.: Radiation induced binding of 2,2,6,6-tetramethyl-4-piperidone-N-oxyl (TAN) to DNA. IV. Congrès Intern. Radiobiol. et Physico-chimie Rayon, Evian, Abstract, p. 156 (1970)
NASIM, A., AUERBACH, C.: The origin of complete and mosaic mutants from mutagenic treatment of single cells. Mutation Res. 4, 1-14 (1967)
NEARY, G.J.: Chromosome aberrations and the theory of RBE. I. General considerations. Intern. J. Radiat. Biol. 9, 477-502 (1965)
NEARY, G.J.: Chromosome aberrations and the theory of RBE. II. Evidence from track-segment experiments with protons and alpha particles. Intern. J. Radiat. Biol. 11, 209-223 (1966)
NEARY, G.J.: Chromosome aberrations and the theory of RBE. III. Evidence from experiments with soft X-rays, and a consideration of the effects of hard X-rays. Intern. J. Radiat. Biol. 12, 317-345 (1967)
NEWCOMBE, H.B.: The genetic effects of ionizing radiations. Advan. Genetics 16, 239-303 (1971)
NICOLAIEFF, A., CHEVALLIER, M.R.: An electron microscopic study of the relation between length and transforming activity of *H. influenzae* DNA. Mol. Gen. Genet. 109, 269-277 (1970)
OGAWA, H., TOMIZAWA, J.-I.: Breakage of polynucleotide strands by disintegration of radiophosphorus atoms in DNA molecules and their repair. I. Single strand breakage by transmutation. J. Mol. Biol. 30, 1-6 (1967)

OKADA, S.: Radiation Biochemistry, Vol. I. London-New York: Academic Press, 1970

ONO, T., OKADA, S.: Estimation in vivo of DNA strand breaks and their rejoining in thymus and liver of mouse. Intern. J. Radiat. Biol. 25, 291-302 (1974)

ORMEROD, M.G.: The measurement of radiation-induced strand breaks in the DNA of mammalian cells. In: Physico-Chemical Properties of Nucleic Acids. Duchesne, J. (ed.), Vol. III, pp. 139-159. London: Academic Press, 1973

ORMEROD, M.G.: Radiation-induced strand breaks in the DNA of mammalian cells. In: The Biology of Radiation Carcinogenesis. Yuhas, J.M., Tennant, R.W., Regan, J.D. (eds.), pp. 67-72. New York: Raven Press, 1975

ORMEROD, M.G., LEHMANN, A.R.: Artefacts arising from the sedimentation of high molecular weight DNA on sucrose gradients. Biochim. Biophys. Acta 247, 369-372 (1971a)

ORMEROD, M.G., LEHMANN, A.R.: The release of high molecular weight DNA from a mammalian cell (L5178Y). Attachment of the DNA to the nuclear membrane. Biochim. Biophys. Acta 228, 331-343 (1971b)

ORMEROD, M.G., STEVENS, U.: The rejoining of X-ray induced strand breaks in the DNA of a murine lymphoma cell (L5178Y). Biochim. Biophys. Acta 232, 72-82 (1971)

OTTOLENGHI, S., LANYON, W.G., PAUL, J., WILLIAMSON, R., WEATHERALL, D.J., CLEGG, J.B., PRITCHARD, J., POOTRAKUL, S., BOON, W.H.: The severe form of α thalassaemia is caused by a haemoglobin gene deletion. Nature (London) 251, 389-392 (1974)

PAINTER, R.B.: Thymidine incorporation as a measure of DNA-synthesis in irradiated cell cultures. Intern. J. Radiat. Biol. 13, 279-281 (1967)

PAINTER, R.B., CLEAVER, J.E.: Repair replication, unscheduled DNA synthesis and the repair of mammalian DNA. Radiat. Res. 37, 451-466 (1969)

PAINTER, R.B., WOLFF, S.: Apparent absence of repair replication in *Vicia faba* after X-irradiation. Mutation Res. 19, 133-137 (1973)

PAINTER, R.B., YOUNG, B.R.: Repair replication in mammalian cells after X-irradiation. Mutation Res. 14, 225-235 (1972)

PALCIC, B., SKARSGARD, L.D.: The effect of oxygen on DNA single-strand breaks produced by ionizing radiation in mammalian cells. Intern. J. Radiat. Biol. 21, 417-433 (1972a)

PALCIC, B., SKARSGARD, L.D.: DNA single-strand breaks produced in mammalian cells by ionizing radiation after treatment with 2,4-dinitrophenol. Intern. J. Radiat. Biol. 21, 535-544 (1972b)

PATERSON, M.C., BOYLE, J.M., SETLOW, R.B.: Ultraviolet and X-ray-induced responses of a deoxyribonucleic acid polymerase-deficient mutant of *Escherichia coli*. J. Bacteriol. 107, 61-67 (1971)

PATERSON, M.C., ROOZEN, K.J., SETLOW, R.B.: Gamma-ray induced chain breaks and alkali-labile bonds in lambda dv DNA of *Escherichia coli* minicells irradiated under aerobic and anoxic conditions. Intern. J. Radiat. Biol. 23, 495-508 (1973)

PATERSON, M.C., SETLOW, R.B.: Endonucleolytic activity from *Micrococcus luteus* that acts on gamma-ray-induced damage in plasmid DNA of *Escherichia coli* minicells. Proc. Nàtl. Acad. Sci. 69, 2927-2931 (1972)

PATERSON, M.C., SMITH, B.P., LOHMAN, P.H.M., ANDERSON, A.K., FISHMAN, L.: Defective excision repair of γ-ray damaged DNA in human (ataxia telangiectasia) fibroblasts. Nature (London) 260, 444-446 (1976)

PAUL, I.J., ZIMMERMAN, A.M.: RNA synthesis and polyribosome metabolism in X-irradiated synchronized *Tetrahymena*. Radiat. Res. 61, 144-157 (1975)

PAULY, H.: Synthesis of protein, RNA, DNA and induced enzyme in *Bacterium cadaveris* after X-irradiation. Intern. J. Radiat. Biol. 6, 221-231 (1963)

PAYES, B.: Enzymatic repair of X-ray-damaged DNA. I. Labelling of the precursor of a malonaldehyde-like material in X-irradiated DNA and the enzymatic excision of labelled lesions. Biochim. Biophys. Acta 366, 251-260 (1974)

PETES, T.D., FANGMAN, W.L.: Sedimentation properties of yeast chromosomal DNA. Proc. Natl. Acad. Sci. 69, 1188-1191 (1972)

PETTIJOHN, D.E., HECHT, R.: RNA molecules bound to the folded bacterial genome stabilize DNA folds and segregate domains of supercoiling. Cold Spring Harbor Symp. Quant. Biol. 38, 31-41 (1974)

PHILLIPS, S.L., PERSON, S., NEWTON, H.P.: Characterization of genetic coding changes in bacteria produced by ionizing radiation and by the radioactive decay of incorporated ^{3}H-labelled compounds. Intern. J. Radiat. Biol. 21, 159-166 (1972)

POLLARD, E.C.: The effect of ionizing radiation on the molecular biology of *Escherichia coli*. Curr. Top. Radiat. Res. 6, 53-127 (1970)

POLLARD, E.C., DAVIS, S.A.: The action of ionizing radiation on transcription (and translation) in several strains of *E. coli*. Radiat. Res. 41, 375-399 (1970)

POLLARD, E.C., RANDALL, E.P.: Studies on the inducible inhibitor of radiation-induced DNA degradation of *Escherichia coli*. Radiat. Res. 55, 265-279 (1973)

PRAKASH, L., SHERMAN, F.: Mutagen specifity: reversion of iso-1-cytochrome *c* mutants of yeast. J. Mol. Biol. 79, 65-82 (1973)

PRESCOTT, D.M.: The structure and replication of eukaryotic chromosomes. Advan. Cell Biol. 1, 57-117 (1970)

PRITCHARD, N.J., BLAKE, A., PEACOCKE, A.R.: Modified intercalation model for the interaction of aminoacridines and DNA. Nature (London) 212, 1360-1361 (1966)

RASMUSSEN, R.E., PAINTER, R.B.: Evidence for repair of ultra-violet damaged deoxyribonucleic acid in cultured mammalian cells. Nature (London) 203, 1360-1362 (1964)

READ, J.: Radiation Biology of *Vicia faba* in Relation to the General Problem. Oxford: Blackwell Scientific Publ., 1959

REGAN, J.D., SETLOW, R.B.: DNA repair in human progeroid cells. Biochem. Biophys. Res. Commun. 59, 858-864 (1974a)

REGAN, J.D., SETLOW, R.B.: Two forms of repair in the DNA of human cells damaged by chemical carcinogens and mutagens. Cancer Res. 34, 3318-3325 (1974b)

REGAN, J.D., SETLOW, R.B., LEY, R.D.: Normal and defective repair of undamaged DNA in human cells: A sensitive assay utilizing the photolysis of bromodeoxyuridine. Proc. Natl. Acad. Sci. 68, 708-712 (1971)

REGAN, J.D., TROSKO, J.E., CARRIER, W.L.: Evidence for excision of ultraviolet induced pyrimidine dimers from the DNA of human cells in vitro. Biophys. J. 8, 319-325 (1968)

REILLY, B.E., SPIZIZEN, J.: Bacteriophage deoxyribonucleate infection of competent *Bacillus subtilis*. J. Bacteriol. 89, 782-788 (1965)

RESNICK, M.A.: The repair of double-strand breaks in chromosomal DNA of yeast. In: Molecular Mechanisms for Repair of DNA. Setlow, R.B., Hanawalt, P.C. (eds.), pp. 547-559. New York: Plenum Press, 1975

REVELL, S.H.: The breakage-and-reunion theory and the exchange theory for chromosomal aberrations induced by ionizing radiations: a short history. Advan. Radiat. Biol. 4, 367-416 (1974)

RIEGER, R., MICHAELIS, A., GREEN, M.M.: A Glossary of Genetics and Cytogenetics. Berlin-Heidelberg-New York: Springer, 1968

ROOTS, R., SMITH, K.C.: Rejoining of DNA single-strand breaks in mammalian cells incubated in buffer or in medium after aerobic or anaerobic X-irradiation. Intern. J. Radiat. Biol. 27, 595-602 (1975)

ROTTLÄNDER, E., TRAUTNER, T.A.: Genetic and transfection studies with *B. subtilis* phage SP50 1. Phage mutants with restricted growth on *B. subtilis* strain 168. Mol. Gen. Genet. 108, 47-60 (1970)
RUBENSTEIN, I., LEIGHTON, S.B.: The influence of rotor speed on the sedimentation behaviour in sucrose gradients of high molecular weight DNAs. Biophys. Soc. Abstracts 11, 209a (1971)
RUPP, W.D., HOWARD-FLANDERS, P.: Discontinuities in the DNA synthesized in an excision defective strain of *E. coli* following ultraviolet irradiation. J. Mol. Biol. 31, 291-304 (1968)
RUSSELL, L.B.: Definition of functional units in a small chromosomal segment of the mouse and its use in interpreting the nature of radiation-induced mutations. Mutation Res. 11, 107-123 (1971)
RUSSELL, W.L.: Evidence from mice concerning the nature of the mutation process. Proc. XI Intern. Congr. Genetics, The Hague, Vol. II, pp. 257-264. Oxford, London, Edinburgh, New York, Paris, Frankfurt: Pergamon Press, 1964
RUSSELL, W.L.: Studies in mammalian radiation genetics. Nucleonics 23, 53-56 (1965)
RUSSELL, W.L.: The genetic effects of radiation. Peaceful Uses At. Energy, 4th, 13, 487-500 (1972)
RYDBERG, B.: The rate of strand separation in alkali of DNA of irradiated mammalian cells. Radiat. Res. 61, 274-287 (1975)
SAITO, H., MIURA, K.I.: Preparation of transforming DNA by phenol treatment. Biochim. Biophys. Acta 72, 619-629 (1963)
SANDERS, J.L., DALRYMPLE, G.V., BAKER, M.L., MOSS, A.J., NASH, J.C.: Cyclic labeling of mammalian cell DNA after X-irradiation; labeling with specific precursors. Proc. Soc. Exp. Biol. Med. 132, 362-366 (1969)
SAUERBIER, W., MILETTE, R.L., HACKETT, P.B., Jr.: The effects of ultraviolet irradiation on the transcription of T4 DNA. Biochim. Biophys. Acta 209, 368-386 (1970)
SAWADA, S., OKADA, S.: Rejoining of single-strand breaks of DNA in cultured mammalian cells. Radiat. Res. 41, 145-162 (1970)
SAWADA, S., OKADA, S.: Effects of BUdr-labelling on radiation-induced DNA breakage and subsequent rejoining in cultured mammalian cells. Intern. J. Radiat. Biol. 21, 599-602 (1972)
SCHACHMAN, H.K.: Ultracentrifugation in Biochemistry. New York-London: Academic Press, 1959
SCHAEFFER, J., MERZ, T.: A comparison of unscheduled DNA synthesis, D_o, cell recovery, and chromosome number in X-irradiated mammalian cell lines. Radiat. Res. 47, 426-436 (1971)
SCHANS, G.P. VAN DER, ATEN, J.B.T., BLOK, J.: Determination of molecular weight distribution of DNA by means of sedimentation in a sucrose gradient. Anal. Biochem. 32, 14-30 (1969)
SCHANS, G.P. VAN DER, BLEICHRODT, J.F.: Contribution of various types of damage to inactivation of bacteriophage PM2 DNA in frozen solution by gamma radiation. Intern. J. Radiat. Biol. 26, 121-126 (1974)
SCHANS, G.P. VAN DER, BLEICHRODT, J.F., BLOK, J.: Contribution of various types of damage to inactivation of a biologically active double-stranded circular DNA by γ-irradiation. Intern. J. Radiat. Biol. 23, 133-150 (1973)
SCHANS, G.P. VAN DER, BLOK, J.: The influence of oxygen and sulphhydryl compounds on the production of breaks in bacteriophage DNA by gamma rays. Intern. J. Radiat. Biol. 17, 25-38 (1970)
SCHEID, W., TRAUT, H.: Ultraviolet-microscopical studies on achromatic lesions ("gaps") induced by X-rays in the chromosomes of *Vicia faba*. Mutation Res. 10, 159-161 (1970)
SCHEID, W., TRAUT, H.: Visualization by scanning electron microscopy of achromatic lesions ("gaps") induced by X-rays in chromosomes of *Vicia faba*. Mutation Res. 11, 253-255 (1971a)

SCHEID, W., TRAUT, H.: On the nature of achromatic lesions ("gaps") induced by X-rays in chromosomes of *Vicia faba*. Z. Naturforsch. 26b, 1384-1385 (1971b)
SCHMIDT, A., HOTZ, G.: The occurence of double-strand breaks in coli-phage T1 - DNA by iodine-125 decay. Intern. J. Radiat. Biol. 24, 307-313 (1973)
SCHÖN, M.: Die Strahlenempfindlichkeit der einzelnen Entwicklungsschritte im Vermehrungszyklus des Bakteriophagen ΦX 174 nach ^{60}Co-γ-Bestrahlung. Thesis, Univ. Karlsruhe, 1973
SEARLE, A.G.: Mutation induction in mice. Advan. Radiat. Biol. 4, 131-207 (1974)
SEDGWICK, S.G., BRIDGES, B.A.: Survival, mutation, and capacity to repair single-strand DNA breaks after gamma irradiation in different *Exr*$^-$ strains of *Escherichia coli*. Mol. Gen. Genetics 119, 93-102 (1972)
SERRES, F.J. DE: Detection of the genetic effects of environmental chemical pollutants with a two-component heterokaryon of Neurospora. EMS Newslett. 4, 24-29 (1971)
SERRES, F.J. DE, MALLING, H.V.: Identification of the genetic alterations in specific locus mutants at the molecular level. Japan J. Genetics 44, 106-113 (1969)
SETLOW, J.K., RANDOLPH, M.L., BOLING, M.E., MATTINGLY, A., PRICE, G., GORDON, M.P.: Repair of DNA in *Haemophilus influenzae*. II. Excision, repair of single-strand breaks, defects in transformation, and host-cell modification in UV-sensitive mutants. Cold Spring Harbor Symp. Quant. Biol. 33, 209-218 (1968)
SETLOW, R.B., CARRIER, W.L.: The disappearance of thymine dimers from DNA, an error-correcting mechanism. Proc. Natl. Acad. Sci. 51, 226-231 (1964)
SETLOW, R.B., HART, R.W.: In: Molecular Mechanisms for Repair of DNA. Hanawalt, P., Setlow, R.B. (eds.). pp. 719-724. New York: Plenum Press, 1975
SETLOW, R.B., SETLOW, J.K.: Effects of radiation on polynucleotides. Ann. Rev. Biophys. Bioeng. 1, 293-346 (1972)
SHARP, J.D., FREIFELDER, D.: Interruption of DNA injection by irradiation of phage λ. Virology 43, 166-175 (1971)
SHIPLEY, W.U., ELKIND, M.M.: DNA damage and repair following irradiation: The effect of 5-bromodeoxyuridine in cultured Chinese hamster cells. Radiat. Res. 48, 86-94 (1971)
SIMPSON, J.R., NAGLE, W.A., BICK, M.D., BELLI, J.A.: Molecular nature of mammalian cell DNA in alkaline sucrose gradients. Proc. Natl. Acad. Sci. 70, 3660-3664 (1973)
SINCLAIR, W.K.: Cyclic X-ray responses in mammalian cells in vitro. Radiat. Res. 33, 620-643 (1968)
SMITH, K.C.: The inhibition of DNA repair processes. Proc. 11th Intern. Cancer Congr., Florence (1974)
SOBELS, F.H. (ed.): Repair from Genetic Radiation Damage and Differential Radiosensitivity in Germ Cells. Oxford-London-New York-Paris: Pergamon Press, 1963
SPIEGLER, P., NORMAN, A.: Unscheduled DNA synthesis at low X-ray doses. Mutation Res. 10, 379-387 (1970)
SPIEGLER, P., NORMAN, A.: Kinetics of unscheduled DNA synthesis induced by ionizing radiation in human lymphocytes. Radiat. Res. 39, 400-412 (1969)
SPIZIZEN, J.: Transformation of biochemically deficient strains of *Bacillus subtilis* by deoxyribonucleate. Proc. Natl. Acad. Sci. 44, 1062-1078 (1958)
STADLER, L.J., ROMAN, H.: The effect of X-rays upon mutation of the gene *A* in maize. Genetics 33, 273-303 (1948)
STENT, G.S., FUERST, C.R.: Inactivation of bacteriophages by decay of incorporated radioactive phosphorus. J. Gen. Physiol. 38, 441-458 (1955)
STEPHAN, G.: Repair of damage induced in ΦX RF DNA by ^{60}Co gamma rays. Intern. J. Radiat. Biol. 23, 533-542 (1973)

STRAUSS, N.: Further evidence concerning the configuration of transforming DNA during entry into *Bacillus subtilis*. J. Bacteriol. 91, 702-708 (1966)
STRAUSS, N.: Transformation of *Bacillus subtilis* using hybrid DNA molecules constructed by annealing resolved complementary strands. Genetics 66, 583-593 (1970)
STREFFER, CH., SCHAFFERUS, S.: The induction of liver enzymes by cortisol after combined treatment of mice with X-irradiation and inhibitors of protein biosynthesis. Intern. J. Radiat. Biol. 20, 301-313 (1971)
STUDIER, F.W.: Sedimentation studies of the size and shape of DNA. J. Mol. Biol. 11, 373-390 (1965)
STUY, J.H.: Fate of transforming DNA in the *Haemophilus influenzae* transformation system. J. Mol. Biol. 13, 554-570 (1965)
STUY, J.H.: Bacterial transformation. In: Uptake of Informative Molecules by Living Cells. Amsterdam-London: North Holland, 1972
SUHADI, F., KITAYAMA, S., OKAZAWA, Y., MATSUYAMA, A.: Isolation and some radiobiological properties of mutants of Micrococcus radiodurans sensitive to ionizing radiation. Radiat. Res. 49, 197-212 (1972)
SUMMERS, W.C., SZYBALSKI, W.: γ-Irradiation of deoxyribonucleic acid in dilute solutions. I. A sensitive method for detection of single-strand breaks in polydisperse DNA samples. J. Mol. Biol. 26, 107-123 (1967)
II. Molecular mechanisms responsible for inactivation of phage, its transfecting DNA and of bacterial transforming activity. J. Mol. Biol. 26, 227-235 (1967)
SWINEHART, J.L., CERUTTI, P.A.: Gamma-ray induced thymine damage in the DNA in coliphage and in *E. coli*. Intern. J. Radiat. Biol. 27, 83-94 (1975)
SZYBALSKI, W.: X-ray sensitization by halogenpyrimidines. Cancer Chemother. Rep. 58, 539-557 (1974)
SZYBALSKI, W., OPARA-KUBINSKA, Z.: Radiobiological and phyciso-chemical properties of 5-bromodeoxyuridine-labeled transforming DNA as related to the nature of the critical radiosensitive structures. Cellular Radiation Biology, Collection of Papers Presented at the Eighteenth Annual Symposium on Fundamental Cancer Research (1964), Published for the Univ. Texas, M.D. Anderson Hospital and Tumor Inst. 1965 by the Williams and Wilkins Co., Baltimore, Maryland
TANOOKA, H.: Modification of the inactivation of bacterial spores and their transforming DNA by ionizing radiation. J. Genetics 40, 229-241 (1964)
TANOOKA, H., HUTCHINSON, F.: Modification of the inactivation by ionizing radiation of the transforming activity of DNA in spores and dry cells. Radiat. Res. 24, 43-56 (1965)
TAYLOR, A.M.R., HARNDEN, D.G., ARLETT, C.F., HARCOURT, S.A., LEHMANN, A.R., STEVENS, S., BRIDGES, B.A.: Ataxia telangiectasia, a human mutation with abnormal radiation sensitivity. Nature (London) 258, 427-429 (1975)
TAYLOR, J.M., DOZY, A., KAN, Y.W., VARMUS, H.E., LIE-INJO, L.E., GANESAN, J., TODD, D.: Genetic lesion in homozygous α thalassaemia (*Hydrops fetalis*). Nature (London) 251, 392-393 (1974)
TAYLOR, W.D., GINOZA, W.: Correlation of γ-ray inactivation and strand scission in the replicative form of ΦX 174 bacteriophage DNA. Proc. Natl. Acad. Sci. 58, 1753-1758 (1967)
TERANO, H., TANOOKA, H., KADOTA, H.: Germination-induced repair of single-strand breaks of DNA in irradiated *Bacillus subtilis* spores. Biochem. Biophys. Res. Commun. 37, 66-71 (1969)
TESSMAN, I.: Some unusual properties of the nucleic acid of bacteriophage S13 and ΦX 174. Virology 7, 263-275 (1959)
THORSETT, G.O., HUTCHINSON, F.: Effects on bacterial transformation of single strand breaks in DNA produced by DNase I and γ-rays. Biochim. Biophys. Acta 238, 67-74 (1971)

TICHY, P., RYTIV, V., KOHOUTOVA, M.: Genetic transformation and transfection of *Bacillus subtilis* Spheroplasts. Folia. Microbiol. Prague 13, 510 (1968)
TIMOFÉEFF-RESSOVSKY, N.W., ZIMMER, K.G.: Biophysik I. Das Trefferprinzip in der Biologie. Leipzig: S. Hirzel, 1947
TOMIZAWA, J.-I., OGAWA, H.: Breakage of polynucleotide strands by disintegration of radiophosphorus atoms in DNA molecules and their repair. II. Simultaneous breakage of both strands. J. Mol. Biol. 30, 7-15 (1967)
TOWN, C.D., SMTIH, K.C., KAPLAN, H.S.: DNA polymerase required for rapid repair of X-ray-induced DNA strand breaks in vivo. Science 172, 851-854 (1971)
TOWN, C.D., SMITH, K.C., KAPLAN, H.S.: Repair of X-ray damage to bacterial DNA. Curr. Top. Radiat. Res. 8, 351-399 (1973)
TRAUTNER, T.A., SPATZ, H.CH.: Transfection in *B. subtilis*. Curr. Top. in Microbiology and Immunology 62, 61-88 (1973)
ULLRICH, M., HAGEN, U.: Base liberation and concomitant reactions in irradiated DNA solutions. Intern. J. Radiat. Biol. 19, 507-517 (1971)
UNSCEAR (U.N. Sci. Comm. on the Effects of Atomic Radiation) Ionizing Radiation: Levels and Effects. Vol. II: Effects. New York: United Nations, 1972
VINCENT, R.A., SHERIDAN, R.B., HUANG, F.C.: DNA strand breakage repair in Ataxia Telangiectasia fibroblast-like cells. Mutation Res. 33, 357-366 (1975)
VINOGRAD, J., BRUNNER, R., RADLOFF, R.: Bandforming centerpieces for the analytical ultracentrifuge. Biopolymers 3, 481-489 (1965)
VOGEL, F., RÖHRBORN, G., SCHLEIERMACHER, E., SCHROEDER, T.: Strahlengenetik der Säuger. In: Fortschritte der allgemeinen und klinischen Humangenetik, Vol. I, pp. 1-80. Becker, P.E., Lenz, W., Vogel, F., Wendt, G.G. (eds.). Stuttgart: Thieme, 1969
VOSS, R., FALK, R.: The nature of reverse mutations in *Drosophila melanogaster*. Mutation Res. 20, 221-234 (1973)
WATANABE, I., OKADA, S.: DNA synthesis during the first post-irradiation life cycle of lethally irradiated cultured mammalian cells (L5178Y). Radiat. Res. 35, 202-212 (1968)
WATKINS, D.K.: High oxygen effect for the release of enzymes from isolated mammalian lysosomes after treatment with ionizing radiation. Advan. Biol. Med. Phys. 13, 289-305 (1970)
WATKINS, D.K.: A differential effect of ionizing radiation on the synthesis of DNA in ascites cells in stationary and multiplying phases. Radiat. Res. 51, 615-630 (1972)
WATSON, J.D.: The properties of X-ray inactivated bacteriophage. I. Inactivation by direct effect. J. Bacteriol. 60, 697-718 (1950)
WATSON, J.D.: The properties of X-ray inactivated bacteriophage. II. Inactivation by indirect effect. J. Bacteriol. 63, 473-485 (1952)
WEISS, J.J.: Chemical effects of ionizing radiation on nucleic acid and related compounds. Prog. Nucleic Acid Res. Mol. Biol. 3, 103-144 (1964)
WEISS, J.J., WHEELER, C.M.: Some studies on γ-irradiated DNA as primer in RNA synthesis. Biochim. Biophys. Acta 145, 68-75 (1967)
WEISSMAN, M., SCHINDLER, H., FEHER, G.: Determination of molecular weights by fluctuation spectroscopy: Application to DNA. Proc. Natl. Acad. Sci. 73, 2776-2780 (1976)
WHEELER, K.T., LETT, J.T.: Formation and rejoining of DNA strand breaks in irradiated neurons in vivo. Radiat. Res. 52, 59-67 (1972)
WHEELER, K.T., LETT, J.T.: On the possibility that DNA repair is related to age in nondividing cells. Proc. Natl. Acad. Sci. 71, 1862-1865 (1974)
WHEELER, K.T., LINN, J.D.: Differential sensitivity of DNA strand-break-rejoining mechanisms to split doses of X-rays. Intern. J. Radiat. Biol. 26, 411-420 (1974)

WHEELER, K.T., SHERIDAN, R.E., PAUTLER, E.L., LETT, J.T.: In vivo restitution of the DNA structure in gamma-irradiated rabbit retinas. Radiat. Res. 53, 414-427 (1973)

WHEELER, K.T., DE WITT, J., LETT, J.T.: A marker for mammalian DNA sedimentation. Radiat. Res. 57, 365-378 (1974)

WHITE, M.J.D.: Animal Cytology and Evolution. Cambridge: Univ. Press, 1973

WILKINS, R.J.: Does *Escherichia coli* possess a DNA excision repair system for gamma-ray damage? Nature (New Biol.) 244, 269-271 (1973a)

WILKINS, R.J.: Endonuclease-sensitive sites in the DNA of irradiated bacteria; a rapid and sensitive assay. Biochim. Biophys. Acta 312, 33-37 (1973b)

WOLFF, S.: The effects of ultraviolet radiation on genes and chromosomes. Radiation Res., pp. 455-465. Proc. 3rd Intern. Congr. Radiation Res. Silini, G. (ed.). Amsterdam: North-Holland, 1967a

WOLFF, S.: Radiation genetics. Ann. Rev. Genetics 1, 221-244 (1967b)

WOLFF, S.: Strandedness of chromosomes. Intern. Rev. Cytol. 25, 279-296 (1969)

WOLFF, S., CLEAVER, J.E.: Absence of DNA repair replication after chemical mutagen damage in *Vicia faba*. Mutation Res. 20, 71-76 (1973)

WORCEL, A., BURGI, E., ROBINTON, J., CARLSON, C.L.: Studies on the folded chromosome of *Escherichia coli*. Cold Spring Harbor Symp. Quant. Biol. 38, 43-51 (1974)

YOUNGS, D.A., BERNSTEIN, I.A.: Involvement of the *recB-recC* nuclease (exonuclease V) in the process of X-ray-induced deoxyribonucleic acid degradation in radiosensitive strains of *Escherichia coli* K-12. J. Bacteriol. 113, 901-906 (1973)

YOUNGS, D.A., SMITH, K.C.: X-ray sensitivity and repair capacity of a *polA/exrA* strain of *Escherichia coli* K-12. J. Bacteriol. 114, 121-127 (1973a)

YOUNGS,D.A., SMITH, K.C.: Sensitivity to X-radiation of strains of *Escherichia coli* K-12 which lack DNA polymerase II. Mol. Gen. Genetics 122, 287-290 (1973b)

YOUNGS, D.A., SMITH, K.C.: The yield and repair of X-ray-induced single-strand breaks in the DNA of *Escherichia coli* K-12 cells. Radiat. Res. 68, 148-154 (1976)

ZIMM, B.H.: Anomalies in sedimentation. IV. Decrease in sedimentation coefficients of chains at high fields. Biophys. Chem. 1, 279-291 (1974)

ZIMMER, K.G.: Studies on Quantitative Radiation Biology. Edinburgh and London: Oliver and Boyd, 1961

ZIMMERMANN, F., KRÖGER, H., HAGEN, U., KECK, K.: The effect of X-irradiation on the priming ability of DNA in the RNA polymerase system. Biochim. Biophys. Acta 87, 160-162 (1964)

Appendix

1. Dose Units

1.1 Exposure

The exposure, X, is the quotient of ΔQ by Δm where ΔQ is the sum of the electrostatic charges on all the ions of one sign produced in air when all the electrons (positrons and negatrons) liberated by photons in a volume element of air whose mass is Δm are completely stopped in air.

$$X = \frac{\Delta Q}{\Delta m}$$

The special unit of exposure is the roentgen (R):

$$1 \text{ R} = 2.58 \cdot 10^{-4} \text{ C/kg of air.}$$

1.2 Absorbed Dose

The absorbed dose, D, is the quotient of ΔE_d by Δm, where ΔE_d is the energy imparted by ionizing radiation to a volume element and Δm is the mass of the matter in that volume element

$$D = \frac{\Delta E_d}{\Delta m}$$

The special unit of absorbed dose is the rad:

$$1 \text{ rad} = 10^{-2} \text{ J/kg} = 100 \text{ erg/g.}$$

1.3 Conversion Factors

Each ion in a pair produced by high energy photons carries a charge of plus or minus $1.602 \cdot 10^{-19}$ C. Thus, an exposure of one roentgen corresponds to a production of $1.61 \cdot 10^{15}$ ion pairs.

Since the average energy required to produce one ion pair is 33.7 eV, air exposed to one roentgen aquires an energy amount of 0.00869 J/kg. The absorbed dose D_a in air exposed to X roentgen is then

$$D_a = 0.869 \cdot X = f_a \cdot X \text{ rad}$$

where a refers to air. Thus, $f_a = 0.869$ is the conversion factor when the irradiated matter is air. For other material an absorbed dose D_a in air is related to the absorbed dose D_m in matter by the equation.

$$D_m = D_a \cdot \frac{(\mu_{en}/\rho)_m}{(\mu_{en}/\rho)_a} = 0.869 \frac{(\mu_{en}/\rho)_m}{(\mu_{en}/\rho)_a} \cdot X$$

provided that the exposures were performed under identical conditions. In this expression, ρ is the density. Thus, the ratio between the mass energy-absorption coefficients (μ_{en}/ρ) for the matter and air multiplied by f_a yields the conversion factor. The mass energy-coefficients depend on the quality of the photon spectrum.

Since matter is composed of different subunits with different absorption features, its mass energy coefficient can be written as:

$$(\mu_{en}/\rho)_m = \sum_i f_i \cdot (\mu_{en}/\rho)_i$$

where f_i is the fraction by weight of the i'th subunit with mass energy coefficient $(\mu_{en}/\rho)_i$. Therefore, D_m can be written as:

$$D_m = \sum_i f_i \cdot D_i \cdot X$$

where $D_i = 0.869 \, (\mu_{en}/\rho)_i/(\mu_{en}/\rho)_a$

The table lists the absorbed doses D_i per unit exposure for photon energies in the range 0.1-3.0 MeV, for some atoms which constitute the main part of organic compounds.

Conversion factors rad/roentgen ($D_i = 0.869 \frac{(\mu_{en}/\rho)_m}{(\mu_{en}/\rho)_a}$)

Photon energy (MeV)	Constituents H	C	N	O	S	P
0.1	1.53	0.801	0.835	0.869	2.25	1.86
0.3	1.72	0.869	0.869	0.869	0.907	0.869
0.6	1.72	0.869	0.869	0.869	0.869	0.840
1.0	1.72	0.869	0.869	0.869	0.863	0.838
1.5	1.72	0.869	0.869	0.869	0.865	0.838
2.0	1.72	0.869	0.869	0.869	0.872	0.846
3.0	1.69	0.866	0.869	0.865	0.894	0.865

In the study of radiation damage to nucleic acids, its building blocks e.g., the bases are sometimes studied separately. For example when thymine ($C_5N_2H_6O_2$) is exposed to say 10^6 R from a ^{60}Co γ-source, the absorbed dose D_T be estimated from the Equation above in the following way:

$$D_T = (1.73 \cdot f_H + 0.869 \cdot f_C + 0.869 \cdot f_N + 0.869 \cdot f_O) \cdot 10^6 \text{ rad}$$
$$= \underline{0.910 \cdot 10^6 \text{ rad}}$$

Thus, the conversion factor here is close to unity. This absorbed energy corresponds to about 5.7 10^{22} eV/kg. Since 1 kg of thymine contains about 4.8 10^{24} molecules it can be shown that approximately four in 10,000 molecules become ionized. Conversely, an absorbed dose of about 0.3 Mrad is required to produce one ion pair in a DNA molecule with M.W. 10^6.

1.4 Radiation-Chemical Yield (G-Value)

The number of molecules formed or destroyed per 100 eV absorbed energy is called G-value (see I.2). The absorbed dose is measured in rad. The conversion factor is:

$$1 \text{ rad} = 10^{-2} \text{ J/kg} = 6.24 \cdot 10^{13} \text{ eV/g}$$

Deriving from the definition of the G-value, the G unit is a frequently used measure of radiation-chemical yield. 1 G unit corresponds to 1.0364 $\cdot$ 10^{-7} mol material formed or destroyed per joule energy absorbed.

Source

International Commission on Radiation Units and Measurements: Linear Energy Transfer. ICRU Report 16. Washington 1970.
BERGER, R.T.: The X- or gamma-ray energy absorption or transfer coefficient: Tabulations and discussion.Radiat. Res. 15, 1, 29 (1961).

2. List of Abbreviations

2.1 Chemical Names

1. For abbreviations of nucleic acid constituents see Table 1.1 on p. 32.

2. Pu, pu : Purine
   Py, pu : Pyrimidine

3. A-T
   A-U : Watson-Crick base pairs
   G-C

4. ADP : Adenosine-5'-diphosphate
   TDP : Thymidine-5'-diphosphate
   PolyA : Polyadenylic acid
   PolyU : Polyuridylic acid
   TpT : Thymidylyl (3'→5') thymidine

5. DNA : Deoxyribonucleic acid
   RNA : Ribonucleic acid
   NaDNA : DNA-sodium salt
   LiDNA : DNA-lithium salt

6. Me_2SO-d_6 : perdeuterated dimethylsulfoxide
   NEM : N-ethylmaleimide
   TAN : 2,2,6,6-tetramethyl-4-piperidone
   CMEC : N-cyclobutyl-N'-β-[4-methylmorpholinium] ethylcarbodiimide

2.2 Experimental and Theoretical Methods

NMR	: Nuclear magnetic resonance
^{1}H-NMR	: Proton magnetic resonance
CD	: Circular dichroism
ORD	: Optical rotatory dispersion
GC	: Gas chromatography
MS	: Mass spectrometry
ESR	: Electron spin resonance spectroscopy, (X-band: 8-12 GHz; Q-band: 26-40 GHz)
ENDOR	: Electron nuclear double resonance
HMO	: Hückel-type molecular orbital calculations
INDO	: Semi-empirical calculations using intermediate neglect of differential overlap
SCF	: Self-consistent-field method for molecular orbital calculations

2.3 Symbols, Entities and Units

OD	: Optical density
T_m	: "melting" temperature
$[\eta]$	: Intrinsic viscosity
$S^o_{20,w}$	: Sedimentation coefficient $[10^{-13}$ S]
B_1; B_2	: Single or double strand break per nucleotide
C	: Crosslink per nucleotide
Dalton	: Atomic weight unit
Mn	: Number average molecular weight
Mw	: Weight average molecular weight
Ms	: Molecular weight determined from the band within a CsCl-density gradient
Msη	: Molecular weight determined by sedimentation and viscosity measurement
LET	: Linear energy transfer
SN1	: Unimolecular nucleophilic substitution
SN2	: Bimolecular nucleophilic substitution

Subject Index

MIX
Papier aus verantwortungsvollen Quellen
Paper from responsible sources
FSC® C105338

If you have any concerns about our products,
you can contact us on
ProductSafety@springernature.com

In case Publisher is established outside the EU,
the EU authorized representative is:
Springer Nature Customer Service Center GmbH
Europaplatz 3, 69115 Heidelberg, Germany

Printed by Libri Plureos GmbH
in Hamburg, Germany